An Outline for the Study of Calculus
to accompany
The Calculus with Analytic Geometry, Fourth Edition
Volume II

AN OUTLINE FOR THE STUDY OF CALCULUS
to accompany

Louis Leithold's **THE CALCULUS** with Analytic Geometry
FOURTH EDITION

JOHN H. MINNICK

De Anza College

(Edited by Louis Leithold)

VOLUME II

HARPER & ROW, PUBLISHERS, New York
Cambridge, Philadelphia, San Francisco,
London, Mexico City, São Paulo, Sydney

1817

Sponsoring Editor: Don Shauger
Project Editor: Cynthia Indriso
Senior Production Manager: Kewal K. Sharma
Compositor: Kingsport Press
Printer and Binder: The Murray Printing Company
Art Studio: J & R Services
Cover Artist: Roman Opalka, Paris, France. Courtesy of John Weber Gallery, New York

An Outline for the Study of Calculus to accompany The Calculus with Analytic Geometry,
Fourth Edition—Volume II
Copyright © 1981 by John H. Minnick

Library of Congress Cataloging in Publication Data
Minnick, John Harper, Date–
    An outline for the study of calculus, to
accompany The calculus with analytic geometry,
fourth edition.
    1. Calculus—Outlines, syllabi, etc.   I. Leithold,
Louis.   II. Leithold, Louis. The calculus with
analytic geometry.   4th ed.   III. Title.
QA303.M686   515   81–2711
ISBN 0–06–044543–2 (v. 1)   AACR2
ISBN 0–06–044544–0 (v. 2)
ISBN 0–06–044545–9 (v. 3)

Contents

Preface

Each section of the student guide contains all of the most important theorems and definitions from the fourth edition of Leithold's *The Calculus with Analytic Geometry*. Often these are followed by a discussion that elaborates the concepts and presents a summary of problem-solving techniques. Every fourth exercise from the text (4, 8, 12, etc.) is completely solved with detailed explanations, including graphs. In the appendix of this volume there is a test for each chapter, followed by solutions for the tests.

Some exercises are more easily solved by using a computer. This guide contains several general flowcharts that show how to use a computer. Each flowchart is followed by a sample program, written in BASIC, that illustrates the solution of a particular exercise. The computer solutions are found in Chapters 6, 15, and 20.

This student guide will be published in three volumes. Volume I contains Chapters 1–7, Volume II contains Chapters 8–15, and Volume III contains Chapters 16–20.

John H. Minnick

CHAPTER 8 | Inverse Functions, Logarithmic Functions, and Exponential Functions

8.1 THE INVERSE OF A FUNCTION

8.1.1 Definition   A function f is said to be *one-to-one* if and only if whenever x_1 and x_2 are any two distinct numbers in the domain of f, then $f(x_1) \neq f(x_2)$.

If $y = f(x)$ and the function f is one-to-one, then for each replacement of x from the domain of f there corresponds exactly one value of y, and for each replacement of y from the range of f there corresponds exactly one value of x. For the graph of a one-to-one function each vertical line intersects the graph in at most one point, and each horizontal line intersects the graph in at most one point.

8.1.2 Theorem   A function that is either increasing on an interval or decreasing on an interval is one-to-one on the interval.

8.1.3 Definition   If f is a one-to-one function that is the set of ordered pairs (x, y), then there is a function f^{-1}, called the *inverse* of f, where f^{-1} is the set of ordered pairs (y, x) defined by

$$x = f^{-1}(y) \qquad \text{if and only if} \qquad y = f(x)$$

The domain of f^{-1} is the range of f and the range of f^{-1} is the domain of f.

8.1.4 Theorem   If f is a one-to-one function having f^{-1} as its inverse, then f^{-1} is a one-to-one function having f as its inverse. Furthermore,

$$f^{-1}(f(x)) = x \qquad \text{for } x \text{ in the domain of } f$$
$$f(f^{-1}(x)) = x \qquad \text{for } x \text{ in the domain of } f^{-1}$$

The graphs of the functions f and f^{-1} are reflections of each other with respect to the line $y = x$, the line that bisects the first and third quadrants. ·

8.1.5 Theorem   Suppose the function f has the closed interval $[a, b]$ as its domain, Then

(i) if f is continuous and increasing on $[a, b]$, f has an inverse f^{-1} that is defined on $[f(a), f(b)]$;

(ii) if f is continuous and decreasing on $[a, b]$, f has an inverse f^{-1} that is defined on $[f(b), f(a)]$.

Exercises 8.1

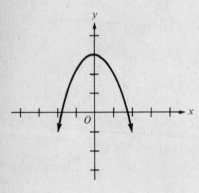

Figure 8. 1. 4

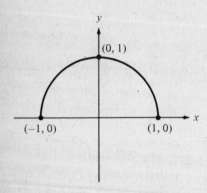

Figure 8. 1. 8

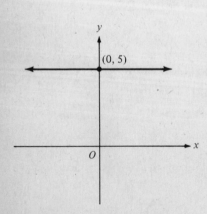

Figure 8. 1. 12

In Exercises 1–14 determine whether the given function is one-to-one. Draw a sketch of the graph of the function.

4. $f(x) = 3 - x^2$

SOLUTION: Because $f(1) = 2$ and $f(-1) = 2$, we have $1 \neq -1$, but $f(1) = f(-1)$. Thus, by Definition 8.1.1, the function f is not one-to-one. Note that in the graph of f, which is the parabola shown in Fig. 8. 1. 4, there are horizontal lines that intersect the graph in two points.

8. $g(x) = \sqrt{1 - x^2}$

SOLUTION: Because $g(-1) = g(1) = 0$, by Definition 8.1.1 the function g is not one-to-one. The graph of g is the semicircle shown in Fig. 8. 1. 8. There are horizontal lines that intersect the graph in two points.

12. $g(x) = 5$

SOLUTION: Because $g(-x) = g(x) = 5$, by Definition 8.1.1 the function g is not one-to-one. The graph of g is the horizontal line shown in Fig. 8. 1. 12.

In Exercises 15–36 determine whether the given function has an inverse. If the inverse exists, find it and state its domain and range and draw sketches of the graphs of the function and its inverse on the same set of axes. If the function does not have an inverse, show that a horizontal line intersects the graph of the function in more than one point.

16. $f(x) = 3x + 6$

SOLUTION: Because $f'(x) = 3 > 0$, the function f is increasing on $(-\infty, +\infty)$. By Theorem 8.1.2, f is therefore one-to-one and thus f has an inverse. To find the inverse of f we let $y = f(x)$ and solve the equation for x in terms of y.

$$y = 3x + 6$$

$$y - 6 = 3x$$

$$x = \frac{y - 6}{3}$$

We have

$$f^{-1}(y) = \frac{y - 6}{3}$$

and replacing y with x we obtain

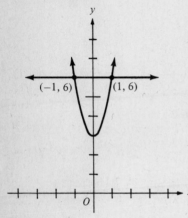

Figure 8. 1. 16

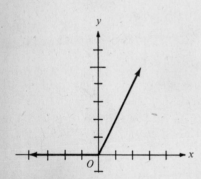

Figure 8. 1. 20

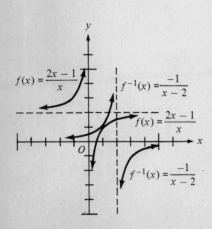

Figure 8. 1. 24

$$f^{-1}(x) = \frac{x-6}{3}$$

which defines the inverse function f^{-1}. Both the domain and range of f^{-1} consist of the set of all real numbers. In Fig. 8. 1. 16 we show sketches of the graphs of f and f^{-1}. Note that the graphs are reflections of each other with respect to the line $y = x$.

20. $F(x) = 3(x^2 + 1)$

SOLUTION: Because $F(-x) = F(x)$, the function F is not one-to-one and thus F does not have an inverse function. In Fig. 8. 1. 20 we show the graph, which is a parabola, and the horizontal line $y = 6$, which intersects the graph in two points, $(1, 6)$ and $(-1, 6)$.

24. $f(x) = |x| + x$

SOLUTION: If $x < 0$, then $|x| = -x$. Thus, when $x < 0$, $f(x) = -x + x = 0$, and so f is not one-to-one and f does not have an inverse function. In Fig. 8. 1. 24 we show the graph of f. The x-axis, which is a horizontal line, intersects the graph in an unlimited number of points.

28. $f(x) = \dfrac{2x - 1}{x}$

SOLUTION: The domain of f is $\{x \mid x \neq 0\}$. Because

$$f'(x) = \frac{1}{x^2} > 0 \qquad \text{if } x \neq 0$$

then f is increasing on $(-\infty, 0)$ and f is increasing on $(0, +\infty)$. Thus, f is one-to-one and has an inverse. Let $y = f(x)$.

$$y = \frac{2x - 1}{x}$$

$$xy = 2x - 1$$

$$xy - 2x = -1$$

$$x(y - 2) = -1$$

$$x = \frac{-1}{y - 2}$$

Thus,

$$f^{-1}(y) = \frac{-1}{y - 2}$$

and

$$f^{-1}(x) = \frac{-1}{x - 2}$$

The domain of f^{-1} is $\{x \mid x \neq 2\}$. Because the domain of f is also the range of f^{-1}, then the range of f^{-1} is $\{y \mid y \neq 0\}$. In Fig. 8. 1. 28 we show sketches of the graphs of f and f^{-1}.

32. $f(x) = (2x - 1)^2, \ x \leq \frac{1}{2}$

Figure 8. 1. 28

SOLUTION: The domain of f is given and is $(-\infty, \frac{1}{2}]$. Because

$$f'(x) = 4(2x - 1)$$

and $2x - 1 \leq 0$ if $x \leq \frac{1}{2}$, then f is decreasing on $(-\infty, \frac{1}{2}]$. Thus, f is one-to-one and f has an inverse function. Let $y = f(x)$.

$$y = (2x - 1)^2$$
$$\sqrt{y} = \sqrt{(2x - 1)^2}$$
$$\sqrt{y} = |2x - 1|$$

Because $2x - 1 \leq 0$, then $|2x - 1| = -(2x - 1)$. Thus, we have

$$\sqrt{y} = -(2x - 1)$$
$$-\sqrt{y} = 2x - 1$$
$$x = \frac{1 - \sqrt{y}}{2}$$

Hence

$$f^{-1}(y) = \frac{1 - \sqrt{y}}{2}$$

and

$$f^{-1}(x) = \frac{1 - \sqrt{x}}{2}$$

The domain of f^{-1} is $[0, +\infty)$, and the range of f^{-1} is $(-\infty, \frac{1}{2}]$, because $(-\infty, \frac{1}{2}]$ is the domain of f. See Fig. 8. 1. 32 for sketches of the graphs of f and f^{-1}.

36. $G(x) = \sqrt{4x^2 - 9}$, $x \geq \frac{3}{2}$

SOLUTION: We are given that the domain of f is $[\frac{3}{2}, +\infty)$.

$$G'(x) = \frac{1}{2}(4x^2 - 9)^{-1/2}(8x)$$

$$= \frac{4x}{\sqrt{4x^2 - 9}}$$

Because $G'(x) > 0$ if $x > \frac{3}{2}$, then G is increasing on $[\frac{3}{2}, +\infty)$. Hence, G is one-to-one and G has an inverse function. Let $y = G(x)$.

$$y = \sqrt{4x^2 - 9}$$

$$y^2 = 4x^2 - 9$$

$$x^2 = \frac{y^2 + 9}{4}$$

$$|x| = \frac{\sqrt{y^2 + 9}}{2}$$

Because $x \geq \frac{3}{2}$, then $|x| = x$. Thus,

$$x = \frac{\sqrt{y^2 + 9}}{2}$$

Hence

$$G^{-1}(y) = \frac{\sqrt{y^2 + 9}}{2}$$

and

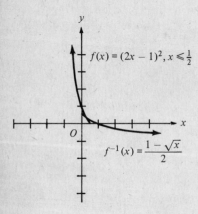

$f(x) = (2x - 1)^2, x \leq \frac{1}{2}$

$f^{-1}(x) = \frac{1 - \sqrt{x}}{2}$

Figure 8. 1. 32

Figure 8. 1. 36

$$G^{-1}(x) = \frac{\sqrt{x^2 + 9}}{2}$$

The domain of G^{-1} is the set of all real numbers, and the range of G^{-1} is $[\frac{3}{2},$ $+\infty)$, which is also the domain of G. In Fig. 8. 1. 36 we show sketches of the graphs of G and G^{-1}.

40. (a) Prove that the given function f has an inverse,
(b) find $f^{-1}(x)$, and **(c)** verify the equations of Theorem 8.1.4 for f and f^{-1}.
$$f(x) = (x + 2)^3$$

SOLUTION:

(a) $f'(x) = 3(x + 2)^2 > 0$ if $x \neq -2$. Thus, f is one-to-one and f^{-1} exists.
(b) Let

$$y = (x + 2)^3$$
$$\sqrt[3]{y} = x + 2$$
$$x = -2 + \sqrt[3]{y}$$
$$f^{-1}(y) = -2 + \sqrt[3]{y}$$
$$f^{-1}(x) = -2 + \sqrt[3]{x}$$

(c) We have

$$f^{-1}(f(x)) = f^{-1}((x + 2)^3)$$
$$= -2 + \sqrt[3]{(x + 2)^3}$$
$$= -2 + (x + 2) = x$$

and

$$f(f^{-1}(x)) = f(-2 + \sqrt[3]{x})$$
$$= [(-2 + \sqrt[3]{x}) + 2]^3$$
$$= (\sqrt[3]{x})^3 = x$$

44. (a) Show that the given function $f(x) = 2x^2 - 6$ is not one-to-one and hence does not have an inverse.
(b) Restrict the domain and obtain two one-to-one functions f_1 and f_2 having the given function value.
(c) Find $f_1^{-1}(x)$ and $f_2^{-1}(x)$ and state the domains of f_1^{-1} and f_2^{-1}.
(d) Draw sketches of the graphs of f_1 and f_1^{-1} on the same set of axes.
(e) Draw sketches of the graphs of f_2 and f_2^{-1} on the same set of axes.

SOLUTION:

(a) Because $f(-1) = f(1) = -4$, f is not one-to-one.
(b) Let

$$f_1(x) = 2x^2 - 6 \qquad \text{if } x \geq 0$$

and

$$f_2(x) = 2x^2 - 6 \qquad \text{if } x \leq 0$$

(c) If $x \geq 0$, and $y = 2x^2 - 6$, then

$$x^2 = \frac{y + 6}{2}$$

$$x = \sqrt{\frac{y + 6}{2}}$$

$$f_1^{-1}(y) = \sqrt{\frac{y + 6}{2}}$$

$$f_1^{-1}(x) = \sqrt{\frac{x+6}{2}}$$

The domain of f_1^{-1} is $[-6, +\infty)$.
If $x \leq 0$, and

$$y = 2x^2 - 6$$

then

$$x^2 = \frac{y+6}{2}$$

$$x = -\sqrt{\frac{y+6}{2}}$$

$$f_2^{-1}(y) = -\sqrt{\frac{y+6}{2}}$$

$$f_2^{-1}(x) = -\sqrt{\frac{x+6}{2}}$$

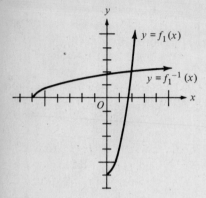

Figure 8. 1. 44a

The domain of f_2^{-1} is $[-6, +\infty)$.
(d) Figure 8. 1. 44a shows sketches of the graphs of f_1 and f_1^{-1}.
(e) Figure 8. 1. 44b shows sketches of the graphs of f_2 and f_2^{-1}.

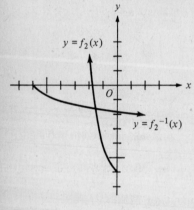

Figure 8. 1. 44b

48. Determine the value of the constant k so that the function defined by $f(x) = (x + 5)/(x + k)$ will be its own inverse.

SOLUTION: We have the fact that

$$f(f^{-1}(x)) = x \tag{1}$$

if $f = f^{-1}$, then from (1) we get

$$f(f(x)) = x$$

In particular, if $x = 0$, we have

$$f(f(0)) = 0 \tag{2}$$

We are given that

$$f(x) = \frac{x+5}{x+k}$$

Thus,

$$f(0) = \frac{5}{k}$$

and

$$f(f(0)) = f\left(\frac{5}{k}\right)$$

$$= \frac{\dfrac{5}{k} + 5}{\dfrac{5}{k} + k} \tag{3}$$

From (2) and (3) we conclude that

$$\frac{\dfrac{5}{k}+5}{\dfrac{5}{k}+k}=0$$

$$\frac{5}{k}+5=0$$

$$k=-1$$

8.2 INVERSE FUNCTION THEOREM AND DERIVATIVE OF THE INVERSE OF A FUNCTION

If a function f is either increasing on an interval $[a,\ b]$ or decreasing on the interval $[a,\ b]$, then f is said to be *monotoic* on the interval $[a,\ b]$.

8.2.3 Theorem

Suppose that the function f is continuous and monotonic on the closed interval $[a,\ b]$, and let $y = f(x)$. If f is differentiable on $[a,\ b]$ and $f'(x) \neq 0$ for any x in $[a,\ b]$, then the derivative of the inverse function f^{-1}, defined by $x = f^{-1}(y)$ is given by

$$D_y x = \frac{1}{D_x y}$$

8.2.4 Theorem

Suppose the function f is continuous and monotonic on a closed interval $[a,\ b]$ containing the number c, and let $f(c) = d$. If $f'(c)$ exists and $f'(c) \neq 0$, then $(f^{-1})'(d)$ exists and

$$(f^{-1})'(d) = \frac{1}{f'(c)}$$

Exercises 8.2

In Exercises 1–10 let $y = f(x)$ and $x = f^{-1}(y)$ and verify that $D_y x = 1/D_x y$.

4. $f(x) = 8x^3$

SOLUTION: If $y = 8x^3$, then

$$D_x y = 24x^2$$

Furthermore,

$$x = \sqrt[3]{\tfrac{1}{8}y} = \tfrac{1}{2}y^{1/3}$$

Thus,

$$D_y x = \tfrac{1}{6}y^{-2/3}$$

Replacing y by $8x^3$, we obtain

$$D_y x = \tfrac{1}{6}(8x^3)^{-2/3}$$

$$= \tfrac{1}{6}(\tfrac{1}{4}x^{-2})$$

$$= \frac{1}{24x^2}$$

$$= \frac{1}{D_x y}$$

8. $f(x) = \sqrt[5]{x}$

SOLUTION: If $y = \sqrt[5]{x}$, *then*

$$D_x y = \tfrac{1}{5} x^{-4/5}$$

$$= \frac{1}{5 x^{4/5}}$$

Furthermore,

$$x = y^5$$

Hence,

$$D_y x = 5 y^4$$

$$= 5(\sqrt[5]{x})^4$$

$$= 5 x^{4/5}$$

$$= \frac{1}{D_x y}$$

In Exercises 11–24 find $(f^{-1})'(d)$ for the given f and d.

12. $f(x) = x^5 + 2; \ d = 1$

SOLUTION: We apply Theorem 8.2.4. First, we find c so that $f(c) = d$. We have

$$c^5 + 2 = 1$$
$$c^5 = -1$$
$$c = -1$$

Because $f'(x) = 5x^4 > 0$ if $x \neq 0$, then f is monotonic and continuous on every closed interval. Furthermore, $f'(c) = f'(-1) = 5 \neq 0$. By Theorem 8.2.4 we conclude that

$$(f^{-1})'(1) = \frac{1}{f'(-1)} = \frac{1}{5}$$

16. $f(x) = 4x^3 + 2x; \ d = 6$

SOLUTION: If $f(c) = d$, then

$$4c^3 + 2c = 6$$
$$4c^3 + 2c - 6 = 0$$
$$2c^3 + c - 3 = 0$$

We use synthetic division to factor.

$$(c - 1)(2c^2 + 2c + 3) = 0$$

Therefore, $c = 1$. Because $f'(x) = 12x^2 + 2$, then $f'(x) > 0$ for all x, and f is monotonic and continuous on any closed interval. Therefore, by Theorem 8.2.4 we have

$$(f^{-1})'(6) = \frac{1}{f'(1)} = \frac{1}{14}$$

20. $f(x) = x^2 - 6x + 7, \ x \leq 3; \ d = 0$

SOLUTION: If $f(x) = 0$, then

$$x^2 - 6x + 7 = 0$$
$$x^2 - 6x + 9 = 2$$
$$(x - 3)^2 = 2$$
$$x - 3 = \pm\sqrt{2}$$
$$x = 3 \pm \sqrt{2}$$

Because we are given that $x \leq 3$, we take $c = 3 - \sqrt{2}$. Now $f'(x) = 2x - 6$, and $x \leq 3$, so $f'(x) \leq 0$. Therefore, f is monotonic and continuous on a closed interval that contains c. For example, f is monotonic and continuous on the closed interval $[1, 2]$. Furthermore, $f'(c) = f'(3 - \sqrt{2}) = 2(3 - \sqrt{2}) - 6 = -2\sqrt{2}$. Therefore, by Theorem 8.2.4,

$$(f^{-1})'(0) = \frac{1}{f'(3 - \sqrt{2})}$$

$$= \frac{1}{-2\sqrt{2}} = -\frac{\sqrt{2}}{4}$$

24. $f(x) = \displaystyle\int_x^2 t \, dt, \ x < 0; \ d = -6$

SOLUTION:

$$\int_x^2 t \, dt = \frac{1}{2} t^2 \Big]_x^2$$

$$= 2 - \frac{1}{2} x^2$$

Thus, we have

$$f(x) = 2 - \tfrac{1}{2}x^2, \quad x < 0$$
$$f'(x) = -x, \quad x < 0$$

If $f(x) = -6$, and $x < 0$, then

$$-6 = 2 - \tfrac{1}{2}x^2$$
$$-8 = -\tfrac{1}{2}x^2$$
$$x^2 = 16$$
$$x = -4$$

Thus, we take $c = -4$. Because $f'(x) = -x > 0$ if $x < 0$, then f is monotonic and continuous on a closed interval containing -4. By Theorem 8.2.4 we conclude that

$$(f^{-1})'(-6) = \frac{1}{f'(-4)} = \frac{1}{4}$$

28. Given $f(x) = 6 - x - x^3$.
(a) On the same set of axes draw sketches of the graphs of the functions f and f^{-1}.
(b) Find the slope of the tangent line to the graph of f at the point $(2, -4)$.
(c) Find the slope of the tangent line to the graph of f^{-1} at the point $(-4, 2)$.

SOLUTION:
(a) $f'(x) = -1 - 3x^2$ and $f''(x) = -6x$. Because $f'(x) < 0$ for all x, the function f is decreasing on its domain. Because $f''(x)$ changes sign at $x = 0$, the point $(0, 6)$ is a point of inflection. Furthermore, the graph is concave downward for $x > 0$ and concave upward for $x < 0$, and the

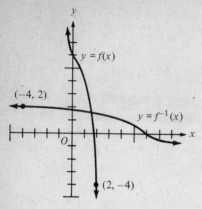

Figure 8. 2. 28

slope of the inflectional tangent is -1. In Fig. 8. 2. 28 we show sketches of the graphs of f and f^{-1}, where the sketch of f is obtained from the above information and the sketch of f^{-1} is made by using the fact that the graphs of f and f^{-1} are reflections of each other with respect to the line $y = x$.

(b) Because $f'(2) = -1 - 3(2^2) = -13$, the slope of the tangent line to the graph of f at the point $(2, -4)$ is -13.

(c) By Theorem 8.2.4,

$$(f^{-1})'(-4) = \frac{1}{f'(2)} = -\frac{1}{13}$$

Thus, $-\frac{1}{13}$ is the slope of the tangent line to the graph of f^{-1} at the point $(-4, 2)$.

32. Given $9y^2 - 8x^3 = 0$.

  (a) Solve the equation for y in terms of x and express y as one or more functions of x.

  (b) For each of the functions obtained in (a) determine whether the function has an inverse, and if it does, determine the domain of the inverse function.

  (c) Use implicit differentiation to find $D_x y$ and $D_y x$ and determine the values of x and y for which $D_x y$ and $D_y x$ are reciprocals.

SOLUTION:

  (a) Solving for y, we obtain $y = \pm \frac{2}{3}\sqrt{2}x^{3/2}$. Let f_1 and f_2 be the functions defined as follows:

$$f_1(x) = \frac{2}{3}\sqrt{2}x^{3/2} \tag{1}$$
$$f_2(x) = -\frac{2}{3}\sqrt{2}x^{3/2} \tag{2}$$

  (b) From (1) we have

$$f_1'(x) = \sqrt{2}x^{1/2}$$

Hence, $f_1'(x) > 0$ for all $x > 0$. Thus, f_1 has an inverse. The domain of f_1^{-1} is the range of f_1, which is $[0, +\infty)$. From (2) we have $f_2'(x) = -\sqrt{2}x^{1/2}$. Thus, $f_2'(x) < 0$ for all $x > 0$, and f_2 has an inverse. The domain of f_2^{-1} is the range of f_2, which is $(-\infty, 0]$.

  (c) Because $D_x y = dy/dx$, we take the differential of both sides of the given equation. This results in

$$18y \, dy - 24x^2 \, dx = 0 \tag{3}$$

Solving (3) for dy/dx, we obtain

$$D_x y = \frac{dy}{dx} = \frac{4x^2}{3y} \qquad \text{if } y \neq 0$$

Solving (3) for dx/dy, we obtain

$$D_y x = \frac{dx}{dy} = \frac{3y}{4x^2} \qquad \text{if } x \neq 0$$

Therefore, $D_x y$ is the reciprocal of $D_y x$ for all x and y for which $x \neq 0$ and $y \neq 0$.

36. Use the formula of Exercise 35 to show that

$$(f^{-1})''(x) = -\frac{f''(f^{-1}(x))}{[f'(f^{-1}(x))]^3}$$

SOLUTION: In Exercise 35 we are given that

$$(f^{-1})'(x) = \frac{1}{f'(f^{-1}(x))} \tag{1}$$

We differentiate with respect to x on both sides of Eq. (1), and apply the chain rule on the right side. Thus,

$$(f^{-1})''(x) = \frac{-1}{[f'(f^{-1}(x))]^2} [D_x f'(f^{-1}(x))]$$

$$= \frac{-1}{[f'(f^{-1}(x))]^2} [f''(f^{-1}(x)) D_x f^{-1}(x)] \tag{2}$$

Because $D_x f^{-1}(x) = (f^{-1})'(x)$, we may substitute from Eq. (1) into Eq. (2) and obtain

$$(f^{-1})''(x) = \frac{-1}{[f'(f^{-1}(x))]^2} [f''(f^{-1}(x))] \left[\frac{1}{f'(f^{-1}(x))} \right]$$

$$= -\frac{f''(f^{-1}(x))}{[f'(f^{-1}(x))]^3} .$$

8.3 THE NATURAL LOGARITHMIC FUNCTION

8.3.1 Definition The *natural logarithmic function* is the function defined by

$$\ln x = \int_1^x \frac{1}{t} \, dt \qquad x > 0$$

8.3.2 Theorem If u is a differentiable function of x and $u(x) > 0$, then

$$D_x(\ln u) = \frac{1}{u} \cdot D_x u$$

8.3.3 Theorem $\ln 1 = 0$

8.3.4 Theorem If a and b are any positive numbers, then

$$\ln(ab) = \ln a + \ln b$$

8.3.5 Theorem If a and b are any positive numbers, then

$$\ln \frac{a}{b} = \ln a - \ln b$$

8.3.6 Theorem If a is any positive number and r is any rational number, then

$$\ln a^r = r \ln a$$

Note the distinction between $\ln(x^2)$ and $(\ln x)^2$. By Theorem 8.3.6, $\ln(x^2) = \ln x^2 = 2 \cdot \ln x$. However, $(\ln x)^2 = (\ln x)(\ln x) \neq 2 \cdot \ln x$. Another symbol for $(\ln x)^2$ is $\ln^2 x$. In general, $\ln^r x = (\ln x)^r \neq r \cdot \ln x$. And $\ln x^r = \ln(x^r) = r \cdot \ln x$, if r is a rational number.

We have the following facts about the natural logarithmic function and its graph.

(i) The domain is the set of all positive numbers.
(ii) The range is the set of all real numbers.
(iii) The function is increasing on its entire domain.
(iv) The function is continuous at all numbers in its domain.

(v) The graph of the function is concave downward at all points.

(vi) The graph of the function is asymptotic to the negative side of the y axis through the fourth quadrant.

Moreover, the following two limits are important.

$$\lim_{x \to +\infty} \ln x = +\infty$$

$$\lim_{x \to 0^+} \ln x = -\infty$$

Exercises 8.3

In Exercises 1–22 differentiate the given function and simplify the result.

4. $f(x) = \ln(8 - 2x)$

SOLUTION: We apply Theorem 8.3.2 with $u = 8 - 2x$.

$$f'(x) = \frac{1}{8 - 2x} D_x(8 - 2x)$$

$$= \frac{-2}{8 - 2x}$$

$$= \frac{1}{x - 4}$$

8. $f(x) = \ln\sqrt{1 + 4x^2}$

SOLUTION: First, we use Theorem 8.3.6 to simplify $f(x)$ before differentiating.

$$f(x) = \ln(1 + 4x^2)^{1/2}$$
$$= \tfrac{1}{2} \ln(1 + 4x^2)$$

Thus,

$$f'(x) = \frac{1}{2} \cdot \frac{1}{1 + 4x^2} (8x)$$

$$= \frac{4x}{4x^2 + 1}$$

12. $f(x) = x \ln x$

SOLUTION: We apply the derivative of a product rule.

$$f'(x) = x \cdot D_x(\ln x) + \ln x \cdot D_x x$$

$$= x\left(\frac{1}{x}\right) + \ln x$$

$$= 1 + \ln x$$

16. $f(x) = \ln[(5x - 3)^4(2x^2 + 7)^3]$

SOLUTION: First, we apply Theorems 8.3.4 and 8.3.6 to express $f(x)$ as a sum.

$$f(x) = \ln(5x - 3)^4 + \ln(2x^2 + 7)^3$$
$$= 4 \ln(5x - 3) + 3 \ln(2x^2 + 7)$$

Then

$$f'(x) = 4\left(\frac{1}{5x-3}\right)(5) + 3\left(\frac{1}{2x^2+7}\right)(4x)$$

$$= \frac{20}{5x-3} + \frac{12x}{2x^2+7}$$

$$= \frac{20(2x^2+7) + 12x(5x-3)}{(5x-3)(2x^2+7)}$$

$$= \frac{100x^2 - 36x + 140}{(5x-3)(2x^2+7)}$$

20. $f(x) = \sqrt[3]{\ln x^3}$

SOLUTION: By Theorem 8.3.6, $\ln x^3 = 3 \ln x$. Thus,

$$f(x) = \sqrt[3]{\ln x^3}$$
$$= \sqrt[3]{3 \ln x}$$
$$= \sqrt[3]{3}\,(\ln x)^{1/3}$$

Therefore,

$$f'(x) = \sqrt[3]{3}\,(\tfrac{1}{3})(\ln x)^{-2/3}\, D_x(\ln x)$$

$$= \tfrac{1}{3}\sqrt[3]{3}\,(\ln x)^{-2/3}\left(\frac{1}{x}\right)$$

$$= \frac{\sqrt[3]{3}}{3x(\ln x)^{2/3}}$$

In Exercises 23–28 find $D_x y$ by implicit differentiation.

24. $\ln\left(\dfrac{y}{x}\right) + xy = 1$

SOLUTION: First, we use Theorem 8.3.5 to simplify the given equation. Thus, we have

$$\ln y - \ln x + xy = 1$$

Differentiating on both sides with respect to x, we obtain

$$\frac{1}{y}\, D_x y - \frac{1}{x} + x\, D_x y + y = 0$$

Multiplying on both sides by xy, we have

$$x\, D_x y - y + x^2 y\, D_x y + xy^2 = 0$$

$$D_x y(x + x^2 y) = y - xy^2$$

$$D_x y = \frac{y - xy^2}{x + x^2 y}$$

28. $x \ln y + y \ln x = xy$

SOLUTION: Differentiating on both sides with respect to x, we obtain

$$\frac{x}{y}\, D_x y + \ln y + \frac{y}{x} + (\ln x)\, D_x y = x \cdot D_x y + y$$

We multiply on both sides by xy.

$$x^2 D_x y + xy \ln y + y^2 + xy(\ln x) D_x y = x^2 y D_x y + xy^2$$

$$(x^2 + xy \ln x - x^2 y) D_x y = xy^2 - y^2 - xy \ln y$$

$$D_x y = \frac{xy^2 - y^2 - xy \ln y}{x^2 + xy \ln x - x^2 y}$$

$$= \frac{y(xy - y - x \ln y)}{x(x - xy + y \ln x)}$$

In Exercises 30–37 draw a sketch of the graph of the curve having the given equation.

32. $y = \ln|x|$

SOLUTION: Because $|x| = |-x|$, the graph of $y = \ln|x|$ is symmetric with respect to the y axis. If $x > 0$, then $|x| = x$; thus the given curve is the graph of $y = \ln x$ if $x > 0$. We use Fig. 8. 3. 4 (in the text) and symmetry to draw a sketch of $y = \ln|x|$. The sketch is shown in Fig. 8. 3. 32.

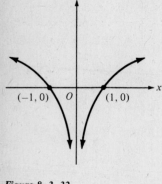

Figure 8. 3. 32

(−1, 0)   O   (1, 0)

36. $y = x \ln x$

SOLUTION: The domain is $(0, +\infty)$. We use a table of natural logarithms or a hand calculator to compute the entries in Table 36. To find $\ln(\frac{1}{2})$, from a table we use

$$\ln(\tfrac{1}{2}) = \ln 1 - \ln 2 = -\ln 2 = -0.7$$

In a similar manner we calculate $\ln(\frac{1}{3})$, $\ln(\frac{1}{4})$, and so on. We plot the points (x, y) from Table 36 and draw smooth curve through them. The sketch is shown in Fig. 8. 3. 36.

Table 36

x	2	1	$\frac{1}{2}$	$\frac{1}{3}$	$\frac{1}{4}$	$\frac{1}{10}$	$\frac{1}{100}$
$\ln x$	0.7	0	−0.7	−1.1	−1.4	−2.3	−4.6
$y = x \ln x$	1.4	0	−0.3	−0.4	−0.3	−0.2	−0.05

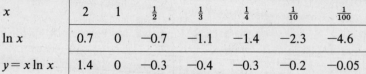

Figure 8. 3. 36

40. Find an equation of the normal line to the graph of $y = x \ln x$ that is perpendicular to the line with equation $x - y + 7 = 0$.

SOLUTION: If the normal line to the graph of the equation $y = x \ln x$ is perpendicular to the line $x - y + 7 = 0$, then the tangent line to the graph of the equation $y = x \ln x$ is parallel to the line $x - y + 7 = 0$. Because the slope-intercept form of the equation of the given line is $y = x + 7$, we have $m = 1$. Thus, we find the point on the given curve where the tangent line has slope 1. We have

$$y = x \ln x \qquad\qquad (1)$$

$$D_x y = x\left(\frac{1}{x}\right) + \ln x$$

$$= 1 + \ln x$$

If $D_x y = 1$, then

$$1 = 1 + \ln x$$
$$0 = \ln x$$
$$x = 1$$

Substituting $x = 1$ into Eq. (1), we obtain $y = 0$, and thus conclude that the required line intersects the curve at the point (1, 0). Furthermore, because the slope of the tangent line at that point is 1, then the slope of the normal line is -1, and an equation of the normal line is

$$y - 0 = (-1)(x - 1)$$
$$x + y - 1 = 0$$

44. A manufacturer of electric generators began operations on January 1, 1971. During the first year there were no sales because the company concentrated on product development and research. After the first year the sales increased steadily according to the equation $y = x \ln x$, where x is the number of years during which the company has been operating and y is sales volume in millions of dollars.　　　**(a)** Draw a sketch of the graph of the equation. Determine the rate at which the sales were increasing on　　　**(b)** January 1, 1975, and　　　**(c)** January 1, 1981.

SOLUTION:

(a) We are given that

$$y = x \ln x \qquad \text{if } x \geq 1$$

The graph is shown in Fig. 8. 3. 44.

(b) On Jan. 1, 1975 we have $x = 4$. Furthermore,

$$D_x y = x \left(\frac{1}{x}\right) + \ln x$$

$$= 1 + \ln x$$

and

$$D_x y \bigg]_{x=4} = 1 + \ln 4$$

$$= 1 + 1.39$$

$$= 2.39$$

Thus, sales are increasing at the rate of \$2.39 million per year on January 1, 1975.

(c) On January 1, 1981 we have $x = 10$. When $x = 10$, then $D_x y = 1 + \ln 10 = 3.30$. Thus, sales are increasing at the rate of \$3.30 million per year on January 1, 1981.

48. Use the result of Exercise 46 to prove that

$$\lim_{x \to 0} \frac{\ln(1 + x)}{x} = 1$$

SOLUTION: In Exercise 46 we are given that

$$1 - \frac{1}{x} < \ln x < x - 1$$

Replacing x with $1 + x$, we have

$$1 - \frac{1}{1 + x} < \ln(1 + x) < (1 + x) - 1$$

$$\frac{x}{1 + x} < \ln(1 + x) < x \qquad (1)$$

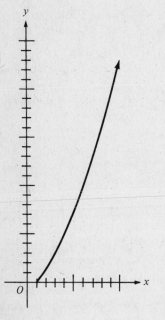

Figure 8. 3. 44

If $x > 0$ and we divide each expression in (1) by x, we obtain

$$\frac{1}{1+x} < \frac{\ln(1+x)}{x} < 1 \tag{2}$$

Because

$$\lim_{x \to 0} \frac{1}{1+x} = 1 \quad \text{and} \quad \lim_{x \to 0} 1 = 1 \tag{3}$$

by the squeeze theorem and (2) we conclude that

$$\lim_{x \to 0^+} \frac{\ln(1+x)}{x} = 1$$

If $x < 0$ and we divide each expression in (1) by x, we reverse the sense of the inequality and obtain

$$\frac{1}{1+x} > \frac{\ln(1+x)}{x} > 1 \tag{4}$$

By the squeeze theorem and (3) and (4), we have

$$\lim_{x \to 0^-} \frac{\ln(1+x)}{x} = 1$$

Therefore,

$$\lim_{x \to 0} \frac{\ln(1+x)}{x} = 1$$

8.4 LOGARITHMIC DIFFERENTIATION AND INTEGRALS YIELDING THE NATURAL LOGARITHMIC FUNCTION

8.4.1 Theorem If u is a differentiable function of x,

$$D_x(\ln|u|) = \frac{1}{u} D_x u$$

8.4.2 Theorem $\displaystyle\int \frac{1}{u} \, du = \ln|u| + C$

Exercises 8.4

4. Use Theorem 8.4.1 to find $D_x y$ if $y = \sin(\ln|2x + 1|)$

SOLUTION:

$$D_x y = \cos(\ln|2x + 1|) \, D_x \ln|2x + 1|$$

$$= \cos(\ln|2x + 1|) \left(\frac{1}{2x + 1}\right)(2)$$

$$= \frac{2 \cos(\ln|2x + 1|)}{2x + 1}$$

In Exercises 5–14 find $D_x y$ by logarithmic differentiation.

8. $y = (5x - 4)(x^2 + 3)(3x^3 - 5)$

SOLUTION:

$$|y| = |5x - 4| \, |x^2 + 3| \, |3x^3 - 5|$$
$$\ln|y| = \ln(|5x - 4| \, |x^2 + 3| \, |3x^3 - 5|)$$
$$= \ln|5x - 4| + \ln|x^2 + 3| + \ln|3x^3 - 5|$$

Differentiating on both sides implicitly with respect to x and applying Theorem 8.4.1, we obtain

$$\left(\frac{1}{y}\right) D_x y = \frac{5}{5x - 4} + \frac{2x}{x^2 + 3} + \frac{9x^2}{3x^3 - 5}$$

$$= \frac{5(x^2 + 3)(3x^3 - 5) + 2x(5x - 4)(3x^3 - 5) + 9x^2(5x - 4)(x^2 + 3)}{(5x - 4)(x^2 + 3)(3x^3 - 5)}$$

$$= \frac{90x^5 - 60x^4 + 180x^3 - 183x^2 + 40x - 75}{(5x - 4)(x^2 + 3)(3x^3 - 5)}$$

Because $y = (5x - 4)(x^2 + 3)(3x^3 - 5)$, if we multiply on both sides by y, we obtain

$$D_x y = 90x^5 - 60x^4 + 180x^3 - 183x^2 + 40x - 75$$

12. $y = \dfrac{\sqrt{1 - x^2}}{(x + 1)^{2/3}}$

SOLUTION: Because $y \geq 0$, we may take the logarithm of each member of the given equation. Thus,

$$\ln y = \ln \frac{\sqrt{1 - x^2}}{(x + 1)^{2/3}}$$

$$= \ln \sqrt{1 - x^2} - \ln(x + 1)^{2/3}$$

$$= \tfrac{1}{2} \ln(1 - x^2) - \tfrac{2}{3} \ln(x + 1)$$

Differentiating on both sides with respect to x, we have

$$\frac{1}{y} D_x y = \frac{1}{2} \cdot \frac{-2x}{1 - x^2} - \frac{2}{3} \cdot \frac{1}{x + 1}$$

$$= \frac{-x}{1 - x^2} - \frac{2}{3(x + 1)}$$

$$= -\frac{x + 2}{3(1 - x)(1 + x)}$$

Multiplying on both sides by y, we have

$$D_x y = -\frac{x + 2}{3(1 - x)(1 + x)} \cdot y$$

Replacing y by its given value, we obtain

$$D_x y = -\frac{x + 2}{3(1 - x)(1 + x)} \cdot \frac{\sqrt{1 - x^2}}{(x + 1)^{2/3}}$$

$$= -\frac{x + 2}{3(1 - x)^{1/2}(1 + x)^{7/6}}$$

In Exercises 15–30 evaluate the indefinite integral.

16. $\displaystyle\int \frac{dx}{7x + 10}$

SOLUTION: Let $u = 7x + 10$. Then $du = 7\,dx$, and by Theorem 8.4.2 we have

$$\int \frac{dx}{7x + 10} = \int \frac{\frac{1}{7}\,du}{u}$$

$$= \frac{1}{7}\int \frac{du}{u}$$

$$= \frac{1}{7}\ln|u| + C$$

$$= \frac{1}{7}\ln|7x + 10| + C$$

20. $\displaystyle\int \frac{2x - 1}{x(x - 1)}\,dx$

SOLUTION: $x(x - 1) = x^2 - x$. Let $u = x^2 - x$ and $du = (2x - 1)\,dx$. Thus,

$$\int \frac{2x - 1}{x(x - 1)}\,dx = \int \frac{(2x - 1)\,dx}{x^2 - x}$$

$$= \int \frac{du}{u}$$

$$= \ln|u| + C$$

$$= \ln|x^2 - x| + C$$

$$= \ln|x(x - 1)| + C$$

24. $\displaystyle\int \frac{5 - 4y^2}{3 + 2y}\,dy$

SOLUTION: We divide the numerator of the given fraction by its denominator. Because

$$\frac{5 - 4y^2}{3 + 2y} = -2y + 3 - \frac{4}{3 + 2y}$$

then

$$\int \frac{5 - 4y^2}{3 + 2y}\,dy = \int (-2y + 3)\,dy - 4\int \frac{dy}{3 + 2y}$$

$$= -y^2 + 3y - 4\int \frac{dy}{3 + 2y} \tag{1}$$

If $u = 3 + 2y$, then $du = 2\,dy$, and

$$\int \frac{dy}{3 + 2y} = \int \frac{\frac{1}{2}\,du}{u}$$

$$= \tfrac{1}{2}\ln|u| + \overline{C}$$

$$= \tfrac{1}{2}\ln|3 + 2y| + \overline{C} \tag{2}$$

Substituting from (2) into (1), we have

$$\int \frac{5-4y^2}{3+2y}\, dy = -y^2 + 3y - 4(\tfrac{1}{2}\ln|3+2y| + \overline{C})$$

$$= -y^2 + 3y - 2\ln|3+2y| + C$$

where $C = -4\overline{C}$.

28. $\displaystyle \int \frac{(2+\ln^2 x)\, dx}{x(1-\ln x)}$

SOLUTION: Let $u = \ln x$. Then $du = dx/x$. Thus,

$$\int \frac{(2+\ln^2 x)\, dx}{x(1-\ln x)} = \int \frac{2+\ln^2 x}{1-\ln x}\cdot \frac{dx}{x}$$

$$= \int \frac{2+u^2}{1-u}\, du \tag{1}$$

Dividing the numerator by the denominator, we obtain

$$\int \frac{2+u^2}{1-u}\, du = \int (-u-1)\, du + 3\int \frac{du}{1-u}$$

$$= -\frac{1}{2}u^2 - u + 3\int \frac{du}{1-u} \tag{2}$$

To find the integral in (2), we let $v = 1 - u$. Then $dv = -du$. Thus

$$\int \frac{du}{1-u} = -\int \frac{dv}{v} = -\ln|v| + C = -\ln|1-u| + C \tag{3}$$

Substituting from (3) into (2), we obtain

$$\int \frac{2+u^2}{1-u}\, du = -\frac{1}{2}u^2 - u - 3\ln|1-u| + C \tag{4}$$

Substituting from (4) into (1) and replacing u by $\ln x$, we get

$$\int \frac{(2+\ln^2 x)\, dx}{x(1-\ln x)} = -\frac{1}{2}\ln^2 x - \ln x - 3\ln|1-\ln x| + C$$

In Exercises 31–36 evaluate the definite integral.

32. $\displaystyle \int_4^5 \frac{x}{4-x^2}\, dx$

SOLUTION: Let $u = 4 - x^2$, and $du = -2x\, dx$. When $x = 4$, then $u = -12$; when $x = 5$, then $u = -21$. Thus,

$$\int_4^5 \frac{x}{4-x^2}\, dx = \int_{-12}^{-21} \frac{-\frac{1}{2}\, du}{u}$$

$$= -\frac{1}{2}\ln|u| \Big]_{-12}^{-21}$$

$$= -\frac{1}{2}(\ln 21 - \ln 12)$$

$$= -\frac{1}{2}\ln \frac{21}{12} = -\frac{1}{2}\ln \frac{7}{4}$$

36. $\displaystyle \int_2^4 \frac{\ln x}{x}\, dx$

SOLUTION: Let $u = \ln x$, and $du = (1/x)\,dx$. When $x = 2$, then $u = \ln 2$; when $x = 4$, then $u = \ln 4$.

$$\int_2^4 \frac{\ln x}{x}\,dx = \int_{\ln 2}^{\ln 4} u\,du$$

$$= \frac{1}{2}u^2 \Big]_{\ln 2}^{\ln 4}$$

$$= \frac{1}{2}(\ln^2 4 - \ln^2 2)$$

40. If $f(x) = (x + 2)/(x - 3)$, find the average value of f on the interval $[4, 6]$.

SOLUTION: The average value of f on the interval $[4, 6]$ is given by

$$\text{A.V.} = \frac{1}{6 - 4}\int_4^6 \frac{x + 2}{x - 3}\,dx$$

$$= \frac{1}{2}\int_4^6 \left(1 + \frac{5}{x - 3}\right)dx$$

$$= \frac{1}{2}\left[x + 5\ln|x - 3|\right]_4^6$$

$$= \frac{1}{2}[(6 + 5\ln 3) - (4 + 5\ln 1)]$$

$$= 1 + \frac{5}{2}\ln 3$$

44. Prove that $\displaystyle\lim_{x \to +\infty} \frac{\ln x}{x} = 0$.

SOLUTION: Because $x \to +\infty$, we may assume that $x \geq 1$. Because $\ln 1 = 0$ and the ln function is increasing, we have

$$\ln x \geq 0 \qquad \text{if } x \geq 1 \tag{1}$$

Furthermore, if $t \geq 1$, then $\sqrt{t} \leq t$, so

$$\frac{1}{\sqrt{t}} \geq \frac{1}{t} \qquad \text{if } t \geq 1$$

Therefore, by Theorem 6.3.8,

$$\int_1^x \frac{1}{\sqrt{t}}\,dt \geq \int_1^x \frac{1}{t}\,dt \qquad \text{if } x \geq 1 \tag{2}$$

Now

$$\int_1^x \frac{1}{\sqrt{t}}\,dt = \int_1^x t^{-1/2}\,dt$$

$$= 2t^{1/2}\Big]_1^x$$

$$= 2(x^{1/2} - 1) \tag{3}$$

and

$$\int_1^x \frac{1}{t}\,dt = \ln x \tag{4}$$

Substituting from (3) and (4) into (2), we have

$$2(x^{1/2} - 1) \geq \ln x \qquad \text{if } x \geq 1 \tag{5}$$

Combining (1) and (5), we have

$$0 \leq \ln x \leq 2(x^{1/2} - 1) \qquad \text{if } x \geq 1$$

Dividing by x if $x \geq 1$, we obtain

$$0 \leq \frac{\ln x}{x} \leq 2\left(\frac{1}{x^{1/2}} - \frac{1}{x}\right) \qquad \text{if } x \geq 1$$

Because

$$\lim_{x \to +\infty} 2\left(\frac{1}{x^{1/2}} - \frac{1}{x}\right) = 0$$

by the squeeze theorem we conclude that

$$\lim_{x \to +\infty} \frac{\ln x}{x} = 0$$

8.5 THE EXPONENTIAL FUNCTION

8.5.1 Definition The *exponential function* is the inverse of the natural logarithmic function, and it is defined by

$$\exp(x) = y \quad \text{if and only if } x = \ln y$$

8.5.2 Definition If a is any positive number and x is any real number, we define

$$a^x = \exp(x \ln a)$$

8.5.3 Theorem If a is any positive number and x is any real number,

$$\ln a^x = x \ln a$$

8.5.4 Definition The number e is defined by the formula

$$e = \exp 1$$

8.5.5 Theorem $\ln e = 1$

8.5.6 Theorem For all values of x,

$$\exp(x) = e^x$$

8.5.7 Theorem If a and b are any real numbers, then

$$e^a \cdot e^b = e^{a+b}$$

8.5.8 Theorem If a and b are any real numbers, then

$$e^a \div e^b = e^{a-b}$$

8.5.9 Theorem If a and b are any real numbers, then

$$(e^a)^b = e^{ab}$$

8.5.10 Theorem If u is a differentiable function of x,

$$D_x(e^u) = e^u \, D_x u$$

8.5.11 Theorem $\int e^u \, du = e^u + C$

The following limits are important.

$$\lim_{x \to +\infty} e^x = +\infty$$

$$\lim_{x \to -\infty} e^x = 0$$

$$\lim_{h \to 0} (1 + h)^{1/h} = e$$

Because the exponential function and the natural logarithmic function are inverse functions, we have

$$\exp(\ln x) = x \quad \text{and} \quad \ln(\exp x) = x$$

or, equivalently,

$$e^{\ln x} = x \quad \text{and} \quad \ln e^x = x$$

Exercises 8.5

In Exercises 1–16 find dy/dx.

4. $y = e^{x^2 - 3}$

SOLUTION: We apply Theorem 8.5.10 with $u = x^2 - 3$.

$$\frac{dy}{dx} = e^{x^2 - 3} D_x(x^2 - 3)$$

$$= 2xe^{x^2 - 3}$$

8. $y = \dfrac{e^x}{x}$

SOLUTION: We apply the derivative of a quotient rule.

$$\frac{dy}{dx} = \frac{xe^x - e^x \cdot 1}{x^2}$$

$$= \frac{e^x(x - 1)}{x^2}$$

12. $y = \ln \dfrac{e^{4x} - 1}{e^{4x} + 1}$

SOLUTION: Before differentiating we apply Theorem 8.3.5.

$$y = \ln(e^{4x} - 1) - \ln(e^{4x} + 1)$$

We differentiate, applying Theorems 8.3.2 and 8.5.10. Thus,

$$\frac{dy}{dx} = \frac{1}{e^{4x} - 1} D_x(e^{4x} - 1) - \frac{1}{e^{4x} + 1} D_x(e^{4x} + 1)$$

$$= \frac{4e^{4x}}{e^{4x} - 1} - \frac{4e^{4x}}{e^{4x} + 1}$$

$$= 4e^{4x}\left(\frac{1}{e^{4x} - 1} - \frac{1}{e^{4x} + 1}\right)$$

$$= 4e^{4x}\frac{(e^{4x} + 1) - (e^{4x} - 1)}{(e^{4x} - 1)(e^{4x} + 1)}$$

$$= 4e^{4x} \frac{2}{(e^{4x})^2 - 1}$$

$$= \frac{8e^{4x}}{e^{8x} - 1}$$

16. $y = e^{x/\sqrt{4+x^2}}$

SOLUTION: First, we take the natural logarithm on each side

$$\ln y = \ln e^{x/\sqrt{4+x^2}}$$

$$= x(4 + x^2)^{-1/2} \tag{1}$$

Differentiating on both sides of (1), we obtain

$$\frac{1}{y} \cdot \frac{dy}{dx} = x\left(-\frac{1}{2}\right)(4 + x^2)^{-3/2}(2x) + (4 + x^2)^{-1/2}$$

$$\frac{1}{y} \cdot \frac{dy}{dx} = (4 + x^2)^{-3/2}[-x^2 + (4 + x^2)]$$

$$\frac{dy}{dx} = 4(4 + x^2)^{-3/2}y$$

Substituting the given value for y, we have

$$\frac{dy}{dx} = 4(4 + x^2)^{-3/2}e^{x/\sqrt{4+x^2}}$$

20. Find $D_x y$ by implicit differentiation.

$$ye^{2x} + xe^{2y} = 1$$

SOLUTION: Differentiating on both sides of the given equation with respect to x gives

$$ye^{2x} \cdot 2 + e^{2x}D_x y + xe^{2y} \cdot 2D_x y + e^{2y} \cdot 1 = 0$$

$$(e^{2x} + 2xe^{2y})D_x y = -2ye^{2x} - e^{2y}$$

$$D_x y = -\frac{2ye^{2x} + e^{2y}}{e^{2x} + 2xe^{2y}}$$

In Exercises 21–28 evaluate the indefinite integral.

24. $\int e^{3x}e^{2x} \, dx$

SOLUTION: By Theorem 8.5.7,

$$\int e^{3x}e^{2x} \, dx = \int e^{5x} \, dx \tag{1}$$

Let $u = 5x$. Then $du = 5 \, dx$. Substituting into the right-hand side of (1), we get

$$\int e^{3x}e^{2x} \, dx = \frac{1}{5}\int e^u \, du$$

$$= \frac{1}{5}e^u + C$$

$$= \frac{1}{5}e^{5x} + C$$

28. $\displaystyle\int \frac{dx}{1+e^x}$

SOLUTION: We multiply the numerator and denominator of the given fraction by e^{-x}.

$$\int \frac{dx}{1+e^x} = \int \frac{e^{-x}\,dx}{e^{-x}+1} \tag{1}$$

Let $u = e^{-x} + 1$. Then $du = -e^{-x}\,dx$. With these substitutions in the right-hand side of (1) we get

$$\int \frac{dx}{1+e^x} = -\int \frac{du}{u}$$

$$= -\ln|u| + C$$

$$= -\ln|e^{-x}+1| + C$$

$$= -\ln\left(\frac{1+e^x}{e^x}\right) + C$$

$$= -[\ln(1+e^x) - \ln e^x] + C$$

$$= -\ln(1+e^x) + x + C \tag{2}$$

Note: The obvious choices for u, namely $u = e^x$ and $u = 1 + e^x$, lead to integrals for which we have no integration formula. However, in Section 10.4 we learn techniques for such integrations.

In Exercises 29–36 evaluate the definite integral.

32. $\displaystyle\int_1^e \frac{\ln x}{x}\,dx$

SOLUTION: Let $u = \ln x$ and $du = dx/x$. When $x = 1$, then $u = 0$; when $x = e$, then $u = 1$. Thus,

$$\int_1^e \frac{\ln x}{x}\,dx = \int_0^1 u\,du$$

$$= \frac{1}{2}u^2 \Big]_0^1$$

$$= \frac{1}{2}$$

36. $\displaystyle\int_1^2 \frac{e^x}{e^x + e}\,dx$

SOLUTION: Note that e is a constant, and thus $D_x e = 0$. Let $u = e^x + e$. Then $du = e^x\,dx$; when $x = 1$, $u = 2e$; when $x = 2$, $u = e^2 + e$. Thus,

$$\int_1^2 \frac{e^x}{e^x + e}\,dx = \int_{2e}^{e^2+e} \frac{du}{u}$$

$$= \ln|u| \Big]_{2e}^{e^2+e}$$

$$= \ln(e^2 + e) - \ln(2e)$$

$$= \ln \frac{e^2 + e}{2e}$$

$$= \ln \frac{1}{2}(e+1)$$

40. Find the relative extrema of f, the points of inflection of the graph of f, the intervals on which f is increasing, the intervals on which f is decreasing, where the graph is concave upward, where the graph is concave downward, and the slope of any inflectional tangent. Draw a sketch of the graph of f.

$$f(x) = e^{-x^2}$$

SOLUTION: f is continuous at every real number.

$$f'(x) = e^{-x^2}(-2x) = \frac{-2x}{e^{x^2}}$$

$$f''(x) = e^{-x^2}(-2) + (-2x)e^{-x^2}(-2x) = \frac{-2(1-2x^2)}{e^{x^2}}$$

Both $f'(x)$ and $f''(x)$ are defined for all x. Because $f'(0) = 0$, then 0 is a critical number. If $f''(x) = 0$, then $x = \pm \frac{1}{2}\sqrt{2}$. Table 40a gives the algebraic sign of both $f'(x)$ and $f''(x)$ in each of the intervals determined by the numbers $-\frac{1}{2}\sqrt{2}$, 0, and $\frac{1}{2}\sqrt{2}$. Table 40 b gives the function value and the value of the derivative at each of these numbers. We use the information in the tables to draw a sketch of the graph, which is shown in Fig. 8. 5. 40.

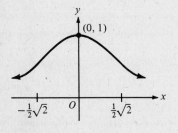

Figure 8. 5. 40

Table 40a

	$-\infty$		$-\frac{1}{2}\sqrt{2}$		0		$\frac{1}{2}\sqrt{2}$		$+\infty$
$f'(x)$	+		+		−		−		
f is	increasing		increasing		decreasing		decreasing		
$f''(x)$	+		−		−		+		
Graph is	concave upward		concave downward		concave downward		concave upward		

Table 40b

x	$f(x)$	$f'(x)$	Description of the Graph
$\frac{1}{2}\sqrt{2} = 0.7$	$e^{-1/2} = 0.6$	$-\left(\frac{2}{e}\right)^{1/2} = -0.9$	Graph has a point of inflection.
0	1	0	f has a relative maximum value.
$-\frac{1}{2}\sqrt{2} = -0.7$	$e^{-1/2} = 0.6$	$\left(\frac{2}{e}\right)^{1/2} = 0.9$	Graph has a point of inflection.

44. Compute the value of e^e to two decimal places.

SOLUTION: We are given that e = 2.718, correct to three decimal places. We apply either the table for powers of e given in the appendix (in the text) or a calculator to obtain

$$e^e = e^{2.178} = 15.15$$

48. Prove that the most general function that is equal to its derivative is given by $f(x) = ke^x$. (*Hint:* Let $y = f(x)$ and solve the differential equation $dy/dx = y$.)

SOLUTION: Suppose that $f(x) = f'(x)$. Then let $y = f(x)$, and we have

$$\frac{dy}{dx} = f'(x) = f(x) = y$$

We solve the differential equation

$$\frac{dy}{dx} = y$$

Separating the variables, we have

$$\frac{dy}{y} = dx$$

$$\int \frac{dy}{y} = \int dx$$

$$\ln|y| = x + c$$

Taking the exponential on both sides, we get

$$|y| = e^{x+c}$$
$$= e^x e^c$$

Thus,

$$y = \begin{cases} e^c e^x & \text{if } y > 0 \\ -e^c e^x & \text{if } y < 0 \end{cases}$$

Let $k = \pm e^c$. Then

$$y = ke^x$$

or, equivalently,

$$f(x) = ke^x$$

52. A simple electric circuit containing no capacitors, a resistance of R ohms, and an inductance of L henrys has the electromotive force cut off when the current is I_0 amperes. The current dies down so that at t sec the current is i amperes and

$$i = I_0 e^{-(R/L)t} \tag{1}$$

Show that the rate of change of the current is proportional to the current.

SOLUTION: I_0, e, R, and L are all constants. The variables are t and i. We find $D_t i$ by differentiating on both sides of (1).

$$D_t i = I_0 e^{-(R/L)t} \cdot D_t(-R/L)t$$
$$= [I_0 e^{-(R/L)t}](-R/L) \tag{2}$$

Substituting from (1) into (2), we obtain

$$D_t i = i(-R/L) \tag{3}$$

Let $k = -R/L$. Then from (3) we have

$$D_t i = ki$$

Therefore, the rate of change of i, the current, is proportional to i.

56. Prove: $\lim\limits_{x \to +\infty} e^x = +\infty$ by showing that for any $N > 0$ there exists an $M > 0$ such that $e^x > N$ whenever $x > M$.

SOLUTION: Because $D_x e^x = e^x > 0$, the exponential function is increasing on $(-\infty, +\infty)$. Thus, whenever $x > M$, then $\exp(x) > \exp(M)$. For any $N > 1$, let $M = \ln N > 0$. Then whenever $x > M$, we have

$$
\begin{aligned}
\exp(x) &> \exp(M) \\
&= \exp(\ln N) \\
&= N
\end{aligned}
$$

or, equivalently, whenever $x > M$, then $e^x > N$. For any N with $0 < N \le 1$, let $M = 1$. Then whenever $x > M$, we have

$$
\begin{aligned}
\exp(x) &> \exp(M) \\
&= \exp(1) \\
&= e \\
&> N
\end{aligned}
$$

or, equivalently, whenever $x > M$, then $e^x > N$.

60. An important function in statistics is the probability density function defined by

$$
f(x) = \frac{1}{\sigma\sqrt{2\pi}} \exp\left[-\frac{(x - \mu)^2}{2\sigma^2} \right]
$$

where σ and μ are constants such that $\sigma > 0$ and μ is any real number.
Find: **(a)** the relative extrema of f; **(b)** the points of inflection of the graph of f; **(c)** $\lim\limits_{x \to +\infty} f(x)$; **(d)** $\lim\limits_{x \to -\infty} f(x)$. **(e)** Draw a sketch of the graph of f.

SOLUTION: We note that $f(x) > 0$ for all x. Also,

(a) $f'(x) = \dfrac{1}{\sigma\sqrt{2\pi}} \exp\left[-\dfrac{(x - \mu)^2}{2\sigma^2} \right] \left[-\dfrac{x - \mu}{\sigma^2} \right]$

$$
= -\frac{x - \mu}{\sigma^2} f(x) \tag{1}
$$

If $f'(x) = 0$, then $-(x - \mu)/\sigma^2 = 0$, or $x = \mu$. Thus, μ is the only critical number of the function f. Furthermore, if $x < \mu$, then $f'(x) > 0$, and if $x > \mu$, then $f'(x) < 0$. Thus the function f has a relative maximum value at the point where $x = \mu$. Because $f(\mu) = 1/(\sigma\sqrt{2\pi})$, then the point $(\mu, 1/(\sigma\sqrt{2\pi}))$ is a relative maximum point on the graph of f. And because $f'(\mu) = 0$, then the tangent line at the relative maximum point is a horizontal line.

(b) We apply the derivative of a product rule to differentiate both sides of Eq. (1). Thus,

$$
f''(x) = -\frac{x - \mu}{\sigma^2} f'(x) - \frac{1}{\sigma^2} f(x) \tag{2}
$$

Substituting from Eq. (1) into Eq. (2), we have

$$
f''(x) = \left[\frac{(x - \mu)^2}{\sigma^4} - \frac{1}{\sigma^2} \right] f(x) \tag{3}
$$

If $f''(x) = 0$, then $(x - \mu)^2 = \sigma^2$, so $x - \mu = \pm\sigma$, or $x = \mu \pm \sigma$. Furthermore, $f''(x)$ changes sign at the points where $x = \mu \pm \sigma$, so these points are points of inflection.

(c) $\lim\limits_{x \to +\infty} f(x) = \lim\limits_{x \to +\infty} \dfrac{1}{\sigma\sqrt{2\pi}} \exp\left[-\dfrac{(x-\mu)^2}{2\sigma^2}\right]$

$$= \dfrac{1}{\sigma\sqrt{2\pi}} \lim\limits_{x \to +\infty} \exp\left[-\dfrac{(x-\mu)^2}{2\sigma^2}\right] \qquad (4)$$

Let $z = -(x-\mu)^2/2\sigma^2$. If $x \to +\infty$, then $z \to -\infty$. Thus,

$$\lim\limits_{x \to +\infty} \exp\left[-\dfrac{(x-\mu)^2}{2\sigma^2}\right] = \lim\limits_{z \to -\infty} e^z = 0 \qquad (5)$$

Substituting from (5) into (4), we obtain

$$\lim\limits_{x \to +\infty} f(x) = 0$$

(d) As in part (c), let $z = -(x-\mu)^2/(2\sigma^2)$. If $x \to -\infty$, then $z \to -\infty$. Thus,

$$\lim\limits_{x \to -\infty} f(x) = \lim\limits_{z \to -\infty} \dfrac{1}{\sigma\sqrt{2\pi}} e^z = 0$$

(e) A sketch of the graph of f is shown in Fig. 8. 5. 60.

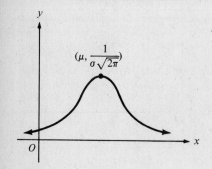

Figure 8. 5. 60

8.6 OTHER EXPONENTIAL AND LOGARITHMIC FUNCTIONS

8.6.1 Definition   If a is any positive number and x is any real number, then the function f defined by

$$f(x) = a^x$$

is called the *exponential function* to the base a.

8.6.2 Theorem   If a is any positive number and u is a differentiable function of x,

$$D_x(a^u) = a^u \ln a\, D_x u$$

8.6.3 Theorem   If a is any positive number other than 1,

$$\int a^u\, du = \dfrac{a^u}{\ln a} + C$$

8.6.4 Definition   If a is any positive number except 1, the *logarithmic function* to the base a is the inverse of the exponential function to the base a, and we write

$$y = \log_a x \text{ if and only if } a^y = x$$

8.6.5 Theorem   If u is a differentiable function of x,

$$D_x(\log_a u) = \dfrac{\log_a e}{u} D_x u$$

8.6.6 Theorem   If n is any real number and the function f is defined by

$$f(x) = x^n \qquad \text{where } x > 0$$

then

$$f'(x) = nx^{n-1}$$

Distinguish carefully between x^a and a^x, where a is a real number. We have

$$D_x(a^x) = a^x \ln a \qquad \text{if } a > 0$$
$$D_x(x^a) = ax^{a-1} \qquad \text{if } x > 0$$

In each of the above differentiation formulas, the number a is a *constant*. Note that $D_x(x^x)$ cannot be found by either of the above formulas. Finding this derivative requires logarithmic differentiation.

The following formula is used to express the logarithmic function to the base a in terms of the natural logarithmic function.

$$\log_a x = \frac{\ln x}{\ln a}$$

A special case of the above is

$$\log_a e = \frac{1}{\ln a}$$

Exercises 8.6

In Exercises 1–24 find the derivative of the given function.

4. $g(x) = 10^{x^2 - 2x}$

SOLUTION: We apply Theorem 8.6.2 with $u = x^2 - 2x$ and $a = 10$. Thus,

$$g'(x) = 10^{x^2 - 2x}(\ln 10) \, D_x(x^2 - 2x)$$
$$= 10^{x^2 - 2x}(\ln 10)(2x - 2)$$
$$= 2(\ln 10)(x - 1)10^{x^2 - 2x}$$

8. $f(x) = (x^3 + 3)2^{-7x}$

SOLUTION:

$$f'(x) = (x^3 + 3) \, D_x(2^{-7x}) + 2^{-7x} \, D_x(x^3 + 3)$$
$$= (x^3 + 3)2^{-7x}(\ln 2)(-7) + 2^{-7x}(3x^2)$$
$$= 2^{-7x}[3x^2 - 7(\ln 2)(x^3 + 3)]$$

12. $f(x) = \log_a[\log_a(\log_a x)]$

SOLUTION: First we apply Theorem 8.6.5 with $u = \log_a(\log_a x)$.

$$f'(x) = \frac{\log_a e}{\log_a(\log_a x)} \cdot D_x[\log_a(\log_a x)] \qquad (1)$$

Then we apply Theorem 8.6.5 again with $u = \log_a x$.

$$D_x[\log_a(\log_a x)] = \frac{\log_a e}{\log_a x} \cdot D_x(\log_a x)$$

$$= \frac{\log_a e}{\log_a x} \cdot \frac{\log_a e}{x} \qquad (2)$$

Substituting from Eq. (2) into Eq. (1), we obtain

$$f'(x) = \frac{\log_a e}{\log_a(\log_a x)} \cdot \frac{\log_a e}{\log_a x} \cdot \frac{\log_a e}{x}$$

$$= \frac{\log_a^3 e}{x(\log_a x)[\log_a(\log_a x)]}$$

16. $f(x) = x^{x^2}$; $x > 0$

SOLUTION: Let $y = f(x)$.

$$y = x^{x^2}$$
$$\ln y = \ln x^{x^2}$$
$$\ln y = x^2 \ln x$$

Differentiating on both sides of the above equation with respect to x, we obtain

$$\frac{1}{y} D_x y = x^2 \cdot \frac{1}{x} + \ln x(2x)$$

$$D_x y = yx(1 + 2 \ln x)$$

$$= x^{x^2} \cdot x(1 + 2 \ln x)$$

$$= x^{x^2+1}(1 + 2 \ln x)$$

20. $g(t) = (\cos t)^t$; $\cos t > 0$

SOLUTION: We apply the method of logarithmic differentiation. Let

$$s = (\cos t)^t \tag{1}$$

Then

$$\ln s = \ln(\cos t)^t$$
$$= t \ln(\cos t)$$

Differentiating on both sides with respect to t, we have

$$\frac{1}{s} D_t s = t D_t \ln(\cos t) + \ln(\cos t) D_t t$$

$$= t \left(\frac{1}{\cos t} \right) D_t(\cos t) + \ln(\cos t)$$

$$= \frac{-t \sin t}{\cos t} + \ln(\cos t)$$

Multiplying on both sides by s and substituting for s from Eq. (1), we obtain

$$D_t s = (\cos t)^t \left[\frac{-t \sin t}{\cos t} + \ln(\cos t) \right]$$

$$= (\cos t)^{t-1}[-t \sin t + \cos t \ln(\cos t)]$$

Thus,

$$g'(t) = (\cos t)^{t-1}[-t \sin t + \cos t \ln(\cos t)]$$

24. $f(x) = (\ln x)^{\ln x}$; $x > 1$

SOLUTION: Because $x > 1$, then $\ln x > 0$. Thus, we may apply the method of logarithmic differentiation. Let $y = f(x)$.

$$y = (\ln x)^{\ln x}$$
$$\ln y = \ln(\ln x)^{\ln x}$$
$$= (\ln x)[\ln(\ln x)]$$

Differentiating with respect to x, we obtain

$$\frac{1}{y} D_x y = (\ln x) D_x[\ln(\ln x)] + [\ln(\ln x)] D_x(\ln x)$$

$$= \ln x \cdot \frac{1}{\ln x} \cdot \frac{1}{x} + [\ln(\ln x)] \cdot \frac{1}{x}$$

$$= \frac{1}{x}[1 + \ln(\ln x)]$$

Thus,

$$D_x y = \frac{y}{x}[1 + \ln(\ln x)]$$

$$= \frac{(\ln x)^{\ln x}[1 + \ln(\ln x)]}{x}$$

In Exercises 25–32 evaluate the indefinite integral.

28. $\displaystyle\int 5^{x^4+2x}(2x^3 + 1)\, dx$

SOLUTION: Let $u = x^4 + 2x$. Then

$$du = (4x^3 + 2)\, dx = 2(2x^3 + 1)\, dx$$

Thus,

$$\int 5^{x^4+2x}(2x^3 + 1)\, dx = \frac{1}{2}\int 5^u\, du$$

$$= \frac{1}{2} \cdot \frac{5^u}{\ln 5} + C$$

$$= \frac{5^{x^4+2x}}{2 \ln 5} + C$$

32. $\displaystyle\int \frac{4^{\ln(1/x)}}{x}\, dx$

SOLUTION: Because $\ln(1/x) = \ln 1 - \ln x = -\ln x$, then

$$\int \frac{4^{\ln(1/x)}}{x}\, dx = \int \frac{4^{-\ln x}}{x}\, dx \qquad (1)$$

Let $u = -\ln x$. Then $du = -dx/x$. With these replacements in the right side of (1), we have

$$\int \frac{4^{\ln(1/x)}}{x}\, dx = -\int 4^u\, du$$

$$= \frac{-4^u}{\ln 4} + C$$

$$= \frac{-4^{-\ln x}}{\ln 4} + C$$

$$= \frac{-1}{4^{\ln x}\ln 4} + C$$

36. Use differentials to find an appropriate value of the given logarithm and express the answer to three decimal places.

$$\log_{10} 1.015$$

SOLUTION: Let f be the function defined by $f(x) = \log_{10} x$. If $y = f(x)$, then

$$\Delta y = f(x + \Delta x) - f(x)$$
$$f(x + \Delta x) = f(x) + \Delta y$$
$$f(x + \Delta x) \approx f(x) + dy \tag{1}$$

Because

$$dy = f'(x) \, \Delta x$$

$$= \frac{\log_{10} e}{x} \Delta x$$

From (1) we have

$$f(x + \Delta x) \approx f(x) + \frac{\log_{10} e}{x} \cdot \Delta x \tag{2}$$

Let $x = 1$ and $\Delta x = .015$ in (2). Then

$$\log_{10} 1.015 \approx \log_{10} 1 + \frac{\log_{10} e}{1} (.015)$$

$$\approx 0 + (.4343)(.015) \approx .006$$

40. Prove the given property if a is any positive number and x and y are any real numbers: $(a^x)^y = a^{xy}$.

SOLUTION: By Definition 8.5.2 and Theorem 8.5.6, we have

$$a^x = e^{x \ln a} \tag{1}$$

Thus, from Eq. (1) and Theorem 8.5.9 we obtain

$$(a^x)^y = (e^{x \ln a})^y$$
$$= e^{(x \ln a)y}$$
$$= e^{xy \ln a} = a^{xy}$$

44. Prove the given property if a is any positive number and x and y are any positive numbers.

$$\log_a(x \div y) = \log_a x - \log_a y$$

SOLUTION: Because $\log_a x = \ln x / \ln a$

$$\log_a(x \div y) = \frac{\ln(x \div y)}{\ln a}$$

$$= \frac{\ln x - \ln y}{\ln a}$$

$$= \frac{\ln x}{\ln a} - \frac{\ln y}{\ln a} = \log_a x - \log_a y$$

48. A particle moves along a straight line according to the equation of motion $s = t^{1/t}$, where s ft is the directed distance of the particle from the starting point at t sec. Find the velocity and acceleration at 2 sec.

SOLUTION: We use logarithmic differentiation. If $t > 0$, we have

$$s = t^{1/t}$$

$$\ln s = \ln t^{1/t}$$

$$\ln s = \frac{\ln t}{t}$$

$$\frac{1}{s}\frac{ds}{dt} = \frac{t\left(\frac{1}{t}\right) - \ln t}{t^2}$$

$$\frac{ds}{dt} = \frac{s(1 - \ln t)}{t^2} \qquad (1)$$

When $t = 2$, then $s = 2^{1/2} = \sqrt{2}$. Thus, from (1) we have

$$\left.\frac{ds}{dt}\right]_{t=2} = \frac{\sqrt{2}(1 - \ln 2)}{4} \qquad (2)$$

Hence, the velocity at 2 sec is $\frac{1}{4}\sqrt{2}(1 - \ln 2)$ ft/sec. Because $ds/dt = v$, from (1) we have

$$vt^2 = s(1 - \ln t)$$

Differentiating on both sides with respect to t, we obtain

$$v(2t) + t^2\frac{dv}{dt} = s\left(-\frac{1}{t}\right) + (1 - \ln t)\frac{ds}{dt}$$

$$\frac{dv}{dt} = \frac{-\frac{s}{t} + (1 - \ln t)v - 2tv}{t^2}$$

$$= \frac{-\frac{s}{t} - v(2t - 1 + \ln t)}{t^2} \qquad (3)$$

When $t = 2$, then $s = \sqrt{2}$, and from Eq. (2) we have $v = \frac{1}{4}\sqrt{2}(1 - \ln 2)$. Substituting these values in Eq. (3), we get

$$\left.\frac{dv}{dt}\right]_{t=2} = \frac{-\frac{1}{2}\sqrt{2} - \frac{1}{4}\sqrt{2}(1 - \ln 2)(3 + \ln 2)}{4}$$

$$= \frac{-\sqrt{2}[2 + (3 - 2\ln 2 - \ln^2 2)]}{16}$$

$$= \frac{-\sqrt{2}(5 - 2\ln 2 - \ln^2 2)}{16}$$

Therefore, when $t = 2$ the acceleration is $-\frac{1}{16}\sqrt{2}(5 - 2\ln 2 - \ln^2 2)$ ft/sec².

52. Find the area of the region bounded by the graph of $y = 3^x$ and the lines $x = 1$ and $y = 1$.

SOLUTION: A sketch of the region and a rectangular element of area is shown in Fig. 8. 6. 52. The width of the element of area is $\Delta_i x$ units and the length is $(3^{\bar{x}_i} - 1)$ units. Thus, if A square units is the area of the region, then

$$A = \lim_{\|\Delta\|\to 0}\sum_{i=1}^{n}(3^{\bar{x}_i} - 1)\,\Delta_i x$$

$$= \int_0^1 (3^x - 1)\,dx$$

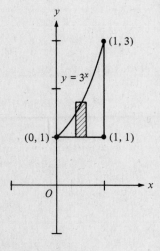

Figure 8. 6. 52

$$= \frac{3^x}{\ln 3} - x \Big]_0^1$$

$$= \left(\frac{3}{\ln 3} - 1\right) - \frac{1}{\ln 3}$$

$$= \frac{2}{\ln 3} - 1$$

The area of the region is $(2/\ln 3 - 1)$ square units.

8.7 LAWS OF GROWTH AND DECAY

If the variable x is a function of time t and if the rate of change of x with respect to time is proportional to x, the amount present, then

$$\frac{dx}{dt} = kx$$

from which we have by separating the variables and integrating on both sides

$$x = Ce^{kt}$$

An important special case of the above is the formula for compound interest. If P dollars is invested at an annual rate of $100\,i$ percent with interest compounded continuously, then the amount after t years is A dollars with

$$A = Pe^{it}$$

Calculate decimal approximations for the answers to the exercises in this section. You may use either a hand-held calculator that has the exponential and natural logarithmic function keys or use the tables in the appendix of the text.

Exercises 8.7

4. Sugar decomposes in water at a rate proportional to the amount still unchanged. If there were 50 lb of sugar present initially and at the end of 5 hr this is reduced to 20 lb, how long will it take until 90% of the sugar is decomposed?

SOLUTION: Let t equal the number of hours that have elapsed since the sugar began to decompose and x the number of pounds of sugar remaining at t hours. Because dx/dt is the rate of change of the amount of sugar present, which is the rate at which the sugar decomposes, we are given that

$$\frac{dx}{dt} = kx$$

Table 4 gives the initial conditions. The complete solution of the differential equation is

$$x = Ce^{kt}$$

Table 4

t	0	5	$\bar{t}$
x	50	20	5

Because $x = 50$ when $t = 0$, then $C = 50$, and we have

$$x = 50e^{kt}$$
$$= 50(e^k)^t \tag{1}$$

Because $x = 20$ when $t = 5$, we have

$$20 = 50(e^k)^5$$
$$\tfrac{2}{5} = (e^k)^5$$
$$e^k = (\tfrac{2}{5})^{1/5}$$

Substituting this value of e^k into Eq. (1), we get

$$x = 50(\tfrac{2}{5})^{t/5} \tag{2}$$

When 90% of the sugar is decomposed, $x = 5$. When $x = 5$, $t = \bar{t}$. Substituting into (2), we get

$$5 = 50(\tfrac{2}{5})^{\bar{t}/5}$$

$$\tfrac{1}{10} = (\tfrac{2}{5})^{\bar{t}/5}$$

$$\ln(\tfrac{1}{10}) = \ln(\tfrac{2}{5})^{\bar{t}/5}$$

$$= \frac{\bar{t}}{5}\ln(\tfrac{2}{5})$$

Thus,

$$\bar{t} = \frac{5 \ln(\tfrac{1}{10})}{\ln(\tfrac{2}{5})}$$

$$= \frac{-5 \ln 10}{\ln 2 - \ln 5} = 12.6$$

Hence, it takes 12.6 hours for 90% of the sugar to decompose.

8. There are 100 gal of brine in a tank and the brine contains 70 lb of dissolved salt. Fresh water runs into the tank at the rate of 3 gal/min, and the mixture, kept uniform by stirring, runs out at the same rate. How many pounds of salt are there in the tank at the end of 1 hr?

SOLUTION: Let x pounds be the weight of the salt that is left in the tank after t minutes. Because the tank contains 100 gal of brine, after t minutes there are $\frac{1}{100}x$ pounds of salt per gallon of brine. Furthermore, because the brine runs out of the tank at the rate of 3 gal/min, then the rate at which the salt leaves the tank is $\frac{3}{100}x$ pounds per minute. Therefore,

$$\frac{dx}{x} = -\frac{3\,dt}{100} \tag{1}$$

We want to find x when $t = 60$, given that $x = 70$ when $t = 0$. From Eq. (1) we obtain

$$\frac{dx}{x} = -\frac{3\,dt}{100}$$

$$\int \frac{dx}{x} = \int \frac{-3\,dt}{100}$$

$$\ln|x| = \frac{-3t}{100} + k$$

$$x = Ce^{-3t/100}$$

Because $x = 70$ when $t = 0$, then $C = 70$. Thus,

$$x = 70e^{-3t/100}$$

Let $x = x_{60}$ when $t = 60$. Then

$$x_{60} = 70e^{-3(60)/100}$$
$$= 70e^{-1.8}$$
$$= 70(0.165) = 11.6$$

There are 11.6 pounds of salt in the tank after 1 hour.

12. When a simple electric circuit, containing no capacitors but having inductance and resistance, has the electromotive force removed, the rate of decrease of the current is proportional to the current. The current is i amperes at t sec after the cutoff, and $i = 40$ when $t = 0$. If the current dies down to 15 amperes in 0.01 sec, find i in terms of t.

SOLUTION: We are given that $di/dt = ki$. Table 12 gives the initial conditions. The complete solution of the differential equation is

$$i = Ce^{kt}$$

Table 12

t	0	0.01
i	40	15

Because $i = 40$ when $t = 0$, then $C = 40$, and thus

$$i = 40e^{kt} \tag{1}$$

We are also given that $i = 15$ when $t = 0.01$. Thus,

$$15 = 40e^{0.01k}$$

$$\tfrac{3}{8} = e^{0.01k}$$

$$\ln(\tfrac{3}{8}) = 0.01k$$

$$k = \frac{\ln 3 - \ln 8}{0.01} = -98.1$$

Substituting this value of k into Eq. (1), we get

$$i = 40e^{-98.1t}$$

16. A tank contains 200 liters of brine in which there are 3 kg of salt per liter. It is desired to dilute this solution by adding brine containing 1 kg of salt per liter, which flows into the tank at the rate of 4 liters/min and runs out at the same rate. When will the tank contain 1.5 kg of salt per liter?

SOLUTION: Let t equal the number of minutes that have elapsed since we began to dilute the solution and x the number of kg of salt in the mixture at t min. If we assume that the mixture is kept uniform by stirring, then at t min there are $x/200$ kg of salt per liter of mixture. Because the mixture runs out at the rate of 4 liters/min, then during each minute $4x/200$, which is $x/50$, kg of salt run out of the tank. Furthermore, because brine containing 1 kg of salt per liter runs into the tank at the rate of 4 liters/min, then during each minute 4 kg of salt run into the tank. Thus, the rate of change of x with respect to time is $4 - x/50$, and we have

$$\frac{dx}{dt} = 4 - \frac{x}{50} \tag{1}$$

Table 16 gives the initial conditions.

Table 16

t	0	$\bar{t}$
x	600	300

From Eq. (1), we have

$$\frac{dx}{dt} = -\frac{x-200}{50}$$

Hence,

$$\frac{dx}{x-200} = -\frac{dt}{50}$$

$$\int \frac{dx}{x-200} = -\int \frac{dt}{50}$$

$$\ln|x-200| = -\frac{1}{50}t + k$$

$$x - 200 = Ce^{-t/50} \tag{2}$$

Because the tank initially has 3 kg of salt per liter and the tank holds 200 liters, then $x = 600$ when $t = 0$. Substituting these values into (2) gives $C = 400$. Thus, from (2) we have

$$x = 200 + 400\,e^{-t/50} \tag{3}$$

When the tank contains 1.5 kg of salt per liter, then $x = 300$, $t = \bar{t}$. Substituting into Eq. (3), we get

$$300 = 200 + 400\,e^{-\bar{t}/50}$$

$$\tfrac{1}{4} = e^{-\bar{t}/50}$$

Taking the reciprocal of each side, we get

$$4 = e^{\bar{t}/50}$$

$$\ln 4 = \frac{\bar{t}}{50}$$

$$\bar{t} = 50 \ln 4 = 69.3$$

Thus, it takes 69.3 minutes to dilute the mixture.

20. If the purchasing power of the dollar is decreasing at the rate of 10% annually, compounded continuously, how long will it take for the purchasing power to be $0.50?

SOLUTION: Let t equal the number of years after the present and x the number of dollars in the purchasing power of a dollar after t years. Then $x = 1$ when $t = 0$, and we want to find t when $x = 0.50$. Because x is decreasing at the rate of 10% of x per year, we have

$$\frac{dx}{dt} = -0.10x \tag{1}$$

Table 20 gives the initial conditions.

Table 20

t	0	$\bar{t}$
x	1	0.50

From Eq. (1) we have

$$\frac{dx}{x} = -0.10 \, dt$$

$$\int \frac{dx}{x} = -0.10 \int dt$$

$$\ln x = -0.10t + k$$

$$x = Ce^{-0.10t}$$

Because $x = 1$ when $t = 0$, then $C = 1$ and

$$x = e^{-0.10t}$$

Substituting $x = 0.50 = \frac{1}{2}$ and $t = \bar{t}$, we get

$$\frac{1}{2} = e^{-0.10\bar{t}}$$

$$2 = e^{0.10\bar{t}}$$

$$\ln 2 = 0.10\bar{t}$$

$$\bar{t} = \frac{\ln 2}{0.10} = 6.93$$

Thus, it will take 6.93 years for the purchasing power to be 50 cents.

Review Exercises for Chapter 8

4. Determine whether the given function has an inverse. If the inverse exists, do the following: **(a)** define it and state its domain and range;   **(b)** draw sketches of the graphs of the function and its inverse on the same set of axes. If the function does not have an inverse, show that a horizontal line intersects the graph of the function in more than one point.

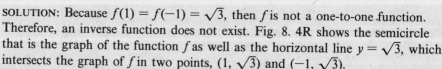

$$f(x) = \sqrt{4 - x^2}$$

SOLUTION: Because $f(1) = f(-1) = \sqrt{3}$, then f is not a one-to-one function. Therefore, an inverse function does not exist. Fig. 8. 4R shows the semicircle that is the graph of the function f as well as the horizontal line $y = \sqrt{3}$, which intersects the graph of f in two points, $(1, \sqrt{3})$ and $(-1, \sqrt{3})$.

8. (a) Prove that the given function f has an inverse,   **(b)** find $f^{-1}(x)$, and   **(c)** verify the equations of Theorem 8.1.4 for f and f^{-1}.

$$f(x) = \frac{2x - 1}{2x + 1}$$

SOLUTION:

(a) $f'(x) = \dfrac{(2x+1)(2) - (2x-1)(2)}{(2x+1)^2}$

$$= \frac{4}{(2x+1)^2}$$

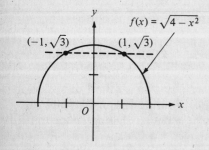

Figure 8. 8. 4

Because $f'(x) > 0$ for all $x \neq -\frac{1}{2}$, then f is increasing on $(-\infty, -\frac{1}{2})$ and increasing on $(-\frac{1}{2}, +\infty)$. Thus, f is increasing on its entire domain, so f is one-to-one and f has an inverse function.

(b) We find $f^{-1}(x)$. Let $y = f(x)$. Then

$$y = \frac{2x-1}{2x+1}$$

$$2xy + y = 2x - 1$$

$$2xy - 2x = -y - 1$$

$$x(2y - 2) = -y - 1$$

$$x = -\frac{y+1}{2(y-1)}$$

Therefore,

$$f^{-1}(y) = -\frac{y+1}{2(y-1)}$$

and

$$f^{-1}(x) = -\frac{x+1}{2(x-1)} = \frac{1+x}{2(1-x)}$$

(c) We verify the equations of Theorem 8.1.4.

$$f^{-1}(f(x)) = f^{-1}\left(\frac{2x-1}{2x+1}\right)$$

$$= \frac{1 + \dfrac{2x-1}{2x+1}}{2\left(1 - \dfrac{2x-1}{2x+1}\right)}$$

$$= \frac{(2x+1) + (2x-1)}{2[(2x+1) - (2x-1)]}$$

$$= \frac{4x}{2(2)} = x$$

and

$$f(f^{-1}(x)) = f\left(\frac{1+x}{2(1-x)}\right)$$

$$= \frac{2\left(\dfrac{1+x}{2(1-x)}\right) - 1}{2\left(\dfrac{1+x}{2(1-x)}\right) + 1}$$

$$= \frac{\dfrac{1+x}{1-x} - 1}{\dfrac{1+x}{1-x} + 1}$$

$$= \frac{(1+x) - (1-x)}{(1+x) + (1-x)}$$

$$= \frac{2x}{2} = x$$

12. Find $(f^{-1})'(d)$ if $f(x) = x^5 + x - 22$ and $d = 12$.

SOLUTION: We apply Theorem 8.2.4. Because

$$f'(x) = 5x^4 + 1$$

then $f'(x) > 0$ for all x, so f is monotonic and continuous on every closed interval. Furthermore, because $f(2) = 12$, we may apply Theorem 8.2.4 with $c = 2$ and $d = 12$. Because $f'(2) = 81$, then

$$(f^{-1})'(12) = \frac{1}{f'(2)} = \frac{1}{81}$$

In Exercises 13–24 differentiate the given function.

16. $f(x) = \dfrac{e^x}{(e^x + e^{-x})^2}$

SOLUTION:

$$f(x) = e^x(e^x + e^{-x})^{-2}$$

$$f'(x) = e^x(-2)(e^x + e^{-x})^{-3}(e^x - e^{-x}) + (e^x + e^{-x})^{-2}e^x$$

$$= e^x(e^x + e^{-x})^{-3}[-2(e^x - e^{-x}) + (e^x + e^{-x})]$$

$$= \frac{e^x(3e^{-x} - e^x)}{(e^x + e^{-x})^3} = \frac{3 - e^{2x}}{(e^x + e^{-x})^3}$$

20. $g(x) = \ln \sqrt{\dfrac{2x+1}{x-3}}$

SOLUTION: First we replace the given expression by an equivalent sum.

$$g(x) = \ln\left(\frac{2x+1}{x-3}\right)^{1/2}$$

$$= \frac{1}{2}\ln\frac{2x+1}{x-3}$$

$$= \frac{1}{2}\left[\ln(2x+1) - \ln(x-3)\right]$$

Hence,

$$g'(x) = \frac{1}{2}\left[\frac{2}{2x+1} - \frac{1}{x-3}\right]$$

$$= \frac{1}{2}\left[\frac{2(x-3) - (2x+1)}{(2x+1)(x-3)}\right] = \frac{-7}{2(2x+1)(x-3)}$$

24. $f(x) = 3^{x^{x^2}}$, $x > 0$

SOLUTION: We apply Theorem 8.6.2 with $a = 3$ and $u = x^{x^2}$.

$$f'(x) = 3^{x^{x^2}}(\ln 3)D_x(x^{x^2}) \tag{1}$$

We use logarithmic differentiation to find the derivative on the right-hand side of (1). Let

$$y = x^{x^2}$$
$$\ln y = \ln x^{x^2}$$
$$= x^2 \ln x$$

Differentiating on both sides with respect to x, we have

$$\frac{1}{y} D_x y = x^2\left(\frac{1}{x}\right) + (2x)\ln x$$

$$= x(1 + 2 \ln x)$$

Thus,

$$D_x y = xy(1 + 2 \ln x)$$
$$= x \cdot x^{x^2}(1 + 2 \ln x)$$
$$= x^{x^2+1}(1 + 2 \ln x) \tag{2}$$

Substituting the derivative given in Eq. (2) into Eq. (1), we obtain

$$f'(x) = 3^{x x^2}(\ln 3)x^{x^2+1}(1 + 2 \ln x)$$

In Exercises 25–32 evaluate the indefinite integral.

28. $\displaystyle\int \frac{10^{\ln x^2}}{x} \, dx$

SOLUTION: First, we have

$$10^{\ln x^2} = 10^{2 \ln x} = (10^2)^{\ln x} = 100^{\ln x}$$

Thus,

$$\int \frac{10^{\ln x^2}}{x} \, dx = \int \frac{100^{\ln x}}{x} \, dx \tag{1}$$

Let $u = \ln x$. Then $du = dx/x$. Substituting in the right-hand side of (1), we obtain

$$\int \frac{10^{\ln x^2}}{x} \, dx = \int 100^u \, du$$

$$= \frac{100^u}{\ln 100} + C = \frac{100^{\ln x}}{\ln 100} + C$$

32. $\displaystyle\int \frac{10^x + 1}{10^x - 1} \, dx$

SOLUTION: We multiply the numerator and denominator of the given fraction by $10^{-x/2}$. Thus,

$$\int \frac{10^x + 1}{10^x - 1} \, dx = \int \frac{10^{x/2} + 10^{-x/2}}{10^{x/2} - 10^{-x/2}} \, dx \tag{1}$$

Let $u = 10^{x/2} - 10^{-x/2}$. Then

$$du = \left[10^{x/2} \cdot \frac{1}{2} - 10^{-x/2}\left(-\frac{1}{2}\right)\right]\ln 10 \, dx$$

$$\frac{2 \, du}{\ln 10} = (10^{x/2} + 10^{-x/2}) \, dx$$

Substituting into the right-hand side of Eq. (1), we have

$$\int \frac{10^x + 1}{10^x - 1} \, dx = \frac{2}{\ln 10} \int \frac{du}{u}$$

$$= \frac{2 \ln|10^{x/2} - 10^{-x/2}|}{\ln 10} + C$$

$$= 2 \log_{10} |10^{x/2} - 10^{-x/2}| + C$$

$$= 2 \log_{10} \left| \frac{10^{x/2} - 10^{-x/2}}{1} \right| \cdot \frac{10^{x/2}}{10^{x/2}} + C$$

$$= 2 \log_{10} |10^x - 1| - 2 \log_{10} 10^{x/2} + C$$

$$= 2 \log_{10} |10^x - 1| - x + C$$

36. Evaluate the definite integral

$$\int_{1/3}^{1/2} \frac{4x^{-3} + 2}{x^{-2} - x} \, dx$$

SOLUTION: Let $u = x^{-2} - x$. Then

$$du = (-2x^{-3} - 1) \, dx$$
$$-2 \, du = (4x^{-3} + 2) \, dx$$

When $x = \frac{1}{3}$, then $u = \frac{26}{3}$; when $x = \frac{1}{2}$, then $u = \frac{7}{2}$. Thus,

$$\int_{1/3}^{1/2} \frac{4x^{-3} + 2}{x^{-2} - x} \, dx = -2 \int_{26/3}^{7/2} \frac{du}{u}$$

$$= -2 \ln u \Big]_{26/3}^{7/2}$$

$$= -2 \left[\ln \frac{7}{2} - \ln \frac{26}{3} \right]$$

$$= -2 \ln \frac{21}{52} = 2 \ln \frac{52}{21}$$

40. If $f(x) = \log_{(e^x)} (x + 1)$, find $f'(x)$.

SOLUTION: Because

$$\log_a x = \frac{\ln x}{\ln a}$$

Then

$$\log_{(e^x)}(x + 1) = \frac{\ln(x + 1)}{\ln e^x} = \frac{\ln(x + 1)}{x}$$

Thus,

$$f(x) = \frac{\ln(x + 1)}{x}$$

and

$$f'(x) = \frac{\dfrac{x}{x + 1} - \ln(x + 1)}{x^2} = \frac{x - (x + 1)\ln(x + 1)}{x^2(x + 1)}$$

44. The area of the region bounded by the curve $y = e^{-x}$, the coordinate axes, and the line $x = b$ ($b > 0$) is a function of b. If f is this function, find $f(b)$. Also find $\lim\limits_{b \to +\infty} f(b)$.

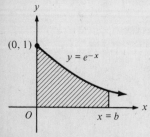

Figure 8. 44R

SOLUTION: A sketch of the region is shown in Fig. 8. 44R. The region is bounded above by the curve $y = e^{-x}$, bounded below by the line $y = 0$, bounded on the left by the line $x = 0$, and bounded on the right by the line $x = b$. Thus,

$$f(b) = \int_0^b e^{-x}\, dx$$

$$= -e^{-x}\Big]_0^b = -e^{-b} + 1$$

Furthermore,

$$\lim_{b \to +\infty} f(b) = \lim_{b \to +\infty} [-e^{-b} + 1]$$

$$= \lim_{b \to -\infty} [-e^b + 1]$$

$$= 0 + 1 = 1$$

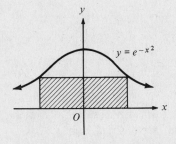

$y = e^{-x^2}$

Figure 8. 48R

48. Prove that if a rectangle is to have its base on the x-axis and two of its vertices on the curve $y = e^{-x^2}$, then the rectangle will have the largest possible area if the two vertices are at the points of inflection of the graph.

SOLUTION: Figure 8. 48R shows the curve $y = e^{-x^2}$ and a rectangle with base on the x-axis and two vertices on the curve. Let $f(x) = e^{-x^2}$. Then the first-quadrant vertex that is on the curve is at the point $(x, f(x))$. Hence, the length of the rectangle is $2x$, and the width is $f(x)$. If A is the area of the rectangle, then

$$A(x) = 2xf(x)$$
$$= 2xe^{-x^2} \qquad \text{if } x > 0$$

We want to find the absolute maximum value of A. Hence, we differentiate A.

$$A'(x) = 2xe^{-x^2}(-2x) + e^{-x^2} \cdot 2$$

$$= \frac{2(1 - 2x^2)}{e^{x^2}}$$

If $A'(x) = 0$, then $x = \pm\frac{1}{2}\sqrt{2}$. We reject the negative square root, because $x > 0$. Because $A'(x) > 0$ if $0 < x < \frac{1}{2}\sqrt{2}$ and $A'(x) < 0$ if $x > \frac{1}{2}\sqrt{2}$, A has a relative maximum value at $x = \frac{1}{2}\sqrt{2}$. Because $\frac{1}{2}\sqrt{2}$ is the only critical number of A in $(0, +\infty)$, then A has an absolute maximum value at $x = \frac{1}{2}\sqrt{2}$. Thus, the first quadrant vertex of the rectangle is at the point where $x = \frac{1}{2}\sqrt{2}$. By symmetry, the second-quadrant vertex is at the point where $x = -\frac{1}{2}\sqrt{2}$. Furthermore, in Exercise 40 of Exercises 8.5 we showed that the curve $y = e^{-x^2}$ has a point of inflection at the points where $x = \pm\frac{1}{2}\sqrt{2}$. Hence, the rectangle will have largest possible area if the vertices are at the points of inflection of the curve.

52. If W in-lb is the work done by a gas expanding against a piston in a cylinder and P lb/in^2 is the pressure of the gas when the volume of the gas is V in^3, show that if V_1 in^3 and V_2 in^3 are the initial and final volumes, respectively, then

$$W = \int_{V_1}^{V_2} P\, dV$$

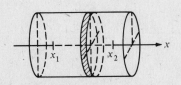

Figure 8. 52R

SOLUTION: Refer to Fig. 8. 52R. Position the cylinder so that its axis is the x-axis with the piston face at the point x when the volume of the cylinder is V. For each partition of the interval $[V_1, V_2]$ there is a corresponding partition of the interval $[x_1, x_2]$. The element of displacement of the piston face is D_i inches with $D_i = \Delta_i x$. Let f and g be the functions such that $P = f(V)$ and $P = g(x)$. Let A in^2 be the area of the piston face. Then the element of force is F_i pounds with $F_i = A \cdot P_i = A \cdot g(\bar{x}_i)$. Hence,

$$W = \lim_{\|\Delta\| \to 0} \sum_{i=1}^{n} F_i D_i$$

$$= \lim_{||\Delta|| \to 0} \sum_{i=1}^{n} A \cdot g(\bar{x}_i) \cdot \Delta_i x \tag{1}$$

Because the element of volume is a right-circular cylinder with base area A in^2 and thickness $\Delta_i x$ in, we have

$$\Delta_i V = A \cdot \Delta_i x \tag{2}$$

Moreover, because $g(x) = P = f(V)$, we have

$$f(\bar{V}_i) = g(\bar{x}_i) \tag{3}$$

Substituting from (2) and (3) into (1), we obtain

$$W = \lim_{||\Delta|| \to 0} \sum_{i=1}^{n} f(\bar{V}_i) \Delta_i V$$

$$= \int_{V_1}^{V_2} f(V) \, dV = \int_{V_1}^{V_2} P \, dV$$

56. A tank contains 60 gal of salt water with 120 lb of dissolved salt. Salt water with 3 lb of salt per gallon flows into the tank at the rate of 2 gal/min, and the mixture, kept uniform by stirring, flows out at the same rate. How long will it be before there are 200 lb of salt in the tank?

SOLUTION: Let x pounds be the weight of the salt that is in the tank after t minutes. Because the tank contains 60 gallons of salt water, then after t minutes the tank contains $x/60$ pounds of salt per gallon. And because the salt water flows out of the tank at the rate of 2 gal/min, then salt is leaving the tank at the rate of $x/30$ pounds per minute. On the other hand, because salt water with 3 lb of salt per gallon flows into the tank at the rate of 2 gal/min, then salt is entering the tank at the rate of 6 pounds per minute. We conclude that the combined effect of these two rates of changes it that $(6 - x/30)$ pounds per minute is the rate of change of x with respect to t. Thus,

$$\frac{dx}{dt} = 6 - \frac{x}{30} \tag{1}$$

We are given that $x = 120$ when $t = 0$, and we must find t when $x = 200$. From Eq. (1) we have

$$\frac{dx}{dt} = \frac{180 - x}{30}$$

$$\frac{dx}{180 - x} = \frac{dt}{30}$$

$$\int \frac{dx}{180 - x} = \int \frac{dt}{30}$$

$$\ln|180 - x| = \frac{t}{30} + k$$

$$180 - x = Ce^{t/30}$$

$$x = 180 - Ce^{t/30}$$

Because $x = 120$ when $t = 0$, we have $C = 60$, so

$$x = 180 - 60e^{t/30}$$

Substituting $x = 200$, we obtain

$$200 = 180 - 60e^{t/30}$$
$$e^{t/30} = -\tfrac{1}{3} \tag{2}$$

Because the exponential function always has a positive value, we conclude that Eq. (2) has no solution. Thus, the tank will never contain 200 pounds of salt.

60. Prove that if $x > 0$, and $\int_1^x t^{h-1} dt = 1$, then

$$\lim_{h \to 0} x = \lim_{h \to 0} (1 + h)^{1/h}$$

SOLUTION: If $h \neq 0$, then

$$\int_1^x t^{h-1} dt = \frac{t^h}{h} \bigg]_{t=1}^{t=x}$$

$$= \frac{x^h}{h} - \frac{1}{h} \tag{1}$$

By the given equation $\int_1^x t^{h-1} dt = 1$ and Eq. (1), we conclude that if $h \neq 0$, and $1 + h > 0$, then

$$\frac{x^h}{h} - \frac{1}{h} = 1$$

$$x^h = 1 + h$$

$$x = (1 + h)^{1/h}$$

Thus,

$$\lim_{h \to 0} x = \lim_{h \to 0} (1 + h)^{1/h}$$

64. Prove that

$$\lim_{x \to 0} \frac{e^{ax} - 1}{x} = a$$

(*Hint:* Let $f(x) = e^{ax}$ and find $f'(0)$ by two methods.)

SOLUTION: Let $f(x) = e^{ax}$. Then, by definition,

$$f'(0) = \lim_{x \to 0} \frac{f(x) - f(0)}{x - 0}$$

$$= \lim_{x \to 0} \frac{e^{ax} - 1}{x} \tag{1}$$

Furthermore, because

$$f'(x) = a\, e^{ax}$$

we have

$$f'(0) = a \cdot e^0 = a \tag{2}$$

From (1) and (2) we conclude that

$$\lim_{x \to 0} \frac{e^{ax} - 1}{x} = a$$

CHAPTER 9 | Trigonometric and Hyperbolic Functions

9.1 THE TANGENT, COTANGENT, SECANT, AND COSECANT FUNCTIONS

9.1.1 Definition The *tangent* and *secant* functions are defined by

$$\tan x = \frac{\sin x}{\cos x} \qquad \sec x = \frac{1}{\cos x}$$

for all real numbers x for which $\cos x \neq 0$.

The *cotangent* and *cosecant* functions are defined by

$$\cot x = \frac{\cos x}{\sin x} \qquad \csc x = \frac{1}{\sin x}$$

for all real numbers x for which $\sin x \neq 0$.

We have the following differentiation formulas.

9.1.2 Theorem $D_x(\tan u) = \sec^2 u \; D_x u$

9.1.3 Theorem $D_x(\cot u) = -\csc^2 u \; D_x u$

9.1.4 Theorem $D_x(\sec u) = \sec u \tan u \; D_x u$

9.1.5 Theorem $D_x(\csc u) = -\csc u \cot u \; D_x u$

The function values given in Table 1 below are needed frequently. The values should be memorized.

Table 1

x (radian measure)	0	$\frac{1}{6}\pi$	$\frac{1}{4}\pi$	$\frac{1}{3}\pi$	$\frac{1}{2}\pi$
θ (degree measure)	0	30	45	60	90
$\tan x$ or $\tan \theta$	0	$\frac{1}{3}\sqrt{3}$	1	$\sqrt{3}$	not defined

The entries in Table 1 and the following reduction formulas can be used to find the exact function value of an integral multiple of x that appears in the table.

$$\tan(\pi - x) = -\tan x$$
$$\tan(x + k\pi) = \tan x \qquad \text{for any integer } k$$

The following are the most frequently used trigonometric identities in this section.

$$1 + \tan^2 t = \sec^2 t$$
$$1 + \cot^2 t = \csc^2 t$$

As was the case for the sine and cosine functions, sometimes you may find a derivative of a function involving the tangent, cotangent, secant, or cosecant function that appears in a form different from that given in the answer section of the text. To check your answer you should use the identities to show that your answer is equivalent to that given in the text.

Exercises 9.1

In Exercises 3–30 find the derivative of the given function.

4. $g(x) = 3 \cot 4x$

SOLUTION: We apply Theorem 9.1.3 with $u = 4x$. Thus,

$$g'(x) = (3)(-\csc^2 4x)(4)$$
$$= -12 \csc^2 4x$$

8. $g(x) = \ln \csc^2 x$

SOLUTION:

$$g(x) = \ln \csc^2 x = 2 \ln \csc x$$

Thus,

$$g'(x) = 2 \cdot \frac{1}{\csc x} (-\csc x \cot x) = -2 \cot x$$

12. $h(x) = \ln|\cot \frac{1}{2} x|$

SOLUTION:

$$h'(x) = \left(\frac{1}{\cot \frac{1}{2} x}\right) D_x(\cot \frac{1}{2} x)$$

$$= (\tan \tfrac{1}{2} x)(-\tfrac{1}{2} \csc^2 \tfrac{1}{2} x)$$

$$= (-\tfrac{1}{2}) \frac{\sin \frac{1}{2} x}{\cos \frac{1}{2} x} \cdot \frac{1}{\sin^2 \frac{1}{2} x}$$

$$= (-\tfrac{1}{2}) \frac{1}{\cos \frac{1}{2} x} \cdot \frac{1}{\sin \frac{1}{2} x}$$

$$= -\tfrac{1}{2} \sec \tfrac{1}{2} x \cdot \csc \tfrac{1}{2} x$$

16. $h(t) = \sec^2 2t - \tan^2 2t$

SOLUTION: We use the identity

$$1 + \tan^2 x = \sec^2 x$$

Thus,

$$h(t) = (1 + \tan^2 2t) - \tan^2 2t = 1$$

Hence,

$$h'(t) = 0$$

20. $g(t) = 2 \sec \sqrt{t}$

SOLUTION:

$$g'(t) = 2 \sec \sqrt{t} \tan \sqrt{t}\ D_t\sqrt{t}$$
$$= 2 \sec \sqrt{t} \tan \sqrt{t}(\tfrac{1}{2}t^{-1/2})$$
$$= \frac{\sec \sqrt{t} \tan \sqrt{t}}{\sqrt{t}}$$

24. $f(x) = \tfrac{1}{3} \sec^3 2x - \sec 2x$

SOLUTION:

$$f'(x) = (\tfrac{1}{3})(3)\sec^2 2x\ D_x \sec 2x - D_x \sec 2x$$
$$= (\sec^2 2x)(\sec 2x \tan 2x)(2) - (\sec 2x \tan 2x)(2)$$
$$= 2 \sec 2x \tan 2x[\sec^2 2x - 1]$$
$$= 2 \sec 2x \tan 2x(\tan^2 2x)$$
$$= 2 \sec 2x \tan^3 2x$$

28. $G(x) = (\tan x)^x,\ \tan x > 0$

SOLUTION: Let $y = G(x)$. Then

$$y = (\tan x)^x$$

$$\ln y = \ln(\tan x)^x$$

$$\ln y = x \ln(\tan x)$$

$$\frac{1}{y}\ D_x y = x \cdot \frac{1}{\tan x} \cdot \sec^2 x + \ln(\tan x)$$

$$D_x y = y\left[\frac{x \sec^2 x}{\tan x} + \ln(\tan x)\right]$$

$$= (\tan x)^x[x \sec x \csc x + \ln(\tan x)]$$

32. Use logarithmic differentiation to find $D_x y$.

$$y = \frac{\tan^2 x}{\sqrt{1 + \sec^2 x}}$$

SOLUTION: Because the logarithm of a negative number is not defined, we first take the absolute value of both sides of the given equation. Thus the solution given applies when $\tan x < 0$ as well as when $\tan x > 0$.

$$|y| = \frac{|\tan x|^2}{(1 + \sec^2 x)^{1/2}}$$

Next, we take the natural logarithm of both sides and simplify the result.

$$\ln|y| = \ln \frac{|\tan x|^2}{(1 + \sec^2 x)^{1/2}}$$

$$= \ln|\tan x|^2 - \ln(1 + \sec^2 x)^{1/2}$$

$$= 2 \ln|\tan x| - \tfrac{1}{2} \ln(1 + \sec^2 x)$$

Differentiating on both sides with respect to x, we obtain

$$\frac{1}{y} D_x y = \frac{2}{\tan x} D_x \tan x - \frac{\tfrac{1}{2}}{1 + \sec^2 x} D_x(1 + \sec^2 x)$$

$$= \frac{2 \sec^2 x}{\tan x} - \frac{(\tfrac{1}{2})(2 \sec x)(\sec x \tan x)}{1 + \sec^2 x}$$

$$= \frac{2 \sec^2 x(1 + \sec^2 x) - \sec^2 x \tan^2 x}{\tan x(1 + \sec^2 x)}$$

$$= \frac{2 \sec^2 x + 2 \sec^4 x - \sec^2 x(\sec^2 x - 1)}{\tan x(1 + \sec^2 x)}$$

$$= \frac{3 \sec^2 x + \sec^4 x}{\tan x(1 + \sec^2 x)}$$

Thus,

$$D_x y = y \cdot \frac{\sec^2 x(3 + \sec^2 x)}{\tan x(1 + \sec^2 x)}$$

$$= \frac{\tan^2 x}{\sqrt{1 + \sec^2 x}} \cdot \frac{\sec^2 x(3 + \sec^2 x)}{\tan x(1 + \sec^2 x)}$$

$$= \frac{\tan x \sec^2 x(3 + \sec^2 x)}{(1 + \sec^2 x)^{3/2}}$$

36. Find $D_x y$ by implicit differentiation.

$$\cot xy + xy = 0$$

SOLUTION: Differentiating on both sides with respect to x, we get

$$-\csc^2 xy \, D_x xy + D_x xy = 0$$

$$-\csc^2 xy(x \, D_x y + y) + x \, D_x y + y = 0$$

$$(x - x \csc^2 xy)D_x y = y \csc^2 xy - y$$

$$D_x y = \frac{y \csc^2 xy - y}{x - x \csc^2 xy}$$

$$= \frac{y(\csc^2 xy - 1)}{-x(\csc^2 xy - 1)} = -\frac{y}{x}$$

40. Draw a sketch of the graph of the given function.

$$f(x) = \tfrac{1}{2} \sec 2x$$

SOLUTION: Because $\sec(x + 2\pi) = \sec x$, then

$$f(x + \pi) = \tfrac{1}{2} \sec[2(x + \pi)] = \tfrac{1}{2} \sec[2x + 2\pi] = \tfrac{1}{2} \sec 2x = f(x)$$

Thus, f is periodic with period π. We sketch the graph for $0 \le x \le \pi$. For each interval $[n\pi, (n + 1)\pi]$ the graph is congruent to the graph in the interval $[0, \pi]$. Because

$$\lim_{x\to\pi/4} f(x) = \tfrac{1}{2} \lim_{x\to\pi/4} \sec 2x = \pm\infty$$

and

$$\lim_{x\to 3\pi/4} f(x) = \tfrac{1}{2} \lim_{x\to 3\pi/4} \sec 2x = \pm\infty$$

the graph of f has vertical asymptotes at $x = \tfrac{1}{4}\pi$ and $x = \tfrac{3}{4}\pi$. Because the function defined by $y = \sec x$ has a relative minimum value at $x = 0$, then f has a relative minimum value at $2x = 0$ or, equivalently, at $x = 0$. Since $f(0) = \tfrac{1}{2} \sec 0 = \tfrac{1}{2}$, the relative minimum value is $\tfrac{1}{2}$. Furthermore, the function defined by $y = \sec x$ has a relative maximum value at $x = \pi$. Thus, f has a relative maximum value at $2x = \pi$ or $x = \tfrac{1}{2}\pi$. Since $f(\tfrac{1}{2}\pi) = \tfrac{1}{2} \sec \pi = -\tfrac{1}{2}$, the relative maximum value is $-\tfrac{1}{2}$. We use this information to draw a sketch of the graph, which is shown in Fig. 9. 1. 40.

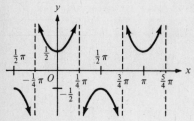

Figure 9. 1. 40

44. Find equations of the tangent line and normal line to the graph of $f(x) = \sec 4x \cdot \csc 2x$ at the point $(\tfrac{1}{4}\pi, -1)$.

SOLUTION:

$$f'(x) = (\sec 4x)D_x \csc 2x + (\csc 2x)D_x \sec 4x$$

$$= (\sec 4x)(-2 \csc 2x \cot 2x) + (\csc 2x)(4 \sec 4x \tan 4x)$$

$$= -2 \sec 4x \csc 2x(\cot 2x - 2 \tan 4x)$$

If $x = \tfrac{1}{4}\pi$, then

$$f'(\tfrac{1}{4}\pi) = -2 \sec \pi \csc \tfrac{1}{2}\pi(\cot \tfrac{1}{2}\pi - 2 \tan \pi)$$

$$= -2(-1)(1)(0 - 0) = 0$$

Therefore, the slope of the tangent line is 0 at the point $(\tfrac{1}{4}\pi, -1)$. An equation of the tangent line is

$$y = -1$$

The normal line at the point $(\tfrac{1}{4}\pi, -1)$ is a vertical line because the tangent line is a horizontal line. An equation of the normal line is

$$x = \tfrac{1}{4}\pi$$

48. If two corridors at right angles to each other are 10 ft and 15 ft wide, respectively, what is the length of the longest steel girder that can be moved horizontally around the corner? Neglect the horizontal width of the girder.

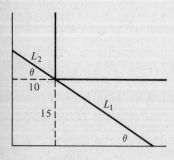

Figure 9. 1. 48

SOLUTION: Refer to Fig. 9. 1. 48. Let L ft be the length of a girder that touches both outside walls and the inside corner, as shown in the figure, and let θ radians be the measure of the angle between the girder and the outside wall of the 15-ft corridor. Then the absolute *minimum* value of L is the length of the longest girder that can be moved around the corner. Because

$$L = L_1 + L_2$$

and

$$L_1 = 15 \csc \theta \quad \text{and} \quad L_2 = 10 \sec \theta$$

then

$$L = 15 \csc \theta + 10 \sec \theta \qquad (1)$$

and $0 < \theta < \frac{1}{2}\pi$.

$$D_\theta L = -15 \csc \theta \cot \theta + 10 \sec \theta \tan \theta \qquad (2)$$

If $D_\theta L = 0$, then

$$0 = -15 \csc \theta \cot \theta + 10 \sec \theta \tan \theta$$

$$0 = -15 \left(\frac{1}{\sin \theta}\right)\left(\frac{\cos \theta}{\sin \theta}\right) + 10 \left(\frac{1}{\cos \theta}\right)\left(\frac{\sin \theta}{\cos \theta}\right)$$

Multiplying on both sides by $\sin^2 \theta \cos^2 \theta$, we get

$$0 = -15 \cos^3 \theta + 10 \sin^3 \theta$$

Dividing both sides by $\cos^3 \theta$, we get

$$0 = -15 + 10 \tan^3 \theta$$

$$\tan^3 \theta = \tfrac{3}{2}$$

$$\tan \theta = \sqrt[3]{\tfrac{3}{2}} = 1.145$$

$$\theta = 0.853$$

Substituting $\theta = 0.853$ in (1) we get

$$L = 15 \csc(0.853) + 10 \sec(0.853) = 35.1$$

We show that 35.1 is the absolute minimum value of L. differentiating on both sides of (2) with respect to θ, we obtain

$$D_\theta^2 L = -15[\csc \theta(-\csc^2 \theta) + \cot \theta(-\csc \theta \cot \theta)] + 10[\sec \theta \sec^2 \theta + \tan \theta \sec \theta \tan \theta]$$

$$= 15[\csc^3 \theta + \csc \theta \cot^2 \theta] + 10[\sec^3 \theta + \tan^2 \theta \sec \theta] \qquad (3)$$

Because $0 < \theta < \frac{1}{2}\theta$, then each term in Eq. (3) is positive. Therefore $D_\theta^2 L > 0$, and thus L has a relative minimum value at $\theta = 0.853$. Since L has only one relative extremum in $(0, \frac{1}{2}\pi)$, the relative extremum is an absolute extremum. We conclude that the longest steel girder that can be moved around the corner is 35.1 feet long.

9.2 AN APPLICATION OF THE TANGENT FUNCTION TO THE SLOPE OF A LINE

9.2.1 Definition   The *angle of inclination* of a line not parallel to the *x*-axis is the smallest angle measured counterclockwise from the positive direction of the *x*-axis to the line. The inclination of a line parallel to the *x*-axis is defined to have measure zero.

9.2.2 Theorem   If α is the degree measure of the angle of inclination of line L, not parallel to the *y*-axis, then the slope m of L is given by

$$m = \tan \alpha°$$

Thus, if $m > 0$ then $0 < \alpha < 90$, and if $m < 0$, then $90 < \alpha < 180$.

9.2.3 Theorem   Let L_1 and L_2 be two nonvertical lines that intersect and are not perpendicular, and let L_2 be the line having the greater angle of inclination. Then if m_1 is the slope of L_1, m_2 is the slope of L_2, and θ is the degree measure of the angle between L_1 and L_2

$$\tan \theta° = \frac{m_2 - m_1}{1 + m_1 m_2}$$

The formula in Theorem 9.2.3 gives the tangent of the angle whose initial side has slope m_1 and whose terminal side has slope m_2, whether or not the angle of inclination of line L_1 is less than the angle of inclination of line L_2. Actually, two lines that intersect form two angles. If m_2 and m_1 are interchanged in the formula 9.2.3, we get the tangent of the second angle. Since the two angles are supplementary, the tangent of one of the angles is the negative of the tangent of the other. If $\tan \theta > 0$, then θ is the acute angle between lines L_1 and L_2, whereas if $\tan \theta < 0$, then θ is the obtuse angle between L_1 and L_2. We have arbitrarily chosen the angle described in 9.2.3 to be *the* angle between two lines.

9.2.4 Definition   The *angle between two curves* at a point of intersection is the angle between the tangent lines to the curves at the point.

Exercises 9.2

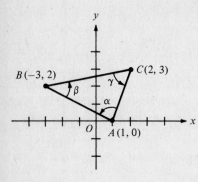

Figure 9. 2. 4

4. Find, to the nearest degree, the measurements of the interior angles of the triangle having vertices $(1, 0)$, $(-3, 2)$, and $(2, 3)$.

SOLUTION: Refer to Fig. 9. 2. 4. We apply Theorem 9.2.3 to find the measurement of angle α. Because line AC contains the initial side of angle α and line AB contains the terminal side of angle α, we let m_1 be the slope of line AC and m_2 be the slope of line AB. Thus,

$$m_1 = \frac{3 - 0}{2 - 1} = 3 \quad \text{and} \quad m_2 = \frac{2 - 0}{-3 - 1} = -\frac{1}{2}$$

Applying Theorem 9.2.3, we have

$$\tan \alpha^\circ = \frac{m_2 - m_1}{1 + m_1 m_2}$$

$$= \frac{-\frac{1}{2} - 3}{1 + (3)(-\frac{1}{2})}$$

$$= \frac{-\frac{7}{2}}{-\frac{1}{2}} = 7$$

We use a table of trigonometric function values or a calculator to determine that

$$\alpha^\circ = 82^\circ$$

To find the measurement of angle β we let m_1 be the slope of line BA and m_2 be the slope of line BC, because line BA contains the initial side of angle β and line BC contains the terminal side of angle β. We have

$$m_1 = \frac{0 - 2}{1 - (-3)} = -\frac{1}{2} \quad \text{and} \quad m_2 = \frac{3 - 2}{2 - (-3)} = \frac{1}{5}$$

Hence,

$$\tan \beta^\circ = \frac{m_2 - m_1}{1 + m_1 m_2}$$

$$= \frac{\frac{1}{5} - (-\frac{1}{2})}{1 + (-\frac{1}{2})(\frac{1}{5})}$$

$$= \frac{\frac{7}{10}}{\frac{9}{10}}$$

$$= \frac{7}{9} = 0.7778$$

and

$$\beta^\circ = 38^\circ$$

Because $\alpha° + \beta° + \gamma° = 180°$, and $\alpha° + \beta° = 120°$, then $\gamma° = 60°$.

8. Use Theorem 9.2.3 to prove that the triangle having vertices $(-3, 0)$, $(-1, 0)$, and $(-2, \sqrt{3})$ is equilateral.

SOLUTION: Refer to Fig. 9. 2. 8. We show that each angle measures 60°. For angle α we take $m_1 = 0$ and $m_2 = \sqrt{3}$, because the slope of the line AB is 0 and the slope of line AC is $\sqrt{3}$. Applying Theorem 9.2.3, we have

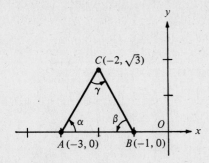

Figure 9. 2. 8

$$\tan \alpha° = \frac{m_2 - m_1}{1 + m_1 m_2}$$

$$= \frac{\sqrt{3} - 0}{1 + (0)(\sqrt{3})} = \sqrt{3}$$

Therefore,

$$\alpha° = 60°$$

For angle β we take $m_1 = -\sqrt{3}$ and $m_2 = 0$, because the slope of line BC is $-\sqrt{3}$ and the slope of line BA is 0. Thus,

$$\tan \beta° = \frac{m_2 - m_1}{1 + m_1 m_2}$$

$$= \frac{0 - (-\sqrt{3})}{1 + (-\sqrt{3})(0)} = \sqrt{3}$$

Hence,

$$\beta° = 60°$$

Because $\alpha° + \beta° + \gamma° = 180°$, and $\alpha° + \beta° = 120°$, then $\gamma° = 60°$. Thus, triangle ABC is equilateral.

12. Find the tangents of the measurements of the interior angles of the triangle having vertices at $(-3, -2)$, $(-6, 3)$, and $(5, 1)$ and check by applying the result of Exercise 11.

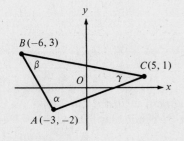

Figure 9. 2. 12

SOLUTION: Let $A = (-3, -2)$, $B = (-6, 3)$, and $C = (5, 1)$. Let α, β, and γ be the measures of the interior angles A, B, and C, respectively, of triangle ABC, as illustrated in Fig. 9. 2. 12. We have

slope of line $AB = -\frac{5}{3}$

slope of line $BC = -\frac{2}{11}$

slope of line $AC = \frac{3}{8}$

We use Theorem 9.2.3 to find $\tan \alpha$. Because line AC is the initial side of angle α, we let $m_1 = \frac{3}{8}$, and because line AB is the terminal side of angle α, we let $m_2 = -\frac{5}{3}$. Then,

$$\tan \alpha = \frac{-\frac{5}{3} - \frac{3}{8}}{1 + \frac{3}{8}(-\frac{5}{3})} = -\frac{49}{9}$$

Next, we find $\tan \beta$.

Because line AB is the initial side of angle β, we let $m_1 = -\frac{5}{3}$, and because line BC is the terminal side of angle β, we let $m_2 = -\frac{2}{11}$. Thus,

$$\tan \beta = \frac{-\frac{2}{11} + \frac{5}{3}}{1 + (-\frac{5}{3})(-\frac{2}{11})} = \frac{49}{43}$$

And we find $\tan \gamma$. Because line BC is the initial side of angle C, we let $m_1 = -\frac{2}{11}$, and because line AC is the terminal side of angle C, we let $m_2 = \frac{3}{8}$. Thus,

$$\tan \gamma = \frac{\frac{3}{8} + \frac{2}{11}}{1 + (-\frac{2}{11})(\frac{3}{8})} = \frac{49}{82}$$

By Exercise 11 we have

$$\tan \alpha + \tan \beta + \tan \gamma = \tan \alpha \tan \beta \tan \gamma \tag{1}$$

We check the values found above.

$$\tan \alpha + \tan \beta + \tan \gamma = -\frac{49}{9} + \frac{49}{43} + \frac{49}{82} = -\frac{117,649}{31,734}$$

$$\tan \alpha \tan \beta \tan \gamma = \left(-\frac{49}{9}\right)\left(\frac{49}{43}\right)\left(\frac{49}{82}\right) = -\frac{117,649}{31,734}$$

Because Eq. (1) is satisfied, we have a check.

16. For what value of a are the given curves orthogonal? $y = a^x$ and $y = a^{-x}$, where $a > 1$.

SOLUTION: The curves are orthogonal if and only if the lines tangent to the curves at their point of intersection are perpendicular. We find the point of intersection of the curves. Eliminating y from the given equations, we have

$$a^x = a^{-x}$$

$$a^{2x} = 1$$

Thus,

$$2x = 0$$

$$x = 0$$

and so $(0, 1)$ is the point of intersection of the curves. See Fig. 9. 2. 16 for a sketch of the curves. Let $f(x) = a^x$ and $g(x) = a^{-x}$.

$$f'(x) = a^x \ln a \quad \text{and} \quad g'(x) = -a^{-x} \ln a$$

Hence,

$$f'(0) = \ln a \quad \text{and} \quad g'(0) = -\ln a$$

Therefore, $m_1 = \ln a$ is the slope of the line tangent to the curve $y = a^x$ at the point $(0, 1)$, and $m_2 = -\ln a$ is the slope of the line tangent to the curve $y = a^{-x}$ at the point $(0, 1)$. If $m_1 m_2 = -1$, the lines are perpendicular. Therefore, we take

$$(\ln a)(-\ln a) = -1$$
$$\ln^2 a = 1$$
$$\ln a = \pm 1$$

Thus, either $a = e$ or $a = e^{-1}$. If $a = e$, then $f(x) = e^x$ and $g(x) = e^{-x}$. Whereas, if $a = e^{-1}$, then $f(x) = e^{-x}$ and $g(x) = e^x$. We conclude that the two solutions are equivalent.

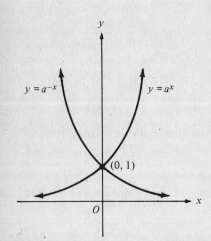

Figure 9. 2. 16

9.3 INTEGRALS INVOLVING THE TANGENT, COTANGENT, SECANT, AND COSECANT

Following are the integration formulas for the functions of this section.

9.3.1 Theorem

$$\int \tan u \, du = \ln|\sec u| + C$$

9.3.2 Theorem
$$\int \cot u \, du = -\ln|\csc u| + C$$
$$= \ln|\sin u| + C$$

9.3.3 Theorem
$$\int \sec u \, du = \ln|\sec u + \tan u| + C$$

9.3.4 Theorem
$$\int \csc u \, du = \ln|\csc u - \cot u| + C$$

9.3.5 Theorem
$$\int \sec^2 u \, du = \tan u + C$$

9.3.6 Theorem
$$\int \sec u \tan u \, du = \sec u + C$$

9.3.7 Theorem
$$\int \csc^2 u \, du = -\cot u + C$$

9.3.8 Theorem
$$\int \csc u \cot u \, du = -\csc u + C$$

To find an integral of the form $\int \tan^m u \sec^n u \, du$ we consider three types:

TYPE 1 The factor sec u appears with even exponent. That is, n is a positive even integer.

1. Factor out $\sec^2 u \, du$.
2. Use the identity $\sec^2 u = 1 + \tan^2 u$ to replace the remaining even power of sec u by an expression involving powers of tan u.
3. Let $v = \tan u$ and $dv = \sec^2 u \, du$.
4. Integrate the resulting algebraic function.

TYPE 2 (See Exercise 20) The factor tan u appears with odd exponent. That is, m is a positive odd integer.

1. Factor out $\sec u \tan u \, du$.
2. Use the identity $\tan^2 u = \sec^2 u - 1$ to replace the remaining even power of tan u by an expression involving powers of sec u.
3. Let $v = \sec u$ and $dv = \sec u \tan u \, du$.
4. Integrate the resulting algebraic function.

TYPE 3 The factor tan u appears with integral exponent larger than 1, and the factor sec u does not appear. That is, the integral is of the form $\int \tan^m u \, du$, where m is an integer greater than 1

1. Factor out $\tan^2 u$, replace $\tan^2 u$ by $\sec^2 u - 1$, and write the given integral as the sum of two integrals.
2. One of the integrals found in step 1 is of the type described in Type 1. Use the method of Type 1 to find the integral.
3. The other integral found in step 1 is either $\int du$ or is of the type described in Type 3. If the latter is true, repeat steps 1, 2, and 3 until the integral is found.

If an integral of the form $\int \tan^m u \sec^n u \, du$ is not Type 1, Type 2, or Type 3, we may not be able to find the integral at this time. In Section 10.2 we have a method called "integration by parts" that enables us to find additional integrals of types not covered here. Sometimes we can find the integral by expressing the tangent and secant functions in terms of the sine and cosine functions and then using the methods of Section 5.3. (See Exercise 28.)

To find an integral of the form $\int \cot^m u \csc^n u \, du$, we consider the same three types described above, where tan u is replaced by cot u and sec u is replaced by csc u. And we must remember that when $v = \cot u$, then $dv = -\csc^2 u \, du$; when $v = \csc u$, then $dv = -\csc u \cot u \, du$.

Sometimes an integral is that described by both Type 1 and Type 2. Then we can use either method to find the integral. In this case, the integrals found differ from each other by a constant. (See Exercise 48.) If the answer that you find to an exercise is not in the same form as that given in the answer section of the text, you should use the identities to show that the two results are either equivalent or that they differ by a constant.

Exercises 9.3

In Exercises 1–34 evaluate the indefinite integral.

4. $\int \csc^2 4t \, dt$

SOLUTION: We let $u = 4t$, $du = 4 \, dt$, and apply Theorem 9.3.7.

$$\int \csc^2 4t \, dt = \tfrac{1}{4} \int \csc^2 u \, du$$

$$= -\tfrac{1}{4} \cot u + C$$

$$= -\tfrac{1}{4} \cot 4t + C$$

8. $\int e^x \csc^2 e^x \, dx$

SOLUTION: We let $u = e^x$ and $du = e^x \, dx$. Thus,

$$\int e^x \csc^2 e^x \, dx = \int \csc^2 u \, du$$

$$= -\cot u + C$$

$$= -\cot e^x + C$$

12. $\int \sec x \tan x \tan(\sec x) \, dx$

SOLUTION: Let $u = \sec x$. Then $du = \sec x \tan x \, dx$. Thus,

$$\int \sec x \tan x \tan(\sec x) \, dx = \int \tan u \, du$$

$$= \ln|\sec u| + C$$

$$= \ln|\sec(\sec x)| + C$$

16. $\int \csc^4 x \, dx$

SOLUTION: We use the method of Type 1.

$$\int \csc^4 x \, dx = \int (1 + \cot^2 x)\csc^2 x \, dx \tag{1}$$

Let $u = \cot x$. Then $du = -\csc^2 x\, dx$. With these substitutions on the right side of (1), we obtain

$$\int \csc^4 x\, dx = \int (1 + u^2)(-du)$$

$$= -u - \tfrac{1}{3}u^3 + C$$

$$= -\cot x - \tfrac{1}{3}\cot^3 x + C$$

20. $\displaystyle \int \tan^5 x \sec^3 x\, dx$

SOLUTION: We use the method of Type 2.

$$\int \tan^5 x \sec^3 x\, dx = \int \tan^4 x \sec^2 x(\sec x \tan x\, dx)$$

$$= \int (\sec^2 x - 1)^2 \sec^2 x(\sec x \tan x\, dx) \qquad (1)$$

Let $u = \sec x$. Then $du = \sec x \tan x\, dx$ and, by substituting in the right-hand side of (1), we get

$$\int \tan^5 x \sec^3 x\, dx = \int (u^2 - 1)^2 u^2\, du$$

$$= \int (u^6 - 2u^4 + u^2)\, du$$

$$= \tfrac{1}{7}u^7 - \tfrac{2}{5}u^5 + \tfrac{1}{3}u^3 + C$$

$$= \tfrac{1}{7}\sec^7 x - \tfrac{2}{5}\sec^5 x + \tfrac{1}{3}\sec^3 x + C$$

24. $\displaystyle \int \frac{dx}{1 + \cos x}$

SOLUTION: We use the identity

$$\cos^2 t = \tfrac{1}{2}(1 + \cos 2t)$$

with t replaced by $\tfrac{1}{2}x$. Thus

$$\cos^2 \tfrac{1}{2}x = \tfrac{1}{2}(1 + \cos x)$$

Hence,

$$\int \frac{dx}{1 + \cos x} = \int \frac{dx}{2\cos^2 \tfrac{1}{2}x}$$

$$= \int \sec^2 \tfrac{1}{2}x\, (\tfrac{1}{2}\, dx)$$

$$= \tan \tfrac{1}{2}x + C$$

28. $\displaystyle \int \frac{\tan^4 y}{\sec^5 y}\, dy$

SOLUTION: Because the integral is not Type 1, 2, or 3, we replace the given expression by an expression involving the sine and cosine functions.

$$\int \frac{\tan^4 y}{\sec^5 y}\, dy = \int \frac{\sin^4 y}{\cos^4 y} \cdot \cos^5 y\, dy$$

$$= \int \sin^4 y \cos y \, dy$$

$$= \tfrac{1}{5} \sin^5 y + C$$

where we let $u = \sin y$ and $du = \cos y \, dy$ to find the above integral.

32. $\displaystyle \int \frac{\sin^2 \pi x}{\cos^6 \pi x} \, dx$

SOLUTION:

$$\int \frac{\sin^2 \pi x}{\cos^6 \pi x} \, dx = \int \frac{\sin^2 \pi x}{\cos^2 \pi x} \cdot \frac{1}{\cos^4 \pi x} \, dx$$

$$= \int \tan^2 \pi x \sec^4 \pi x \, dx \tag{1}$$

We use the method of Type 1. From (1) we get

$$\int \frac{\sin^2 \pi x}{\cos^6 \pi x} \, dx = \int \tan^2 \pi x (1 + \tan^2 \pi x) \sec^2 \pi x \, dx \tag{2}$$

Let $u = \tan \pi x$ and $du = \pi \sec^2 \pi x \, dx$ on the right side of (2). We have

$$\int \frac{\sin^2 \pi x}{\cos^6 \pi x} \, dx = \frac{1}{\pi} \int u^2 (1 + u^2) \, du$$

$$= \frac{1}{\pi} \left(\tfrac{1}{3} u^3 + \tfrac{1}{5} u^5 \right) + C$$

$$= \frac{1}{\pi} \left(\tfrac{1}{3} \tan^3 \pi x + \tfrac{1}{5} \tan^5 \pi x \right) + C$$

In Exercises 35–40 evaluate the definite integral.

36. $\displaystyle \int_{\pi/8}^{\pi/4} 3 \csc 2x \, dx$

SOLUTION: Let $u = 2x$ and $du = 2 \, dx$. When $x = \tfrac{1}{8}\pi$, then $u = \tfrac{1}{4}\pi$; when $x = \tfrac{1}{4}\pi$, then $u = \tfrac{1}{2}\pi$.

$$\int_{\pi/8}^{\pi/4} 3 \csc 2x \, dx = \frac{3}{2} \int_{\pi/4}^{\pi/2} \csc u \, du$$

$$= \frac{3}{2} \ln|\csc u - \cot u| \Big]_{\pi/4}^{\pi/2}$$

$$= \frac{3}{2} \left(\ln\left|\csc \frac{1}{2}\pi - \cot \frac{1}{2}\pi\right| - \ln\left|\csc \frac{1}{4}\pi - \cot \frac{1}{4}\pi\right| \right)$$

$$= \frac{3}{2} (\ln 1 - \ln|\sqrt{2} - 1|)$$

$$= -\frac{3}{2} \ln(\sqrt{2} - 1)$$

40. $\displaystyle \int_{\pi/6}^{\pi/4} \cot^3 w \, dw$

SOLUTION: We use the method of Type 3.

$$\int_{\pi/6}^{\pi/4} \cot^3 w \, dw = \int_{\pi/6}^{\pi/4} (\csc^2 w - 1)\cot w \, dw$$

$$= \int_{\pi/6}^{\pi/4} \cot w \csc^2 w \, dw - \int_{\pi/6}^{\pi/4} \cot w \, dw \qquad (1)$$

In the first integral on the right side of (1) let $u = \cot w$. Then $du = -\csc^2 w \, dw$; when $w = \frac{1}{6}\pi$, $u = \sqrt{3}$; when $w = \frac{1}{4}\pi$, $u = 1$. Thus,

$$\int_{\pi/6}^{\pi/4} \cot w \csc^2 w \, dw = -\int_{\sqrt{3}}^{1} u \, du$$

$$= -\frac{1}{2} u^2 \Big]_{\sqrt{3}}^{1} = 1 \qquad (2)$$

In the second integral on the right side of (1), use the integration formula 9.3.2. Thus,

$$\int_{\pi/6}^{\pi/4} \cot w \, dw = \ln|\sin w| \Big]_{\pi/6}^{\pi/4}$$

$$= \ln(\tfrac{1}{2}\sqrt{2}) - \ln(\tfrac{1}{2})$$

$$= \ln \frac{\tfrac{1}{2}\sqrt{2}}{\tfrac{1}{2}}$$

$$= \ln\sqrt{2} = \tfrac{1}{2}\ln 2 \qquad (3)$$

By substituting from (2) and (3) into (1), we obtain

$$\int_{\pi/6}^{\pi/4} \cot^3 w \, dw = 1 - \tfrac{1}{2}\ln 2$$

44. Prove

$$\int \cot x \csc^n x \, dx = -\frac{\csc^n x}{n} + C \qquad \text{if } n \neq 0$$

SOLUTION: We use the method of Type 2.

$$\int \cot x \csc^n x \, dx = \int \csc^{n-1} x(\csc x \cot x) \, dx \qquad (1)$$

Let $u = \csc x$. Then $du = -\csc x \cot x \, dx$. Substituting in the right side of (1), we have

$$\int \cot x \csc^n x \, dx = -\int u^{n-1} \, du$$

$$= -\frac{u^n}{n} + C$$

$$= -\frac{\csc^n x}{n} + C \qquad \text{if } n \neq 0$$

48. We can integrate $\int \sec^2 x \tan x \, dx$ in two ways as follows:

$$\int \sec^2 x \tan x \, dx = \int \tan x(\sec^2 x \, dx)$$

$$= \tfrac{1}{2}\tan^2 x + C$$

and

$$\int \sec^2 x \tan x \, dx = \int \sec x (\sec x \tan x \, dx)$$

$$= \tfrac{1}{2} \sec^2 x + C$$

Explain the difference in the appearance of the two answers.

SOLUTION: Because $1 + \tan^2 x = \sec^2 x$, then $\sec^2 x - \tan^2 x = 1$. Thus,

$$\tfrac{1}{2} \sec^2 x - \tfrac{1}{2} \tan^2 x = \tfrac{1}{2}$$

and the answers differ by the constant $\tfrac{1}{2}$. This agrees with Theorem 5.2.2.

52. Find the volume of the solid of revolution generated if the region bounded by the curve $y = 3 \csc^3 x$, the x-axis, and the lines $x = \tfrac{1}{6}\pi$ and $x = \tfrac{1}{2}\pi$ is revolved about the x-axis.

SOLUTION: Figure 9. 3. 52 shows a sketch of the given region and a plane section of the solid of revolution. The element of area is a rectangle with width $\Delta_i x$ units. The element of volume is a circular disk with element of thickness $\Delta_i h = \Delta_i x$ units and radius $R_i = 3 \csc^3 \bar{x}_i$ units. If V cubic units is the volume of the solid of revolution, then

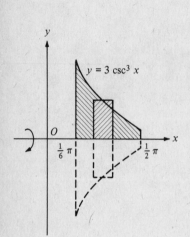

Figure 9. 3. 52

$$V = \lim_{\|\Delta\| \to 0} \sum_{i=1}^{n} \pi R_i^2 \, \Delta_i h$$

$$= \int_{\pi/6}^{\pi/2} \pi (9 \csc^6 x) \, dx$$

$$= 9\pi \int_{\pi/6}^{\pi/2} (1 + \cot^2 x)^2 \csc^2 x \, dx \qquad (1)$$

Let $u = \cot x$ and $du = -\csc^2 x \, dx$. When $x = \tfrac{1}{6}\pi$, then $u = \sqrt{3}$; when $x = \tfrac{1}{2}\pi$, then $u = 0$. Substituting into Eq. (1), we obtain

$$V = -9\pi \int_{\sqrt{3}}^{0} (1 + u^2)^2 \, du$$

$$= 9\pi \int_{0}^{\sqrt{3}} (1 + 2u^2 + u^4) \, du$$

$$= 9\pi \left[u + \frac{2}{3} u^3 + \frac{1}{5} u^5 \right]_{0}^{\sqrt{3}}$$

$$= 9\pi \left[\sqrt{3} + 2\sqrt{3} + \frac{1}{5} (9\sqrt{3}) \right]$$

$$= \frac{216\sqrt{3}\pi}{5}$$

The volume of the solid of revolution is $\tfrac{216}{5} \sqrt{3}\pi$ cubic units.

9.4 INVERSE TRIGONOMETRIC FUNCTIONS

The inverse trigonometric functions are defined as follows.

9.4.1 Definition

The *inverse sine function*, denoted by $\sin^{-1}$ is defined as follows:

$$y = \sin^{-1} x \text{ if and only if } x = \sin y \text{ and } -\tfrac{1}{2}\pi \le y \le \tfrac{1}{2}\pi$$

9.4.2 Definition

The *inverse cosine function*, denoted by $\cos^{-1}$, is defined as follows:

$$y = \cos^{-1} x \text{ if and only if } x = \cos y \text{ and } 0 \le y \le \pi$$

9.4.3 Definition The *inverse tangent function,* denoted by $\tan^{-1}$, is defined as follows:

$$y = \tan^{-1} x \text{ if and only if } x = \tan y \text{ and } -\tfrac{1}{2}\pi < y < \tfrac{1}{2}\pi$$

9.4.4 Definition The *inverse cotangent function,* denoted by $\cot^{-1}$, is defined by

$$\cot^{-1} x = \tfrac{1}{2}\pi - \tan^{-1} x, \text{ where } x \text{ is any real number}$$

9.4.5 Definition The *inverse secant function,* denoted by $\sec^{-1}$, is defined by

$$\sec^{-1} x = \cos^{-1}\left(\frac{1}{x}\right) \quad \text{for } |x| \geq 1$$

9.4.6 Definition The *inverse cosecant function,* denoted by $\csc^{-1}$, is defined by

$$\csc^{-1} x = \sin^{-1}\left(\frac{1}{x}\right) \quad \text{for } |x| \geq 1$$

We summarize the domain and range of each function.

Function defined by:	Domain	Range
$y = \sin^{-1} x$	$\{x|-1 \leq x \leq 1\}$	$\{y|-\tfrac{1}{2}\pi \leq y \leq \tfrac{1}{2}\pi\}$
$y = \cos^{-1} x$	$\{x|-1 \leq x \leq 1\}$	$\{y|0 \leq y \leq \pi\}$
$y = \tan^{-1} x$	$\{x|-\infty < x < +\infty\}$	$\{y|-\tfrac{1}{2}\pi < y < \tfrac{1}{2}\pi\}$
$y = \cot^{-1} x$	$\{x|-\infty < x < +\infty\}$	$\{y|0 < y < \pi\}$
$y = \sec^{-1} x$	$\{x|x \geq 1 \text{ or } x \leq -1\}$	$\{y|0 \leq y \leq \pi, \ y \neq \tfrac{1}{2}\pi\}$
$y = \csc^{-1} x$	$\{x|x \geq 1 \text{ or } x \leq -1\}$	$\{y|-\tfrac{1}{2}\pi \leq y \leq \tfrac{1}{2}\pi, \ y \neq 0\}$

We note that if f is any one of the six trigonometric functions and x is a *positive* number in the domain of f^{-1}, then the number in the range of f^{-1} is between 0 and $\tfrac{1}{2}\pi$. That is,

$$0 \leq f^{-1}(x) \leq \tfrac{1}{2}\pi \quad \text{if } x > 0$$

However, when $-x$ is a *negative* number in the domain of f^{-1} we have the following results.

$-\tfrac{1}{2}\pi \leq \sin^{-1}(-x) < 0$	$\tfrac{1}{2}\pi < \cos^{-1}(-x) \leq \pi$
$-\tfrac{1}{2}\pi < \tan^{-1}(-x) < 0$	$\tfrac{1}{2}\pi < \cot^{-1}(-x) < \pi$
$-\tfrac{1}{2}\pi \leq \csc^{-1}(-x) < 0$	$\tfrac{1}{2}\pi < \sec^{-1}(-x) \leq \pi$

The following identities are helpful when finding the function value of a negative number.

$\sin^{-1}(-x) = -\sin^{-1} x$	$\cos^{-1}(-x) = \pi - \cos^{-1} x$
$\tan^{-1}(-x) = -\tan^{-1} x$	$\cot^{-1}(-x) = \pi - \cot^{-1} x$
$\csc^{-1}(-x) = -\csc^{-1} x$	$\sec^{-1}(-x) = \pi - \sec^{-1} x$

Exercises 9.4

4. Find the value of each of the following:
   (a) $\sin^{-1}(\tfrac{1}{2})$ (b) $\sin^{-1}(-\tfrac{1}{2})$ (c) $\cos^{-1}(\tfrac{1}{2})$ (d) $\cos^{-1}(-\tfrac{1}{2})$
   (e) $\sec^{-1}(2)$ (f) $\csc^{-1}(-2)$

SOLUTION:
   (a) Let $t = \sin^{-1}(\tfrac{1}{2})$. Then $\sin t = \tfrac{1}{2}$ and $0 \leq t \leq \tfrac{1}{2}\pi$. Thus, $t = \tfrac{1}{6}\pi$. That is, $\sin^{-1}(\tfrac{1}{2}) = \tfrac{1}{6}\pi$.

    (b) $\sin^{-1}(-\tfrac{1}{2}) = -\sin^{-1}(\tfrac{1}{2}) = -\tfrac{1}{6}\pi$ [By part (a)]

    (c) Let $t = \cos^{-1}(\tfrac{1}{2})$. Then $\cos t = \tfrac{1}{2}$ and $0 \le t \le \tfrac{1}{2}\pi$. Thus $t = \tfrac{1}{3}\pi$. That is, $\cos^{-1}(\tfrac{1}{2}) = \tfrac{1}{3}\pi$.

    (d) $\cos^{-1}(-\tfrac{1}{2}) = \pi - \cos^{-1}(\tfrac{1}{2}) = \pi - \tfrac{1}{3}\pi = \tfrac{2}{3}\pi$ [By part (c)]

    (e) $\sec^{-1}(2) = \cos^{-1}(\tfrac{1}{2}) = \tfrac{1}{3}\pi$ [By part (c)]

    (f) $\csc^{-1}(-2) = \sin^{-1}(-\tfrac{1}{2}) = -\tfrac{1}{6}\pi$ [By part (b)]

8. Given: $y = \cot^{-1}(-\tfrac{1}{2})$. Find the exact value of each of the following:

    (a) $\sin y$ **(b)** $\cos y$ **(c)** $\tan y$ **(d)** $\sec y$ **(e)** $\csc y$

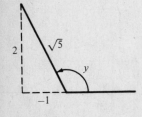

Figure 9. 4. 8

SOLUTION: We have $\cot y = -\tfrac{1}{2}$, and $\tfrac{1}{2}\pi < y < \pi$. Figure 9. 4. 8 shows an angle having radian measure y that satisfies these requirements. From the figure we see that:

    (a) $\sin y = \tfrac{2}{5}\sqrt{5}$

    (b) $\cos y = -\tfrac{1}{5}\sqrt{5}$

    (c) $\tan y = -2$

    (d) $\sec y = -\sqrt{5}$

    (e) $\csc y = \tfrac{1}{2}\sqrt{5}$

12. Find the exact value of **(a)** $\cos[\tan^{-1}(-3)]$ and **(b)** $\tan[\sec^{-1}(-3)]$.

SOLUTION:

    (a) Let $t = \tan^{-1}(-3)$. Then $\tan t = -3$ and $-\tfrac{1}{2}\pi < t < 0$. Furthermore, $\sec^2 t = 1 + \tan^2 t = 1 + (-3)^2 = 10$. Because $\sec t > 0$ if $-\tfrac{1}{2}\pi < t < 0$, then $\sec t = \sqrt{10}$. Therefore,

$$\cos[\tan^{-1}(-3)] = \cos t$$

$$= \frac{1}{\sec t} = \frac{1}{\sqrt{10}}$$

    (b) Let $t = \sec^{-1}(-3)$. Then $t = \cos^{-1}(-\tfrac{1}{3})$, so $\cos t = -\tfrac{1}{3}$ and $\tfrac{1}{2}\pi < t < \pi$. Furthermore, $\sin^2 t = 1 - \cos^2 t = 1 - (-\tfrac{1}{3})^2 = \tfrac{8}{9}$. Because $\sin t > 0$ if $\tfrac{1}{2}\pi < t < \pi$, then $\sin t = \sqrt{\tfrac{8}{9}} = \tfrac{2}{3}\sqrt{2}$. Therefore,

$$\tan[\sec^{-1}(-3)] = \tan t$$

$$= \frac{\sin t}{\cos t}$$

$$= \frac{\tfrac{2}{3}\sqrt{2}}{-\tfrac{1}{3}} = -2\sqrt{2}$$

16. Find the exact value of each of the following:

    (a) $\cos^{-1}[\sin(-\tfrac{1}{6}\pi)]$ **(b)** $\sin^{-1}[\cos(-\tfrac{1}{6}\pi)]$ **(c)** $\cos^{-1}[\sin \tfrac{2}{3}\pi]$

    (d) $\sin^{-1}[\cos \tfrac{2}{3}\pi]$

SOLUTION:

    (a) $\cos^{-1}[\sin(-\tfrac{1}{6}\pi)] = \cos^{-1}[-\tfrac{1}{2}]$
                  $= \pi - \cos^{-1}(\tfrac{1}{2})$
                  $= \pi - \tfrac{1}{3}\pi = \tfrac{2}{3}\pi$

    (b) $\sin^{-1}[\cos(-\tfrac{1}{6}\pi)] = \sin^{-1}[\tfrac{1}{2}\sqrt{3}] = \tfrac{1}{3}\pi$

    (c) $\cos^{-1}[\sin \tfrac{2}{3}\pi] = \cos^{-1}[\tfrac{1}{2}\sqrt{3}] = \tfrac{1}{6}\pi$

    (d) $\sin^{-1}[\cos \tfrac{2}{3}\pi] = \sin^{-1}[-\tfrac{1}{2}]$
                  $= -\sin^{-1}(\tfrac{1}{2}) = -\tfrac{1}{6}\pi$

In Exercises 17–26 find the exact value of the given quantity.

20. $\cos[\sin^{-1}(-\frac{1}{2}) + \sin^{-1}\frac{1}{4}]$

SOLUTION: Let $s = \sin^{-1}(-\frac{1}{2})$ and $t = \sin^{-1}\frac{1}{4}$. Then

$$\cos[\sin^{-1}(-\tfrac{1}{2}) + \sin^{-1}\tfrac{1}{4}] = \cos[s + t]$$
$$= \cos s \cdot \cos t - \sin s \cdot \sin t \qquad (1)$$

Because $s = \sin^{-1}(-\frac{1}{2})$, then

$$\sin s = -\tfrac{1}{2} \qquad (2)$$

and $-\frac{1}{2}\pi < s < 0$. Thus $\cos s > 0$, and

$$\cos s = \sqrt{1 - \sin^2 s} = \sqrt{1 - (-\tfrac{1}{2})^2} = \tfrac{1}{2}\sqrt{3} \qquad (3)$$

Because $t = \sin^{-1}\frac{1}{4}$, then

$$\sin t = \tfrac{1}{4} \qquad (4)$$

and $0 < t < \frac{1}{2}\pi$. Thus, $\cos t > 0$, and

$$\cos t = \sqrt{1 - \sin^2 t} = \sqrt{1 - (\tfrac{1}{4})^2} = \tfrac{1}{4}\sqrt{15} \qquad (5)$$

Substituting from (2), (3), (4), and (5) into the right-hand side of (1), we obtain

$$\cos[\sin^{-1}(-\tfrac{1}{2}) + \sin^{-1}\tfrac{1}{4}] = (\tfrac{1}{2}\sqrt{3})(\tfrac{1}{4}\sqrt{15}) - (-\tfrac{1}{2})(\tfrac{1}{4})$$
$$= \frac{3\sqrt{5} + 1}{8}$$

24. $\tan[\sec^{-1}\frac{5}{3} + \csc^{-1}(-\frac{13}{12})]$

SOLUTION: Let $s = \sec^{-1}\frac{5}{3}$ and $t = \csc^{-1}(-\frac{13}{12})$. Then

$$\tan[\sec^{-1}\tfrac{5}{3} + \csc^{-1}(-\tfrac{13}{12})] = \tan[s + t] = \frac{\tan s + \tan t}{1 - \tan s \cdot \tan t} \qquad (1)$$

Because $s = \sec^{-1}\frac{5}{3}$, then $\sec s = \frac{5}{3}$ and $0 < s < \frac{1}{2}\pi$. Thus, $\tan s > 0$ and

$$\tan s = \sqrt{\sec^2 s - 1} = \sqrt{(\tfrac{5}{3})^2 - 1} = \tfrac{4}{3} \qquad (2)$$

Because $t = \csc^{-1}(-\frac{13}{12})$, then $\csc t = -\frac{13}{12}$ and $-\frac{1}{2}\pi < t < 0$. Thus, $\cot t < 0$ and

$$\cot t = -\sqrt{\csc^2 t - 1} = -\sqrt{(\tfrac{13}{12})^2 - 1} = -\tfrac{5}{12}$$

Hence,

$$\tan t = \frac{1}{\cot t} = -\frac{12}{5} \qquad (3)$$

Substituting from (2) and (3) into the right-hand side of (1), we obtain

$$\tan[\sec^{-1}\tfrac{5}{3} + \csc^{-1}(-\tfrac{13}{12})] = \frac{\tfrac{4}{3} - \tfrac{12}{5}}{1 - (\tfrac{4}{3})(-\tfrac{12}{5})}$$
$$= -\tfrac{16}{63}$$

28. Prove

$$2 \tan^{-1}\tfrac{1}{3} - \tan^{-1}(-\tfrac{1}{7}) = \tfrac{1}{4}\pi$$

SOLUTION: Because $\tan^{-1}(-\frac{1}{7}) = -\tan^{-1}\frac{1}{7}$, then

$$2 \tan^{-1}\tfrac{1}{3} - \tan^{-1}(-\tfrac{1}{7}) = 2 \tan^{-1}\tfrac{1}{3} + \tan^{-1}\tfrac{1}{7} \qquad (1)$$

With $s = \tan^{-1}\frac{1}{3}$ and $t = \tan^{-1}\frac{1}{7}$ on the right-hand side of (1), we have

$$2 \tan^{-1}\tfrac{1}{3} - \tan^{-1}(-\tfrac{1}{7}) = 2s + t \qquad (2)$$

We must show that $2s + t = \frac{1}{4}\pi$. First, we show that $0 < 2s + t < \pi$. We have

$$s = \tan^{-1} \tfrac{1}{3} < \tan^{-1} 1 = \tfrac{1}{4}\pi$$

Thus,

$$0 < 2s < \tfrac{1}{2}\pi \tag{3}$$

and

$$t = \tan^{-1} \tfrac{1}{7} < \tfrac{1}{2}\pi$$

Thus,

$$0 < t < \tfrac{1}{2}\pi \tag{4}$$

By adding the corresponding members of (3) and (4) we obtain

$$0 < 2s + t < \pi \tag{5}$$

Next, we show that $\tan(2s + t) = 1$. Now

$$\tan(2s + t) = \frac{\tan 2s + \tan t}{1 - \tan 2s \cdot \tan t} \tag{6}$$

Because $s = \tan^{-1} \tfrac{1}{3}$, then $\tan s = \tfrac{1}{3}$. Thus,

$$\tan 2s = \frac{2 \tan s}{1 - \tan^2 s}$$

$$= \frac{2(\tfrac{1}{3})}{1 - (\tfrac{1}{3})^2} = \tfrac{3}{4} \tag{7}$$

Because $t = \tan^{-1} \tfrac{1}{7}$, then

$$\tan t = \tfrac{1}{7} \tag{8}$$

Substituting from (7) and (8) into (6), we have

$$\tan(2s + t) = \frac{\tfrac{3}{4} + \tfrac{1}{7}}{1 - \tfrac{3}{4} \cdot \tfrac{1}{7}} = 1 \tag{9}$$

From (9), we conclude that

$$2s + t = \tfrac{1}{4}\pi + k\pi \tag{10}$$

From (10) and (5), we conclude that

$$2s + t = \tfrac{1}{4}\pi \tag{11}$$

Thus, from (11) and (2), we have

$$2 \tan^{-1} \tfrac{1}{3} - \tan^{-1}(-\tfrac{1}{7}) = \tfrac{1}{4}\pi$$

In Exercises 29–36, draw a sketch of the graph of the given equation.

32. $y = 2 \tan^{-1} x$

SOLUTION: Let $f(x) = 2 \tan^{-1} x$. Because $\tan^{-1} x$ is defined for all real x, the domain of f is $(-\infty, +\infty)$. Because $-\tfrac{1}{2}\pi < \tan^{-1} x < \tfrac{1}{2}\pi$, then $-\pi < 2 \tan^{-1} x < \pi$, and the range of f is $(-\pi, \pi)$.

$$\lim_{x \to +\infty} f(x) = 2 \lim_{x \to +\infty} \tan^{-1} x$$

$$= 2(\tfrac{1}{2}\pi) = \pi$$

Thus, the line $y = \pi$ is a horizontal asymptote for the graph of f.

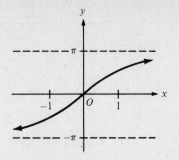

Figure 9. 4. 32

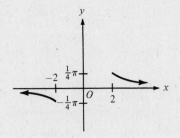

Figure 9. 4. 36

$$\lim_{x \to -\infty} f(x) = 2 \lim_{x \to -\infty} \tan^{-1} x$$

$$= 2(-\tfrac{1}{2}\pi)$$

$$= -\pi$$

Thus, the line $y = -\pi$ is a horizontal asymptote for the graph of f. Because $f(0) = 2 \tan^{-1} 0 = 0$, the graph contains the origin. A sketch of the graph is shown in Fig. 9. 4. 32.

36. $y = \tfrac{1}{2} \csc^{-1} \tfrac{1}{2} x$

SOLUTION: Let $f(x) = \tfrac{1}{2} \csc^{-1} \tfrac{1}{2} x$. Because $\csc^{-1} x$ is defined for $|x| \geq 1$, then f is defined for $|\tfrac{1}{2} x| \geq 1$, or equivalently, for $x \geq 2$ or $x \leq -2$. Thus, the domain of f is $(-\infty, -2] \cup [2, +\infty)$.

$$\lim_{x \to \pm\infty} f(x) = \tfrac{1}{2} \lim_{x \to \pm\infty} \csc^{-1}(\tfrac{1}{2} x) = 0$$

Thus, the x-axis is a horizontal asymptote for the graph of f. Because

$$0 < |\csc^{-1} x| \leq \tfrac{1}{2}\pi$$

then

$$0 < |\tfrac{1}{2} \csc^{-1} \tfrac{1}{2} x| \leq \tfrac{1}{4}\pi$$

Thus, the range of f is $[-\tfrac{1}{4}\pi, 0) \cup (0, \tfrac{1}{4}\pi]$. Furthermore, $f(2) = \tfrac{1}{2} \csc^{-1}(1) = \tfrac{1}{4}\pi$ and $f(-2) = \tfrac{1}{2} \csc^{-1}(-1) = -\tfrac{1}{4}\pi$. A sketch of the graph of f is shown in Fig. 9. 4. 36.

9.5 DERIVATIVES OF THE INVERSE TRIGONOMETRIC FUNCTIONS

9.5.1 Theorem   If u is a differentiable function of x,

$$D_x(\sin^{-1} u) = \frac{1}{\sqrt{1 - u^2}} D_x u$$

9.5.2 Theorem   If u is a differentiable function of x,

$$D_x(\cos^{-1} u) = -\frac{1}{\sqrt{1 - u^2}} D_x u$$

9.5.3 Theorem   If u is a differentiable function of x,

$$D_x(\tan^{-1} u) = \frac{1}{1 + u^2} D_x u$$

9.5.4 Theorem   If u is a differentiable function of x,

$$D_x(\cot^{-1} u) = -\frac{1}{1 + u^2} D_x u$$

9.5.5 Theorem   If u is a differentiable function of x,

$$D_x(\sec^{-1} u) = \frac{1}{|u|\sqrt{u^2 - 1}} D_x u$$

9.5.6 Theorem   If u is a differentiable function of x,

$$D_x(\csc^{-1} u) = -\frac{1}{|u|\sqrt{u^2 - 1}} D_x u$$

Exercises 9.5

In Exercises 1–26 find the derivative of the given function.

4. $f(x) = \csc^{-1} 2x$

SOLUTION:

$$f'(x) = -\frac{1}{|2x|\sqrt{(2x)^2 - 1}} \cdot D_x(2x)$$

$$= \frac{-1}{|x|\sqrt{4x^2 - 1}}$$

8. $f(y) = \cot^{-1} e^y$

SOLUTION: We apply Theorem 9.5.4 with $u = e^y$

$$f'(y) = -\frac{1}{1 + (e^y)^2} D_y e^y$$

$$= \frac{-e^y}{1 + e^{2y}}$$

12. $f(x) = \ln(\tan^{-1} 3x)$

SOLUTION:

$$f'(x) = \frac{1}{\tan^{-1} 3x} D_x(\tan^{-1} 3x)$$

$$= \frac{1}{\tan^{-1} 3x} \cdot \frac{1}{1 + (3x)^2} D_x 3x$$

$$= \frac{3}{(1 + 9x^2)\tan^{-1} 3x}$$

16. $g(s) = \cos^{-1} s + \dfrac{s}{1 - s^2}$

SOLUTION:

$$g'(s) = -\frac{1}{\sqrt{1 - s^2}} + \frac{(1 - s^2) - s(-2s)}{(1 - s^2)^2}$$

$$= \frac{-(1 - s^2)^{3/2} + 1 + s^2}{(1 - s^2)^2}$$

20. $F(x) = \sec^{-1} \sqrt{x^2 + 4}$

SOLUTION:

$$F'(x) = \frac{1}{|\sqrt{x^2 + 4}|\sqrt{(x^2 + 4) - 1}} \cdot D_x \sqrt{x^2 + 4}$$

$$= \frac{1}{(x^2 + 4)^{1/2}(x^2 + 3)^{1/2}} \cdot \frac{x}{(x^2 + 4)^{1/2}}$$

$$= \frac{x}{(x^2 + 4)(x^2 + 3)^{1/2}}$$

24. $f(x) = \sin^{-1} x + \cos^{-1} x$

SOLUTION:

$$f'(x) = \frac{1}{\sqrt{1-x^2}} + \frac{-1}{\sqrt{1-x^2}} = 0$$

Because $f'(x) = 0$, we conclude that $f(x)$ is a constant. Indeed, $f(x) = \sin^{-1} x + \cos^{-1} x = \frac{1}{2}\pi$.

28. Find $D_x y$ by implicit differentiation.

$$\ln(\sin^2 3x) = e^x + \cot^{-1} y$$

SOLUTION: Because $\ln x^r = r \ln x$, we may simplify the given equation as follows:

$$2 \ln(\sin 3x) = e^x + \cot^{-1} y$$

Differentiating on both sides with respect to x, we obtain

$$\frac{2}{\sin 3x} (\cos 3x)3 = e^x - \frac{1}{1+y^2} D_x y$$

$$6 \cot 3x = e^x - \frac{D_x y}{1+y^2}$$

$$D_x y = (1+y^2)(e^x - 6 \cot 3x)$$

32. A ladder 25 ft long is leaning against a vertical wall. If the bottom of the ladder is pulled horizontally away from the wall so that the top is sliding down at 3 ft/sec, how fast is the measure of the angle between the ladder and the ground changing when the bottom of the ladder is 15 ft from the wall?

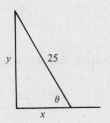

Figure 9. 5. 32

SOLUTION: Refer to Fig. 9. 5. 32,
     Let

   $t =$ the number of seconds that have elapsed since the ladder began to be pulled away from the wall
   $x =$ the number of feet in the distance between the wall and the bottom of the ladder at t sec
   $y =$ the number of feet in the distance between the floor and the top of the ladder at t sec
   $\theta =$ the radian measure of the angle between the ladder and the ground at t sec

We are given that $D_t y = -3$. We want to find $D_t \theta$ when $x = 15$. Because $x^2 + y^2 = 625$, then $y = 20$ when $x = 15$. We have

$$\theta = \sin^{-1} \frac{y}{25}$$

$$D_t \theta = \frac{1}{\sqrt{1 - \left(\dfrac{y}{25}\right)^2}} \cdot \frac{1}{25} D_t y$$

Substituting $D_t y = -3$, we obtain

$$D_t \theta = \frac{-3}{\sqrt{625 - y^2}}$$

Then,

$$D_t\theta]_{y=20} = \frac{-3}{\sqrt{625-20^2}} = -\frac{1}{5}$$

We conclude that the angle is decreasing at the rate of $\frac{1}{5}$ radian per second when the bottom of the ladder is 15 feet from the wall.

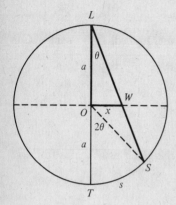

Figure 9. 5. 36

36. A woman is walking at the rate of 5 ft/sec along the diameter of a circular courtyard. A light at one end of a diameter perpendicular to her path casts a shadow on the circular wall. How far is the woman from the center of the courtyard when the speed of her shadow along the wall is 9 ft/sec?

SOLUTION: Refer to Fig. 9. 5. 36. The light is at point L, the center of the circle is at point O, and line segment LT is a diameter of the circle. At a certain instant the woman is at point W moving away from point O along the diameter perpendicular to LT. At a certain instant the light casts a shadow on point S. Thus, the shadow moves along arc TS away from point T. We define the variables.
Let

$t =$ the number of seconds that have elapsed since the woman began to walk
$x =$ the number of feet in the length of line segment OW at t sec
$s =$ the number of feet in the length of arc TS at t sec
$\theta =$ the radian measure of angle TLS at t sec

We are given that $D_t x = 5$. We want to find x when $D_t s = 9$. Thus, we must find an equation that expresses the functional relationship between x and s. If a is the radius of the circle, we have

$$2\theta = \frac{s}{a}$$

because 2θ is the radian measure of a central angle in a circle with radius a that intercepts an arc of length s. Thus

$$\theta = \frac{s}{2a} \tag{1}$$

Furthermore, from right triangle LOW, we have

$$\theta = \tan^{-1}\left(\frac{x}{a}\right) \tag{2}$$

From (1) and (2) we get

$$\tan^{-1}\left(\frac{x}{a}\right) = \frac{s}{2a}$$

Differentiating on both sides with respect to t, we obtain

$$\frac{1}{1+\left(\frac{x}{a}\right)^2} \cdot \frac{1}{a} D_t x = \frac{D_t s}{2a}$$

$$\frac{a^2}{a^2+x^2} D_t x = \frac{1}{2} D_t s$$

Substituting $D_t x = 5$ and $D_t s = 9$, we get

$$\frac{5a^2}{a^2+x^2} = \frac{9}{2}$$

Solving for x, we have

$$x = \frac{1}{3}a$$

Thus, when the speed of the woman's shadow is 9 ft/sec, her distance from the center of the courtyard is one-third of the radius of the courtyard.

40. Given: $f(x) = \tan^{-1}(1/x) - \cot^{-1} x$

(a) Show that $f'(x) = 0$ for all x in the domain of f.
(b) Prove that there is no constant C for which $f(x) = C$ for all x in its domain.
(c) Why doesn't the statement in part (b) contradict Theorem 5.2.2?

SOLUTION:

(a) $f'(x) = \dfrac{1}{1 + \left(\dfrac{1}{x}\right)^2} D_x\left(\dfrac{1}{x}\right) - \dfrac{-1}{1 + x^2}$

$= \dfrac{x^2}{x^2 + 1} \cdot \dfrac{-1}{x^2} + \dfrac{1}{1 + x^2}$

$= \dfrac{-1}{1 + x^2} + \dfrac{1}{1 + x^2} = 0$

(b) We apply Definition 9.4.4 to obtain

$f(x) = \tan^{-1}\left(\dfrac{1}{x}\right) - \left(\dfrac{\pi}{2} - \tan^{-1} x\right)$

$= \tan^{-1}\left(\dfrac{1}{x}\right) + \tan^{-1} x - \dfrac{\pi}{2}$

For $x = 1$ we have
$f(1) = \tan^{-1} 1 + \tan^{-1} 1 - \dfrac{\pi}{2}$

$= \dfrac{\pi}{4} + \dfrac{\pi}{4} - \dfrac{\pi}{2} = 0$

For $x = -1$ we have

$f(-1) = \tan^{-1}(-1) + \tan^{-1}(-1) - \dfrac{\pi}{2}$

$= -\dfrac{\pi}{4} - \dfrac{\pi}{4} - \dfrac{\pi}{2} = -\pi$

Because $f(1) \neq f(-1)$, there is no constant C for which $f(x) = C$ for all x in the domain of f.

(c) The given function f is not defined at $x = 0$. For every $x > 0$ we can show that $f(x) = 0$, and for every $x < 0$ we can show that $f(x) = -\pi$. In part (b) we illustrated this for the special cases $x = 1$ and $x = -1$. Any interval that contains both positive and negative values of x must also contain $x = 0$. Thus, there is no *interval* such that f is defined for all x in the interval and for which f is not constant. Thus, the function f does not contradict Theorem 5.2.2.

9.6 INTEGRALS YIELDING INVERSE TRIGONOMETRIC FUNCTIONS

9.6.1 Theorem
$$\int \frac{du}{\sqrt{1 - u^2}} = \sin^{-1} u + C$$

$$\int \frac{du}{1 + u^2} = \tan^{-1} u + C$$

$$\int \frac{du}{u\sqrt{u^2 - 1}} = \sec^{-1}|u| + C$$

From the above we may also derive the following special cases.

9.6.2 Theorem $$\int \frac{du}{\sqrt{a^2 - u^2}} = \sin^{-1} \frac{u}{a} + C \qquad \text{where } a > 0$$

$$\int \frac{du}{a^2 + u^2} = \frac{1}{a} \tan^{-1} \frac{u}{a} + C \qquad \text{where } a \neq 0$$

$$\int \frac{du}{u\sqrt{u^2 - a^2}} = \frac{1}{a} \sec^{-1} \left| \frac{u}{a} \right| + C \qquad \text{where } a > 0$$

Exercises 9.6

In Exercises 1–25 evaluate the indefinite integral.

4. $\displaystyle\int \frac{dt}{\sqrt{1 - 16t^2}}$

SOLUTION: We apply Theorem 9.6.1 with $u = 4t$ and $du = 4\, dt$.

$$\int \frac{dt}{\sqrt{1 - 16t^2}} = \int \frac{\frac{1}{4}\, du}{\sqrt{1 - u^2}}$$

$$= \tfrac{1}{4} \sin^{-1} u + C$$

$$= \tfrac{1}{4} \sin^{-1} 4t + C$$

8. $\displaystyle\int \frac{x\, dx}{x^4 + 16}$

SOLUTION: Let $u = x^2$. Then $du = 2x\, dx$. Thus, by Theorem 9.6.2 we have

$$\int \frac{x\, dx}{x^4 + 16} = \tfrac{1}{2} \int \frac{du}{u^2 + 4^2}$$

$$= \tfrac{1}{2} \cdot \tfrac{1}{4} \tan^{-1} \tfrac{1}{4} u + C$$

$$= \tfrac{1}{8} \tan^{-1} \tfrac{1}{4} x^2 + C$$

12. $\displaystyle\int \frac{du}{u\sqrt{16u^2 - 9}}$

SOLUTION:

$$\int \frac{du}{u\sqrt{16u^2 - 9}} = \frac{1}{4} \int \frac{du}{u\sqrt{u^2 - \frac{9}{16}}}$$

$$= \frac{1}{4} \cdot \frac{1}{\frac{3}{4}} \sec^{-1} \left| \frac{u}{\frac{3}{4}} \right| + C$$

$$= \frac{1}{3} \sec^{-1} \left| \frac{4}{3} u \right| + C$$

16. $\displaystyle\int \frac{ds}{\sqrt{2s-s^2}}$

SOLUTION: We complete the square under the radical sign. Because

$$2s - s^2 = -(s^2 - 2s + 1) + 1$$
$$= 1 - (s-1)^2$$

then

$$\int \frac{ds}{\sqrt{2s-s^2}} = \int \frac{ds}{\sqrt{1-(s-1)^2}}$$
$$= \sin^{-1}(s-1) + C$$

20. $\displaystyle\int \frac{dx}{2x^2 + 2x + 3}$

SOLUTION:

$$\int \frac{dx}{2x^2 + 2x + 3} = \frac{1}{2} \int \frac{dx}{x^2 + x + \frac{3}{2}}$$

$$= \frac{1}{2} \int \frac{dx}{(x^2 + x + \frac{1}{4}) + \frac{5}{4}}$$

$$= \frac{1}{2} \int \frac{dx}{(x + \frac{1}{2})^2 + (\frac{1}{2}\sqrt{5})^2}$$

$$= \frac{1}{2} \cdot \frac{1}{\frac{1}{2}\sqrt{5}} \tan^{-1} \frac{x + \frac{1}{2}}{\frac{1}{2}\sqrt{5}} + C$$

$$= \frac{1}{\sqrt{5}} \tan^{-1} \frac{2x + 1}{\sqrt{5}} + C$$

24. $\displaystyle\int \frac{2\,dt}{(t-3)\sqrt{t^2 - 6t + 5}}$

SOLUTION: First we complete the square under the radical sign.

$$\int \frac{2\,dt}{(t-3)\sqrt{t^2 - 6t + 5}} = 2 \int \frac{dt}{(t-3)\sqrt{(t-3)^2 - 4}}$$

Then we let $u = t - 3$, $du = dt$, and apply Theorem 9.6.2.

$$\int \frac{2\,dt}{(t-3)\sqrt{t^2 - 6t + 5}} = 2 \int \frac{du}{u\sqrt{u^2 - 2^2}}$$

$$= 2(\tfrac{1}{2}) \sec^{-1} |\tfrac{1}{2} u| + C$$

$$= \sec^{-1} \left| \frac{t-3}{2} \right| + C$$

In Exercises 26–33 evaluate the definite integral.

28. $\displaystyle\int_0^{\sqrt{3}} \frac{x\,dx}{\sqrt{1+x^2}}$

SOLUTION: Let $u = 1 + x^2$. Then $du = 2x\,dx$; when $x = 0$, $u = 1$; when $x = \sqrt{3}$, $u = 4$. Thus,

$$\int_0^{\sqrt{3}} \frac{x\,dx}{\sqrt{1+x^2}} = \frac{1}{2}\int_1^4 u^{-1/2}\,du$$

$$= u^{1/2}\Big]_1^4 = 1$$

32. $\displaystyle\int_{1/\sqrt{2}}^1 \frac{dx}{x\sqrt{4x^2-1}}$

SOLUTION: Let $u = 2x$. Then $du = 2\,dx$. When $x = 1/\sqrt{2}$, $u = \sqrt{2}$; when $x = 1$, $u = 2$. Thus,

$$\int_{1/\sqrt{2}}^1 \frac{dx}{x\sqrt{4x^2-1}} = \int_{\sqrt{2}}^2 \frac{\frac{1}{2}\,du}{\frac{1}{2}u\sqrt{u^2-1}}$$

$$= \sec^{-1}|u|\,\Big]_{\sqrt{2}}^2$$

$$= \sec^{-1}2 - \sec^{-1}\sqrt{2}$$

$$= \cos^{-1}\frac{1}{2} - \cos^{-1}\frac{1}{2}\sqrt{2}$$

$$= \frac{1}{3}\pi - \frac{1}{4}\pi = \frac{1}{12}\pi$$

36. Find the area of the region bounded by the curves $x^2 = 4ay$ and $y = 8a^3/(x^2 + 4a^2)$.

SOLUTION: Figure 9. 6. 36 shows a sketch of the region if $a > 0$. To find the intersection of the two curves, we eliminate y from the given pair of equations. Thus,

$$\frac{x^2}{4a} = \frac{8a^3}{x^2 + 4a^2}$$

$$x^4 + 4a^2x^2 - 32a^4 = 0$$

$$(x^2 + 8a^2)(x^2 - 4a^2) = 0$$

$$x^2 - 4a^2 = 0$$

$$x = \pm 2a$$

We ignore $x^2 + 8a^2 = 0$ because there is no real solution of this equation. The region is bounded above by the curve $y = 8a^3/(x^2 + 4a^2)$, bounded below by

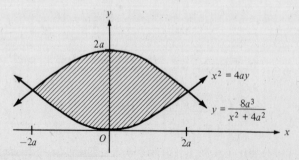

Figure 9. 6. 36

the curve $y = x^2/4a$, bounded on the left by the line $x = -2a$, and bounded on the right by the line $x = 2a$. Thus,

$$A = \int_{-2a}^{2a} \left[\frac{8a^3}{x^2 + 4a^2} - \frac{x^2}{4a} \right] dx$$

$$= 8a^3 \int_{-2a}^{2a} \frac{dx}{x^2 + (2a)^2} - \frac{1}{4a} \int_{-2a}^{2a} x^2 \, dx$$

$$= 8a^3 \cdot \frac{1}{2a} \left[\tan^{-1} \frac{x}{2a} \right]_{-2a}^{2a} - \frac{1}{12a} \left[x^3 \right]_{-2a}^{2a}$$

$$= 4a^2 [\tan^{-1} 1 - \tan^{-1}(-1)] - \frac{1}{12a} [8a^3 - (-8a^3)]$$

$$= 4a^2 \left[\frac{\pi}{4} - \left(-\frac{\pi}{4} \right) \right] - \frac{16a^3}{12a} = 2\pi a^2 - \frac{4}{3} a^2$$

Thus, the area is $2\pi a^2 - \frac{4}{3} a^2$ square units.

40. Prove the given formula by showing that the derivative of the right side is equal to the integrand. Is the formula equivalent to formula (1) in the text? Why?

$$\int \frac{du}{\sqrt{1 - u^2}} = -\cos^{-1} u + C$$

SOLUTION: Let $f(u) = -\cos^{-1} u$. Then

$$f'(u) = -\frac{-1}{\sqrt{1 - u^2}} = \frac{1}{\sqrt{1 - u^2}}$$

Because $f'(u)$ is equal to the given integrand, the given formula is proved. Formula (1) in the text is

$$\int \frac{du}{\sqrt{1 - u^2}} = \sin^{-1} u + C$$

Because $\sin^{-1} u + \cos^{-1} u = \frac{1}{2}\pi$ for all u in the closed interval $[-1, 1]$, the two formulas differ by the constant $\frac{1}{2}\pi$ and thus the two formulas are equivalent.

44. A particle is moving in a straight line according to the equation of motion $s = 5 - 10 \sin^2 2t$, where s cm is the directed distance of the particle from the origin at t sec. Use the result of part (b) of Exercise 43 to show that the motion is simple harmonic. Find the amplitude and period of this motion.

SOLUTION: Exercise 43 states that if $s = a \cos kt$, then the particle has simple harmonic motion. We are given that

$$s = 5 - 10 \sin^2 2t$$
$$= 5(1 - 2 \sin^2 2t) \tag{1}$$

Because $\cos 2x = 1 - 2 \sin^2 x$, then with $x = 2t$ we have

$$\cos 4t = 1 - 2 \sin^2 2t \tag{2}$$

Substituting from (2) into (1), we have

$$s = 5 \cos 4t$$

Therefore, the motion is simple harmonic. The amplitude is $a = 5$ and the period is $2\pi/k = 2\pi/4 = \frac{1}{2}\pi$.

9.7 THE HYPERBOLIC FUNCTIONS

9.7.1 Definition   The *hyperbolic sine function* is defined by

$$\sinh x = \frac{e^x - e^{-x}}{2}$$

The domain and range are the set of all real numbers.

9.7.2 Definition   The *hyperbolic cosine function* is defined by

$$\cosh x = \frac{e^x + e^{-x}}{2}$$

The domain is the set of all real numbers, and the range is the set of all numbers in the interval $[1, +\infty)$.

9.7.3 Theorem   If u is a differentiable function of x,

$$D_x(\sinh u) = \cosh u \, D_x u$$
$$D_x(\cosh u) = \sinh u \, D_x u$$

9.7.4 Definition   The *hyperbolic tangent function, hyperbolic cotangent function, hyperbolic secant function*, and *hyperbolic cosecant function* are defined, respectively, as follows:

$$\tanh x = \frac{\sinh x}{\cosh x} = \frac{e^x - e^{-x}}{e^x + e^{-x}}$$

$$\coth x = \frac{\cosh x}{\sinh x} = \frac{e^x + e^{-x}}{e^x - e^{-x}}$$

$$\operatorname{sech} x = \frac{1}{\cosh x} = \frac{2}{e^x + e^{-x}}$$

$$\operatorname{csch} x = \frac{1}{\sinh x} = \frac{2}{e^x - e^{-x}}$$

9.7.5 Theorem   If u is a differentiable function of x,

$$D_x(\tanh u) = \operatorname{sech}^2 u \, D_x u$$
$$D_x(\coth u) = -\operatorname{csch}^2 u \, D_x u$$
$$D_x(\operatorname{sech} u) = -\operatorname{sech} u \tanh u \, D_x u$$
$$D_x(\operatorname{csch} u) = -\operatorname{csch} u \coth u \, D_x u$$

9.7.6 Theorem

$$\int \sinh u \, du = \cosh u + C$$

$$\int \cosh u \, du = \sinh u + C$$

$$\int \operatorname{sech}^2 u \, du = \tanh u + C$$

$$\int \operatorname{csch}^2 u \, du = -\coth u + C$$

$$\int \operatorname{sech} u \tanh u \, du = -\operatorname{sech} u + C$$

$$\int \operatorname{csch} u \coth u \, du = -\operatorname{csch} u + C$$

The following are the most frequently used identities for the hyperbolic functions.

$$\cosh^2 x - \sinh^2 x = 1$$
$$1 - \tanh^2 x = \operatorname{sech}^2 x$$
$$1 - \coth^2 x = -\operatorname{csch}^2 x$$
$$\sinh 2x = 2 \sinh x \cosh x$$
$$\cosh 2x = \cosh^2 x + \sinh^2 x$$
$$= 2 \sinh^2 x + 1$$
$$= 2 \cosh^2 x - 1$$

Exercises 9.7

In Exercises 1–10 prove the identities.

4. (a) $\sinh 2x = 2 \sinh x \cosh x$
(b) $\cosh 2x = \cosh^2 x + \sinh^2 x = 2 \sinh^2 x + 1 = 2 \cosh^2 x - 1$

SOLUTION:
(a) From Eq. (11) in the text we have

$$\sinh(x + y) = \sinh x \cosh y + \cosh x \sinh y \tag{1}$$

Replacing y by x in (1), we obtain

$$\sinh(x + x) = \sinh x \cosh x + \cosh x \sinh x$$
$$\sinh 2x = 2 \sinh x \cosh x$$

(b) From Exercise 2 we have

$$\cosh(x + y) = \cosh x \cosh y + \sinh x \sinh y \tag{2}$$

Replacing y by x in (2), we obtain

$$\cosh(x + x) = \cosh x \cosh x + \sinh x \sinh x$$
$$\cosh 2x = \cosh^2 x + \sinh^2 x \tag{3}$$

From Eq. (6) in the text we obtain

$$\cosh^2 x = 1 + \sinh^2 x \tag{4}$$

Substituting from (4) into (3), we get

$$\cosh 2x = 1 + 2 \sinh^2 x$$

Solving (4) for $\sinh^2 x$ and substituting into (3), we obtain

$$\cosh 2x = 2 \cosh^2 x - 1$$

8. $\sinh 3x = 3 \sinh x + 4 \sinh^3 x$

SOLUTION: From Eq. (11) in the text, Exercise 4, and Eq. (6) in the text, we obtain

$$\sinh 3x = \sinh(2x + x)$$
$$= \sinh 2x \cosh x + \cosh 2x \sinh x$$
$$= (2 \sinh x \cosh x)\cosh x + (2 \sinh^2 x + 1)\sinh x$$
$$= 2 \sinh x \cosh^2 x + 2 \sinh^3 x + \sinh x$$
$$= 2 \sinh x(1 + \sinh^2 x) + 2 \sinh^3 x + \sinh x$$
$$= 3 \sinh x + 4 \sinh^3 x$$

12. Prove that the hyperbolic tangent function is an odd function and the hyperbolic secant function is an even function.

SOLUTION: From Definition 9.7.4,

$$\tanh x = \frac{e^x - e^{-x}}{e^x + e^{-x}}$$

Replacing x with $-x$, we obtain

$$\tanh(-x) = \frac{e^{-x} - e^x}{e^{-x} + e^x}$$

$$= -\frac{e^x - e^{-x}}{e^x + e^{-x}}$$

$$= -\tanh x$$

Therefore, the hyperbolic tangent function is an odd function. From Definition 8.7.4,

$$\text{sech } x = \frac{2}{e^x + e^{-x}}$$

Replacing x with $-x$, we obtain

$$\text{sech}(-x) = \frac{2}{e^{-x} + e^x}$$

$$= \frac{2}{e^x + e^{-x}} = \text{sech } x$$

which proves that the hyperbolic secant function is an even function.

16. Prove the following differentiation formulas.
    (a) $D_x(\text{sech } x) = -\text{sech } x \tanh x$ **(b)** $D_x(\text{csch } x) = -\text{csch } x \coth x$

SOLUTION:

(a) $D_x(\text{sech } x) = D_x\left(\dfrac{1}{\cosh x}\right)$

$$= \frac{-1}{\cosh^2 x} \cdot D_x(\cosh x)$$

$$= \frac{-\sinh x}{\cosh^2 x}$$

$$= -\frac{1}{\cosh x} \cdot \frac{\sinh x}{\cosh x}$$

$$= -\text{sech } x \tanh x$$

(b) $D_x(\text{csch } x) = D_x\left(\dfrac{1}{\sinh x}\right)$

$$= \frac{-1}{\sinh^2 x} \cdot D_x(\sinh x)$$

$$= \frac{-\cosh x}{\sinh^2 x}$$

$$= -\frac{1}{\sinh x} \cdot \frac{\cosh x}{\sinh x}$$

$$= -\text{csch } x \coth x$$

In Exercises 17–30 find the derivative of the given function.

20. $f(x) = \coth(\ln x)$

SOLUTION: Applying Theorem 9.7.5 with $u = \ln x$, we obtain

$$f'(x) = -\text{csch}^2(\ln x) \, D_x \ln x$$

$$= \frac{-\text{csch}^2(\ln x)}{x}$$

24. $h(x) = \coth \dfrac{1}{x}$

SOLUTION:

$$h'(x) = -\text{csch}^2 \frac{1}{x} \cdot D_x \left(\frac{1}{x}\right)$$

$$= \frac{\text{csch}^2 \dfrac{1}{x}}{x^2}$$

28. $f(x) = \ln(\coth 3x - \text{csch } 3x)$

SOLUTION:

$$f'(x) = \frac{1}{\coth 3x - \text{csch } 3x}[-\text{csch}^2 3x + \text{csch } 3x \coth 3x] \cdot 3$$

$$= \frac{3 \text{ csch } 3x(\coth 3x - \text{csch } 3x)}{\coth 3x - \text{csch } 3x} = 3 \text{ csch } 3x$$

32. Prove: $\displaystyle\int \text{sech } u \, du = 2 \tan^{-1} e^u + C$

SOLUTION:

$$\int \text{sech } u \, du = \int \frac{2 \, du}{e^u + e^{-u}}$$

$$= \int \frac{2e^u \, du}{e^u(e^u + e^{-u})} = \int \frac{2e^u \, du}{e^{2u} + 1} \tag{1}$$

Let $z = e^u$. Then $dz = e^u \, du$. Thus, from (1) we get

$$\int \text{sech } u \, du = \int \frac{2 \, dz}{z^2 + 1}$$

$$= 2 \tan^{-1} z + C$$

$$= 2 \tan^{-1} e^u + C$$

In Exercises 34–43 evaluate the indefinite integral.

36. $\displaystyle\int \tanh^2 3x \, dx$

SOLUTION:

$$\int \tanh^2 3x \, dx = \int (1 - \text{sech}^2 3x) \, dx$$

$$= x - \tfrac{1}{3} \tanh 3x + C$$

40. $\int x \sinh^2 x^2 \, dx$

SOLUTION: Let $u = x^2$ and $du = 2x \, dx$. Then

$$\int x \sinh^2 x^2 \, dx = \tfrac{1}{2} \int \sinh^2 u \, du \tag{1}$$

In Exercise 4(b) we proved that

$$\cosh 2x = 2 \sinh^2 x + 1$$

Thus,

$$\sinh^2 u = \tfrac{1}{2}(\cosh 2u - 1) \tag{2}$$

Substituting from (2) into (1), we obtain

$$\int x \sinh^2 x^2 \, dx = \tfrac{1}{2} \int \tfrac{1}{2}(\cosh 2u - 1) \, du$$

$$= \tfrac{1}{4}(\tfrac{1}{2} \sinh 2u - u) + C$$

$$= \tfrac{1}{8} \sinh 2x^2 - \tfrac{1}{4}x^2 + C$$

44. Evaluate the definite integral

$$\int_0^{\ln 2} \tanh z \, dz$$

SOLUTION:

$$\int_0^{\ln 2} \tanh z \, dz = \int_0^{\ln 2} \frac{\sinh z}{\cosh z} \, dz \tag{1}$$

We let $u = \cosh z$ and $du = \sinh z \, dz$. When $z = \ln 2$, then $e^z = 2$, and $e^{-z} = \tfrac{1}{2}$, so

$$\cosh z = \frac{e^z + e^{-z}}{2}$$

$$= \frac{2 + \tfrac{1}{2}}{2} = \frac{5}{4}$$

Thus, $u = \tfrac{5}{4}$ when $z = \ln 2$. When $z = 0$, then $u = \cosh 0 = \tfrac{1}{2}(e^0 + e^{-0}) = 1$. With the indicated substitutions in Eq. (1) we obtain

$$\int_0^{\ln 2} \tanh z \, dz = \int_1^{5/4} \frac{du}{u}$$

$$= \ln|u| \Big]_1^{5/4}$$

$$= \ln \frac{5}{4} - \ln 1 = \ln \frac{5}{4}$$

48. Draw a sketch of the graph of the hyperbolic cotangent function. Find equations of the asymptotes.

SOLUTION: Because

$$\coth x = \frac{e^x + e^{-x}}{e^x - e^{-x}}$$

then

$$\coth(-x) = \frac{e^{-x} + e^x}{e^{-x} - e^x} = -\frac{e^x + e^{-x}}{e^x - e^{-x}} = -\coth x$$

Therefore, the hyperbolic cotangent function is an odd function. Hence, the graph of the function is symmetric with respect to the origin. Because

$$\coth x = \frac{e^x + e^{-x}}{e^x - e^{-x}} \cdot \frac{e^{-x}}{e^{-x}} = \frac{1 + e^{-2x}}{1 - e^{-2x}}$$

then

$$\lim_{x \to +\infty} \coth x = \lim_{x \to +\infty} \frac{1 + e^{-2x}}{1 - e^{-2x}} = 1$$

Thus, the line $y = 1$ is a horizontal asymptote, and by symmetry the line $y = -1$ is a horizontal asymptote. Because

$$\coth x = \frac{e^x + e^{-x}}{e^x - e^{-x}} \cdot \frac{e^x}{e^x} = \frac{e^{2x} + 1}{e^{2x} - 1}$$

then

$$\lim_{x \to 0^+} \coth x = \lim_{x \to 0^+} \frac{e^{2x} + 1}{e^{2x} - 1} = +\infty$$

Thus, the y-axis is a vertical asymptote. We plot a few points, use the asymptotes and symmetry, and draw a sketch of the graph, which is shown in Fig. 9. 7. 48.

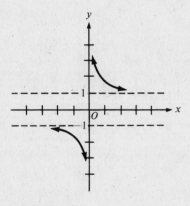

Figure 9. 7. 48

52. Find the volume of the solid of revolution if the region bounded by the catenary $y = a \cosh(x/a)$, where $a > 0$, the y-axis, the x-axis, and the line $x = x_1$, where $x_1 > 0$, is revolved about the x-axis.

SOLUTION: Figure 9. 7. 52 shows a sketch of the region. Let $f(x) = a \cosh(x/a)$. Then the element of volume is a circular disk with thickness $\Delta_i x$ units and radius $f(\bar{x}_i)$ units. Thus,

$$V = \lim_{\|\Delta\| \to 0} \sum_{i=1}^{n} \pi [f(\bar{x}_i)]^2 \Delta_i x$$

$$= \int_0^{x_1} \pi [f(x)]^2 \, dx$$

$$= a^2 \int_0^{x_1} \pi \cosh^2 \left(\frac{x}{a}\right) dx \tag{1}$$

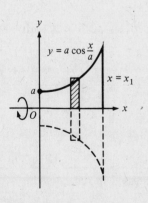

Figure 9. 7. 52

We use the result of Exercise 4(b). Thus,

$$\cosh^2 \left(\frac{x}{a}\right) = \frac{1}{2} \left(1 + \cosh \frac{2x}{a}\right)$$

Substituting in (1) we have

$$V = \frac{1}{2} a^2 \pi \int_0^{x_1} \left(1 + \cosh \frac{2x}{a}\right) dx$$

$$= \frac{1}{2} a^2 \pi \left[x + \frac{1}{2} a \sinh \frac{2x}{a}\right]_0^{x_1}$$

$$= \frac{1}{2} a^2 \pi \left(x_1 + \frac{1}{2} a \sinh \frac{2x_1}{a}\right)$$

$$= \frac{1}{4} a^2 \pi \left(2x_1 + a \sinh \frac{2x_1}{a} \right)$$

Thus, the volume is $\frac{1}{4}a^2\pi[2x_1 + a\sinh(2x_1/a)]$ cubic units.

56. Prove that the hyperbolic cosine function is continuous on its entire domain but is not monotonic on its entire domain. Find the intervals on which the function is increasing and the intervals on which the function is decreasing.

SOLUTION: Let $f(x) = \cosh x$. Then $f'(x) = \sinh(x)$. Because f' is defined for all real numbers, then f, the hyperbolic cosine function, is continuous for all real numbers.

$$\sinh(x) = \frac{e^x - e^{-x}}{2} \cdot \frac{e^x}{e^x} = \frac{e^{2x} - 1}{2e^x} \tag{1}$$

Because $e^{2x} > 1$ if $x > 0$ and $e^{2x} < 1$ if $x < 0$, we conclude from Eq. (1) that

$$f'(x) = \sinh(x) > 0 \qquad \text{if } x > 0$$

and

$$f'(x) = \sinh(x) < 0 \qquad \text{if } x < 0$$

Thus, f is increasing on $[0, +\infty)$ and decreasing on $(-\infty, 0]$, and hence f is not monotonic.

9.8 THE INVERSE HYPERBOLIC FUNCTIONS

9.8.1 Definition The *inverse hyperbolic sine function*, denoted by $\sinh^{-1}$, is defined as follows:

$$y = \sinh^{-1} x \text{ if and only if } x = \sinh y$$

9.8.2 Definition The *inverse hyperbolic cosine function*, denoted by $\cosh^{-1}$, is defined as follows:

$$y = \cosh^{-1} x \text{ if and only if } x = \cosh y \text{ and } y \geq 0$$

9.8.3 Definition The *inverse hyperbolic tangent function*, the *inverse hyperbolic cotangent function*, and the *inverse hyperbolic cosecant function*, denoted respectively by $\tanh^{-1}$, $\coth^{-1}$, and csch^{-1}, are defined as follows:

$$y = \tanh^{-1} x \text{ if and only if } x = \tanh y$$
$$y = \coth^{-1} x \text{ if and only if } x = \coth y$$
$$y = \text{csch}^{-1} x \text{ if and only if } x = \text{csch } y$$

9.8.4 Definition The *inverse hyperbolic secant function*, denoted by sech^{-1}, is defined as follows:

$$y = \text{sech}^{-1} x \text{ if and only if } x = \text{sech } y \text{ and } y \geq 0$$

Each of the inverse hyperbolic functions may be expressed in terms of the natural logarithmic function.

$$\sinh^{-1} x = \ln(x + \sqrt{x^2 + 1}) \qquad x \text{ any real number}$$

$$\cosh^{-1} x = \ln(x + \sqrt{x^2 - 1}) \qquad x \geq 1$$

$$\tanh^{-1} x = \frac{1}{2} \ln \frac{1 + x}{1 - x} \qquad |x| < 1$$

$$\coth^{-1} x = \frac{1}{2} \ln \frac{x + 1}{x - 1} \qquad |x| > 1$$

Following are the differentiation formulas for the inverse hyperbolic functions.

9.8.5 Theorem

$$D_x(\sinh^{-1} u) = \frac{1}{\sqrt{u^2 + 1}} \, D_x u$$

$$D_x(\cosh^{-1} u) = \frac{1}{\sqrt{u^2 - 1}} \, D_x u \qquad u > 1$$

$$D_x(\tanh^{-1} u) = \frac{1}{1 - u^2} \, D_x u \qquad |u| < 1$$

$$D_x(\coth^{-1} u) = \frac{1}{1 - u^2} \, D_x u \qquad |u| > 1$$

$$D_x(\operatorname{sech}^{-1} u) = -\frac{1}{u\sqrt{1 - u^2}} \, D_x u \qquad 0 < u < 1$$

$$D_x(\operatorname{csch}^{-1} u) = -\frac{1}{|u|\sqrt{1 + u^2}} \, D_x u \qquad u \neq 0$$

Exercises 9.8

In Exercises 3–10, prove the indicated formula of this section.

4. Formula (10): $\tanh^{-1} x = \dfrac{1}{2} \ln \dfrac{1 + x}{1 - x} \qquad |x| < 1$

SOLUTION: Let $y = \tanh^{-1} x$. Then $x = \tanh y$, and thus $|x| < 1$ because the range of the hyperbolic tangent function is the interval $(-1, 1)$. Furthermore,

$$x = \frac{e^y - e^{-y}}{e^y + e^{-y}} \cdot \frac{e^y}{e^y}$$

$$x = \frac{e^{2y} - 1}{e^{2y} + 1}$$

$$x(e^{2y} + 1) = e^{2y} - 1$$

$$1 + x = e^{2y}(1 - x)$$

$$\frac{1 + x}{1 - x} = e^{2y} \tag{1}$$

$$\ln \frac{1 + x}{1 - x} = 2y$$

$$y = \frac{1}{2} \ln \frac{1 + x}{1 - x}$$

Thus,

$$\tanh^{-1} x = \frac{1}{2} \ln \frac{1 + x}{1 - x} \qquad \text{if } |x| < 1$$

8. Formula (18): $D_x(\coth^{-1} u) = \dfrac{1}{1 - u^2} \, D_x u \qquad |u| > 1$

SOLUTION: Let $y = \coth^{-1} u$. Then

$$\coth y = u \quad \text{and} \quad |u| > 1 \tag{1}$$

because the range of the hyperbolic cotangent function is $(-\infty, -1) \cup (1, +\infty)$. Differentiating on both sides of (1) with respect to x, we obtain

$$-\text{csch}^2 \, y \, D_x y = D_x u$$

Thus,

$$D_x y = \frac{1}{-\text{csch}^2 \, y} \, D_x u$$

$$= \frac{1}{1 - \coth^2 \, y} \, D_x u \qquad (2)$$

Substituting from Eq. (1) into (2), we get

$$D_x y = \frac{1}{1 - u^2} \, D_x u \qquad \text{if } |u| > 1$$

or, equivalently,

$$D_x(\coth^{-1} u) = \frac{1}{1 - u^2} \, D_x u \qquad \text{if } |u| > 1$$

12. Express $\cosh^{-1} 3$ in terms of a natural logarithm.

SOLUTION: From Formula (9) of the text we have

$$\cosh^{-1} x = \ln(x + \sqrt{x^2 - 1}) \qquad \text{if } x \geq 1$$

Thus,

$$\cosh^{-1} 3 = \ln(3 + 2\sqrt{2})$$

In Exercises 15–32 find the derivative of the given function.

16. $G(x) = \cosh^{-1} \frac{1}{3} x$

SOLUTION:

$$G'(x) = \frac{1}{\sqrt{\frac{1}{9}x^2 - 1}} \cdot \frac{1}{3} = \frac{1}{\sqrt{x^2 - 9}}$$

20. $f(r) = \text{csch}^{-1} \frac{1}{2} r^2$

SOLUTION:

$$f'(r) = -\frac{1}{|\frac{1}{2}r^2|\sqrt{1 + \frac{1}{4}r^4}} \cdot r = \frac{-4}{r\sqrt{r^4 + 4}}$$

24. $g(x) = \tanh^{-1}(\sin 3x)$

SOLUTION:

$$g'(x) = \frac{1}{1 - \sin^2 3x} \cdot 3 \cos 3x$$

$$= \frac{3 \cos 3x}{\cos^2 3x} = 3 \sec 3x$$

28. $f(t) = \sinh^{-1} e^{2t}$

SOLUTION:

$$f'(t) = \frac{1}{\sqrt{(e^{2t})^2 + 1}} \, D_t e^{2t} = \frac{2e^{2t}}{\sqrt{e^{4t} + 1}}$$

32. $H(x) = \ln\sqrt{x^2 - 1} - x \tanh^{-1} x$

SOLUTION:

$$H(x) = \tfrac{1}{2} \ln(x^2 - 1) - x \tanh^{-1} x$$

Thus,

$$H'(x) = \frac{x}{x^2 - 1} - \left[x \cdot \frac{1}{1 - x^2} + \tanh^{-1} x \right] = \frac{2x}{x^2 - 1} - \tanh^{-1} x$$

36. Draw a sketch of the graph of the inverse hyperbolic cosecant function.

SOLUTION: Let

$$y = \operatorname{csch}^{-1} x \tag{1}$$

Then

$$x = \operatorname{csch} y$$

$$\frac{1}{x} = \sinh y \tag{2}$$

Because $\sinh(-y) = -\sinh y$, from (2) we conclude that the graph of (1) is symmetric with respect to the origin. Furthermore, from (2) we have

$$y = \sinh^{-1}\left(\frac{1}{x}\right)$$

Thus, by Eq. (8) of the text, we have

$$y = \ln\!\left(\frac{1}{x} + \sqrt{\frac{1}{x^2} + 1}\right)$$

Because

$$\lim_{x \to +\infty} \ln\!\left(\frac{1}{x} + \sqrt{\frac{1}{x^2} + 1}\right) = \ln 1 = 0$$

we conclude that the x-axis is a horizontal asymptote of the graph of (1). Because

$$\lim_{x \to 0^+} \ln\!\left(\frac{1}{x} + \sqrt{\frac{1}{x^2} + 1}\right) = +\infty$$

we conclude that the y-axis is a vertical asymptote of the graph of (1). We plot a few points and use the asymptotes and symmetry to draw a sketch of the graph, which is shown in Fig. 9. 8. 36.

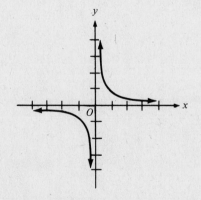

Figure 9. 8. 36

Review Exercises for Chapter 9

In Exercises 1–27 find the derivative of the given function.

4. $f(w) = w \tan \dfrac{1}{w}$

SOLUTION:

$$f'(w) = w\left(\sec^2 \frac{1}{w}\right) D_w\left(\frac{1}{w}\right) + \left(\tan \frac{1}{w}\right) D_w w$$

$$= w\left(\sec^2 \frac{1}{w}\right)\left(\frac{-1}{w^2}\right) + \tan \frac{1}{w}$$

$$= -\frac{1}{w} \sec^2 \frac{1}{w} + \tan \frac{1}{w}$$

8. $g(x) = \cot 3x\sqrt{\sin 2x}$

SOLUTION:

$$g'(x) = (\cot 3x)(\tfrac{1}{2})(\sin 2x)^{-1/2} D_x(\sin 2x) + (\sin 2x)^{1/2} D_x(\cot 3x)$$

$$= \tfrac{1}{2} \cot 3x(\sin 2x)^{-1/2}(2 \cos 2x) + (\sin 2x)^{1/2}(-3 \csc^2 3x)$$

$$= (\sin 2x)^{1/2}\left[\cot 3x \cdot \frac{\cos 2x}{\sin 2x} - 3 \csc^2 3x\right]$$

$$= \sqrt{\sin 2x} \,(\cot 3x \cot 2x - 3\csc^2 3x)$$

12. $h(x) = e^x(\cosh x + \sinh x)$

SOLUTION:

$$h'(x) = e^x(\sinh x + \cosh x) + (\cosh x + \sinh x)e^x$$

$$= 2e^x(\sinh x + \cosh x)$$

16. $f(t) = \sec^3 t + 3^{\sec t}$

SOLUTION:

$$f'(t) = 3 \sec^2 t \, D_t(\sec t) + 3^{\sec t}(\ln 3)D_t(\sec t)$$

$$= (3 \sec^2 t + 3^{\sec t} \ln 3)\sec t \tan t$$

20. $F(x) = \cot^3 x \csc x$

SOLUTION:

$$F'(x) = \cot^3 x(-\csc x \cdot \cot x) + \csc x(3 \cot^2 x)(-\csc^2 x)$$

$$= -\cot^2 x \csc x(\cot^2 x + 3 \csc^2 x)$$

24. $f(x) = \tan^{-1}(\tan^{-1} x)$

SOLUTION:

$$f'(x) = \frac{1}{1 + (\tan^{-1} x)^2} D_x(\tan^{-1} x)$$

$$= \frac{1}{(1 + x^2)[1 + (\tan^{-2} x)^2]}$$

28. The gudermannian is the function defined by gd $x = \tan^{-1}(\sinh x)$. Show that $D_x(\text{gd } x) = \text{sech } x$.

SOLUTION:

$$D_x(\text{gd } x) = \frac{1}{1 + \sinh^2 x} D_x(\sinh x)$$

$$= \frac{\cosh x}{1 + \sinh^2 x}$$

$$= \frac{\cosh x}{\cosh^2 x} = \frac{1}{\cosh x} = \operatorname{sech} x$$

32. Evaluate the following limit if it exists.

$$\lim_{x \to 0} \frac{\csc 3x}{\cot x}$$

SOLUTION: We apply Theorem 2.9.2,

$$\lim_{t \to 0} \frac{\sin t}{t} = 1$$

Thus,

$$\lim_{x \to 0} \frac{\csc 3x}{\cot x} = \lim_{x \to 0} \frac{\sin x}{\cos x \sin 3x}$$

$$= \left(\lim_{x \to 0} \frac{1}{\cos x} \right) \left(\lim_{x \to 0} \frac{\sin x}{\sin 3x} \right)$$

$$= (1) \lim_{x \to 0} \frac{\dfrac{\sin x}{x}}{\dfrac{3 \sin 3x}{3x}}$$

$$= \frac{\displaystyle\lim_{x \to 0} \frac{\sin x}{x}}{3 \displaystyle\lim_{3x \to 0} \frac{\sin 3x}{3x}} = \frac{1}{3}$$

36. Find $D_x y$ by implicit differentiation: $\tan^{-1}(x + 3y) = \ln y$.

SOLUTION: We differentiate implicitly with respect to x on both sides of the given equation.

$$\frac{1}{1 + (x + 3y)^2} (1 + 3D_x y) = \frac{1}{y} D_x y$$

$$y + 3yD_x y = [1 + (x + 3y)^2] D_x y$$

$$y = [1 + (x + 3y)^2 - 3y] D_x y$$

$$D_x y = \frac{y}{1 + (x + 3y)^2 - 3y}$$

In Exercises 37–52 evaluate the indefinite integral.

40. $\int (\csc 2t - \cot 2t)\, dt$

SOLUTION: Let $u = 2t$ and $du = 2dt$. Then

$$\int (\csc 2t - \cot 2t)\, dt = \frac{1}{2} \int (\csc u - \cot u)\, du$$

$$= \frac{1}{2} [\ln|\csc u - \cot u| - \ln|\sin u|] + C$$

$$= \frac{1}{2} \ln \left| \frac{\csc u - \cot u}{\sin u} \right| + C$$

$$= \tfrac{1}{2} \ln \left| \frac{1 - \cos u}{\sin^2 u} \right| + C$$

$$= \tfrac{1}{2} \ln \left| \frac{1 - \cos u}{1 - \cos^2 u} \right| + C$$

$$= \tfrac{1}{2} \ln \left| \frac{1 - \cos u}{(1 - \cos u)(1 + \cos u)} \right| + C$$

$$= \tfrac{1}{2} \ln \left| \frac{1}{1 + \cos u} \right| + C$$

$$= \tfrac{1}{2} \ln |1 + \cos u|^{-1} + C$$

$$= -\tfrac{1}{2} \ln |1 + \cos u| + C$$

$$= -\tfrac{1}{2} \ln |1 + \cos 2t| + C$$

44. $\displaystyle\int \frac{dx}{\sqrt{7 + 5x - 2x^2}}$

SOLUTION: We complete the square under the radical sign.

$$7 + 5x - 2x^2 = -2(x^2 - \tfrac{5}{2}x + \tfrac{25}{16}) + 7 + \tfrac{25}{8}$$
$$= -2(x - \tfrac{5}{4})^2 + \tfrac{81}{8}$$
$$= 2[\tfrac{81}{16} - (x - \tfrac{5}{4})^2]$$

Thus,

$$\int \frac{dx}{\sqrt{7 + 5x - 2x^2}} = \int \frac{dx}{\sqrt{2[\tfrac{81}{16} - (x - \tfrac{5}{4})^2]}}$$

$$= \frac{1}{\sqrt{2}} \int \frac{dx}{\sqrt{(\tfrac{9}{4})^2 - (x - \tfrac{5}{4})^2}}$$

$$= \frac{1}{\sqrt{2}} \sin^{-1} \frac{x - \tfrac{5}{4}}{\tfrac{9}{4}} + C$$

$$= \frac{1}{\sqrt{2}} \sin^{-1} \frac{4x - 5}{9} + C$$

48. $\int 2^x \sec^6(2^x)\, dx$

SOLUTION: Let $u = 2^x$ and $du = 2^x(\ln 2)\, dx$. Thus,

$$\int 2^x \sec^6(2^x)\, dx = \frac{1}{\ln 2} \int \sec^6 u\, du$$

$$= \frac{1}{\ln 2} \int (1 + \tan^2 u)^2 \sec^2 u\, du \tag{1}$$

Now let $v = \tan u$ and $dv = \sec^2 u\, du$. We have

$$\int (1 + \tan^2 u)^2 \sec^2 u\, du = \int (1 + v^2)^2\, dv$$

$$= \int (1 + 2v^2 + v^4)\, dv$$

$$= v + \tfrac{2}{3} v^3 + \tfrac{1}{5} v^5 + \overline{C}$$

$$= \tan u + \tfrac{2}{3} \tan^3 u + \tfrac{1}{5} \tan^5 u + \overline{C} \tag{2}$$

Substituting from Eq. (2) into Eq. (1) and replacing u by 2^x, we obtain

$$\int 2^x \sec^6(2^x)\, dx = \frac{1}{\ln 2}\left[\tan(2^x) + \tfrac{2}{3}\tan^3(2^x) + \tfrac{1}{5}\tan^5(2^x) + C\right]$$

where $C = \overline{C}/\ln 2$.

52. $\displaystyle \int \frac{\cosh^3 x}{\sqrt{\sinh x}}\, dx$

SOLUTION: Let $u = \sinh x$ and $du = \cosh x\, dx$.

$$\int \frac{\cosh^3 x}{\sqrt{\sinh x}}\, dx = \int \frac{(1+\sinh^2 x)\cosh x\, dx}{\sqrt{\sinh x}}$$

$$= \int \frac{(1+u^2)\, du}{\sqrt{u}}$$

$$= \int (u^{-1/2} + u^{3/2})\, du$$

$$= 2u^{1/2} + \tfrac{2}{5}u^{5/2} + C$$

$$= \tfrac{2}{5}u^{1/2}(5 + u^2) + C$$

$$= \tfrac{2}{5}\sqrt{\sinh x}(5 + \sinh^2 x) + C$$

56. Evaluate the definite integral

$$\int_{-1}^{1} \frac{2x+6}{x^2+2x+5}\, dx$$

SOLUTION: First, we complete the square in the denominator.

$$\int_{-1}^{1} \frac{2x+6}{x^2+2x+5}\, dx = \int_{-1}^{1} \frac{2(x+1)+4}{(x+1)^2+4}\, dx \tag{1}$$

Next, we let $u = x + 1$ and $du = dx$. When $x = -1$, then $u = 0$; when $x = 1$, then $u = 2$. Substituting into Eq. (1), we obtain

$$\int_{-1}^{1} \frac{2x+6}{x^2+2x+5}\, dx = \int_{0}^{2} \frac{(2u+4)\, du}{u^2+4}$$

$$= \int_{0}^{2} \frac{2u\, du}{u^2+4} + \int_{0}^{2} \frac{4\, du}{u^2+4}$$

$$= \ln(u^2+4)\Big]_0^2 + 4(\tfrac{1}{2})\tan^{-1}\left(\frac{u}{2}\right)\Big]_0^2$$

$$= (\ln 8 - \ln 4) + 2(\tan^{-1} 1 - \tan^{-1} 0)$$

$$= \ln\frac{8}{4} + 2\left(\frac{\pi}{4}\right)$$

$$= \ln 2 + \frac{1}{2}\pi$$

60. Evaluate

$$\int_{-\pi/3}^{\pi/3} |\sec x - 2|\, dx$$

SOLUTION: Because $\sec x \le 2$ if $-\tfrac{1}{3}\pi \le x \le \tfrac{1}{3}\pi$, then $|\sec x - 2| = 2 - \sec x$ if $-\tfrac{1}{3}\pi \le x \le \tfrac{1}{3}\pi$. Hence,

$$\int_{-\pi/3}^{\pi/3} |\sec x - 2|\, dx = \int_{-\pi/3}^{\pi/3} (2 - \sec x)\, dx$$

$$= 2\int_{0}^{\pi/3} (2 - \sec x)\, dx$$

$$= 2\Big[2x - \ln|\sec x + \tan x|\Big]_{0}^{\pi/3}$$

$$= 2\Big[\frac{2\pi}{3} - \ln\Big|\sec \frac{\pi}{3} + \tan \frac{\pi}{3}\Big|\Big]$$

$$= \frac{4\pi}{3} - 2\ln(2 + \sqrt{3})$$

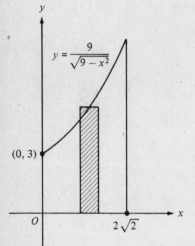

$$y = \frac{9}{\sqrt{9 - x^2}}$$

(0, 3)

O $2\sqrt{2}$

Figure 9. 64R

64. Find the area of the region bounded by the curve $y = 9/\sqrt{9 - x^2}$, the two coordinate axes, and the line $x = 2\sqrt{2}$.

SOLUTION: Figure 9. 64R shows a sketch of the region and an element of area. Let $f(x) = 9/\sqrt{9 - x^2}$. The width of the element of area is $\Delta_i x$ units and the length is $f(\bar{x}_i)$ units. Therefore, if A square units is the area of the region, then

$$A = \lim_{\|\Delta\| \to 0} \sum_{i=1}^{n} f(\bar{x}_i)\, \Delta_i x$$

$$= \int_{0}^{2\sqrt{2}} \frac{9}{\sqrt{9 - x^2}}\, dx$$

$$= 9\, \sin^{-1}\Big(\frac{x}{3}\Big)\Big]_{0}^{2\sqrt{2}}$$

$$= 9\, \sin^{-1}\Big(\frac{2\sqrt{2}}{3}\Big)$$

The area of the region is $9\, \sin^{-1}(\tfrac{2}{3}\sqrt{2})$ square units.

68. Prove

(a) $\lim_{x \to +\infty} \coth x = 1$

(b) $\lim_{x \to +\infty} \operatorname{csch} x = 0$

SOLUTION:

(a) $\lim_{x \to +\infty} \coth x = \lim_{x \to +\infty} \dfrac{e^x + e^{-x}}{e^x - e^{-x}}$

$$= \lim_{x \to +\infty} \frac{(e^x + e^{-x})e^{-x}}{(e^x - e^{-x})e^{-x}}$$

$$= \lim_{x \to +\infty} \frac{1 + e^{-2x}}{1 - e^{-2x}}$$

$$= \frac{1 + \lim_{x \to +\infty} e^{-2x}}{1 - \lim_{x \to +\infty} e^{-2x}}$$

$$= \frac{1 + 0}{1 - 0} = 1$$

(b) $\displaystyle\lim_{x\to+\infty} \text{csch } x = \lim_{x\to+\infty} \frac{2}{e^x - e^{-x}}$

$$= \lim_{x\to+\infty} \frac{2e^{-x}}{1 - e^{-2x}}$$

$$= \frac{2\cdot 0}{1 - 0} = 0$$

72. A helicopter leaves the ground at a point 800 ft from an observer and rises vertically at 25 ft/sec. Find the time rate of change of the measure of the observer's angle of elevation of the helicopter when the helicopter is 600 ft above the ground.

SOLUTION: Refer to Fig. 9. 72R. The observer is at point Q, the helicopter leaves the ground at point P, and after t sec the helicopter is at point H, which is x ft from point P. Thus, x ft is the distance that the helicopter rises in t sec, and we are given that

$$\frac{dx}{dt} = 25$$

Let θ be the radian measure of the observers's angle of elevation after t sec. We want to find $d\theta/dt$ when $x = 600$. Because $\tan \theta = x/800$, then

$$\theta = \tan^{-1}\frac{x}{800}$$

We differentiate on both sides with respect to t. Thus,

$$\frac{d\theta}{dt} = \frac{1}{1 + \left(\dfrac{x}{800}\right)^2}\left(\frac{1}{800}\right)\frac{dx}{dt}$$

Because $dx/dt = 25$, we have

$$\frac{d\theta}{dt} = \frac{\frac{1}{32}}{1 + \left(\dfrac{x}{800}\right)^2}$$

When $x = 600$, we obtain

$$\left.\frac{d\theta}{dt}\right]_{x=600} = \frac{\frac{1}{32}}{1 + (\frac{3}{4})^2} = \frac{1}{50}$$

The observer's angle of elevation is increasing at the rate of $\frac{1}{50}$ radians per second when the helicopter is 600 ft above the ground.

76. If an equation of motion is $s = \cos 2t + 2 \sin 2t$, prove that the motion is simple harmonic.

SOLUTION: We show that a, the measure of the acceleration, is always proportional to s, the measure of the displacement from a fixed point, and that the acceleration and displacement are oppositely directed. We have

$$s = \cos 2t + 2 \sin 2t$$

$$v = \frac{ds}{dt} = -2 \sin 2t + 4 \cos 2t$$

$$a = \frac{dv}{dt} = -4 \cos 2t - 8 \sin 2t$$

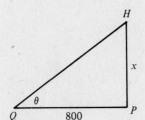

Figure 9. 72R

Thus,

$$a = -4(\cos 2t + 2 \sin 2t) = -4s$$

Because $a = -4s$, then a is proportional to s, and because the constant -4 is negative, then the acceleration and the displacement are oppositely directed. Thus, the motion is simple harmonic.

CHAPTER 10 Techniques of Integration

10.1 INTRODUCTION   You should memorize the following standard integration formulas.

$$\int du = u + C$$

$$\int a \, du = au + C \qquad \text{where } a \text{ is any constant}$$

$$\int [f(u) + g(u)] \, du = \int f(u) \, du + \int g(u) \, du$$

$$\int u^n \, du = \frac{u^{n+1}}{n+1} + C \qquad n \neq -1$$

$$\int \frac{du}{u} = \ln|u| + C$$

$$\int a^n \, du = \frac{a^n}{\ln a} + C$$

$$\int e^n \, du = e^n + C$$

$$\int \sin u \, du = -\cos u + C$$

$$\int \cos u \, du = \sin u + C$$

$$\int \sec^2 u \, du = \tan u + C$$

$$\int \csc^2 u \, du = -\cot u + C$$

$$\int \sec u \tan u \, du = \sec u + C$$

$$\int \csc u \cot u \, du = -\csc u + C$$

$$\int \tan u \, du = \ln|\sec u| + C$$

$$\int \cot u \, du = \ln|\sin u| + C$$

$$\int \sec u \, du = \ln|\sec u + \tan u| + C$$

$$\int \csc u \, du = \ln|\csc u - \cot u| + C$$

$$\int \frac{du}{\sqrt{a^2 - u^2}} = \sin^{-1}\frac{u}{a} + C \qquad \text{where } a > 0$$

$$\int \frac{du}{a^2 + u^2} = \frac{1}{a}\tan^{-1}\frac{u}{a} + C$$

$$\int \frac{du}{u\sqrt{u^2 - a^2}} = \frac{1}{a}\sec^{-1}\left|\frac{u}{a}\right| + C \qquad \text{where } a > 0$$

The integration formulas for the hyperbolic functions are also important, but they are not used as frequently as the above standard formulas.

$$\int \sinh u \, du = \cosh u + C$$

$$\int \cosh u \, du = \sinh u + C$$

$$\int \operatorname{sech}^2 u \, du = \tanh u + C$$

$$\int \operatorname{csch}^2 u \, du = -\coth u + C$$

$$\int \operatorname{sech} u \tanh u \, du = -\operatorname{sech} u + C$$

$$\int \operatorname{csch} u \coth u \, du = -\operatorname{csch} u + C$$

10.2 INTEGRATION BY PARTS

Following is the formula for integration by parts.

$$\int u \, du = uv - \int v \, du$$

To use the formula, we must factor the given integrand in such a way that we can integrate one of the factors—which we designate dv—and differentiate the other factor—which we designate u. And we must choose u and dv in such a way that we can integrate $\int v \, du$. Sometimes, we may have to simplify the given integral by making a change of variable before integrating by parts. (See Exercise 28.) Sometimes we may have to integrate by parts more than once. (See Exercises 12 and 32.)

Exercises 10.2

In Exercises 1–24 evaluate the indefinite integral.

4. $\displaystyle\int x 3^x \, dx$

SOLUTION: We use integration by parts with

$$u = x \qquad dv = 3^x \, dx$$

$$du = dx \qquad v = \frac{3^x}{\ln 3}$$

Thus,

$$\int x 3^x \, dx = \frac{x \, 3^x}{\ln 3} - \frac{1}{\ln 3} \int 3^x \, dx$$

$$= \frac{x \, 3^x}{\ln 3} - \frac{1}{\ln 3} \frac{3^x}{\ln 3} + C$$

$$= \frac{3^x (x \ln 3 - 1)}{\ln^2 3} + C$$

8. $\displaystyle\int x \sec^2 x \, dx$

SOLUTION: We use integration by parts with

$$u = x \qquad dv = \sec^2 x \, dx$$
$$du = dx \qquad v = \tan x$$

Thus,

$$\int x \sec^2 x \, dx = x \tan x - \int \tan x \, dx$$

$$= x \tan x - \ln|\sec x| + C$$

12. $\displaystyle\int x^2 \sin 3x \, dx$

SOLUTION: We let

$$u = x^2 \qquad dv = \sin 3x \, dx$$
$$du = 2x \, dx \qquad v = -\tfrac{1}{3} \cos 3x$$

Thus,

$$\int x^2 \sin 3x \, dx = -\frac{1}{3} x^2 \cos 3x + \frac{2}{3} \int x \cos 3x \, dx \tag{1}$$

For the integral on the right side of (1), we let

$$\bar{u} = x \qquad d\bar{v} = \cos 3x \, dx$$
$$d\bar{u} = dx \qquad \bar{v} = \tfrac{1}{3} \sin 3x$$

Thus,

$$\int x \cos 3x \, dx = \frac{1}{3} x \sin 3x - \frac{1}{3} \int \sin 3x \, dx$$

$$= \frac{1}{3} x \sin 3x + \frac{1}{9} \cos 3x + C \qquad (2)$$

Substituting from (2) into (1), we obtain

$$\int x^2 \sin 3x \, dx = -\frac{1}{3} x^2 \cos 3x + \frac{2}{9} x \sin 3x + \frac{2}{27} \cos 3x + C$$

16. $\displaystyle \int x^3 e^{x^2} \, dx$

SOLUTION: We let

$$u = x^2 \qquad dv = xe^{x^2} \, dx$$
$$du = 2x \, dx \qquad v = \tfrac{1}{2} e^{x^2}$$

Hence,

$$\int x^3 e^{x^2} \, dx = \frac{1}{2} x^2 e^{x^2} - \int x \, e^{x^2} \, dx$$

$$= \frac{1}{2} x^2 e^{x^2} - \frac{1}{2} e^{x^2} + C$$

$$= \frac{1}{2} e^{x^2} (x^2 - 1) + C$$

20. $\displaystyle \int \frac{e^{2x}}{\sqrt{1 - e^x}} \, dx$

SOLUTION: Integration by parts is not required. We let $u = 1 - e^x$. Then $du = -e^x \, dx$ and $e^x = 1 - u$. With these substitutions we obtain

$$\int \frac{e^{2x}}{\sqrt{1 - e^x}} \, dx = \int \frac{e^x (e^x \, dx)}{\sqrt{1 - e^x}}$$

$$= \int \frac{(1 - u)(-du)}{\sqrt{u}}$$

$$= \int (u^{1/2} - u^{-1/2}) \, du$$

$$= \tfrac{2}{3} u^{3/2} - 2 u^{1/2} + C$$

$$= \tfrac{2}{3} u^{1/2} (u - 3) + C$$

$$= \tfrac{2}{3} \sqrt{1 - e^x} (1 - e^x - 3) + C$$

$$= -\tfrac{2}{3} \sqrt{1 - e^x} (2 + e^x) + C$$

24. $\displaystyle \int \tanh^{-1} x \, dx$

SOLUTION: We use integration by parts with

$$u = \tanh^{-1} x \qquad dv = dx$$

$$du = \frac{dx}{1 - x^2} \qquad v = x$$

Thus,

$$\int \tanh^{-1} x \, dx = x \tanh^{-1} x - \int \frac{x \, dx}{1 - x^2}$$

$$= x \tanh^{-1} x + \tfrac{1}{2} \ln|1 - x^2| + C$$

In Exercises 25–34 evaluate the definite integral.

28. $\displaystyle\int_0^{\pi^2/2} \cos\sqrt{2x} \, dx$

SOLUTION: Before integrating by parts we let $y = \sqrt{2x}$. Thus $x = \tfrac{1}{2} y^2$ and $dx = y \, dy$. When $x = 0$, $y = 0$; when $x = \tfrac{1}{2} \pi^2$, $y = \pi$. Hence,

$$\int_0^{\pi^2/2} \cos\sqrt{2x} \, dx = \int_0^{\pi} y \cos y \, dy \qquad (1)$$

Now we integrate by parts. Let

$$u = y \qquad dv = \cos y \, dy$$

$$du = dy \qquad v = \sin y$$

With integration by parts on the right-hand side of (1) we obtain

$$\int_0^{\pi^2/2} \cos\sqrt{2x} \, dx = \left[y \sin y \right]_0^{\pi} - \int_0^{\pi} \sin y \, dy$$

$$= [\pi \sin \pi - 0 \sin 0] + \left[\cos y \right]_0^{\pi}$$

$$= 0 + [-1 - 1] = -2$$

32. $\displaystyle\int_0^1 x \sin^{-1} x \, dx$

SOLUTION: First we find $\int \sin^{-1} x \, dx$. Let

$$u = \sin^{-1} x \qquad dv = dx$$

$$du = \frac{dx}{\sqrt{1 - x^2}} \qquad v = x$$

Hence,

$$\int \sin^{-1} x \, dx = x \sin^{-1} x - \int \frac{x \, dx}{\sqrt{1 - x^2}}$$

$$= x \sin^{-1} x + \sqrt{1 - x^2} + C \qquad (1)$$

Therefore, we may use integration by parts on the given integral as follows. Let

$$\bar{u} = x \qquad d\bar{v} = \sin^{-1} x \, dx$$

$$d\bar{u} = dx \qquad \bar{v} = x \sin^{-1} x + \sqrt{1 - x^2} \qquad \text{(by (1))}$$

Hence,

$$\int_0^1 x \sin^{-1} x \, dx = \left[x^2 \sin^{-1} x + x\sqrt{1-x^2} \right]_0^1 - \int_0^1 (x \sin^{-1} x + \sqrt{1-x^2}) \, dx$$

$$= \frac{1}{2}\pi - \int_0^1 x \sin^{-1} x \, dx - \int_0^1 \sqrt{1-x^2} \, dx$$

Adding $\int_0^1 x \sin^{-1} x \, dx$ to both sides, we obtain

$$2\int_0^1 x \sin^{-1} x \, dx = \frac{1}{2}\pi - \int_0^1 \sqrt{1-x^2} \, dx \qquad (2)$$

The value of the definite integral on the right side of (2) is the measure of the area of the region under the curve $y = \sqrt{1-x^2}$ between the line $x = 0$ and the line $x = 1$. Because this region is the first quadrant part of the circle $x^2 + y^2 = 1$, we conclude that

$$\int_0^1 \sqrt{1-x^2} \, dx = \frac{1}{4}\pi \qquad (3)$$

Substituting from (3) into (2), we obtain

$$2\int_0^1 x \sin^{-1} x \, dx = \frac{1}{2}\pi - \frac{1}{4}\pi = \frac{1}{4}\pi$$

Thus,

$$\int_0^1 x \sin^{-1} x \, dx = \frac{1}{8}\pi$$

In Section 10.3 we have a method for finding $\int_0^1 \sqrt{1-x^2} \, dx$ by using the fundamental theorem of the calculus.

36. Find the volume of the solid generated by revolving about the x-axis the region bounded by the curve $y = \ln x$, the x-axis, and the line $x = e^2$.

SOLUTION: Figure 10. 2. 36 shows a sketch of the region and a plane section of the solid of revolution. The element of volume is a circular disk with thickness $\Delta_i x$ units and radius $\ln \bar{x}_i$ units. If V cubic units is the volume of the solid of revolution, then

$$V = \lim_{\|\Delta\| \to 0} \sum_{i=1}^{n} \pi (\ln \bar{x}_i)^2 \, \Delta_i x$$

$$= \int_1^{e^2} \pi \ln^2 x \, dx \qquad (1)$$

To find the indefinite integral we use integration by parts with

$$u = \ln^2 x \qquad dv = dx$$

$$du = \frac{2 \ln x}{x} \, dx \qquad v = x$$

Hence,

$$\int \ln^2 x \, dx = x \ln^2 x - 2 \int \ln x \, dx \qquad (2)$$

Now we integrate by parts again with

$$\bar{u} = \ln x \qquad d\bar{v} = dx$$

$$d\bar{u} = \frac{1}{x} \, dx \qquad \bar{v} = x$$

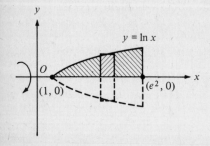

Figure 10. 2. 36

Thus,

$$\int \ln x \, dx = x \ln x - \int dx$$

$$= x \ln x - x \qquad (3)$$

We have taken $C = 0$ in the indefinite integral because we intend to use the result to evaluate a definite integral. Substituting from (3) into (2) and then into (1), we obtain

$$V = \pi \left[x \ln^2 x - 2x \ln x + 2x \right]_1^{e^2}$$

$$= \pi [(e^2 \ln^2 e^2 - 2e^2 \ln e^2 + 2e^2) - (\ln^2 1 - 2 \ln 1 + 2)]$$

$$= \pi [4e^2 - 4e^2 + 2e^2 - 2]$$

$$= 2\pi(e^2 - 1)$$

The volume of the solid of revolution is $2\pi(e^2 - 1)$ cubic units.

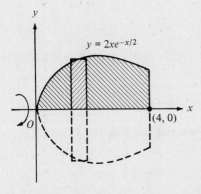

$y = 2xe^{-x/2}$

$(4, 0)$

Figure 10. 2. 40

40. Find the volume of the solid of revolution generated by revolving about the x-axis the region bounded by the curve $y = 2xe^{-x/2}$, the x-axis, and the line $x = 4$.

SOLUTION: Let $f(x) = 2xe^{-x/2}$. Because $f'(x) = e^{-x/2}(2 - x)$, then f has a relative maximum value at $x = 2$. Figure 10. 2. 40 shows a sketch of the region and a plane section of the solid of revolution. The element of volume is a circular disk with thickness $\Delta_i x$ units and radius $f(\bar{x}_i)$ units. If V cubic units is the volume of the solid of revolution, then

$$V = \lim_{\|\Delta\| \to 0} \sum_{i=1}^{n} \pi [f(\bar{x}_i)]^2 \, \Delta_i x$$

$$= \int_0^4 \pi(2xe^{-x/2})^2 \, dx$$

$$= 4\pi \int_0^4 x^2 e^{-x} \, dx \qquad (1)$$

We use integration by parts. Let

$$u = x^2 \qquad dv = e^{-x} \, dx$$
$$du = 2x \, dx \qquad v = -e^{-x}$$

Then

$$\int x^2 e^{-x} \, dx = -x^2 e^{-x} + 2 \int xe^{-x} \, dx \qquad (2)$$

We integrate by parts again with

$$\bar{u} = x \qquad d\bar{v} = e^{-x}$$
$$d\bar{u} = dx \qquad \bar{v} = -e^{-x}$$

Thus,

$$\int xe^{-x} \, dx = -xe^{-x} + \int e^{-x} \, dx$$

$$= -xe^{-x} - e^{-x} \qquad (3)$$

Substituting from (3) into (2) and then into (1), we obtain

$$V = 4\pi \left[-x^2 e^{-x} + 2(-xe^{-x} - e^{-x}) \right]_0^4$$

$$= 4\pi[(-16e^{-4} - 8e^{-4} - 2e^{-4}) - (-2e^0)]$$

$$= 4\pi(2 - 26e^{-4})$$

$$= 8\pi(1 - 13e^{-4})$$

The volume of the solid of revolution is $8\pi(1 - 13e^{-4})$ cubic units.

44. A particle is moving along a straight line and s ft is the directed distance of the particle from the origin at t sec. If v ft/sec is the velocity at t sec, $s = 0$ when $t = 0$, and $vs = t \sin t$, find s in terms of t and also s when $t = \frac{1}{2}\pi$.

SOLUTION: Because $v = ds/dt$, we are given that

$$\frac{ds}{dt} s = t \sin t$$

Thus,

$$\int s \, ds = \int t \sin t \, dt$$

$$\tfrac{1}{2}s^2 = \int t \sin t \, dt \tag{1}$$

We use integration by parts for the integral on the right side of (1). Let

$$u = t \qquad dv = \sin t \, dt$$
$$du = dt \qquad v = -\cos t$$

Hence,

$$\int t \sin t \, dt = -t \cos t + \int \cos t \, dt$$

$$= -t \cos t + \sin t + C \tag{2}$$

Substituting from (2) into the right side of (1), we obtain

$$\tfrac{1}{2}s^2 = -t \cos t + \sin t + C$$

Because $s = 0$ when $t = 0$, then $C = 0$. Thus, solving for s we have

$$s = \pm\sqrt{2(\sin t - t \cos t)}$$

If $t = \frac{1}{2}\pi$, then

$$s = \pm\sqrt{2 \sin \tfrac{1}{2}\pi - (\tfrac{1}{2}\pi)\cos \tfrac{1}{2}\pi}$$
$$= \pm\sqrt{2}$$

48. (a) Derive the following formula where r is any real number.

$$\int x^r e^x \, dx = x^r e^x - r \int x^{r-1} e^x \, dx$$

(b) Use the formula derived in (a) to find $\int x^4 e^x \, dx$.

SOLUTION:

(a) We use integration by parts with

$$u = x^r \qquad dv = e^x \, dx$$
$$du = rx^{r-1} \, dx \qquad v = e^x$$

Thus,

$$\int x^r e^x \, dx = x^r e^x - \int rx^{r-1} e^x \, dx$$

$$= x^r e^x - r \int x^{r-1} e^x \, dx$$

and the formula is derived.

(b) First, we apply the formula with $r = 4$. Thus,

$$\int x^4 e^x \, dx = x^4 e^x - 4 \int x^3 e^x \, dx$$

To continue, we apply the formula successively with $r = 3$, $r = 2$, and $r = 1$. Thus,

$$\int x^4 e^x \, dx = x^4 e^x - 4 \int x^3 e^x \, dx$$

$$= x^4 e^x - 4 \left[x^3 e^x - 3 \int x^2 e^x \, dx \right]$$

$$= x^4 e^x - 4x^3 e^x + 12 \left[x^2 e^x - 2 \int x e^x \, dx \right]$$

$$= x^4 e^x - 4x^3 e^x + 12x^2 e^x - 24 \left[x e^x - \int e^x \, dx \right]$$

$$= x^4 e^x - 4x^3 e^x + 12x^2 e^x - 24x e^x + 24 e^x + C$$

$$= e^x (x^4 - 4x^3 + 12x^2 - 24x + 24) + C$$

10.3 INTEGRATION BY TRIGONOMETRIC SUBSTITUTION

Sometimes we can "eliminate" the square root radical in an expression that we wish to integrate by making a trigonometric substitution. We consider three cases.

Case 1: (See Exercise 24) The integrand contains an expression of the form $\sqrt{a^2 - u^2}$, where $a > 0$. We let

$$\theta = \sin^{-1} \frac{u}{a}$$

Then,

$$\sqrt{a^2 - u^2} = a \cos \theta \quad \text{and} \quad du = a \cos \theta \, d\theta$$

Case 2: (See Exercise 4) The integrand contains an expression of the form $\sqrt{a^2 + u^2}$, where $a > 0$. We let

$$\theta = \tan^{-1} \left(\frac{u}{a} \right)$$

Then

$$\sqrt{a^2 + u^2} = a \sec \theta \quad \text{and} \quad du = a \sec^2 \theta \, d\theta$$

Case 3: (See Exercise 8) The integrand contains an expression of the form $\sqrt{u^2 - a^2}$, where $a > 0$. We let

$$\theta = \begin{cases} \sec^{-1} \left(\dfrac{u}{a} \right) & \text{if } u \geq a \\[2mm] 2\pi - \sec^{-1} \left(\dfrac{u}{a} \right) & \text{if } u \leq -a \end{cases}$$

Then

$$\sqrt{u^2 - a^2} = a \tan \theta \quad \text{and} \quad du = a \sec \theta \tan \theta \, d\theta$$

You should memorize the choice for θ in each of the three cases, but do not attempt to memorize the replacement for the expression containing the radical or the replacement for du. Rather, use the trigonometric identities and the differentiation formulas for the trigonometric functions to derive these replacements. The identities needed are the following.

$$1 - \sin^2 \theta = \cos^2 \theta$$
$$1 + \tan^2 \theta = \sec^2 \theta$$
$$\sec^2 \theta - 1 = \tan^2 \theta$$

In Case 3 we let $\theta = 2\pi - \sec^{-1}(u/a)$ when $u < -a$, so that $\tan \theta \geq 0$, and thus $\sqrt{\tan^2 \theta} = \tan \theta$. For a *definite* integral we may need to consider only one of the two possible replacements. (See Exercise 32.)

Exercises 10.3

In Exercises 1–24 evaluate the indefinite integral.

4. $\displaystyle\int \frac{x^2 \, dx}{\sqrt{x^2 + 6}}$

SOLUTION: We use the method of Case 2 with $u = x$ and $a = \sqrt{6}$. Thus, we let

$$\theta = \tan^{-1} \frac{x}{\sqrt{6}}$$

Hence,

$$\tan \theta = \frac{x}{\sqrt{6}} \tag{1}$$

$$x = \sqrt{6} \tan \theta$$

$$dx = \sqrt{6} \sec^2 \theta \, d\theta$$

and

$$\sqrt{x^2 + 6} = \sqrt{6 \tan^2 \theta + 6}$$
$$= \sqrt{6(\tan^2 \theta + 1)}$$
$$= \sqrt{6 \sec^2 \theta}$$
$$= \sqrt{6} \sec \theta \tag{2}$$

We have

$$\int \frac{x^2 \, dx}{\sqrt{x^2 + 6}} = \int \frac{(6 \tan^2 \theta)\sqrt{6} \sec^2 \theta \, d\theta}{\sqrt{6} \sec \theta}$$

$$= 6 \int \tan^2 \theta \sec \theta \, d\theta$$

$$= 6 \int (\sec^2 \theta - 1) \sec \theta \, d\theta$$

$$= 6 \left[\int \sec^3 \theta \, d\theta - \int \sec \theta \, d\theta \right] \tag{3}$$

In Example 6 of Section 10.2 (in the text) we showed that

$$\int \sec^3 \theta \; d\theta = \frac{1}{2} \sec \theta \tan \theta + \frac{1}{2} \ln|\sec \theta + \tan \theta| + C \tag{4}$$

Substituting from (4) into the right side of (3) and using the formula for $\int \sec \theta \; d\theta$, we have

$$\int \frac{x^2 \; dx}{\sqrt{x^2 + 6}} = 3 \sec \theta \tan \theta + 3 \ln|\sec \theta + \tan \theta| - 6 \ln|\sec \theta + \tan \theta| + C_1$$

$$= 3 \sec \theta \tan \theta - 3 \ln|\sec \theta + \tan \theta| + C_1 \tag{5}$$

Substituting the value given for $\sec \theta$ in (2) and the value given for $\tan \theta$ in (1) into (5), we obtain

$$\int \frac{x^2 \; dx}{\sqrt{x^2 + 6}} = 3 \frac{\sqrt{x^2 + 6}}{\sqrt{6}} \cdot \frac{x}{\sqrt{6}} - 3 \ln \left| \frac{\sqrt{x^2 + 6}}{\sqrt{6}} + \frac{x}{\sqrt{6}} \right| + C_1$$

$$= \tfrac{1}{2} x \sqrt{x^2 + 6} - 3 \ln(\sqrt{x^2 + 6} + x) + 3 \ln \sqrt{6} + C_1$$

$$= \tfrac{1}{2} x \sqrt{x^2 + 6} - 3 \ln(\sqrt{x^2 + 6} + x) + C$$

where we let $3 \ln \sqrt{6} + C_1 = C$ on the last step.

8. $\displaystyle \int \frac{dw}{w^2 \sqrt{w^2 - 7}}$

SOLUTION: We use the method of Case 3 with $u = w$ and $a = \sqrt{7}$. Thus, we let

$$\theta = \begin{cases} \sec^{-1} \dfrac{w}{\sqrt{7}} & \text{if } w \geq \sqrt{7} \\[2mm] 2\pi - \sec^{-1} \dfrac{w}{\sqrt{7}} & \text{if } w \leq -\sqrt{7} \end{cases}$$

Then,

$$\sec \theta = \frac{w}{\sqrt{7}} \tag{1}$$

$$w = \sqrt{7} \sec \theta$$

$$dw = \sqrt{7} \sec \theta \tan \theta \; d\theta$$

and

$$\sqrt{w^2 - 7} = \sqrt{7 \sec^2 \theta - 7}$$
$$= \sqrt{7(\sec^2 \theta - 1)}$$
$$= \sqrt{7 \tan^2 \theta}$$
$$= \sqrt{7} \tan \theta \tag{2}$$

Thus,

$$\int \frac{dw}{w^2 \sqrt{w^2 - 7}} = \int \frac{\sqrt{7} \sec \theta \tan \theta \; d\theta}{(7 \sec^2 \theta)(\sqrt{7} \tan \theta)}$$

$$= \frac{1}{7} \int \cos \theta \; d\theta$$

$$= \frac{1}{7} \sin \theta + C \tag{3}$$

$$= \frac{1}{7} \left[\frac{\sin \theta}{\cos \theta} \cdot \cos \theta \right] + C$$

$$= \frac{1}{7} \left[\tan \theta \cdot \frac{1}{\sec \theta} \right] + C \tag{4}$$

Substituting the values for $\sec \theta$ and $\tan \theta$ given in (1) and (2) into (4), we obtain

$$\int \frac{dw}{w^2\sqrt{w^2-7}} = \frac{1}{7}\left[\frac{\sqrt{w^2-7}}{\sqrt{7}} \cdot \frac{\sqrt{7}}{w}\right] + C$$

$$= \frac{\sqrt{w^2-7}}{7w} + C \qquad (5)$$

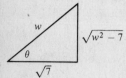

Figure 10. 3. 8

ALTERNATE SOLUTION: After deriving Eq. (3), we draw the sketch shown in Fig. 10. 3. 8, which shows a right triangle containing the angle θ that satisfies Eq. (1). From the triangle we have

$$\sin \theta = \frac{\sqrt{w^2-7}}{w}$$

With this replacement for $\sin \theta$ in (3), we obtain (5).

12. $\displaystyle\int \frac{dx}{x^4\sqrt{16+x^2}}$

SOLUTION: We use the method of Case 2 with $u = x$ and $a = \sqrt{16} = 4$. Thus, we let

$$\theta = \tan^{-1}\frac{x}{4}$$

Then

$$\tan \theta = \frac{x}{4}$$

$$x = 4 \tan \theta$$

$$dx = 4 \sec^2 \theta \, d\theta$$

and

$$\sqrt{16+x^2} = \sqrt{16 + 16\tan^2 \theta}$$
$$= \sqrt{16(1+\tan^2 \theta)}$$
$$= \sqrt{16 \sec^2 \theta} = 4 \sec \theta$$

With these substitutions we obtain

$$\int \frac{dx}{x^4\sqrt{16+x^2}} = \int \frac{4 \sec^2 \theta \, d\theta}{(256 \tan^4 \theta)(4 \sec \theta)}$$

$$= \frac{1}{256}\int \frac{\sec \theta \, d\theta}{\tan^4 \theta}$$

$$= \frac{1}{256}\int \frac{\cos^3 \theta \, d\theta}{\sin^4 \theta}$$

$$= \frac{1}{256}\int \frac{(1-\sin^2 \theta)\cos \theta \, d\theta}{\sin^4 \theta} \qquad (1)$$

Now we let $u = \sin \theta$ and $du = \cos \theta \, d\theta$. Substituting in the right-hand side of Eq. (1), we have

$$\int \frac{dx}{x^4\sqrt{16+x^2}} = \frac{1}{256}\int \frac{(1-u^2)\,du}{u^4}$$

$$= \frac{1}{256}\int (u^{-4} - u^{-2})\,du$$

$$= \frac{1}{256}\left(-\frac{1}{3}u^{-3} + u^{-1}\right) + C$$

$$= \frac{1}{256}\left(\frac{-1}{3\sin^3\theta} + \frac{1}{\sin\theta}\right) + C$$

$$= \frac{\csc\theta}{256}\left(-\frac{\csc^2\theta}{3} + 1\right) + C \tag{2}$$

Because $\tan\theta = x/4$, by Fig. 10. 3. 12 we have

$$\csc\theta = \frac{\sqrt{16+x^2}}{x}$$

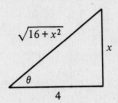

Figure 10. 3. 12

Substituting into the right-hand side of Eq. (2), we obtain

$$\int \frac{dx}{x^4\sqrt{16+x^2}} = \frac{\sqrt{16+x^2}}{256x}\left(-\frac{16+x^2}{3x^2} + 1\right) + C$$

$$= \frac{\sqrt{16+x^2}\,(2x^2 - 16)}{768x^3} + C$$

$$= \frac{(x^2 - 8)\sqrt{16+x^2}}{384x^3} + C$$

16. $\displaystyle\int \frac{dx}{\sqrt{4x - x^2}}$

SOLUTION: We complete the square

$$4x - x^2 = -(x^2 - 4x + 4) + 4$$
$$= 4 - (x - 2)^2$$

Thus,

$$\int \frac{dx}{\sqrt{4x - x^2}} = \int \frac{dx}{\sqrt{4 - (x - 2)^2}} \tag{1}$$

We let $u = x - 2$. Then $du = dx$. Thus, (1) becomes

$$\int \frac{dx}{\sqrt{4x - x^2}} = \int \frac{du}{\sqrt{4 - u^2}}$$

$$= \sin^{-1}\frac{u}{2} + C = \sin^{-1}\left(\frac{x-2}{2}\right) + C$$

20. $\displaystyle\int \frac{e^{-x}\,dx}{(9e^{-2x} + 1)^{3/2}}$

SOLUTION: First, we let $u = 3e^{-x}$. Then $du = -3e^{-x}\,dx$. Thus,

$$\int \frac{e^{-x}\,dx}{(9e^{-2x} + 1)^{3/2}} = -\frac{1}{3}\int \frac{du}{(u^2 + 1)^{3/2}} \tag{1}$$

Next, we use the method of Case 2. Let

$$\theta = \tan^{-1} u$$
$$\tan\theta = u \tag{2}$$
$$du = \sec^2\theta\,d\theta$$

and

$$(u^2 + 1)^{3/2} = (\tan^2\theta + 1)^{3/2}$$
$$= (\sec^2\theta)^{3/2}$$
$$= \sec^3\theta \tag{3}$$

Thus, from (1) we obtain

$$\int \frac{e^{-x}\,dx}{(9e^{-2x}+1)^{3/2}} = -\frac{1}{3}\int \frac{\sec^2\theta\,d\theta}{\sec^3\theta}$$

$$= -\frac{1}{3}\sin\theta + C$$

$$= -\frac{1}{3}\left(\frac{\sin\theta}{\cos\theta}\cdot\cos\theta\right) + C$$

$$= -\frac{1}{3}\left(\tan\theta\cdot\frac{1}{\sec\theta}\right) + C \qquad (4)$$

From (3) we have $\sec\theta = \sqrt{u^2+1}$, and from (2) we have $\tan\theta = u$. Substituting these values into (4), we obtain

$$\int \frac{e^{-x}\,dx}{(9e^{-2x}+1)^{3/2}} = -\frac{1}{3}\left(u\cdot\frac{1}{\sqrt{u^2+1}}\right) + C$$

$$= -\frac{1}{3}\frac{3e^{-x}}{\sqrt{9e^{-2x}+1}} + C$$

$$= \frac{-1}{\sqrt{9+e^{2x}}} + C$$

24. $\displaystyle\int \frac{\sqrt{16-e^{2x}}\,dx}{e^x}$

SOLUTION: We use the method of Case 1 with $u = e^x$ and $a = 4$. Thus, we let

$$\theta = \sin^{-1}\left(\frac{e^x}{4}\right) \qquad (1)$$

Then

$$e^x = 4\sin\theta \qquad (2)$$
$$e^x\,dx = 4\cos\theta\,d\theta$$

and

$$\sqrt{16-e^{2x}} = \sqrt{16-16\sin^2\theta}$$
$$= \sqrt{16(1-\sin^2\theta)}$$
$$= \sqrt{16\cos^2\theta}$$
$$= 4\cos\theta \qquad (3)$$

By first multiplying the numerator and denominator of the given fraction by e^x, we may then make the indicated substitutions in the given integral to obtain

$$\int \frac{\sqrt{16-e^{2x}}\,dx}{e^x} = \int \frac{\sqrt{16-e^{2x}}\,e^x\,dx}{e^{2x}}$$

$$= \int \frac{(4\cos\theta)(4\cos\theta\,d\theta)}{(4\sin\theta)^2}$$

$$= \int \cot^2\theta\,d\theta$$

$$= \int (\csc^2\theta - 1)\,d\theta$$

$$= -\cot\theta - \theta + C \qquad (4)$$

From Eqs. (2) and (3), we have

$$\cot \theta = \frac{\cos \theta}{\sin \theta} = \frac{\sqrt{16 - e^{2x}}}{e^x} \qquad (5)$$

Substituting from Eqs. (5) and (1) into (4), we get

$$\int \frac{\sqrt{16 - e^{2x}}\, dx}{e^x} = -\frac{\sqrt{16 - e^{2x}}}{e^x} - \sin^{-1}\left(\frac{e^x}{4}\right) + C$$

In Exercises 25–32, evaluate the definite integral.

28. $\displaystyle\int_0^1 \frac{x^2\, dx}{\sqrt{4 - x^2}}$

SOLUTION: We use the method of Case 1 with $u = x$ and $a = 2$. Let

$$\theta = \sin^{-1} \frac{x}{2}$$

Then

$$x = 2 \sin \theta$$
$$dx = 2 \cos \theta \, d\theta$$

and

$$\begin{aligned}
\sqrt{4 - x^2} &= \sqrt{4 - 4 \sin^2 \theta} \\
&= 2\sqrt{1 - \sin^2 \theta} \\
&= 2\sqrt{\cos^2 \theta} \\
&= 2 \cos \theta
\end{aligned}$$

Furthermore, when $x = 0$, $\theta = 0$; when $x = 1$, $\theta = \frac{1}{6}\pi$. Thus,

$$\begin{aligned}
\int_0^1 \frac{x^2\, dx}{\sqrt{4 - x^2}} &= \int_0^{\pi/6} \frac{(4 \sin^2 \theta)(2 \cos \theta \, d\theta)}{2 \cos \theta} \\
&= 4 \int_0^{\pi/6} \frac{1}{2}(1 - \cos 2\theta)\, d\theta \\
&= 2 \left[\theta - \tfrac{1}{2} \sin 2\theta \right]_0^{\pi/6} \\
&= 2[(\tfrac{1}{6}\pi - \tfrac{1}{2} \sin \tfrac{1}{3}\pi) - (0 - \tfrac{1}{2} \sin 0)] \\
&= \tfrac{1}{3}\pi - \tfrac{1}{2}\sqrt{3}
\end{aligned}$$

32. $\displaystyle\int_4^8 \frac{dw}{(w^2 - 4)^{3/2}}$

SOLUTION: We use the method of Case 3. Because w is in the closed interval $[4, 8]$, we let

$$\theta = \sec^{-1}\left(\frac{w}{2}\right)$$

Thus,

$$w = 2 \sec \theta$$
$$dw = 2 \sec \theta \tan \theta \, d\theta$$

and

$$\begin{aligned}
(w^2 - 4)^{3/2} &= (4 \sec^2 \theta - 4)^{3/2} \\
&= [4(\sec^2 \theta - 1)]^{3/2} \\
&= (4 \tan^2 \theta)^{3/2} \\
&= 8 \tan^3 \theta
\end{aligned}$$

Furthermore, when $w = 4$, $\theta = \frac{1}{3}\pi$; when $w = 8$, $\theta = \sec^{-1} 4$. Thus,

$$\int_4^8 \frac{dw}{(w^2 - 4)^{3/2}} = \int_{\pi/3}^{\sec^{-1} 4} \frac{2 \sec \theta \tan \theta \ d\theta}{8 \tan^3 \theta}$$

$$= \frac{1}{4} \int_{\pi/3}^{\sec^{-1} 4} \frac{\cos \theta \ d\theta}{\sin^2 \theta} \tag{1}$$

Now we let $u = \sin \theta$ and $du = \cos \theta \ d\theta$. When $\theta = \frac{1}{3}\pi$, $u = \frac{1}{2}\sqrt{3}$; when $\theta = \sec^{-1} 4$, then $\theta = \cos^{-1} \frac{1}{4}$ and $\cos \theta = \frac{1}{4}$. Thus, when $\theta = \sec^{-1} 4$,

$$\begin{aligned}
u &= \sin \theta \\
&= \sqrt{1 - \cos^2 \theta} \\
&= \sqrt{1 - (\tfrac{1}{4})^2} = \tfrac{1}{4}\sqrt{15}
\end{aligned}$$

With these substitutions in (1), we have

$$\int_4^8 \frac{dw}{(w^2 - 4)^{3/2}} = \frac{1}{4} \int_{\frac{1}{2}\sqrt{3}}^{\frac{1}{4}\sqrt{15}} u^{-2} \ du$$

$$= -\frac{1}{4} \left[u^{-1} \right]_{\frac{1}{2}\sqrt{3}}^{\frac{1}{4}\sqrt{15}}$$

$$= -\frac{1}{4} \left[\frac{4}{\sqrt{15}} - \frac{2}{\sqrt{3}} \right]$$

$$= -\frac{1}{4} \left[\frac{4}{15}\sqrt{15} - \frac{2}{3}\sqrt{3} \right]$$

$$= -\frac{1}{15}\sqrt{15} + \frac{1}{6}\sqrt{3}$$

36. Find the volume of the solid generated by revolving the region bounded by the curve $y = \sqrt{x^2 - 9}/x^2$, the x-axis, and the line $x = 5$ about the y-axis.

SOLUTION: Let f be the function defined by

$$f(x) = \frac{\sqrt{x^2 - 9}}{x^2}$$

We note that $f(x)$ is defined for $|x| \geq 3$. A sketch of the graph of the region is shown in Fig. 10. 3. 36. The element of area is a rectangle with width $\Delta_i x$ units. We rotate the element of area about the y-axis. This results in an element of volume that is a cylindrical shell with the number of units in the thickness given by $\Delta_i x$, the number of units in the mean radius given by $\bar{x}_i$, and the number of units in the altitude given by $f(\bar{x}_i)$. Furthermore, the region is bounded on the left by the line $x = 3$ and bounded on the right by the line $x = 5$. Thus

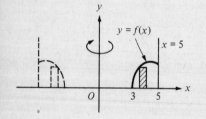

Figure 10. 3. 36

$$V = \lim_{\|\Delta\| \to 0} \sum_{i=1}^n 2\pi \bar{x}_i f(\bar{x}_i) \ \Delta_i x$$

$$= 2\pi \int_3^5 x f(x) \ dx$$

$$= 2\pi \int_3^5 \frac{\sqrt{x^2 - 9}}{x} \ dx \tag{1}$$

We use the method of Case 3 with $a = 3$. Because $x \geq 3$, we let

$$\theta = \sec^{-1} \frac{x}{3} \tag{2}$$

Thus,

$$x = 3 \sec \theta$$
$$dx = 3 \sec \theta \tan \theta \, d\theta$$

and

$$\sqrt{x^2 - 9} = \sqrt{9 \sec^2 \theta - 9}$$
$$= 3 \tan \theta \tag{3}$$

Thus,

$$\int \frac{\sqrt{x^2 - 9}}{x} \, dx = \int \frac{(3 \tan \theta)(3 \sec \theta \tan \theta \, d\theta)}{3 \sec \theta}$$

$$= 3 \int (\sec^2 \theta - 1)$$

$$= 3(\tan \theta - \theta) + C$$

$$= 3\left(\frac{1}{3} \sqrt{x^2 - 9} - \sec^{-1} \frac{1}{3} x \right) + C \tag{4}$$

where we substituted from (3) and (2) on the last step. From (1) and (4), we obtain

$$V = 2\pi \left[\sqrt{x^2 - 9} - 3 \sec^{-1} \frac{1}{3} x \right]_3^5$$

$$= 2\pi \left[\left(4 - 3 \sec^{-1} \frac{5}{3} \right) - (0 - 3 \sec^{-1} 1) \right]$$

$$= 2\pi \left(4 - 3 \cos^{-1} \frac{3}{5} \right)$$

Thus, the volume is $2\pi(4 - 3 \cos^{-1} \frac{3}{5})$ cubic units.

10.4 INTEGRATION OF RATIONAL FUNCTIONS BY PARTIAL FRACTIONS WHEN THE DENOMINATOR HAS ONLY LINEAR FACTORS

Case 1: The factors of the denominator are all linear and none is repeated. Find constants A, B, C, etc., such that

$$\frac{P(x)}{(x + a)(x + b)} = \frac{A}{x + a} + \frac{B}{x + b}$$

or

$$\frac{P(x)}{(x + a)(x + b)(x + c)} = \frac{A}{x + a} + \frac{B}{x + b} + \frac{C}{x + c}$$

and so on.

Case 2: There is a linear factor in the denominator that is repeated. Find constants A, B, C, etc., such that

$$\frac{P(x)}{(x + a)^2} = \frac{A}{(x + a)^2} + \frac{B}{x + a}$$

$$\frac{P(x)}{(x + a)^3} = \frac{A}{(x + a)^3} + \frac{B}{(x + a)^2} + \frac{C}{x + a}$$

and so on.

You may find the constants (A, B, C, etc.) either by making arbitrarily chosen replacements for x and solving for A, B, C, and so on, or by expressing each member of the equation as a polynomial and then equating coefficients of like powers of x.

When the denominator contains linear factors some of which are repeated and some of which are not repeated, use a combination of the methods of Case

1 and Case 2. Remember to reduce an improper fraction to the sum of a polynomial and a proper fraction by dividing the numerator by the denominator before attempting to find partial fractions.

Although the following integration formulas can be easily derived by using the method of this section, they occur frequently enough to be worth memorizing.

$$\int \frac{du}{u^2 - a^2} = \frac{1}{2a} \ln \left| \frac{u - a}{u + a} \right| + C$$

$$\int \frac{du}{a^2 - u^2} = \frac{1}{2a} \ln \left| \frac{u + a}{u - a} \right| + C$$

Exercises 10.4

In Exercises 1–20 evaluate the indefinite integral.

4. $\displaystyle\int \frac{(4x - 2)\, dx}{x^3 - x^2 - 2x}$

SOLUTION: First, we factor the denominator $x^3 - x^2 - 2x = x(x + 1)(x - 2)$. Because none of the factors is repeated, we use the method of Case 1. That is, we find constants A, B, and C such that

$$\frac{4x - 2}{x(x + 1)(x - 2)} = \frac{A}{x} + \frac{B}{x + 1} + \frac{C}{x - 2} \tag{1}$$

Eliminating the fractions, we have

$$4x - 2 = A(x + 1)(x - 2) + Bx(x - 2) + Cx(x + 1) \tag{2}$$

We choose replacements for x in (2) that enable us to find A, B, and C. If $x = 0$, then (2) becomes

$$-2 = -2A \quad \text{or} \quad A = 1$$

If $x = -1$, then $x + 1 = 0$, and (2) becomes

$$-6 = 3B \quad \text{or} \quad B = -2$$

If $x = 2$, then $x - 2 = 0$, and from (2) we obtain

$$6 = 6C \quad \text{or} \quad C = 1$$

With $A = 1$, $B = -2$, and $C = 1$ in (1) we obtain

$$\frac{4x - 2}{x(x + 1)(x - 2)} = \frac{1}{x} - \frac{2}{x + 1} + \frac{1}{x - 2}$$

Thus,

$$\int \frac{(4x - 2)\, dx}{x^3 - x^2 - 2x} = \int \frac{(4x - 2)\, dx}{x(x + 1)(x - 2)}$$

$$= \int \frac{dx}{x} - \int \frac{2\, dx}{x + 1} + \int \frac{dx}{x - 2}$$

$$= \ln|x| - 2 \ln|x + 1| + \ln|x - 2| + C$$

$$= \ln|x(x - 2)| - \ln(x + 1)^2 + C$$

$$= \ln \frac{|x(x - 2)|}{(x + 1)^2} + C$$

8. $\int \dfrac{x^2 + x + 2}{x^2 - 1}\, dx$

SOLUTION: First, we divide the numerator by the denominator to replace the given improper fraction by the sum of a polynomial and a proper fraction.

$$\int \frac{x^2 + x + 2}{x^2 - 1}\, dx = \int \left(1 + \frac{x + 3}{x^2 - 1} \right) dx$$

$$= x + \int \frac{x + 3}{(x + 1)(x - 1)}\, dx \tag{1}$$

We find partial fractions for the integrand that remains. Let

$$\frac{x + 3}{(x + 1)(x - 1)} = \frac{A}{x + 1} + \frac{B}{x - 1} \tag{2}$$

Eliminating fractions, we obtain

$$x + 3 = A(x - 1) + B(x + 1) \tag{3}$$

If $x = -1$, then from (3) we get

$$2 = -2A$$
$$A = -1$$

If $x = 1$, then from (3) we have

$$4 = 2B$$
$$B = 2$$

Substituting the values for A and B into Eq. (2), we get

$$\frac{x + 3}{(x + 1)(x - 1)} = \frac{-1}{x + 1} + \frac{2}{x - 1} \tag{4}$$

And substituting from Eq. (4) into Eq. (1), we obtain

$$\int \frac{x^2 + x + 2}{x^2 - 1}\, dx = x + \int \frac{-dx}{x + 1} + \int \frac{2\, dx}{x - 1}$$

$$= x - \ln|x + 1| + 2\ln|x - 1| + C$$

$$= x + \ln \frac{(x - 1)^2}{|x + 1|} + C$$

12. $\int \dfrac{3x^2 - x + 1}{x^3 - x^2}\, dx$.

SOLUTION: We factor the denominator. Because $x^3 - x^2 = x^2(x - 1)$, we have a repeated factor, and thus we use the method of Case 2. That is, we find constants A, B, and C such that

$$\frac{3x^2 - x + 1}{x^2(x - 1)} = \frac{A}{x^2} + \frac{B}{x} + \frac{C}{x - 1} \tag{1}$$

Eliminating the fractions, we have

$$3x^2 - x + 1 = A(x - 1) + Bx(x - 1) + Cx^2 \tag{2}$$

If $x = 0$, from (2) we get

$$1 = -A \quad \text{or} \quad A = -1$$

If $x = 1$, then $x - 1 = 0$, and from (2) we obtain

$$3 = C$$

If $x = 2$, from (2) we have

$$11 = A + 2B + 4C$$

Because $A = -1$ and $C = 3$, the above yields

$$11 = -1 + 2B + 12 \quad \text{or} \quad B = 0$$

Thus, from (1) with $A = -1$, $B = 0$, and $C = 3$, we have

$$\frac{3x^2 - x + 1}{x^2(x - 1)} = -\frac{1}{x^2} + \frac{3}{x - 1}$$

Therefore,

$$\int \frac{3x^2 - x + 1}{x^3 - x^2}\, dx = \int \frac{3x^2 - x + 1}{x^2(x - 1)}\, dx$$

$$= -\int \frac{dx}{x^2} + \int \frac{3\, dx}{x - 1}$$

$$= \frac{1}{x} + 3\ln|x - 1| + C$$

16. $\displaystyle \int \frac{(5x^2 - 11x + 5)\, dx}{x^3 - 4x^2 + 5x - 2}$

SOLUTION: We use trial and error with synthetic division to factor the denominator.

$$x^3 - 4x^2 + 5x - 2 = (x - 1)^2(x - 2)$$

We find constants A, B, and C such that

$$\frac{5x^2 - 11x + 5}{(x - 1)^2(x - 2)} = \frac{A}{(x - 1)^2} + \frac{B}{x - 1} + \frac{C}{x - 2} \tag{1}$$

Thus,

$$5x^2 - 11x + 5 = A(x - 2) + B(x - 1)(x - 2) + C(x - 1)^2 \tag{2}$$

If $x = 1$ in Eq. (2), we get

$$-1 = -A \quad \text{or} \quad A = 1$$

If $x = 2$ in Eq. (2), we obtain

$$3 = C$$

If $x = 0$ in Eq. (2), we have

$$5 = -2A + 2B + C$$

Because $A = 1$ and $C = 3$, the above gives

$$5 = -2 + 2B + 3 \quad \text{or} \quad B = 2$$

From (1) with $A = 1$, $B = 2$, and $C = 3$, we have

$$\frac{5x^2 - 11x + 5}{(x - 1)^2(x - 2)} = \frac{1}{(x - 1)^2} + \frac{2}{x - 1} + \frac{3}{x - 2}$$

Therefore,

$$\int \frac{(5x^2 - 11x + 5)\, dx}{x^3 - 4x^2 + 5x - 2} = \int \frac{(5x^2 - 11x + 5)\, dx}{(x - 1)^2(x - 2)}$$

$$= \int \frac{dx}{(x - 1)^2} + \int \frac{2\, dx}{x - 1} + \int \frac{3\, dx}{x - 2}$$

$$= -\frac{1}{x-1} + 2\ln|x-1| + 3\ln|x-2| + C$$

$$= -\frac{1}{x-1} + \ln|(x-1)^2(x-2)^3| + C$$

20. $\displaystyle\int \frac{dx}{16x^4 - 8x^2 + 1}$

SOLUTION:

$$\int \frac{dx}{16x^4 - 8x^2 + 1} = \frac{1}{16}\int \frac{dx}{x^4 - \frac{1}{2}x^2 + \frac{1}{16}}$$

$$= \frac{1}{16}\int \frac{dx}{(x-\frac{1}{2})^2(x+\frac{1}{2})^2} \tag{1}$$

We let

$$\frac{1}{(x-\frac{1}{2})^2(x+\frac{1}{2})^2} = \frac{A}{(x-\frac{1}{2})^2} + \frac{B}{x-\frac{1}{2}} + \frac{C}{(x+\frac{1}{2})^2} + \frac{D}{x+\frac{1}{2}} \tag{2}$$

Then

$$1 = A(x+\tfrac{1}{2})^2 + B(x-\tfrac{1}{2})(x+\tfrac{1}{2})^2 + C(x-\tfrac{1}{2})^2 + D(x+\tfrac{1}{2})(x-\tfrac{1}{2})^2 \tag{3}$$

If $x = \frac{1}{2}$ in (3), we have

$$1 = A$$

If $x = -\frac{1}{2}$ in (3), we get

$$1 = C$$

Next, we let $x = 0$, $A = 1$, and $C = 1$ in (3), obtaining

$$1 = \tfrac{1}{4} - \tfrac{1}{8}B + \tfrac{1}{4} + \tfrac{1}{8}D$$
$$4 = -B + D \tag{4}$$

And then we let $x = 1$, $A = 1$, and $C = 1$ in (3), which yields

$$1 = \tfrac{9}{4} + \tfrac{9}{8}B + \tfrac{1}{4} + \tfrac{3}{8}D$$
$$-4 = 3B + D \tag{5}$$

Solving (4) and (5) simultaneously, we get

$$B = -2 \quad \text{and} \quad D = 2$$

Substituting the values found for A, B, C, and D into Eq. (2), we have

$$\frac{1}{(x-\frac{1}{2})^2(x+\frac{1}{2})^2} = \frac{1}{(x-\frac{1}{2})^2} - \frac{2}{x-\frac{1}{2}} + \frac{1}{(x+\frac{1}{2})^2} + \frac{2}{x+\frac{1}{2}}$$

Thus,

$$\int \frac{dx}{(x-\frac{1}{2})^2(x+\frac{1}{2})^2} = \int \frac{dx}{(x-\frac{1}{2})^2} - 2\int \frac{dx}{x-\frac{1}{2}} + \int \frac{dx}{(x+\frac{1}{2})^2} + 2\int \frac{dx}{x+\frac{1}{2}}$$

$$= -(x-\tfrac{1}{2})^{-1} - 2\ln|x-\tfrac{1}{2}| - (x+\tfrac{1}{2})^{-1}$$
$$+ 2\ln|x+\tfrac{1}{2}| + C'$$

$$= -\left[\frac{1}{x-\frac{1}{2}} + \frac{1}{x+\frac{1}{2}}\right] + 2\ln\left|\frac{x+\frac{1}{2}}{x-\frac{1}{2}}\right| + C'$$

$$= -\frac{8x}{(2x-1)(2x+1)} + 2\ln\left|\frac{2x+1}{2x-1}\right| + C' \tag{6}$$

Substituting from (6) into (1), we obtain

$$\int \frac{dx}{16x^4 - 8x^2 + 1} = -\frac{x}{2(2x-1)(2x+1)} + \frac{1}{8}\ln\left|\frac{2x+1}{2x-1}\right| + C$$

In Exercises 21–28 evaluate the definite integral.

24. $\int_1^4 \frac{(2x^2 + 13x + 18)\,dx}{x^3 + 6x^2 + 9x}$

SOLUTION: $x^3 + 6x^2 + 9x = x(x+3)^2$. Thus, we find constants A, B, and C such that

$$\frac{2x^2 + 13x + 18}{x^3 + 6x^2 + 9x} = \frac{A}{x} + \frac{B}{(x+3)^2} + \frac{C}{x+3} \tag{1}$$

We have

$$2x^2 + 13x + 18 = A(x+3)^2 + Bx + Cx(x+3) \tag{2}$$

If $x = 0$, from (2) we have

$$18 = 9A \quad \text{or} \quad A = 2$$

If $x = -3$, from (2) we have

$$-3 = -3B \quad \text{or} \quad B = 1$$

If $x = -2$, from (2) we get

$$0 = A - 2B - 2C$$

Because $A = 2$ and $B = 1$, the above gives $C = 0$. From (1) we have

$$\frac{2x^2 + 13x + 18}{x^3 + 6x^2 + 9x} = \frac{2}{x} + \frac{1}{(x+3)^2}$$

Thus,

$$\int_1^4 \frac{(2x^2 + 13x + 18)\,dx}{x^3 + 6x^2 + 9x} = 2\int_1^4 \frac{dx}{x} + \int_1^4 \frac{dx}{(x+3)^2}$$

$$= 2\ln x\Big]_1^4 - \left[\frac{1}{x+3}\right]_1^4$$

$$= 2(\ln 4 - \ln 1) - \left(\frac{1}{7} - \frac{1}{4}\right)$$

$$= \ln 16 + \frac{3}{28}$$

28. $\int_0^4 \frac{x^2\,dx}{2x^3 + 9x^2 + 12x + 4}$

SOLUTION: We use synthetic division to factor the denominator. Thus,

$$\frac{x^2}{2x^3 + 9x^2 + 12x + 4} = \frac{x^2}{(x+2)^2(2x+1)}$$

$$= \frac{A}{(x+2)^2} + \frac{B}{x+2} + \frac{C}{2x+1} \tag{1}$$

Eliminating the fractions, we obtain

$$x^2 = A(2x+1) + B(x+2)(2x+1) + C(x+2)^2 \tag{2}$$

If $x = -2$, then $x + 2 = 0$, and from (2) we have

$$4 = -3A \quad \text{or} \quad A = -\tfrac{4}{3}$$

If $x = -\tfrac{1}{2}$, then $2x + 1 = 0$, and from (2) we have

$$\tfrac{1}{4} = \tfrac{9}{4}C \quad \text{or} \quad C = \tfrac{1}{9}$$

If $x = 0$, $A = -\tfrac{4}{3}$ and $C = \tfrac{1}{9}$, then from (2) we obtain

$$0 = -\tfrac{4}{3} + 2B + \tfrac{4}{9} \quad \text{or} \quad B = \tfrac{4}{9}$$

Thus, with $A = -\tfrac{4}{3}$, $B = \tfrac{4}{9}$, and $C = \tfrac{1}{9}$, Eq. (1) becomes

$$\frac{x^2}{2x^3 + 9x^2 + 12x + 4} = \frac{-\tfrac{4}{3}}{(x+2)^2} + \frac{\tfrac{4}{9}}{x+2} + \frac{\tfrac{1}{9}}{2x+1}$$

Hence

$$\int_0^4 \frac{x^2 \, dx}{2x^3 + 9x^2 + 12x + 4} = -\frac{4}{3} \int_0^4 \frac{dx}{(x+2)^2} + \frac{4}{9} \int_0^4 \frac{dx}{x+2} + \frac{1}{18} \int_0^4 \frac{2 \, dx}{2x+1}$$

$$= \frac{4}{3} \left[(x+2)^{-1} \right]_0^4 + \frac{4}{9} \left[\ln(x+2) \right]_0^4 + \frac{1}{18} \left[\ln(2x+1) \right]_0^4$$

$$= \frac{4}{3} \left[\frac{1}{6} - \frac{1}{2} \right] + \frac{4}{9} \left[\ln 6 - \ln 2 \right] + \frac{1}{18} \left[\ln 9 - \ln 1 \right]$$

$$= -\frac{4}{9} + \frac{4}{9} \ln 3 + \frac{1}{18} \ln 3^2$$

$$= -\frac{4}{9} + \frac{5}{9} \ln 3$$

32. Find the volume of the solid of revolution if the region in the first quadrant bounded by the curve $(x + 2)^2 y = 4 - x$ is revolved about the x-axis.

SOLUTION: Solving the given equation for y, we get

$$y = f(x) = \frac{4 - x}{(x+2)^2}$$

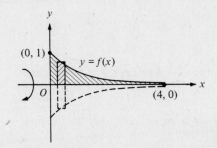

Figure 10. 4. 32

Figure 10. 4. 32 shows a sketch of the given region and a plane section of the solid of revolution. The element of volume is a circular disk with thickness $\Delta_i x$ units and radius $f(\bar{x}_i)$ units. If V cubic units is the volume of the solid of revolution, then

$$V = \lim_{\|\Delta\| \to 0} \sum_{i=1}^{n} \pi [f(\bar{x}_i)]^2 \, \Delta_x x$$

$$= \pi \int_0^4 \frac{(4 - x)^2}{(x+2)^4} \, dx \tag{1}$$

We let

$$\frac{(4 - x)^2}{(x+2)^4} = \frac{A}{(x+2)^4} + \frac{B}{(x+2)^3} + \frac{C}{(x+2)^2} + \frac{D}{x+2} \tag{2}$$

Then

$$(4 - x)^2 = A + B(x+2) + C(x+2)^2 + D(x+2)^3$$

If $x = -2$, then we get $A = 6^2 = 36$. Replacing A with 36 and multiplying out, we obtain

$$x^2 - 8x + 16 = 36 + Bx + 2B + Cx^2 + 4Cx + 4C + Dx^3 + 6Dx^2 + 12Dx$$
$$+ 8D$$
$$= Dx^3 + (C + 6D)x^2 + (B + 4C + 12D)x + (36 + 2B + 4C$$
$$+ 8D)$$

Because the coefficients of like powers of x must be equal, we have

$$0 = D$$
$$1 = C + 6D$$
$$-8 = B + 4C + 12D$$
$$16 = 36 + 2B + 4C + 8D$$

Solving this system, we obtain $B = -12$, $C = 1$, and $D = 0$. Substituting for A, B, C, and D into Eq. (2), we have

$$\frac{(4-x)^2}{(x+2)^4} = \frac{36}{(x+2)^4} - \frac{12}{(x+2)^3} + \frac{1}{(x+2)^2} \tag{3}$$

Substituting from Eq. (3) into Eq. (1), we have

$$V = \pi \left[\int_0^4 \frac{36\ dx}{(x+2)^4} - \int_0^4 \frac{12\ dx}{(x+2)^3} + \int_0^4 \frac{dx}{(x+2)^2} \right]$$

$$= \pi \left[\frac{-12}{(x+2)^3} + \frac{6}{(x+2)^2} - \frac{1}{x+2} \right]_0^4$$

$$= \pi \left[\left(\frac{-12}{6^3} + \frac{6}{6^2} - \frac{1}{6} \right) - \left(\frac{-12}{2^3} + \frac{6}{2^2} - \frac{1}{2} \right) \right]$$

$$= \frac{4}{9} \pi$$

The volume of the solid of revolution is $\frac{4}{9}\pi$ cubic units.

36. A particle is moving along a straight line so that if v ft/sec is the velocity of the particle at t sec, then

$$v = \frac{t+3}{t^2 + 3t + 2}$$

Find the distance traveled by the particle from the time when $t = 0$ to the time when $t = 2$.

SOLUTION: Let s ft be the distance of the particle from a fixed point on the line at t sec. Because $ds/dt = v$, we are given that

$$ds = \frac{(t+3)\ dt}{t^2 + 3t + 2}$$

Thus,

$$s = \int \frac{(t+3)\ dt}{t^2 + 3t + 2} \tag{1}$$

Let

$$\frac{t+3}{t^2 + 3t + 2} = \frac{t+3}{(t+1)(t+2)}$$

$$= \frac{A}{t+1} + \frac{B}{t+2}$$

Then

$$t + 3 = A(t + 2) + B(t + 1)$$

If $t = -1$, then $2 = A$. If $t = -2$, then $1 = -B$ or $B = -1$. Thus,

$$\frac{t + 3}{t^2 + 3t + 2} = \frac{2}{t + 1} - \frac{1}{t + 2}$$

and from (1) we have

$$s = 2 \int \frac{dt}{t + 1} - \int \frac{dt}{t + 2}$$

$$= 2 \ln|t + 1| - \ln|t + 2| + C$$

$$= \ln \frac{(t + 1)^2}{|t + 2|} + C$$

When $t = 2$, we have

$$s = \ln \tfrac{9}{4} + C$$

When $t = 0$, we have

$$s = \ln \tfrac{1}{2} + C$$

Thus,

$$s(2) - s(0) = \ln \tfrac{9}{4} - \ln \tfrac{1}{2}$$
$$= \ln \tfrac{9}{2}$$

The distance traveled is $\ln \tfrac{9}{2}$ feet.

10.5 INTEGRATION OF RATIONAL FUNCTIONS BY PARTIAL FRACTIONS WHEN THE DENOMINATOR CONTAINS QUADRATIC FACTORS

Case 3: The denominator contains quadratic factors that are not repeated. For each quadratic factor $x^2 + a$ in the denominator, find constants A and B and a partial fraction of the form

$$\frac{Ax + B}{x^2 + a}$$

For each quadratic factor of the form $x^2 + px + q$, when $p \neq 0$, find a partial fraction of the form

$$\frac{A(2x + p) + B}{x^2 + px + q}$$

Case 4: The denominator contains quadratic factors that are repeated. For each quadratic factor $(x^2 + a)^2$, find partial fractions

$$\frac{Ax + B}{(x^2 + a)^2} + \frac{Cx + D}{x^2 + a}$$

For each quadratic factor $(x^2 + a)^3$, find partial fractions

$$\frac{Ax + B}{(x^2 + a)^3} + \frac{Cx + D}{(x^2 + a)^2} + \frac{Ex + F}{x^2 + a}$$

And so on. For each quadratic factor $(x^2 + px + q)^2$, where $p \neq 0$, find partial fractions

$$\frac{A(2x + p) + B}{(x^2 + px + q)^2} + \frac{C(2x + p) + D}{x^2 + px + q}$$

And so on.

Exercises 10.5

In Exercises 1–20 evaluate the indefinite integral

4. $\displaystyle\int \frac{(x^2 - 4x - 4)\,dx}{x^3 - 2x^2 + 4x - 8}$

SOLUTION:

$$\int \frac{(x^2 - 4x - 4)\,dx}{x^3 - 2x^2 + 4x - 8} = \int \frac{(x^2 - 4x - 4)\,dx}{(x^2 + 4)(x - 2)} \tag{1}$$

We find constants A, B, and C such that

$$\frac{x^2 - 4x - 4}{(x^2 + 4)(x - 2)} = \frac{Ax + B}{x^2 + 4} + \frac{C}{x - 2} \tag{2}$$

$$x^2 - 4x - 4 = (Ax + B)(x - 2) + C(x^2 + 4) \tag{3}$$

If $x = 2$, from Eq. (3) we have

$$-8 = 8C \quad \text{or} \quad C = -1$$

From (3), with $C = -1$, we have

$$x^2 - 4x - 4 = Ax^2 + Bx - 2Ax - 2B - x^2 - 4$$
$$= (A - 1)x^2 + (B - 2A)x + (-2B - 4)$$

Equating coefficients of like powers of x, we obtain

$$1 = A - 1 \quad \text{or} \quad A = 2$$
$$-4 = -2B - 4 \quad \text{or} \quad B = 0$$

Thus, from (2) we have

$$\frac{x^2 - 4x - 4}{(x^2 + 4)(x - 2)} = \frac{2x}{x^2 + 4} - \frac{1}{x - 2} \tag{4}$$

Hence, from (1) and (4) we get

$$\int \frac{(x^2 - 4x - 4)\,dx}{x^3 - 2x^2 + 4x - 8} = \int \frac{2x\,dx}{x^2 + 4} - \int \frac{dx}{x - 2}$$

$$= \ln(x^2 + 4) - \ln|x - 2| + C$$

$$= \ln \frac{x^2 + 4}{|x - 2|} + C$$

8. $\displaystyle\int \frac{dx}{9x^4 + x^2}$

SOLUTION:

$$\int \frac{dx}{9x^4 + x^2} = \int \frac{dx}{x^2(9x^2 + 1)} \tag{1}$$

We find constants A, B, C, and D such that

$$\frac{1}{x^2(9x^2 + 1)} = \frac{A}{x^2} + \frac{B}{x} + \frac{Cx + D}{9x^2 + 1} \tag{2}$$

$$1 = A(9x^2 + 1) + Bx(9x^2 + 1) + (Cx + D)x^2$$

If $x = 0$, we get $1 = A$. Replacing A with 1 and multiplying out, we obtain

$$1 = 9x^2 + 1 + 9Bx^3 + Bx + Cx^3 + Dx^2$$
$$1 = (9B + C)x^3 + (9 + D)x^2 + Bx + 1$$

The coefficients of x^3, x^2, and x must be 0. Thus,

$$0 = 9B + C$$
$$0 = 9 + D$$
$$0 = B$$

and we have $B = 0$, $C = 0$, and $D = -9$. Substituting the values for A, B, C, and D into Eq. (2), we get

$$\frac{1}{x^2(9x^2 + 1)} = \frac{1}{x^2} - \frac{9}{9x^2 + 1} \tag{3}$$

And substituting from Eq. (3) into Eq. (1), we obtain

$$\int \frac{dx}{9x^4 + x^2} = \int \frac{dx}{x^2} - \int \frac{9\,dx}{9x^2 + 1}$$

$$= -\frac{1}{x} - \int \frac{9\,dx}{(3x)^2 + 1} \tag{4}$$

For the integral that remains in (4) we let $u = 3x$ and $du = 3\,dx$. Then

$$\int \frac{9\,dx}{(3x)^2 + 1} = \int \frac{3\,du}{u^2 + 1}$$

$$= 3\,\tan^{-1} u + \overline{C}$$

$$= 3\,\tan^{-1} 3x + \overline{C} \tag{5}$$

Substituting from (5) into (4), we obtain

$$\int \frac{dx}{9x^4 + x^2} = -\frac{1}{x} - 3\,\tan^{-1} 3x + C$$

where $C = -\overline{C}$.

12. $\displaystyle\int \frac{(2x^3 + 9x)\,dx}{(x^2 + 3)(x^2 - 2x + 3)}$

SOLUTION: We find constants A, B, C, and D such that

$$\frac{2x^3 + 9x}{(x^2 + 3)(x^2 - 2x + 3)} = \frac{Ax + B}{x^2 + 3} + \frac{C(2x - 2) + D}{x^2 - 2x + 3}$$

Note that the coefficient of C is the derivative of the denominator of the fraction in which C appears. From the above equation we obtain

$$2x^3 + 9x = (Ax + B)(x^2 - 2x + 3) + [C(2x - 2) + D](x^2 + 3)$$
$$= Ax^3 - 2Ax^2 + 3Ax + Bx^2 - 2Bx + 3B + 2Cx^3 + 6Cx - 2Cx^2$$
$$\quad - 6C + Dx^2 + 3D$$
$$= (A + 2C)x^3 + (-2A + B - 2C + D)x^2 + (3A - 2B + 6C)x$$
$$\quad + (3B - 6C + 3D)$$

Equating coefficients of like powers of x, we obtain

$$A \qquad\quad + 2C \qquad\quad = 2 \tag{1}$$
$$-2A + \quad B - 2C + \quad D = 0 \tag{2}$$
$$3A - 2B + 6C \qquad\quad = 9 \tag{3}$$
$$3B - 6C + 3D = 0 \tag{4}$$

We solve the above system as follows. First, we eliminate A from Eqs. (2) and (3) by adding a multiple of Eq. (1) to Eq. (2) and a multiple of Eq. (1) to Eq. (3).

$$A \quad\quad + 2C \quad\quad = 2 \tag{1}$$
$$B + 2C + D \;= 4 \quad\quad \text{[Eq. (2) plus two times Eq. (1)]} \tag{5}$$
$$-2B \quad\quad\quad = 3 \quad\quad \text{[Eq. (3) minus three times Eq. (1)]} \tag{6}$$
$$3B - 6C + 3D = 0 \tag{4}$$

Next, we eliminate B from Eq. (6) and Eq. (4) by adding a multiple of Eq. (5) to Eq. (6) and a multiple of Eq. (5) to Eq. (4).

$$A \quad + 2C \quad\quad = \quad 2 \tag{1}$$
$$B + \;2C + \;D = \quad 4 \tag{5}$$
$$4C + 2D = \;11 \quad\quad \text{[Eq. (6) plus two times Eq. (5)]} \tag{7}$$
$$-12C \quad\quad = -12 \quad\quad \text{[Eq. (4) minus three times Eq. (5)]} \tag{8}$$

Next, solve Eq. (8) for C and interchange Eq. (7) and (8).

$$A \quad + 2C \quad\quad = \;2 \tag{1}$$
$$B + 2C + \;D = \;4 \tag{5}$$
$$C \quad\quad = \;1 \tag{8}$$
$$4C + 2D = 11 \tag{7}$$

Now, we eliminate C from Eq. (7) by adding a multiple of Eq. (8) to Eq. (7).

$$A \quad + 2C \quad\quad = 2 \tag{1}$$
$$B + 2C + \;D = 4 \tag{5}$$
$$C \quad\quad = 1 \tag{8}$$
$$2D = 7 \quad\quad \text{[Eq. (7) minus four times Eq. (8)]} \tag{9}$$

We solve Eq. (9) for D, obtaining $D = \frac{7}{2}$. Substituting the values for D and C into (5) and solving for B, we get $B = -\frac{3}{2}$. Substituting the value for C into Eq. (1) and solving for A, we get $A = 0$. Thus, $A = 0$, $B = -\frac{3}{2}$, $C = 1$, and $D = \frac{7}{2}$, and returning to our original equation we have

$$\frac{2x^3 + 9x}{(x^2 + 3)(x^2 - 2x + 3)} = \frac{-\frac{3}{2}}{x^2 + 3} + \frac{(2x - 2) + \frac{7}{2}}{x^2 - 2x + 3}$$

Thus,

$$\int \frac{(2x^3 + 9x)\,dx}{(x^2 + 3)(x^2 - 2x + 3)} = -\frac{3}{2} \int \frac{dx}{x^2 + 3} + \int \frac{(2x - 2)\,dx}{x^2 - 2x + 3} + \frac{7}{2} \int \frac{dx}{(x - 1)^2 + 2}$$

$$= -\frac{3}{2} \cdot \frac{1}{\sqrt{3}} \tan^{-1}\!\left(\frac{x}{\sqrt{3}}\right) + \ln(x^2 - 2x + 3)$$

$$+ \frac{7}{2} \cdot \frac{1}{\sqrt{2}} \tan^{-1}\!\left(\frac{x - 1}{\sqrt{2}}\right) + C$$

$$= -\frac{1}{2}\sqrt{3}\, \tan^{-1}\!\left(\frac{x}{\sqrt{3}}\right) + \ln(x^2 - 2x + 3)$$

$$+ \frac{7}{4}\sqrt{2}\, \tan^{-1}\!\left(\frac{x - 1}{\sqrt{2}}\right) + C$$

16. $\displaystyle \int \frac{e^{5x}\,dx}{(e^{2x} + 1)^2}$

SOLUTION: Let $u = e^x$. Then $du = e^x\,dx$. Thus,

$$\int \frac{e^{5x}\,dx}{(e^{2x} + 1)^2} = \int \frac{u^4\,du}{(u^2 + 1)^2} \tag{1}$$

Because the fraction is improper, we divide the numerator by the denominator. Thus, from (1) we get

$$\int \frac{e^{5x}\, dx}{(e^{2x}+1)^2} = \int du - \int \frac{(2u^2+1)\, du}{(u^2+1)^2} \tag{2}$$

We find constants A, B, C, and D such that

$$\frac{2u^2+1}{(u^2+1)^2} = \frac{Au+B}{(u^2+1)^2} + \frac{Cu+D}{u^2+1} \tag{3}$$

$$2u^2+1 = Au+B+(Cu+D)(u^2+1)$$

$$= Cu^3 + Du^2 + (A+C)u + (B+D)$$

Equating coefficients of like powers of u, we obtain

$$A=0 \qquad B=-1 \qquad C=0 \qquad D=2$$

Thus, from (3) we have

$$\frac{2u^2+1}{(u^2+1)^2} = -\frac{1}{(u^2+1)^2} + \frac{2}{u^2+1}$$

Substituting this into (2), we obtain

$$\int \frac{e^{5x}\, dx}{(e^{2x}+1)^2} = \int du + \int \frac{du}{(u^2+1)^2} - 2\int \frac{du}{u^2+1}$$

$$= u + \int \frac{du}{(u^2+1)^2} - 2\tan^{-1} u \tag{4}$$

For the remaining integral on the right-hand side of (4), we let $\theta = \tan^{-1} u$. Then

$$u = \tan\theta \tag{5}$$

$$(u^2+1)^2 = \sec^4\theta \tag{6}$$

$$du = \sec^2\theta\, d\theta$$

Thus,

$$\int \frac{du}{(u^2+1)^2} = \int \frac{\sec^2\theta\, d\theta}{\sec^4\theta}$$

$$= \int \cos^2\theta\, d\theta$$

$$= \frac{1}{2}\int (1+\cos 2\theta)\, d\theta$$

$$= \frac{1}{2}\left(\theta + \frac{1}{2}\sin 2\theta\right) + C$$

$$= \frac{1}{2}(\theta + \sin\theta\, \cos\theta) + C \tag{7}$$

From (6) we get $\sec\theta = \sqrt{u^2+1}$. Thus,

$$\cos\theta = \frac{1}{\sqrt{u^2+1}}$$

Furthermore, from (5) we have

$$\sin\theta = \tan\theta\, \cos\theta = \frac{u}{\sqrt{u^2+1}}$$

Substituting into (7), we obtain

$$\int \frac{du}{(u^2+1)^2} = \frac{1}{2}\left[\tan^{-1} u + \frac{u}{(u^2+1)}\right] + C \tag{8}$$

Substituting from (8) into (4), we obtain

$$\int \frac{e^{5x}\, dx}{(e^{2x}+1)^2} = u + \frac{1}{2}\tan^{-1} u + \frac{u}{2(u^2+1)} - 2\tan^{-1} u + C$$

$$= e^x - \frac{3}{2}\tan^{-1} e^x + \frac{e^x}{2(e^{2x}+1)} + C$$

20. $\displaystyle\int \frac{(6w^4 + 4w^3 + 9w^2 + 24w + 32)\, dw}{(w^3 + 8)(w^2 + 3)}$

SOLUTION: Because $w^3 + 8 = (w+2)(w^2 - 2w + 4)$, we let

$$\frac{6w^4 + 4w^3 + 9w^2 + 24w + 32}{(w^3+8)(w^2+3)} = \frac{A}{w+2} + \frac{Bw+C}{w^2+3} + \frac{D(2w-2)+E}{w^2-2w+4} \tag{1}$$

Eliminating fractions, we have

$$6w^4 + 4w^3 + 9w^2 + 24w + 32 = A(w^2+3)(w^2-2w+4)$$
$$+ (Bw+C)(w+2)(w^2-2w+4)$$
$$+ [D(2w-2)+E](w+2)(w^2+3)$$

If $w = -2$, then we have $84 = A(7)(12)$, and so $A = 1$. Replacing A with 1 and multiplying out, we obtain

$$6w^4 + 4w^3 + 9w^2 + 24w + 32 = (w^4 - 2w^3 + 7w^2 - 6w + 12)$$
$$+ (Bw^4 + Cw^3 + 8Bw + 8C)$$
$$+ (2Dw^4 + 4Dw^3 + 6Dw^2 + 12Dw - 2Dw^3$$
$$- 4Dw^2 - 6Dw - 12D + Ew^3 + 2Ew^2$$
$$+ 3Ew + 6E)$$
$$= (1 + B + 2D)w^4 + (-2 + C + 2D + E)w^3$$
$$+ (7 + 2D + 2E)w^2 + (-6 + 8B + 6D + 3E)w$$
$$+ 12 + 8C - 12D + 6E$$

Equating like coefficients of w, we get the system

$$\begin{array}{llll}
6 = 1 + B + 2D & \Longleftrightarrow & B + 2D & = 5 \\
4 = -2 + C + 2D + E & \Longleftrightarrow & C + 2D + E = 6 \\
9 = 7 + 2D + 2E & \Longleftrightarrow & D + E = 1 \\
24 = -6 + 8B + 6D + 3E & \Longleftrightarrow & 8B + 6D + 3E = 30 \\
32 = 12 + 8C - 12D + 6E & \Longleftrightarrow & 4C - 6D + 3E = 10
\end{array}$$

The system may be solved by the method of Exercise 12. The resulting solution is $B = 3$, $C = 4$, $D = 1$, and $E = 0$. Substituting for A, B, C, D, and E into Eq. (1), we have

$$\int \frac{(6w^4 + 4w^3 + 9w^2 + 24w + 32)\, dw}{(w^3+8)(w^2+3)} = \int \frac{dw}{w+2} + \int \frac{(3w+4)\, dw}{w^2+3} + \int \frac{(2w-2)\, dw}{w^2-2w+4}$$

$$= \ln|w+2| + \frac{3}{2}\ln(w^2+3) + \frac{4}{\sqrt{3}}\tan^{-1}\!\left(\frac{w}{\sqrt{3}}\right) + \ln(w^2 - 2w + 4) + C$$

$$= \ln|(w+2)(w^2-2w+4)(w^2+3)^{3/2}| + \frac{4}{\sqrt{3}}\tan^{-1}\!\left(\frac{w}{\sqrt{3}}\right) + C$$

$$= \ln|(w^3-8)(w^2+3)^{3/2}| + \frac{4}{\sqrt{3}}\tan^{-1}\!\left(\frac{w}{\sqrt{3}}\right) + C$$

In Exercises 21–29 evaluate the definite integral.

24. $\displaystyle\int_0^1 \frac{9\, dx}{8x^3 + 1}$

SOLUTION:

$$\frac{9}{8x^3+1} = \frac{9}{(2x+1)(4x^2-2x+1)}$$

$$= \frac{A}{2x+1} + \frac{B(8x-2)+C}{4x^2-2x+1} \tag{1}$$

Thus,

$$9 = A(4x^2-2x+1) + [B(8x-2)+C](2x+1)$$
$$= 4Ax^2 - 2Ax + A + 16Bx^2 - 4Bx + 2Cx + 8Bx - 2B + C$$
$$= (4A+16B)x^2 + (-2A+4B+2C)x + (A-2B+C)$$

Equating coefficients of like powers of x, we obtain

$$\begin{aligned} 4A + 16B &= 0 \\ -2A + 4B + 2C &= 0 \\ A - 2B + C &= 9 \end{aligned}$$

Solving this system as in Exercise 12, we obtain

$$A = 3 \qquad B = -\tfrac{3}{4} \qquad C = \tfrac{9}{2}$$

Thus, from (1) we have

$$\frac{9}{8x^3+1} = \frac{3}{2x+1} + \frac{-\tfrac{3}{4}(8x-2)+\tfrac{9}{2}}{4x^2-2x+1}$$

Then,

$$\int_0^1 \frac{9\,dx}{8x^3+1} = 3\int_0^1 \frac{dx}{2x+1} - \frac{3}{4}\int_0^1 \frac{(8x-2)\,dx}{4x^2-2x+1} + \frac{9}{2}\int_0^1 \frac{dx}{4x^2-2x+1}$$

$$= \frac{3}{2}\left[\ln(2x+1)\right]_0^1 - \frac{3}{4}\left[\ln(4x^2-2x+1)\right]_0^1 + \frac{9}{8}\int_0^1 \frac{dx}{(x-\tfrac{1}{4})^2 + \tfrac{3}{16}}$$

$$= \frac{3}{2}\ln 3 - \frac{3}{4}\ln 3 + \frac{9}{8}\cdot\frac{4}{\sqrt{3}}\left[\tan^{-1}\frac{4}{\sqrt{3}}\left(x-\frac{1}{4}\right)\right]_0^1$$

$$= \frac{3}{4}\ln 3 + \frac{3}{2}\sqrt{3}\left[\tan^{-1}\sqrt{3} - \tan^{-1}\left(-\frac{1}{3}\sqrt{3}\right)\right]$$

$$= \frac{3}{4}\ln 3 + \frac{3}{2}\sqrt{3}\left[\frac{1}{3}\pi + \frac{1}{6}\pi\right]$$

$$= \frac{3}{4}(\ln 3 + \sqrt{3}\,\pi)$$

28. $\displaystyle\int_{\pi/6}^{\pi/2} \frac{\cos x\,dx}{\sin x + \sin^3 x}$

SOLUTION: First, we let $u = \sin x$. Then $du = \cos x\,dx$; when $x = \tfrac{1}{6}\pi$, $u = \tfrac{1}{2}$; when $x = \tfrac{1}{2}\pi$, $u = 1$. Thus

$$\int_{\pi/6}^{\pi/2} \frac{\cos x\,dx}{\sin x + \sin^3 x} = \int_{1/2}^1 \frac{du}{u(u^2+1)} \tag{1}$$

We let

$$\frac{1}{u(u^2+1)} = \frac{A}{u} + \frac{Bu+C}{u^2+1}$$

$$1 = (A+B)u^2 + Cu + A$$

Thus, $A = 1$, $B = -1$, and $C = 0$, and we have

$$\frac{1}{u(u^2 + 1)} = \frac{1}{u} - \frac{u}{u^2 + 1} \tag{2}$$

From (1) and (2) we get

$$\int_{\pi/6}^{\pi/2} \frac{\cos x \, dx}{\sin x + \sin^3 x} = \int_{1/2}^{1} \frac{du}{u} - \int_{1/2}^{1} \frac{u \, du}{u^2 + 1}$$

$$= \ln u \Big]_{1/2}^{1} - \frac{1}{2}\Big[\ln(u^2 + 1)\Big]_{1/2}^{1}$$

$$= \ln 2 - \frac{1}{2}\Big[\ln 2 - \ln \frac{5}{4}\Big] = \frac{1}{2}\ln\frac{5}{2}$$

32. Find the volume of the solid of revolution generated by revolving the region bounded by the curve $y(x^3 + 8) = 4$, the x-axis, the y-axis, and the line $x = 1$ about the y-axis.

SOLUTION: Figure 10. 5. 32 shows a sketch of the region. Let f be the function defined by

$$f(x) = \frac{4}{x^3 + 8}$$

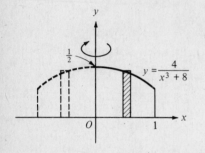

Figure 10. 5. 32

The element of volume is a cylindrical shell with units of thickness given by $\Delta_i x$ units of mean radius given by $\bar{x}_i$, and altitude $f(\bar{x}_i)$ units. Thus,

$$V = \lim_{\|\Delta\| \to 0} \sum_{i=1}^{n} 2\pi \bar{x}_i f(\bar{x}_i) \, \Delta_i x$$

$$= 8\pi \int_0^1 \frac{x \, dx}{x^3 + 8} \tag{1}$$

We let

$$\frac{x}{x^3 + 8} = \frac{x}{(x + 2)(x^2 - 2x + 4)}$$

$$= \frac{A}{x + 2} + \frac{B(2x - 2) + C}{x^2 - 2x + 4} \tag{2}$$

Eliminating the fractions, we have

$$x = A(x^2 - 2x + 4) + [B(2x - 2) + C](x + 2)$$
$$= Ax^2 - 2Ax + 4A + 2Bx^2 - 2Bx + Cx + 4Bx - 4B + 2C$$
$$= (A + 2B)x^2 + (-2A + 2B + C)x + (4A - 4B + 2C)$$

Equating coefficients of like powers of x, we obtain

$$A + 2B \qquad = 0$$
$$-2A + 2B + \quad C = 1$$
$$4A - 4B + 2C = 0$$

Solving this system, we obtain $A = -\frac{1}{6}$, $B = \frac{1}{12}$, and $C = \frac{1}{2}$. From (2) we get

$$\frac{x}{x^3 + 8} = \frac{-\frac{1}{6}}{x + 2} + \frac{\frac{1}{12}(2x - 2) + \frac{1}{2}}{x^2 - 2x + 4}$$

Thus,

$$\int_0^1 \frac{x \, dx}{x^3 + 8} = -\frac{1}{6}\int_0^1 \frac{dx}{x + 2} + \frac{1}{12}\int_0^1 \frac{(2x - 2) \, dx}{x^2 - 2x + 4} + \frac{1}{2}\int_0^1 \frac{dx}{(x - 1)^2 + 3}$$

$$= -\frac{1}{6}\Big[\ln(x+2)\Big]_0^1 + \frac{1}{12}\Big[\ln(x^2-2x+4)\Big]_0^1 + \frac{1}{6}\sqrt{3}\left[\tan^{-1}\left(\frac{x-1}{\sqrt{3}}\right)\right]_0^1$$

$$= -\frac{1}{6}(\ln 3 - \ln 2) + \frac{1}{12}(\ln 3 - \ln 4) + \frac{\sqrt{3}}{6}\left[\tan^{-1}0 - \tan^{-1}\left(\frac{-1}{\sqrt{3}}\right)\right]$$

$$= -\frac{1}{12}\ln 3 + \frac{1}{6}\ln 2 - \frac{1}{12}\ln 2^2 + \frac{\sqrt{3}}{6}\left(\frac{\pi}{6}\right)$$

$$= -\frac{1}{12}\ln 3 + \frac{\sqrt{3}\pi}{36} \tag{3}$$

Substituting from (3) into (1), we obtain

$$V = -\tfrac{2}{3}\pi \ln 3 + \tfrac{2}{9}\sqrt{3}\,\pi^2$$

Thus, the volume of the solid of revolution is $-\tfrac{2}{3}\pi \ln 3 + \tfrac{2}{9}\sqrt{3}\,\pi^2$ cubic units.

10.6 INTEGRALS YIELDING INVERSE HYPERBOLIC FUNCTIONS

We have the following integration formulas

10.6.4 Theorem

$$\int \frac{du}{\sqrt{u^2+a^2}} = \sinh^{-1}\frac{u}{a} + C = \ln(u + \sqrt{u^2+a^2}) + C \qquad \text{if } a > 0$$

$$\int \frac{du}{\sqrt{u^2-a^2}} = \cosh^{-1}\frac{u}{a} + C = \ln(u + \sqrt{u^2-a^2}) + C \qquad \text{if } u > a > 0$$

$$\int \frac{du}{a^2-u^2} = \begin{cases} \dfrac{1}{a}\tanh^{-1}\dfrac{u}{a} + C & \text{if } |u| < a \\[2mm] \dfrac{1}{a}\coth^{-1}\dfrac{u}{a} + C & \text{if } |u| > a \end{cases} = \frac{1}{2a}\ln\left|\frac{a+u}{a-u}\right| + C \qquad \begin{array}{l}\text{if } u \neq a \\ \text{and } a \neq 0\end{array}$$

Exercises 10.6

In Exercises 1–16 express the indefinite integral in terms of an inverse hyperbolic function and as a natural logarithm.

4. $\displaystyle\int \frac{dx}{25 - x^2}$

SOLUTION:

$$\int \frac{dx}{25-x^2} = \begin{cases} \dfrac{1}{5}\tanh^{-1}\dfrac{x}{5} + C & \text{if } |x| < 5 \\[2mm] \dfrac{1}{5}\coth^{-1}\dfrac{x}{5} + C & \text{if } |x| > 5 \end{cases}$$

$$= \frac{1}{10}\ln\left|\frac{5+x}{5-x}\right| + C$$

8. $\displaystyle\int \frac{dx}{\sqrt{x^2 - 4x + 1}}$

SOLUTION:

$$\int \frac{dx}{\sqrt{x^2-4x+1}} = \int \frac{dx}{\sqrt{(x-2)^2 - 3}}$$

$$= \cosh^{-1}\frac{x-2}{\sqrt{3}} + C$$

$$= \ln(x - 2 + \sqrt{x^2 - 4x + 1}) + C$$

12. $\displaystyle\int \frac{dx}{\sqrt{25x^2 + 9}}$

SOLUTION: Let $u = 5x$ and $du = 5\,dx$. Then

$$\int \frac{dx}{\sqrt{25x^2 + 9}} = \frac{1}{5}\int \frac{du}{\sqrt{u^2 + 9}}$$

$$= \frac{1}{5}\sinh^{-1}\left(\frac{u}{3}\right) + C$$

$$= \frac{1}{5}\sinh^{-1}\left(\frac{5x}{3}\right) + C$$

$$= \frac{1}{5}\ln(5x + \sqrt{25x^2 + 9}) + C$$

16. $\displaystyle\int \frac{3x\,dx}{\sqrt{x^4 + 6x^2 + 5}}$

SOLUTION: First, we complete the square under the radical sign.

$$\int \frac{3x\,dx}{\sqrt{x^4 + 6x^2 + 5}} = \int \frac{3x\,dx}{\sqrt{(x^2 + 3)^2 - 4}}$$

Then we let $u = x^2 + 3$ and $du = 2x\,dx$. Thus,

$$\int \frac{3x\,dx}{\sqrt{x^4 + 6x^2 + 5}} = \frac{3}{2}\int \frac{du}{\sqrt{u^2 - 4}}$$

$$= \frac{3}{2}\cosh^{-1}\left(\frac{u}{2}\right) + C$$

$$= \frac{3}{2}\cosh^{-1}\left(\frac{x^2 + 3}{2}\right) + C$$

$$= \frac{3}{2}\ln(x^2 + 3 + \sqrt{x^4 + 6x^2 + 5}) + C$$

In Exercises 17–22 evaluate the definite integral and express the answer in terms of a natural logarithm.

20. $\displaystyle\int_1^2 \frac{dx}{\sqrt{x^2 + 2x}}$

SOLUTION:

$$\int_1^2 \frac{dx}{\sqrt{x^2 + 2x}} = \int_1^2 \frac{dx}{\sqrt{(x + 1)^2 - 1}}$$

$$= \ln(x + 1 + \sqrt{x^2 + 2x})\Big]_1^2$$

$$= \ln(3 + 2\sqrt{2}) - \ln(2 + \sqrt{3})$$

$$= \ln\frac{3 + 2\sqrt{2}}{2 + \sqrt{3}}$$

$$= \ln\left(\frac{3 + 2\sqrt{2}}{2 + \sqrt{3}} \cdot \frac{2 - \sqrt{3}}{2 - \sqrt{3}}\right)$$

$$= \ln(6 + 4\sqrt{2} - 3\sqrt{3} - 2\sqrt{6})$$

24. A curve goes through the point $(0, a)$, $a > 0$, and the slope at any point is $\sqrt{y^2/a^2 - 1}$. Prove that the curve is a catenary.

SOLUTION: Because the slope is dy/dx, we are given that

$$\frac{dy}{dx} = \sqrt{\frac{y^2}{a^2} - 1}$$

$$= \frac{1}{a}\sqrt{y^2 - a^2}$$

Thus,

$$\frac{dy}{\sqrt{y^2 - a^2}} = \frac{1}{a}\,dx$$

$$\int \frac{dy}{\sqrt{y^2 - a^2}} = \int \frac{1}{a}\,dx$$

$$\cosh^{-1}\frac{y}{a} = \frac{x}{a} + C \tag{1}$$

Because the curve contains the point $(0, a)$, this point must satisfy (1). Thus

$$\cosh^{-1} 1 = C$$

Hence, $C = 0$, and from (1) we have

$$\cosh^{-1}\frac{y}{a} = \frac{x}{a}$$

$$\frac{y}{a} = \cosh\left(\frac{x}{a}\right)$$

$$y = a\cosh\left(\frac{x}{a}\right)$$

Thus, the curve is a catenary.

10.7 INTEGRATION OF RATIONAL FUNCTIONS OF SINE AND COSINE

If an integrand is a rational function of $\sin x$ and $\cos x$, it can always be reduced to a rational function of z by the substitution

$$z = \tan \tfrac{1}{2} x$$

It follows that

$$\cos x = \frac{1 - z^2}{1 + z^2} \qquad \sin x = \frac{2z}{1 + z^2} \qquad dx = \frac{2\,dz}{1 + z^2}$$

Sometimes we may be able to reduce the integrand to a rational function of u by a suitable choice of u. We illustrate this in Exercise 24.

Exercises 10.7

In Exercises 1–18 evaluate the indefinite integral.

4. $\displaystyle\int \frac{dx}{\sin x - \cos x + 2}$

SOLUTION: We let $z = \tan \frac{1}{2} x$. By using the substitutions indicated at the beginning of this section, we obtain

$$\int \frac{dx}{\sin x - \cos x + 2} = \int \frac{\dfrac{2\,dz}{1+z^2}}{\dfrac{2z}{1+z^2} - \dfrac{1-z^2}{1+z^2} + 2}$$

$$= \int \frac{2\,dz}{3z^2 + 2z + 1}$$

$$= \frac{2}{3} \int \frac{dz}{z + \frac{2}{3} z + \frac{1}{3}}$$

$$= \frac{2}{3} \int \frac{dz}{(z + \frac{1}{3})^2 + \frac{2}{9}}$$

$$= \frac{2}{3} \cdot \frac{3}{\sqrt{2}} \tan^{-1} \frac{z + \frac{1}{3}}{\frac{1}{3}\sqrt{2}} + C$$

$$= \sqrt{2} \tan^{-1} \frac{3 \tan \frac{1}{2} x + 1}{\sqrt{2}} + C$$

8. $\displaystyle \int \frac{dx}{\tan x - 1}$

SOLUTION: We let $z = \tan \frac{1}{2} x$. Because

$$\sin x = \frac{2z}{1+z^2} \quad \text{and} \quad \cos x = \frac{1-z^2}{1+z^2}$$

then

$$\tan x = \frac{\sin x}{\cos x} = \frac{2z}{1-z^2}$$

Thus,

$$\int \frac{dx}{\tan x - 1} = \int \frac{\dfrac{2\,dz}{1+z^2}}{\dfrac{2z}{1-z^2} - 1}$$

$$= \int \frac{2(1-z^2)\,dz}{(z^2+1)(z^2+2z-1)} \tag{1}$$

We use partial fractions to integrate (1).

$$\frac{2(1-z^2)}{(z^2+1)(z^2+2z-1)} = \frac{Az+B}{z^2+1} = \frac{C(2z+2)+D}{z^2+2z-1}$$

$$-2z^2 + 2 = (Az+B)(z^2+2z-1) + [C(2z+2)+D](z^2+1)$$

$$= (A+2C)z^3 + (2A+B+2C+D)z^2$$
$$+ (-A+2B+2C)z + (-B+2C+D)$$

Setting coefficients of like powers of z equal and solving, we obtain

$$A = -1 \qquad B = -1 \qquad C = \tfrac{1}{2} \qquad D = 0$$

Thus,

$$\frac{2(1-z^2)}{(z^2+1)(z^2+2z-1)} = \frac{-z-1}{z^2+1} + \frac{\frac{1}{2}(2z+2)}{z^2+2z-1}$$

From the above and (1) we obtain

$$\int \frac{dx}{\tan x - 1} = -\int \frac{z\,dz}{z^2 + 1} - \int \frac{dz}{z^2 + 1} + \frac{1}{2}\int \frac{(2z + 2)\,dz}{z^2 + 2z - 1}$$

$$= -\frac{1}{2}\ln(z^2 + 1) - \tan^{-1}z + \frac{1}{2}\ln|z^2 + 2z - 1| + C$$

$$= \frac{1}{2}\ln\left|\frac{z^2 + 2z - 1}{z^2 + 1}\right| - \tan^{-1}z + C$$

$$= \frac{1}{2}\ln\left|\frac{2z}{1 + z^2} - \frac{1 - z^2}{1 + z^2}\right| - \tan^{-1}\left[\tan\frac{1}{2}x\right] + C$$

$$= \frac{1}{2}\ln|\sin x - \cos x| - \frac{1}{2}x + C$$

12. $\displaystyle\int \frac{\cos x\,dx}{3\cos x - 5}$

SOLUTION: We let $z + \tan\frac{1}{2}x$. Then

$$\int \frac{\cos x\,dx}{3\cos x - 5} = \int \frac{\dfrac{1 - z^2}{1 + z^2}\cdot \dfrac{2\,dz}{1 + z^2}}{3\cdot \dfrac{1 - z^2}{1 + z^2} - 5}$$

$$= \int \frac{(z^2 - 1)\,dz}{(z^2 + 1)(4z^2 + 1)} \tag{1}$$

We find partial fractions. Let

$$\frac{z^2 - 1}{(z^2 + 1)(4z^2 + 1)} = \frac{Az + B}{z^2 + 1} + \frac{Cz + D}{4z^2 + 1} \tag{2}$$

$$z^2 - 1 = 4Az^3 + 4Bz^2 + Az + B + Cz^3 + Dz^2 + Cz + D$$

$$= (4A + C)z^3 + (4B + D)z^2 + (A + C)z + (B + D)$$

Thus,

$$\begin{aligned} 0 &= 4A + C \\ 0 &= A + C \end{aligned} \quad \text{and} \quad \begin{aligned} 1 &= 4B + D \\ -1 &= B + D \end{aligned}$$

Solving the system of equations, we obtain $A = 0$, $B = \frac{2}{3}$, $C = 0$, and $D = -\frac{5}{3}$. Substituting the constants into Eq. (2) and then substituting from Eq. (2) into Eq. (1), we obtain

$$\int \frac{\cos x\,dx}{3\cos x - 5} = \frac{2}{3}\int \frac{dz}{z^2 + 1} - \frac{5}{3}\int \frac{dz}{4z^2 + 1}$$

$$= \frac{2}{3}\tan^{-1}z - \frac{5}{6}\tan^{-1}(2z) + C$$

$$= \frac{2}{3}\tan^{-1}\left(\tan\frac{1}{2}x\right) - \frac{5}{6}\tan^{-1}\left(2\tan\frac{1}{2}x\right) + C$$

$$= \frac{1}{3}x - \frac{5}{6}\tan^{-1}\left(2\tan\frac{1}{2}x\right) + C$$

16. $\displaystyle\int \frac{dx}{\cot 2x\,(1 - \cos 2x)}$

SOLUTION: We let $z = \tan \frac{1}{2}(2x) = \tan x$. Then

$$\cos 2x = \frac{1-z^2}{1+z^2} \qquad \sin 2x = \frac{2z}{1+z^2} \qquad d(2x) = \frac{2\,dz}{1+z^2}$$

Thus,

$$\int \frac{dx}{\cot 2x\,(1-\cos 2x)} = \frac{1}{2}\int \frac{d(2x)}{\dfrac{\cos 2x}{\sin 2x}(1-\cos 2x)}$$

$$= \frac{1}{2}\int \frac{\dfrac{2\,dz}{1+z^2}}{\dfrac{1-z^2}{2z}\left(1-\dfrac{1-z^2}{1+z^2}\right)}$$

$$= \int \frac{-dz}{z(z+1)(z-1)} \tag{1}$$

We let

$$\frac{-1}{z(z+1)(z-1)} = \frac{A}{z} + \frac{B}{z+1} + \frac{C}{z-1}$$

$$-1 = A(z+1)(z-1) + Bz(z-1) + Cz(z+1)$$

If $z = 0$, we get $A = 1$. If $z = -1$, we get $B = -\frac{1}{2}$. If $z = 1$, we get $C = -\frac{1}{2}$. Thus,

$$\frac{-1}{z(z+1)(z-1)} = \frac{1}{z} - \frac{\frac{1}{2}}{z+1} - \frac{\frac{1}{2}}{z-1}$$

From the above and (1) we obtain

$$\int \frac{dx}{\cot 2x\,(1-\cos 2x)} = \int \frac{dz}{z} - \frac{1}{2}\int \frac{dz}{z+1} - \frac{1}{2}\int \frac{dz}{z-1}$$

$$= \ln|z| - \frac{1}{2}\ln|z+1| - \frac{1}{2}\ln|z-1| + C$$

$$= \frac{1}{2}\ln\left|\frac{z^2}{z^2-1}\right| + C$$

$$= \frac{1}{2}\ln\left|\frac{\tan^2 x}{\tan^2 x - 1}\right| + C$$

In Exercises 19–24 evaluate the definite integral.

20. $\displaystyle\int_0^{\pi/4} \frac{8\,dx}{\tan x + 1}$

SOLUTION: Let $z = \tan \frac{1}{2}x$. Then, as in Exercise 8, we have $\tan x = 2z/(1 - z^2)$. Thus, the indefinite integral is

$$\int \frac{8\,dx}{\tan x + 1} = 8\int \frac{\dfrac{2\,dz}{1+z^2}}{\dfrac{2z}{1-z^2} + 1}$$

$$= 16\int \frac{(z^2-1)\,dz}{(z^2+1)(z^2-2z-1)} \tag{1}$$

We use partial fractions to evaluate the indefinite integral. Let

$$\frac{z^2 - 1}{(z^2 + 1)(z^2 - 2z - 1)} = \frac{Az + B}{z^2 + 1} + \frac{C(2z - 2) + D}{z^2 - 2z - 1} \tag{2}$$

Then

$$\begin{aligned}
z^2 - 1 &= (Az + B)(z^2 - 2z - 1) + [C(2z - 2) + D](z^2 + 1) \\
&= (Az^3 - 2Az^2 - Az + Bz^2 - 2Bz - B) \\
&\quad + (2Cz^3 - 2Cz^2 + Dz^2 + 2Cz - 2C + D) \\
&= (A + 2C)z^3 + (-2A + B - 2C + D)z^2 \\
&\quad + (-A - 2B + 2C)z + (-B - 2C + D)
\end{aligned}$$

Equating coefficients of like powers of z, we have

$$\begin{aligned}
0 &= A + 2C \\
1 &= -2A + B - 2C + D \\
0 &= -A - 2B + 2C \\
-1 &= -B - 2C + D
\end{aligned}$$

Solving the system, we get $A = -\frac{1}{2}$, $B = \frac{1}{2}$, $C = \frac{1}{4}$, and $D = 0$. Substituting these constants into Eq. (2) and substituting the result into Eq. (1), we obtain

$$\int \frac{8\, dx}{\tan x - 1} = 16 \left[\int \frac{(-\frac{1}{2}z + \frac{1}{2})\, dz}{z^2 + 1} + \int \frac{\frac{1}{4}(2z - 2)\, dz}{z^2 - 2z - 1} \right]$$

$$= -4 \int \frac{2z\, dz}{z^2 + 1} + 8 \int \frac{dz}{z^2 + 1} + 4 \int \frac{(2z - 2)\, dz}{z^2 - 2z - 1}$$

$$= -4 \ln(z^2 + 1) + 8 \tan^{-1} z + 4 \ln|z^2 - 2z - 1| + C$$

$$= 4 \ln \left| \frac{z^2 - 2z - 1}{z^2 + 1} \right| + 8 \tan^{-1} z + C$$

Let

$$F(z) = 4 \ln \left| \frac{z^2 - 2z - 1}{z^2 + 1} \right| + 8 \tan^{-1} z \tag{3}$$

To evaluate the definite integral we note that when $x = 0$, then $z = \tan 0 = 0$. Furthermore, by substituting into Eq. (3),

$$\begin{aligned}
F(0) &= 4 \ln|-1| + 8 \tan^{-1} 0 \\
&= 0
\end{aligned}$$

When $x = \frac{1}{4}\pi$, then $z = \tan \frac{1}{8}\pi$. We apply the identity

$$\tan \frac{t}{2} = \frac{1 - \cos t}{\sin t}$$

with $t = \frac{1}{4}\pi$ to find the value of $\tan \frac{1}{8}\pi$. Thus,

$$\tan \frac{1}{8}\pi = \frac{1 - \cos \frac{1}{4}\pi}{\sin \frac{1}{4}\pi}$$

$$= \frac{1 - \dfrac{1}{\sqrt{2}}}{\dfrac{1}{\sqrt{2}}} = \sqrt{2} - 1$$

Thus, $z = \sqrt{2} - 1$, when $x = \frac{1}{4}\pi$. Because $z = \tan \frac{1}{8}\pi$ when $x = \frac{1}{4}\pi$, then we also have $\tan^{-1} z = \frac{1}{8}\pi$ when $x = \frac{1}{4}\pi$. Therefore, by substituting into Eq. (3) when $x = \frac{1}{4}\pi$, we have

$$F(\sqrt{2}-1) = 4\ln\left|\frac{(\sqrt{2}-1)^2 - 2(\sqrt{2}-1)-1}{(\sqrt{2}-1)^2+1}\right| + 8\left(\frac{1}{8}\pi\right)$$

$$= 4\ln\left|\frac{2-2\sqrt{2}+1-2\sqrt{2}+2-1}{2-2\sqrt{2}+1+1}\right| + \pi$$

$$= 4\ln\left|\frac{4(1-\sqrt{2})}{2(2-\sqrt{2})}\right| + \pi$$

$$= 4\ln\left|\frac{2(1-\sqrt{2})}{2-\sqrt{2}}\cdot\frac{2+\sqrt{2}}{2+\sqrt{2}}\right| + \pi$$

$$= 4\ln\left|\frac{2(2-2\sqrt{2}+\sqrt{2}-2)}{4-2}\right| + \pi$$

$$= 4\ln|-\sqrt{2}| + \pi$$

$$= 4\ln\sqrt{2} + \pi$$

$$= \ln(\sqrt{2})^4 + \pi$$

$$= \ln 4 + \pi$$

Therefore,

$$\int_0^{\pi/4}\frac{8\,dx}{\tan x + 1} = F(\sqrt{2}-1) - F(0)$$

$$= \ln 4 + \pi$$

24. Evaluate the definite integral

$$\int_0^{\pi/2}\frac{\sin 2x\,dx}{2+\cos x}$$

SOLUTION: The substitution $z = \tan\frac{1}{2}x$ is not required.

$$\int_0^{\pi/2}\frac{\sin 2x\,dx}{2+\cos x} = \int_0^{\pi/2}\frac{2\sin x\cos x\,dx}{2+\cos x} \qquad (1)$$

Let $u = 2+\cos x$. Then $du = -\sin x\,dx$; when $x=0$, $u=3$; when $x=\frac{1}{2}\pi$, $u=2$. With these replacements on the right-hand side of (1), we get

$$\int_0^{\pi/2}\frac{\sin 2x\,dx}{2+\cos x} = \int_3^2\frac{2(u-2)(-du)}{u}$$

$$= 2\int_2^3 du - 4\int_2^3\frac{du}{u}$$

$$= 2u\Big]_2^3 - 4\ln u\Big]_2^3$$

$$= 2 - 4\ln\frac{3}{2}$$

28. Evaluate the given integral by two methods: **(a)** Let $z=\tan\frac{1}{2}x$; **(b)** let $u=\frac{1}{2}x$ and obtain an integral involving trigonometric functions of u.

$$\int\frac{\tan\frac{1}{2}x}{\sin x}\,dx$$

SOLUTION:

(a) If $z = \tan\frac{1}{2}x$, then

$$\int \frac{\tan \frac{1}{2}x}{\sin x}\, dx = \int \frac{z \cdot \dfrac{2\, dz}{1+z^2}}{\dfrac{2z}{1+z^2}}$$

$$= \int dz$$

$$= z + C$$

$$= \tan \tfrac{1}{2}x + C$$

(b) If $u = \frac{1}{2}x$, then $x = 2u$; so $dx = 2\, du$ and $\sin x = \sin 2u$. Thus,

$$\int \frac{\tan \frac{1}{2}x}{\sin x}\, dx = \int \frac{\tan u(2\, du)}{\sin 2u}$$

$$= \int \frac{2\dfrac{\sin u}{\cos u}\, du}{2 \sin u \cos u}$$

$$= \int \sec^2 u\, du$$

$$= \tan u + C$$

$$= \tan \tfrac{1}{2}x + C$$

10.8 MISCELLANEOUS SUBSTITUTIONS

Exercises 10.8

In Exercises 1–13 evaluate the indefinite integral.

4. $\displaystyle\int x(1+x)^{2/3}\, dx$

SOLUTION: Let $u = (1+x)^{1/3}$. Then $u^3 = 1 + x$, $dx = 3u^2\, du$, $(1+x)^{2/3} = u^2$, and $x = u^3 - 1$. Thus,

$$\int x(1+x)^{2/3}\, dx = \int (u^3 - 1)u^2(3u^2\, du)$$

$$= 3\int (u^7 - u^4)\, du$$

$$= \tfrac{3}{8}u^8 - \tfrac{3}{5}u^5 + C$$

$$= \tfrac{3}{8}(1+x)^{8/3} - \tfrac{3}{5}(1+x)^{5/3} + C$$

8. $\displaystyle\int \frac{dx}{2\sqrt[3]{x} + \sqrt{x}}$

SOLUTION: Because 6 is the least common denominator of the fractional exponents of the given powers of x, which are $x^{1/3}$ and $x^{1/2}$, we let $z = x^{1/6}$. Then $x = z^6$; $dx = 6z^5\, dz$; $\sqrt[3]{x} = z^2$; $\sqrt{x} = z^3$. Thus,

$$\int \frac{dx}{2\sqrt[3]{x} + \sqrt{x}} = \int \frac{6z^5 \, dz}{2z^2 + z^3}$$

$$= \int \frac{6z^3 \, dz}{2 + z}$$

$$= \int (6z^2 - 12z + 24) \, dz - \int \frac{48 \, dz}{z + 2}$$

$$= 2z^3 - 6z^2 + 24z - 48 \ln|z + 2| + C$$

$$= 2x^{1/2} - 6x^{1/3} + 24x^{1/6} - 48 \ln(x^{1/6} + 2) + C$$

12. $\displaystyle \int \frac{dx}{\sqrt{\sqrt{x} + 1}}$

SOLUTION: Let $u = \sqrt{\sqrt{x} + 1}$. Then

$$u^2 = \sqrt{x} + 1$$
$$u^2 - 1 = \sqrt{x}$$
$$u^4 - 2u^2 + 1 = x$$
$$(4u^3 - 4u) \, du = dx$$

Thus,

$$\int \frac{dx}{\sqrt{\sqrt{x} + 1}} = \int \frac{(4u^3 - 4u) \, du}{u}$$

$$= \int (4u^2 - 4) \, du$$

$$= \tfrac{4}{3} u^3 - 4u + C$$

$$= \tfrac{4}{3} u(u^2 - 3) + C$$

$$= \tfrac{4}{3}\sqrt{\sqrt{x} + 1}(\sqrt{x} - 2) + C$$

In Exercises 14–19 evaluate the indefinite integral by using the reciprocal substitution $x = 1/z$.

16. $\displaystyle \int \frac{dx}{x^2\sqrt{1 + 2x + 3x^2}}$

SOLUTION: Let $x = z^{-1}$. Then $dx = -z^{-2} \, dz$. Thus,

$$\int \frac{dx}{x^2\sqrt{1 + 2x + 3x^2}} = \int \frac{-z^{-2} \, dz}{z^{-2}\sqrt{1 + 2z^{-1} + 3z^{-2}}}$$

$$= \int \frac{-dz}{\sqrt{z^{-2}}\sqrt{z^2 + 2z + 3}}$$

$$= \int \frac{-|z| \, dz}{\sqrt{z^2 + 2z + 3}}$$

$$= \begin{cases} -\displaystyle\int \frac{z \, dz}{\sqrt{z^2 + 2z + 3}} & \text{if } z > 0 \\[2ex] \displaystyle\int \frac{z \, dz}{\sqrt{z^2 + 2z + 3}} & \text{if } z < 0 \end{cases} \qquad (1)$$

Now

$$\int \frac{z\,dz}{\sqrt{z^2+2z+3}} = \frac{1}{2}\int \frac{(2z+2)\,dz}{\sqrt{z^2+2z+3}} - \int \frac{dz}{\sqrt{(z+1)^2+2}}$$

$$= \sqrt{z^2+2z+3} - \int \frac{dz}{\sqrt{(z+1)^2+2}} \qquad (2)$$

For the integral remaining on the right-hand side of (2), we let

$$\theta = \tan^{-1}\frac{z+1}{\sqrt{2}}$$

$$z+1 = \sqrt{2}\tan\theta \qquad (3)$$

$$dz = \sqrt{2}\sec^2\theta\,d\theta$$

$$\sqrt{(z+1)^2+2} = \sqrt{2\tan^2\theta+2}$$

$$= \sqrt{2}\sec\theta \qquad (4)$$

Thus,

$$\int \frac{dz}{\sqrt{(z+1)^2+2}} = \int \frac{\sqrt{2}\sec^2\theta\,d\theta}{\sqrt{2}\sec\theta}$$

$$= \int \sec\theta\,d\theta$$

$$= \ln|\sec\theta+\tan\theta|+k \qquad (5)$$

Solving Eq. (3) for $\tan\theta$ and solving Eq. (4) for $\sec\theta$ and substituting the results into Eq. (5), we obtain

$$\int \frac{dz}{\sqrt{(z+1)^2+2}} = \ln\left|\frac{\sqrt{z^2+2z+3}}{\sqrt{2}}+\frac{z+1}{\sqrt{2}}\right|+k$$

$$= \ln|\sqrt{z^2+2z+3}+z+1|+k' \qquad (6)$$

Substituting from (6) into (2), we get

$$\int \frac{z\,dz}{\sqrt{z^2+2z+3}} = \sqrt{z^2+2z+3} - \ln|\sqrt{z^2+2z+3}+z+1|+k'$$

$$= \sqrt{x^{-2}+2x^{-1}+3} - \ln|\sqrt{x^{-2}+2x^{-1}+3}+x^{-1}+1|+k'$$

$$= \frac{\sqrt{1+2x+3x^2}}{|x|} - \ln\left|\frac{\sqrt{1+2x+3x^2}}{|x|}+\frac{1+x}{x}\right|+k' \qquad (7)$$

Substituting from (7) into (1), we obtain

$$\int \frac{dx}{x^2\sqrt{1+2x+3x^2}} =
\begin{cases}
-\dfrac{\sqrt{1+2x+3x^2}}{|x|} + \ln\left|\dfrac{\sqrt{1+2x+3x^2}}{|x|}+\dfrac{1+x}{x}\right|+k' \\
\qquad\qquad\qquad \text{if } x>0 \\[6pt]
\dfrac{\sqrt{1+2x+3x^2}}{|x|} - \ln\left|\dfrac{\sqrt{1+2x+3x^2}}{|x|}+\dfrac{1+x}{x}\right|+k' \\
\qquad\qquad\qquad \text{if } x<0
\end{cases}$$

$$=
\begin{cases}
-\dfrac{\sqrt{1+2x+3x^2}}{x} + \ln\left|\dfrac{\sqrt{1+2x+3x^2}}{x}+\dfrac{1+x}{x}\right|+k' \\
\qquad\qquad\qquad \text{if } x>0 \qquad (8) \\[6pt]
-\dfrac{\sqrt{1+2x+3x^2}}{x} - \ln\left|\dfrac{-\sqrt{1+2x+3x^2}}{x}+\dfrac{1+x}{x}\right|+k' \\
\qquad\qquad\qquad \text{if } x<0
\end{cases}$$

We show that the two equations in (8) are equivalent. Because $|a| = |-a|$, we have

$$-\ln\left|\frac{-\sqrt{1+2x+3x^2}}{x}+\frac{1+x}{x}\right| = -\ln\left|\frac{\sqrt{1+2x+3x^2}-(1+x)}{x}\right|$$

$$= \ln\left|\frac{x}{\sqrt{1+2x+3x^2}-(1+x)}\right|$$

$$= \ln\left|\frac{x[\sqrt{1+2x+3x^2}+(1+x)]}{[\sqrt{1+2x+3x^2}-(1+x)][\sqrt{1+2x+3x^2}+(1+x)]}\right|$$

$$= \ln\left|\frac{x[\sqrt{1+2x+3x^2}+1+x]}{2x^2}\right|$$

$$= \ln\left|\frac{\sqrt{1+2x+3x^2}+1+x}{x}\right| - \ln 2 \qquad (9)$$

Substituting from (9) into the second equation in (8), we get

$$\int \frac{dx}{x^2\sqrt{1+2x+3x^2}} = \begin{cases} -\dfrac{\sqrt{1+2x+3x^2}}{x}+\ln\left|\dfrac{\sqrt{1+2x+3x^2}+1+x}{x}\right|+k' \\ \qquad\qquad \text{if } x>0 \\ -\dfrac{\sqrt{1+2x+3x^2}}{x}+\ln\left|\dfrac{\sqrt{1+2x+3x^2}+1+x}{x}\right| - \ln 2 + k' \\ \qquad\qquad \text{if } x<0 \end{cases}$$

$$= -\frac{\sqrt{1+2x+3x^2}}{x}+\ln\left|\frac{\sqrt{1+2x+3x^2}+1+x}{x}\right|+C$$

20. Use the substitution $\sqrt{x^2+2x-1} = z-x$ to evaluate the given integral.

$$\int \frac{dx}{x\sqrt{x^2+2x-1}}$$

SOLUTION: If $\sqrt{x^2+2x-1} = z-x$, then taking the differential of both sides, we have

$$\tfrac{1}{2}(x^2+2x-1)^{-1/2}(2x+2)\,dx = dz-dx$$

$$\frac{(x+1)\,dx}{\sqrt{x^2+2x-1}} = dz-dx$$

$$\frac{(x+1)\,dx}{z-x} = dz-dx$$

$$\left(\frac{x+1}{z-x}+1\right)dx = dz$$

$$\frac{1+z}{z-x}\,dx = dz$$

$$\frac{dx}{z-x} = \frac{dz}{1+z} \qquad (1)$$

Furthermore, by squaring both sides of the given substitution we obtain

$$x^2+2x-1 = z^2-2zx+x^2$$

$$2x(1+z) = 1+z^2$$

$$x = \frac{1}{2}\left(\frac{1+z^2}{1+z}\right) \qquad (2)$$

Thus, with the given substitution we may apply (1) and (2) to obtain

$$\int \frac{dx}{x\sqrt{x^2 + 2x - 1}} = \int \frac{dx}{x(z - x)}$$

$$= \int \frac{1}{x} \left(\frac{dx}{z - x} \right)$$

$$= \int \frac{2(1 + z)}{1 + z^2} \left(\frac{dz}{1 + z} \right)$$

$$= 2 \int \frac{dz}{1 + z^2}$$

$$= 2 \tan^{-1} z + C$$

$$= 2 \tan^{-1}(x + \sqrt{x^2 + 2x - 1}) + C$$

In Exercises 21–28 evaluate the definite integral.

24. $\int_{16}^{18} \frac{dx}{\sqrt{x} - \sqrt[4]{x^3}}$

SOLUTION: Let $u = x^{1/4}$. Then $\sqrt{x} = u^2$, $\sqrt[4]{x^3} = u^3$, $x = u^4$, $dx = 4u^3\, du$. Furthermore, when $x = 16$, $u = 2$; when $x = 18$, $u = \sqrt[4]{18}$. Thus,

$$\int_{16}^{18} \frac{dx}{\sqrt{x} - \sqrt[4]{x^3}} = \int_{2}^{\sqrt[4]{18}} \frac{4u^3\, du}{u^2 - u^3}$$

$$= -4 \int_{2}^{\sqrt[4]{18}} \frac{u\, du}{u - 1}$$

$$= -4 \left[\int_{2}^{\sqrt[4]{18}} du + \int_{2}^{\sqrt[4]{18}} \frac{du}{u - 1} \right]$$

$$= -4 \left[u \right]_{2}^{\sqrt[4]{18}} - 4 \left[\ln(u - 1) \right]_{2}^{\sqrt[4]{18}}$$

$$= -4(\sqrt[4]{18} - 2) - 4 \ln(\sqrt[4]{18} - 1)$$

$$= 8 - 4\sqrt[4]{18} - 4 \ln(\sqrt[4]{18} - 1)$$

28. $\int_{2}^{11} \frac{x^3\, dx}{\sqrt[3]{x^2 + 4}}$

SOLUTION: Let $u = \sqrt[3]{x^2 + 4}$. Then $u^3 = x^2 + 4$, so $3u^2\, du = 2x\, dx$. Hence, $x\, dx = \frac{3}{2}u^2\, du$ and $x^2 = u^3 - 4$. Furthermore, when $x = 2$, then $u = 2$; when $x = 11$, then $u = 5$. Thus,

$$\int_{2}^{11} \frac{x^3 dx}{\sqrt[3]{x^2 + 4}} = \int_{2}^{5} \frac{(u^3 - 4)(\frac{3}{2}u^2\, du)}{u}$$

$$= \frac{3}{2} \int_{2}^{5} (u^4 - 4u)\, du$$

$$= \frac{3}{2} \left[\frac{1}{5} u^5 - 2u^2 \right]_{2}^{5}$$

$$= \frac{3}{2} \left[(625 - 50) - \left(\frac{32}{5} - 8 \right) \right] = \frac{8,649}{10}$$

Review Exercises for Chapter 10 and Review of Integration

In Exercises 1–62 evaluate the indefinite integral.

4. $\dfrac{dx}{x^2\sqrt{a^2 + x^2}}$

SOLUTION: Let $\theta = \tan^{-1}(x/a)$. Then

$$x = a\tan\theta \tag{1}$$
$$dx = a\sec^2\theta\,d\theta$$
$$\sqrt{a^2 + x^2} = a\sec\theta \tag{2}$$

Thus,

$$\int\frac{dx}{x^2\sqrt{a^2 + x^2}} = \int\frac{a\sec^2\theta\,d\theta}{(a^2\tan^2\theta)(a\sec\theta)}$$

$$= \frac{1}{a^2}\int\cot^2\theta\sec\theta\,d\theta$$

$$= \frac{1}{a^2}\int\cot\theta\csc\theta\,d\theta$$

$$= -\frac{\csc\theta}{a^2} + C \tag{3}$$

Because

$$\frac{\sec\theta}{\tan\theta} = \frac{\dfrac{1}{\cos\theta}}{\dfrac{\sin\theta}{\cos\theta}} = \frac{1}{\sin\theta} = \csc\theta$$

by substitution from Eqs. (1) and (2) we have

$$\csc\theta = \frac{\sec\theta}{\tan\theta}$$

$$= \frac{a\sec\theta}{a\tan\theta}$$

$$= \frac{\sqrt{a^2 + x^2}}{x} \tag{4}$$

And by substitution from (4) into (3) we get

$$\int\frac{dx}{x^2\sqrt{a^2 + x^2}} = -\frac{\sqrt{a^2 + x^2}}{a^2x} + C$$

8. $\displaystyle\int\frac{\sqrt{x+1}+1}{\sqrt{x+1}-1}\,dx$

SOLUTION: Let $u = \sqrt{x+1}$. Then $u^2 = x + 1$, and $dx = 2u\,du$. Hence,

$$\int\frac{\sqrt{x+1}+1}{\sqrt{x+1}-1}\,dx = \int\frac{(u+1)(2u\,du)}{u-1}$$

$$= 2\int(u+2)\,du + 4\int\frac{du}{u-1}$$

$$= u^2 + 4u + 4\ln|u-1| + \overline{C}$$

$$= (x+1) + 4\sqrt{x+1} + 4\ln|\sqrt{x+1} - 1| + \overline{C}$$
$$= x + 4\sqrt{x+1} + 4\ln|\sqrt{x+1} - 1| + C$$

where $C = 1 + \overline{C}$.

12. $\int \cos\theta \cos 2\theta \, d\theta$

SOLUTION: We use the identity

$$\cos a \cos b = \tfrac{1}{2}[\cos(a+b) + \cos(a-b)]$$

Thus,

$$\int \cos\theta \cos 2\theta \, d\theta = \frac{1}{2}\int \cos 3\theta \, d\theta + \frac{1}{2}\int \cos\theta \, d\theta$$

$$= \frac{1}{6}\sin 3\theta + \frac{1}{2}\sin\theta + C$$

16. $\int \dfrac{dx}{\sqrt{e^x - 1}}$

SOLUTION: Let $u = \sqrt{e^x - 1}$. Then

$$u^2 = e^x - 1$$
$$2u\,du = e^x\,dx$$
$$2u\,du = (u^2 + 1)\,dx$$
$$\frac{2u\,du}{u^2 + 1} = dx$$

Thus,

$$\int \frac{dx}{\sqrt{e^x - 1}} = \int \frac{2u\,du}{(u^2 + 1)u}$$

$$= 2\int \frac{du}{1 + u^2}$$

$$= 2\tan^{-1}u + C$$

$$= 2\tan^{-1}\sqrt{e^x - 1} + C$$

20. $\int \dfrac{\sqrt{x^2 - 4}}{x^2}\,dx$

SOLUTION: Let $\theta = \sec^{-1}\frac{1}{2}x$. Then $x = 2\sec\theta$, $dx = 2\sec\theta\tan\theta\,d\theta$, and $\sqrt{x^2 - 4} = 2\tan\theta$. Thus,

$$\int \frac{\sqrt{x^2 - 4}}{x^2}\,dx = \int \frac{(2\tan\theta)(2\sec\theta\tan\theta\,d\theta)}{4\sec^2\theta}$$

$$= \int \frac{\tan^2\theta\,d\theta}{\sec\theta}$$

$$= \int \frac{(\sec^2\theta - 1)\,d\theta}{\sec\theta}$$

$$= \int (\sec\theta - \cos\theta)\,d\theta$$

$$= \ln|\sec\theta + \tan\theta| - \sin\theta + \overline{C}$$

$$= \ln\left|\frac{x}{2} + \frac{\sqrt{x^2-4}}{2}\right| - \frac{\sqrt{x^2-4}}{x} + \overline{C}$$

$$= \ln|x + \sqrt{x^2-4}| - \frac{\sqrt{x^2-4}}{x} + C$$

where $C = \overline{C} - \ln 2$.

24. $\displaystyle\int \frac{4x^2 + x - 2}{x^3 - 5x^2 + 8x - 4}\,dx$

SOLUTION: We use synthetic division to factor the denominator. Thus, $x^3 - 5x^2 + 8x - 4 = (x-1)(x-2)^2$. Let

$$\frac{4x^2 + x - 2}{(x-1)(x-2)^2} = \frac{A}{x-1} + \frac{B}{x-2} + \frac{C}{(x-2)^2}$$

Then

$$4x^2 + x - 2 = A(x-2)^2 + B(x-1)(x-2) + C(x-1)$$

If $x = 1$, we get $3 = A$. If $x = 2$, we have $16 = C$. If $x = 0$, we obtain $-2 = 4A + 2B - C$, and because $A = 3$ and $C = 16$, then $-2 = 12 + 2B - 16$, so $B = 1$. Therefore,

$$\int \frac{4x^2 + x - 2}{x^3 - 5x^2 + 8x - 4}\,dx = \int \frac{3\,dx}{x-1} + \int \frac{dx}{x-2} + \int \frac{16\,dx}{(x-2)^2}$$

$$= 3\ln|x-1| + \ln|x-2| - 16(x-2)^{-1} + C$$

$$= \ln|(x-1)^3(x-2)| - \frac{16}{x-2} + C$$

28. $\displaystyle\int \frac{du}{u^{5/8} - u^{1/8}}$

SOLUTION: Let $z = u^{1/8}$. Then $u = z^8$; $du = 8z^7\,dz$. Thus,

$$\int \frac{du}{u^{5/8} - u^{1/8}} = \int \frac{8z^7\,dz}{z^5 - z}$$

$$= 8\int \frac{z^6\,dz}{z^4 - 1}$$

$$= 8\int z^2\,dz + 8\int \frac{z^2\,dz}{z^4 - 1} \qquad (1)$$

For the second integral on the right of (1), we use partial fractions

$$\frac{z^2}{z^4 - 1} = \frac{z^2}{(z^2+1)(z+1)(z-1)}$$

$$= \frac{Az + B}{z^2 + 1} + \frac{C}{z+1} + \frac{D}{z-1} \qquad (2)$$

$$z^2 = (Az + B)(z^2 - 1) + C(z-1)(z^2+1) + D(z+1)(z^2+1)$$

$$= (A + C + D)z^3 + (B - C + D)z^2 + (-A + C + D)z + (-B - C + D)$$

Equating coefficients of like powers of z, we get

$$A = 0 \qquad B = \tfrac{1}{2} \qquad C = -\tfrac{1}{4} \qquad D = \tfrac{1}{4}$$

Thus, from (2) we have

$$\int \frac{z^2 \, dz}{z^4 - 1} = \frac{1}{2} \int \frac{dz}{z^2 + 1} - \frac{1}{4} \int \frac{dz}{z + 1} + \frac{1}{4} \int \frac{dz}{z - 1}$$

$$= \tfrac{1}{2} \tan^{-1} z - \tfrac{1}{4} \ln|z + 1| + \tfrac{1}{4} \ln|z - 1| + C \qquad (3)$$

Substituting from (3) into (1), we obtain

$$\int \frac{du}{u^{5/8} - u^{1/8}} = \frac{8}{3} z^3 + 4 \tan^{-1} z - 2 \ln|z + 1| + 2 \ln|z - 1| + C$$

$$= \frac{8}{3} u^{3/8} + 4 \tan^{-1} u^{1/8} + 2 \ln \left| \frac{u^{1/8} - 1}{u^{1/8} + 1} \right| + C$$

32. $\displaystyle \int \frac{dx}{x \ln x (\ln x - 1)}$

SOLUTION: Let $u = \ln x$, and $du = dx/x$. Then

$$\int \frac{dx}{x \ln x (\ln x - 1)} = \int \frac{du}{u(u - 1)}$$

$$= \int \frac{du}{u - 1} - \int \frac{du}{u}$$

$$= \ln|u - 1| - \ln|u| + C$$

$$= \ln \left| \frac{u - 1}{u} \right| + C$$

$$= \ln \left| \frac{\ln x - 1}{\ln x} \right| + C$$

36. $\displaystyle \int \frac{x^5 \, dx}{(x^2 - a^2)^3}$

SOLUTION: Let $u = x^2 - a^2$. Then $du = 2x \, dx$ and $x^2 = u + a^2$.

$$\int \frac{x^5 \, dx}{(x^2 - a^2)^3} = \int \frac{(x^2)^2 (x \, dx)}{(x^2 - a^2)^3}$$

$$= \int \frac{(u + a^2)^2 (\tfrac{1}{2} \, du)}{u^3}$$

$$= \frac{1}{2} \int \frac{(u^2 + 2ua^2 + a^4)}{u^3} \, du$$

$$= \frac{1}{2} \int (u^{-1} + 2a^2 u^{-2} + a^4 u^{-3}) \, du$$

$$= \frac{1}{2} \ln|u| - a^2 u^{-1} - \frac{1}{4} a^4 u^{-2} + C$$

$$= \frac{1}{2} \ln|x^2 - a^2| - \frac{a^2}{x^2 - a^2} - \frac{a^4}{4(x^2 - a^2)^2} + C$$

40. $\displaystyle \int \frac{dx}{\sqrt{1 - x + 3x^2}}$

SOLUTION:

$$\int \frac{dx}{\sqrt{1-x+3x^2}} = \frac{1}{\sqrt{3}} \int \frac{dx}{\sqrt{x^2 - \frac{1}{3}x + \frac{1}{3}}}$$

$$= \frac{1}{\sqrt{3}} \int \frac{dx}{\sqrt{(x-\frac{1}{6})^2 + \frac{11}{36}}} \qquad (1)$$

Let

$$\theta = \tan^{-1}\left(\frac{x - \frac{1}{6}}{\frac{1}{6}\sqrt{11}}\right)$$

Then

$$x - \tfrac{1}{6} = \tfrac{1}{6}\sqrt{11}\tan\theta \qquad (2)$$

$$dx = \tfrac{1}{6}\sqrt{11}\sec^2\theta \, d\theta$$

$$\sqrt{(x-\tfrac{1}{6})^2 + \tfrac{11}{36}} = \sqrt{\tfrac{11}{36}\tan^2\theta + \tfrac{11}{36}} = \tfrac{1}{6}\sqrt{11}\sec\theta \qquad (3)$$

Substituting into (1), we obtain

$$\int \frac{dx}{\sqrt{1-x+3x^2}} = \frac{1}{\sqrt{3}} \int \frac{\frac{1}{6}\sqrt{11}\sec^2\theta \, d\theta}{\frac{1}{6}\sqrt{11}\sec\theta}$$

$$= \frac{1}{\sqrt{3}} \ln|\sec\theta + \tan\theta| + C' \qquad (4)$$

Solving (2) for $\tan\theta$, solving (3) for $\sec\theta$, and substituting into (4) we obtain

$$\int \frac{dx}{\sqrt{1-x+3x^2}} = \frac{1}{\sqrt{3}} \ln\left| \frac{\sqrt{(x-\frac{1}{6})^2 + \frac{11}{36}}}{\frac{1}{6}\sqrt{11}} + \frac{x - \frac{1}{6}}{\frac{1}{6}\sqrt{11}} \right| + C'$$

$$= \frac{1}{\sqrt{3}} \ln\left| \frac{\sqrt{(6x-1)^2 + 11} + 6x - 1}{\sqrt{11}} \right| + C'$$

$$= \frac{1}{\sqrt{3}} \ln|2\sqrt{9x^2 - 3x + 3} + 6x - 1| + C$$

44. $\displaystyle\int \frac{dx}{5 + 4\cos 2x}$

SOLUTION: First, let $u = 2x$ and $du = 2\,dx$.

$$\int \frac{dx}{5 + 4\cos 2x} = \frac{1}{2} \int \frac{du}{5 + 4\cos u}$$

Now let $z = \tan\frac{1}{2}u$. Then

$$du = \frac{2\,dz}{1 + z^2} \quad \text{and} \quad \cos u = \frac{1 - z^2}{1 + z^2}$$

Hence,

$$\int \frac{dx}{5 + 4\cos 2x} = \frac{1}{2} \int \frac{\dfrac{2\,dz}{1 + z^2}}{5 + 4\left(\dfrac{1 - z^2}{1 + z^2}\right)}$$

$$= \int \frac{dz}{9 + z^2}$$

$$= \frac{1}{3} \tan^{-1} \frac{z}{3} + C$$

$$= \frac{1}{3} \tan^{-1} \left(\frac{1}{3} \tan \frac{1}{2} u \right) + C$$

$$= \frac{1}{3} \tan^{-1} \left(\frac{1}{3} \tan x \right) + C$$

48. $\displaystyle\int \frac{dx}{x\sqrt{5x - 6 - x^2}}$

SOLUTION: We let $x = z^{-1}$ and $dx = -z^{-2} \, dz$.

$$\int \frac{dx}{x\sqrt{5x - 6 - x^2}} = \int \frac{-z^{-2} \, dz}{z^{-1}\sqrt{5z^{-1} - 6 - z^{-2}}}$$

$$= -\int \frac{dz}{z\sqrt{\dfrac{5z - 6z^2 - 1}{z^2}}}$$

$$= -\int \frac{|z| \, dz}{z\sqrt{5z - 6z^2 - 1}}$$

$$= \begin{cases} -\displaystyle\int \dfrac{dz}{\sqrt{5z - 6z^2 - 1}} & \text{if } z > 0 \\[3mm] \displaystyle\int \dfrac{dz}{\sqrt{5z - 6z^2 - 1}} & \text{if } z < 0 \end{cases} \qquad (1)$$

We evaluate the second integral in (1).

$$\int \frac{dz}{\sqrt{5z - 6z^2 - 1}} = \frac{1}{\sqrt{6}} \int \frac{dz}{\sqrt{\frac{5}{6}z - z^2 - \frac{1}{6}}}$$

$$= \frac{1}{\sqrt{6}} \int \frac{dz}{\sqrt{\frac{1}{144} - \left(z - \frac{5}{12}\right)^2}}$$

$$= \frac{1}{\sqrt{6}} \sin^{-1} \left(\frac{z - \frac{5}{12}}{\frac{1}{12}} \right) + C$$

$$= \frac{1}{\sqrt{6}} \sin^{-1}(12z - 5) + C$$

$$= \frac{1}{\sqrt{6}} \sin^{-1} \left(\frac{12}{x} - 5 \right) + C$$

$$= -\frac{1}{\sqrt{6}} \sin^{-1} \left(\frac{5x - 12}{x} \right) + C \qquad (2)$$

Because $z > 0$ if $x > 0$, and $z < 0$ if $x < 0$, by substitution from (2) into (1), we conclude that

$$\int \frac{dx}{x\sqrt{5x - 6 - x^2}} = \begin{cases} \dfrac{1}{\sqrt{6}} \sin^{-1} \left(\dfrac{5x - 12}{x} \right) + C & \text{if } x > 0 \\[4mm] -\dfrac{1}{\sqrt{6}} \sin^{-1} \left(\dfrac{5x - 12}{x} \right) + C & \text{if } x < 0 \end{cases}$$

52. $\displaystyle\int \frac{dx}{(x^2 + 6x + 34)^2}$

SOLUTION:

$$\int \frac{dx}{(x^2 + 6x + 34)^2} = \int \frac{dx}{[(x+3)^2 + 25]^2} \tag{1}$$

Let

$$\theta = \tan^{-1}\left(\frac{x+3}{5}\right) \tag{2}$$

$$x + 3 = 5 \tan \theta$$

$$dx = 5 \sec^2 \theta \, d\theta$$

$$[(x + 3)^2 + 25]^2 = (25 \tan^2 \theta + 25)^2 = (25 \sec^2 \theta)^2 = 625 \sec^4 \theta$$

With these replacements in (1) we obtain

$$\int \frac{dx}{(x^2 + 6x + 34)^2} = \int \frac{5 \sec^2 \theta \, d\theta}{625 \sec^4 \theta}$$

$$= \frac{1}{125} \int \cos^2 \theta \, d\theta$$

$$= \frac{1}{250} \int (1 + \cos 2\theta) \, d\theta$$

$$= \frac{1}{250} \left(\theta + \frac{1}{2} \sin 2\theta\right) + C$$

$$= \frac{1}{250} (\theta + \sin \theta \cos \theta) + C \tag{3}$$

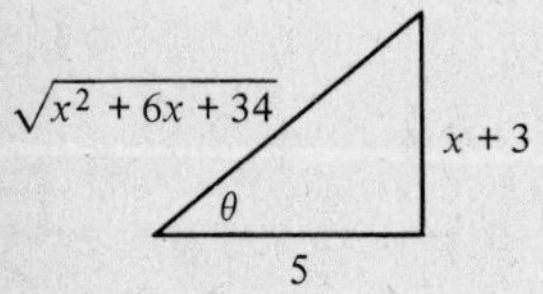

Figure 10. 52R

Substituting from (2) and Fig. 10. 52R into (3), we obtain

$$\int \frac{dx}{(x^2 + 6x + 34)^2} = \frac{1}{250} \left[\tan^{-1}\left(\frac{x+3}{5}\right) + \frac{5(x+3)}{x^2 + 6x + 34}\right] + C$$

$$= \frac{1}{250} \tan^{-1}\left(\frac{x+3}{5}\right) + \frac{x+3}{50(x^2 + 6x + 34)} + C$$

56. $\displaystyle\int \ln(x^2 + 1) \, dx$

SOLUTION: We use integration by parts. Let

$$u = \ln(x^2 + 1) \qquad dv = dx$$

$$du = \frac{2x \, dx}{x^2 + 1} \qquad v = x$$

Then,

$$\int \ln(x^2 + 1) \, dx = x \ln(x^2 + 1) - \int \frac{2x^2 \, dx}{x^2 + 1}$$

$$= x \ln(x^2 + 1) - 2 \int dx + 2 \int \frac{dx}{x^2 + 1}$$

$$= x \ln(x^2 + 1) - 2x + 2 \tan^{-1} x + C$$

60. $\displaystyle\int \frac{dx}{\sqrt{1 + \sqrt[3]{x}}}$

SOLUTION: Let $u = \sqrt{1 + \sqrt[3]{x}}$. Then

$$u^2 = 1 + \sqrt[3]{x}$$
$$x = (u^2 - 1)^3$$
$$dx = 3(u^2 - 1)^2(2u \, du)$$

Thus,

$$\int \frac{dx}{\sqrt{1 + \sqrt[3]{x}}} = \int \frac{6u(u^2 - 1)^2 \, du}{u}$$

$$= 6 \int (u^4 - 2u^2 + 1) \, du$$

$$= \frac{6}{5} u^5 - 4u^3 + 6u + C$$

$$= \frac{2}{5} u(3u^4 - 10u^2 + 15) + C$$

$$= \frac{2}{5} \sqrt{1 + \sqrt[3]{x}} \, [3(1 + \sqrt[3]{x})^2 - 10(1 + \sqrt[3]{x}) + 15] + C$$

$$= \frac{2}{5} \sqrt{1 + \sqrt[3]{x}} \, (3\sqrt[3]{x^2} - 4\sqrt[3]{x} + 8) + C$$

In Exercises 63–94 evaluate the definite integral.

64. $\displaystyle\int_{1/2}^{1} \sqrt{\frac{1 - x}{x}} \, dx$

SOLUTION: Let $\theta = \sin^{-1} \sqrt{x}$. Then $x = \sin^2 \theta$ and $dx = 2 \sin \theta \cos \theta \, d\theta$. When $x = \frac{1}{2}$, then $\theta = \frac{1}{4}\pi$; when $x = 1$, then $\theta = \frac{1}{2}\pi$. Furthermore,

$$\sqrt{\frac{1 - x}{x}} = \sqrt{\frac{1 - \sin^2 \theta}{\sin^2 \theta}} = \sqrt{\frac{\cos^2 \theta}{\sin^2 \theta}} = \frac{\cos \theta}{\sin \theta}$$

Thus,

$$\int_{1/2}^{1} \sqrt{\frac{1 - x}{x}} \, dx = \int_{\pi/4}^{\pi/2} \frac{\cos \theta}{\sin \theta} (2 \sin \theta \cos \theta \, d\theta)$$

$$= \int_{\pi/4}^{\pi/2} 2 \cos^2 \theta \, d\theta$$

$$= \int_{\pi/4}^{\pi/2} (1 + \cos 2\theta) \, d\theta$$

$$= \theta + \frac{1}{2} \sin 2\theta \Big]_{\pi/4}^{\pi/2}$$

$$= \left(\frac{1}{2} \pi + \frac{1}{2} \sin \pi \right) - \left(\frac{1}{4} \pi + \frac{1}{2} \sin \frac{1}{2} \pi \right)$$

$$= \frac{1}{4} \pi - \frac{1}{2}$$

68. $\displaystyle\int_{0}^{\pi/2} \sin^3 t \cos^3 t \, dt$

SOLUTION: Let $u = \sin t$ and $du = \cos t \, dt$. When $t = 0$, then $u = 0$; when $t = \frac{1}{2}\pi$, then $u = 1$. Thus,

$$\int_0^{\pi/2} \sin^3 t \cos^3 t \, dt = \int_0^{\pi/2} \sin^3 t(1 - \sin^2 t)\cos t \, dt$$

$$= \int_0^1 u^3(1 - u^2) \, du$$

$$= \frac{1}{4} u^4 - \frac{1}{6} u^6 \Big]_0^1 = \frac{1}{12}$$

72. $\displaystyle\int_0^2 \frac{(1 - x) \, dx}{x^2 + 3x + 2}$

SOLUTION: Because $x^2 + 3x + 2 = (x + 1)(x + 2)$, we let

$$\frac{1 - x}{x^2 + 3x + 2} = \frac{A}{x + 1} + \frac{B}{x + 2}$$

$$1 - x = A(x + 2) + B(x + 1)$$

If $x = -1$, then $2 = A$; if $x = -2$, than $3 = -B$, so $B = -3$. Thus,

$$\int_0^2 \frac{(1 - x) \, dx}{x^2 + 3x + 2} = \int_0^2 \frac{2 \, dx}{x + 1} - \int_0^2 \frac{3 \, dx}{x + 2}$$

$$= 2 \ln|x + 1| - 3 \ln|x + 2| \Big]_0^2$$

$$= (2 \ln 3 - 3 \ln 4) - (2 \ln 1 - 3 \ln 2)$$

$$= 2 \ln 3 - 3 \ln 4 + 3 \ln 2$$

$$= \ln \frac{(3^2)(2^3)}{4^3} = \ln \frac{9}{8}$$

76. $\displaystyle\int_1^2 (\ln x)^2 \, dx$

SOLUTION: We use integration by parts.

$$u = (\ln x)^2 \qquad dv = dx$$

$$du = \frac{2 \ln x \, dx}{x} \qquad v = x$$

Thus,

$$\int_1^2 (\ln x)^2 \, dx = x(\ln x)^2 \Big]_1^2 - 2 \int_1^2 \ln x \, dx \tag{1}$$

For the integral on the right-hand side of (1), we integrate by parts.

$$\bar{u} = \ln x \qquad d\bar{v} = dx$$

$$d\bar{u} = \frac{dx}{x} \qquad \bar{v} = x$$

Thus, from (1) we obtain

$$\int_1^2 (\ln x)^2 \, dx = 2(\ln 2)^2 - 2 \left[x \ln x - \int dx \right]_1^2$$

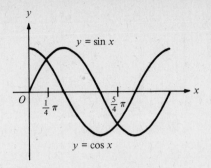

Figure 10. 80R

$$= 2(\ln 2)^2 - 2\left[x\ln x - x\right]_1^2$$

$$= 2(\ln 2)^2 - 2[(2\ln 2 - 2) - (0 - 1)]$$

$$= 2(\ln 2)^2 - 4\ln 2 + 2$$

$$= 2(\ln 2 - 1)^2$$

80. $\displaystyle\int_0^{2\pi} |\sin x - \cos x|\, dx$

SOLUTION: See Fig. 10. 80R. If $0 \le x \le \frac{1}{4}\pi$, then $\sin x \le \cos x$, so $|\sin x - \cos x| = -\sin x + \cos x$. If $\frac{1}{4}\pi \le x \le \frac{5}{4}\pi$, then $\sin x \ge \cos x$, so $|\sin x - \cos x| = \sin x - \cos x$. If $\frac{5}{4}\pi \le x \le 2\pi$, then $\sin x \le \cos x$, so $|\sin x - \cos x| = -\sin x + \cos x$. Therefore,

$$\int_0^{2\pi} |\sin x - \cos x|\, dx = \int_0^{\pi/4} (-\sin x + \cos x)\, dx + \int_{\pi/4}^{5\pi/4} (\sin x - \cos x)\, dx + \int_{5\pi/4}^{2\pi} (-\sin x + \cos x)\, dx$$

$$= \left[\cos x + \sin x\right]_0^{\pi/4} + \left[-\cos x - \sin x\right]_{\pi/4}^{5\pi/4} + \left[\cos x + \sin x\right]_{5\pi/4}^{2\pi}$$

$$= [(\tfrac{1}{2}\sqrt{2} + \tfrac{1}{2}\sqrt{2}) - 1] + [(\tfrac{1}{2}\sqrt{2} + \tfrac{1}{2}\sqrt{2}) - (-\tfrac{1}{2}\sqrt{2} - \tfrac{1}{2}\sqrt{2})] + [1 - (-\tfrac{1}{2}\sqrt{2} - \tfrac{1}{2}\sqrt{2})]$$

$$= (\sqrt{2} - 1) + 2\sqrt{2} + (1 + \sqrt{2}) = 4\sqrt{2}$$

84. $\displaystyle\int_{-\pi/4}^{\pi/4} |\tan^5 x|\, dx$

SOLUTION: Because $|\tan^5(-x)| = |\tan^5 x|$, then

$$\int_{-\pi/4}^{\pi/4} |\tan^5 x|\, dx = 2\int_0^{\pi/4} |\tan^5 x|\, dx$$

$$= 2\int_0^{\pi/4} \tan^5 x\, dx$$

$$= 2\int_0^{\pi/4} (\sec^2 x - 1)\tan^3 x\, dx$$

$$= 2\int_0^{\pi/4} \tan^3 x \sec^2 x\, dx - 2\int_0^{\pi/4} (\sec^2 x - 1)\tan x\, dx$$

$$= \frac{1}{2}\tan^4 x\Big]_0^{\pi/4} - 2\int_0^{\pi/4} \tan x \sec^2 x\, dx + 2\int_0^{\pi/4} \tan x\, dx$$

$$= \frac{1}{2} - \left[\tan^2 x\right]_0^{\pi/4} + 2\ln|\sec x|\Big]_0^{\pi/4}$$

$$= \frac{1}{2} - 1 + 2\ln\sqrt{2}$$

$$= -\frac{1}{2} + \ln 2$$

88. $\displaystyle\int_0^{\pi/12} \frac{dx}{\cos^4 3x}$

SOLUTION:

$$\int_0^{\pi/12} \frac{dx}{\cos^4 3x} = \int_0^{\pi/12} \sec^4 3x\, dx$$

$$= \int_0^{\pi/12} (1 + \tan^2 3x)\sec^2 3x \, dx$$

$$= \left[\frac{1}{3} \tan 3x + \frac{1}{9} \tan^3 3x \right]_0^{\pi/12}$$

$$= \frac{4}{9}$$

92. $\displaystyle\int_0^3 \frac{dr}{(r+2)\sqrt{r+1}}$

SOLUTION: Let $x = \sqrt{r+1}$. Then $x^2 = r + 1$, so $dr = 2x \, dx$ and $r + 2 = x^2 + 1$. When $r = 0$, then $x = 1$; when $r = 3$, then $x = 2$. Therefore,

$$\int_0^3 \frac{dr}{(r+2)\sqrt{r+1}} = \int_1^2 \frac{2x \, dx}{(x^2+1)x}$$

$$= 2\int_1^2 \frac{dx}{1+x^2}$$

$$= 2 \tan^{-1} x \Big]_1^2$$

$$= 2(\tan^{-1} 2 - \tfrac{1}{4}\pi)$$

$$= 2 \tan^{-1} 2 - \tfrac{1}{2}\pi$$

96. Obtain the given result by a hyperbolic function substitution.

$$\int \frac{dx}{(a^2 - x^2)^{3/2}} = \frac{1}{a^2} \sinh\left(\tanh^{-1} \frac{x}{a} \right) + C, \ a > 0$$

SOLUTION: Let $z = \tanh^{-1}(x/a)$. Then

$$x = a \tanh z$$
$$dx = a \operatorname{sech}^2 z \, dz$$
$$(a^2 - x^2)^{3/2} = (a^2 - a^2 \tanh^2 z)^{3/2}$$
$$= [a^2(1 - \tanh^2 z)]^{3/2}$$
$$= (a^2 \operatorname{sech}^2 z)^{3/2}$$
$$= a^3 \operatorname{sech}^3 z$$

Hence,

$$\int \frac{dx}{(a^2 - x^2)^{3/2}} = \int \frac{a \operatorname{sech}^2 z \, dz}{a^3 \operatorname{sech}^3 z}$$

$$= \frac{1}{a^2} \int \cosh z \, dz$$

$$= \frac{1}{a^2} \sinh z + C$$

$$= \frac{1}{a^2} \sinh\left(\tanh^{-1} \frac{x}{a} \right) + C$$

100. Find the length of the arc of the curve $y = \ln x$ from $x = 1$ to $x = e$.

SOLUTION: We apply Theorem 7.6.3. Let $f(x) = \ln x$.

$$f'(x) = \frac{1}{x}$$

$$\sqrt{1 + [f'(x)]^2} = \sqrt{1 + \frac{1}{x^2}}$$

$$= \frac{\sqrt{x^2 + 1}}{x}$$

If L units is the length of the arc, then

$$L = \int_1^e \frac{\sqrt{x^2 + 1}}{x} \, dx$$

Let $u = \sqrt{x^2 + 1}$. Then $u^2 = x^2 + 1$, so $u \, du = x \, dx$ and $x^2 = u^2 - 1$. Multiplying the numerator and denominator of the fraction by x and making the indicated substitutions, we have for the indefinite integral

$$\int \frac{\sqrt{x^2 + 1} \, dx}{x} = \int \frac{\sqrt{x^2 + 1}(x \, dx)}{x^2}$$

$$= \int \frac{u(u \, du)}{u^2 - 1}$$

$$= \int du + \int \frac{du}{u^2 - 1}$$

$$= u + \tfrac{1}{2} \ln \left| \frac{u - 1}{u + 1} \right| + C$$

$$= u + \tfrac{1}{2} \ln \left| \frac{u - 1}{u + 1} \cdot \frac{u - 1}{u - 1} \right| + C$$

$$= u + \tfrac{1}{2} \ln \left| \frac{(u - 1)^2}{u^2 - 1} \right| + C$$

$$= u + \ln|u - 1| - \tfrac{1}{2} \ln|u^2 - 1| + C$$

$$= \sqrt{x^2 + 1} + \ln(\sqrt{x^2 + 1} - 1) - \tfrac{1}{2} \ln x^2 + C$$

$$= \sqrt{x^2 + 1} + \ln(\sqrt{x^2 + 1} - 1) - \ln x + C$$

Let

$$F(x) = \sqrt{x^2 + 1} + \ln(\sqrt{x^2 + 1} - 1) - \ln x$$

Then

$$F(e) = \sqrt{e^2 + 1} + \ln(\sqrt{e^2 + 1} - 1) - 1$$
$$F(1) = \sqrt{2} + \ln(\sqrt{2} - 1)$$

Therefore,

$$L = F(e) - F(1)$$
$$= \sqrt{e^2 + 1} - 1 - \sqrt{2} + \ln(\sqrt{e^2 + 1} - 1) - \ln(\sqrt{2} - 1)$$
$$\approx 2.00$$

The length of the arc is ≈ 2.00 units.

104. A tank is in the shape of the solid of revolution formed by rotating about the x-axis the region bounded by the curve $y = \ln x$, the x-axis, and the lines $x = e$ and $x = e^2$. If the tank is full of water, find the work done in pumping all the water to the top of the tank. Distance is measured in feet. Take the positive x-axis vertically downward.

SOLUTION: Figure 10. 104R shows a sketch of a plane section of the tank. Let $f(x) = \ln x$. The element of volume is a circular disk with units of thickness $\Delta_i x$ and units of radius $f(\bar{x}_i)$. Thus the number of cubic units in the element of volume is

$$\Delta_i V = \pi [f(\bar{x}_i)]^2 \, \Delta_i x$$

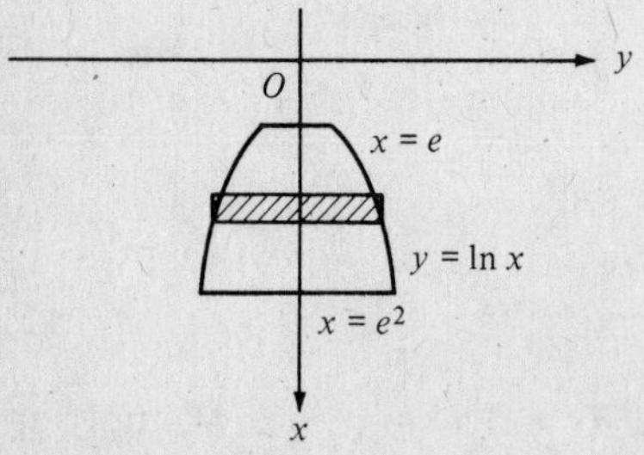

If w is the density of water, then the number of units in the element of force is

$$\begin{aligned} \Delta_i F &= w \, \Delta_i V \\ &= \pi w (\ln \bar{x}_i)^2 \, \Delta_i x \end{aligned} \tag{1}$$

Because the top of the tank is at the point where $x = e$, the number of units in the element of displacement is

$$D_i = \bar{x}_i - e \tag{2}$$

Figure 10. 104R

Furthermore, the tank is bounded by the lines $x = e$ and $x = e^2$. Therefore, by (1) and (2) we have

$$\begin{aligned} W &= \lim_{\|\Delta\| \to 0} \sum_{i=1}^{n} D_i \, \Delta_i F \\ \\ &= \lim_{\|\Delta\| \to 0} \sum_{i=1}^{n} (\bar{x}_i - e) \pi w (\ln \bar{x}_i)^2 \, \Delta_i x \\ \\ &= \pi w \int_{e}^{e^2} (x - e)(\ln x)^2 \, dx \\ \\ &= \pi w \int_{e}^{e^2} x(\ln x)^2 \, dx - \pi w e \int_{e}^{e^2} (\ln x)^2 \, dx \end{aligned} \tag{3}$$

We use integration by parts on the first integral in (3).

$$u = (\ln x)^2 \qquad dv = x \, dx$$

$$du = \frac{2 \ln x \, dx}{x} \qquad v = \tfrac{1}{2} x^2$$

Thus,

$$\int x(\ln x)^2 \, dx = \frac{1}{2} x^2 (\ln x)^2 - \int x \ln x \, dx \tag{4}$$

We integrate by parts again.

$$\bar{u} = \ln x \qquad d\bar{v} = x \, dx$$

$$d\bar{u} = \frac{dx}{x} \qquad \bar{v} = \tfrac{1}{2} x^2$$

Thus, from (4) we have

$$\int x(\ln x)^2 \, dx = \frac{1}{2} x^2 (\ln x)^2 - \frac{1}{2} x^2 \ln x + \frac{1}{4} x^2 + C_1 \tag{5}$$

For the second integral in (3), we integrate by parts as in Exercise 76. Thus

$$\int (\ln x)^2 \, dx = x(\ln x)^2 - 2x \ln x + 2x + C_2 \tag{6}$$

Substituting from (5) and (6) into (3), we obtain

$$W = \pi w \left[\frac{1}{2} x^2 (\ln x)^2 - \frac{1}{2} x^2 \ln x + \frac{1}{4} x^2 \right]_{e}^{e^2} - \pi w e \left[x(\ln x)^2 - 2x \ln x + 2x \right]_{e}^{e^2}$$

$$= \pi w \left[\frac{5}{4} e^4 - \frac{1}{4} e^2 \right] - \pi w e [2 e^2 - e]$$

$$= \pi w \left[\frac{5}{4} e^4 - 2 e^3 + \frac{3}{4} e^2 \right]$$

Thus, the work done is $\pi w (\frac{5}{4} e^4 - 2 e^3 + \frac{3}{4} e^2)$ foot-pounds.

CHAPTER 11 | Additional Applications of Integration

11.1 CENTER OF MASS OF A ROD

If a system of n particles is located on the x-axis at the points $x_1, x_2, \ldots, x_n$ and the mass of the ith particle is given by m_i, then the center of mass for the system is at the point $\bar{x}$ on the x-axis, where

$$\bar{x} = \frac{\displaystyle\sum_{i=1}^{n} m_i x_i}{\displaystyle\sum_{i=1}^{n} m_i} \tag{1}$$

11.1.1 Definition

A rod of length L meters has its left endpoint at the origin. If the number of kilograms per meter in the linear density at a point x meters from the origin is $\rho(x)$, where ρ is continuous on $[0, L]$, then the total *mass* of the rod is M kg, where

$$M = \lim_{\|\Delta\| \to 0} \sum_{i=1}^{n} \rho(\xi_i)\, \Delta_i x = \int_0^L \rho(x)\, dx$$

11.1.2 Definition

A rod of length L meters has its left endpoint at the origin and the number of kilograms per meter in the linear density at a point x meters from the origin is $\rho(x)$, where ρ is continuous on $[0, L]$. The *moment of mass* of the rod with respect to the origin is M_0, where

$$M_0 = \lim_{\|\Delta\| \to 0} \sum_{i=1}^{n} \xi_i \rho(\xi_i)\, \Delta_i x = \int_0^L x\rho(x)\, dx$$

The center of mass of the rod described in Definitions 11.1.1 and 11.1.2 is at the point $\bar{x}$, where

$$\bar{x} = \frac{M_0}{M}$$

Because $M_0 = M\bar{x}$, the integral given in Definition 11.1.2 gives $M \cdot \bar{x}$. The following table gives the units that are used in the three systems of units considered in this book.

System of Units	Force	Mass	Acceleration
Britsh engineering	pound	slug	ft/sec²
mks	newton	kilogram	m/sec²
cgs	dyne	gram	cm/sec²

Exercises 11.1

4. A particle is moving on a horizontal line. Find the force exerted on the particle if it has mass of 22 slugs and acceleration of 4 ft/sec².

SOLUTION: We are given that

$$M = \frac{F}{a}$$

Thus,

$$F = Ma$$

We let $M = 22$ and $a = 4$ and obtain

$$F = (22)(4) = 88$$

Because the mass and acceleration are given in British engineering units, the force is measured in pounds. Thus, the force is 88 pounds.

8. A particle is subjected to a horizontal force of 700 dynes, and the acceleration is 80 cm/sec². Find the mass of the particle.

SOLUTION: The given units are in the cgs system. We let $F = 700$ and $a = 80$, so

$$M = \frac{F}{a} = \frac{700}{80} = 8.75$$

The mass is 8.75 grams.

12. A system of particles is located on the x-axis. The number of kilograms in the mass of each particle and the coordinate of its position are given. Distance is measured in meters. Find the center of mass of the system: $m_1 = 5$ at -7; $m_2 = 3$ at -2; $m_3 = 5$ at 0; $m_4 = 1$ at 2; $m_5 = 8$ at 10.

SOLUTION: We use Eq. (1) with m_i as given for $i = 1, 2, 3, 4, 5$ and $x_1 = -7$, $x_2 = -2$, $x_3 = 0$, $x_4 = 2$, $x_5 = 10$. Thus,

$$\bar{x} = \frac{m_1 x_1 + m_2 x_2 + m_3 x_3 + m_4 x_4 + m_5 x_5}{m_1 + m_2 + m_3 + m_4 + m_5}$$

$$= \frac{5(-7) + 3(-2) + (5 \cdot 0) + (1 \cdot 2) + (8 \cdot 10)}{5 + 3 + 5 + 1 + 8} = \frac{41}{22}$$

Thus the center of mass is $\frac{41}{22}$ meters to the right of the origin.

16. The length of a rod is 3 ft, and the linear density of the rod at a point x ft from one end is $(5 + 2x)$ slugs per foot. Find the total mass of the rod and the center of mass.

SOLUTION: To find the total mass of the rod we apply Definition 11.1.1. We are given that $\rho(x) = 5 + 2x$. If M slugs is the total mass of the rod, then

$$M = \lim_{\|\Delta\| \to 0} \sum_{i=1}^{n} (5 + 2\bar{x}_i)\,\Delta_i x$$

$$= \int_0^3 (5 + 2x)\,dx$$

$$= 5x + x^2 \Big]_0^3 = 24 \tag{1}$$

The total mass of the rod is 24 slugs. To find the center of mass of the rod, we apply Definition 11.1.2. If the center of mass is $\bar{x}$ ft from the end of the rod, then

$$M \cdot \bar{x} = \lim_{\|\Delta\| \to 0} \sum_{i=1}^{n} (5 + 2\bar{x}_i)\bar{x}_i\,\Delta_i x$$

$$= \int_0^3 (5 + 2x)x\,dx$$

$$= \int_0^3 (5x + 2x^2)\,dx$$

$$= \frac{5}{2}x^2 + \frac{2}{3}x^3 \Big]_0^3$$

$$= \frac{45}{2} + 18 = \frac{81}{2} \tag{2}$$

Substituting the value of M, given in (1), into Eq. (2), we obtain

$$24\bar{x} = \tfrac{81}{2}$$
$$\bar{x} = \tfrac{27}{16}$$

The center of mass of the rod is $\frac{27}{16}$ ft from the end of the rod.

20. A rod is 10 ft long, and the measure of the linear density at a point is a linear function of the measure of the distance from the center of the rod. The linear density at each end of the rod is 5 slugs/ft, and at the center the linear density is $3\frac{1}{2}$ slugs/ft. Find the total mass of the rod and the center of mass.

SOLUTION: Position the rod so that it lies along the x-axis with left endpoint at the origin. Let ρ be the linear density function. We must find $\rho(x)$. Because the rod is 10 ft long, its center is at the point where $x = 5$, and thus $|x - 5|$ is the measure of the distance from the point x to the center of the rod, with x in $[0, 10]$. Because the linear density is a linear function of $|x - 5|$, there are constants a and b such that

$$\rho(x) = a + b\,|x - 5| \tag{1}$$

Because the linear density is $\frac{7}{2}$ slugs/ft at the center of the rod, we are given that $\rho(5) = \frac{7}{2}$. From (1) we get $\rho(5) = a$. Thus $a = \frac{7}{2}$. Because the linear density is 5 slugs/ft at the end of the rod, we are given that $\rho(0) = 5$. From (1) we have $\rho(0) = a + 5b$, and because $a = \frac{7}{2}$, then $5 = \frac{7}{2} + 5b$. Hence, $b = \frac{3}{10}$. Therefore, from Eq. (1) with $a = \frac{7}{2}$ and $b = \frac{3}{10}$, we have

$$\rho(x) = \tfrac{7}{2} + \tfrac{3}{10}|x - 5| \tag{2}$$

To find the total mass of the rod we use Definition 11.1.1 and Eq. (2). Thus,

$$M = \lim_{\|\Delta\| \to 0} \sum_{i=1}^{n} \left[\frac{7}{2} + \frac{3}{10}|\bar{x}_i - 5| \right] \Delta_i x$$

$$= \int_0^{10} \left[\frac{7}{2} + \frac{3}{10}|x - 5| \right] dx \tag{3}$$

Because $|x - 5| = -(x - 5)$ if $0 \le x \le 5$ and $|x - 5| = x - 5$ if $5 \le x \le 10$, we separate the integral in (3) into two integrals. Thus,

$$M = \int_0^5 \left[\frac{7}{2} - \frac{3}{10}(x - 5) \right] dx + \int_5^{10} \left[\frac{7}{2} + \frac{3}{10}(x - 5) \right] dx$$

$$= \int_0^5 \left(-\frac{3}{10}x + 5 \right) dx + \int_5^{10} \left(\frac{3}{10}x + 2 \right) dx$$

$$= \left[-\frac{3}{20}x^2 + 5x \right]_0^5 + \left[\frac{3}{20}x^2 + 2x \right]_5^{10}$$

$$= \frac{85}{4} + \frac{85}{4} = \frac{85}{2} \tag{4}$$

Thus the total mass of the rod is $\frac{85}{2}$ slugs.

To find $\bar{x}$, the center of mass, we use Definition 11.1.2 and Eq. (2). Thus

$$M \cdot \bar{x} = \lim_{\|\Delta\| \to 0} \sum_{i=1}^{n} \left[\frac{7}{2} + \frac{3}{10}|\bar{x}_i - 5| \right] \bar{x}_i \, \Delta_i x$$

$$= \int_0^{10} \left[\frac{7}{2} + \frac{3}{10}|x - 5| \right] x \, dx$$

$$= \int_0^5 \left(-\frac{3}{10}x^2 + 5x \right) dx + \int_5^{10} \left(\frac{3}{10}x^2 + 2x \right) dx$$

$$= \left[-\frac{1}{10}x^3 + \frac{5}{2}x^2 \right]_0^5 + \left[\frac{1}{10}x^3 + x^2 \right]_5^{10} = \frac{425}{2} \tag{5}$$

Therefore, substituting from (4) into (5), we get

$$\tfrac{85}{2}\bar{x} = \tfrac{425}{2}$$
$$\bar{x} = 5$$

Thus, the center of mass is at the point where $x = 5$, the center of the rod.

24. The measure of the linear density at a point on a rod varies directly as the fourth power of the measure of the distance of the point from one end. The length of the rod is 2 meters and the linear density is 2 kg/m at the center. If the total mass of the rod is $\frac{64}{5}$ kg, find the center of mass of the rod.

SOLUTION: Let the linear density of the rod at a point that is x meters from one end of the rod be $\rho(x)$ kg/m. We are given that $\rho(1) = 2$ and that

$$\rho(x) = kx^4$$

Thus,

$$2 = k(1^4)$$
$$k = 2$$

so

$$\rho(x) = 2x^4$$

Let the center of mass be at the point that is $\bar{x}$ meters from the end. Because we are given that the total mass of the rod is $\frac{64}{5}$ kg, then by Definition 11.1.2 we conclude that

$$\tfrac{64}{5}x = \lim_{\|\Delta\| \to 0} \sum_{i=1}^{n} (2\bar{x}_i^4)\bar{x}_i\, \Delta_i x$$

$$= \int_0^2 2x^5\, dx$$

$$= \frac{1}{3} x^6 \bigg]_0^2 = \frac{64}{3}$$

Thus,

$$\bar{x} = \tfrac{64}{3} \cdot \tfrac{5}{64} = \tfrac{5}{3}$$

The center of mass is $\frac{5}{3}$ meters from the end of the rod.

28. The total mass of a rod of length L ft is M slugs, and the measure of the linear density at a point x ft from the left end is proportional to the measure of the distance of the point from the right end. Show that the linear density at a point on the rod x ft from the left end is $2M(L - x)/L^2$ slugs/ft.

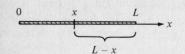

Figure 11. 1. 28

SOLUTION: Position the rod with its left endpoint at the origin and right endpoint at the point where $x = L$, as illustrated in Fig. 11. 1. 28. Let ρ be the linear density function. Because $(L - x)$ ft is the distance to the right endpoint from any point that is x ft from the origin, we are given that

$$\rho(x) = k(L - x) \tag{1}$$

We must find the constant k. By Definition 11.1.1 and Eq. (1), we obtain

$$M = \lim_{\|\Delta\| \to 0} \sum_{i=1}^{n} k(L - \bar{x}_i)\, \Delta_i x$$

$$= \int_0^L k(L - x)\, dx$$

$$= k\left[Lx - \frac{1}{2} x^2 \right]_0^L$$

$$= k\left(L^2 - \frac{1}{2} L^2 \right)$$

$$= \frac{1}{2} kL^2$$

Solving for k, we obtain

$$k = \frac{2M}{L^2}$$

Substituting this value of k into (1) we get the required result

$$\rho(x) = \frac{2M}{L^2} (L - x)$$

11.2 CENTROID OF A PLANE REGION

The formulas given in the text for finding the centroid of a region in the xy plane apply only to the case in which the region is bounded below by the x-axis. When these formulas do not apply, we use the following more general method.

Let R be a region in the xy plane that is bounded above by the curve $y = f(x)$, bounded below by the curve $y = g(x)$, bounded on the left by the line $x =$

a, and bounded on the right by the line $x = b$. To find $(\bar{x}, \bar{y})$, the centroid of R, first use the method of Section 7.1 to find A, the measure of the area of R. Then, because the centroid of the vertical element of area is at the point $(\bar{x}_i, \frac{1}{2}[f(\bar{x}_i) + g(\bar{x}_i)])$, we have

$$A \cdot \bar{x} = \lim_{\|\Delta\| \to 0} \sum_{i=1}^{n} [f(\bar{x}_i) - g(\bar{x}_i)]\bar{x}_i \, \Delta_i x$$

$$= \int_a^b [f(x) - g(x)]x \, dx$$

$$A \cdot \bar{y} = \lim_{\|\Delta\| \to 0} \sum_{i=1}^{n} \frac{1}{2} [f(\bar{x}_i) - g(\bar{x}_i)][f(\bar{x}_i) + g(\bar{x}_i)] \, \Delta_i x$$

$$= \frac{1}{2} \int_a^b [f(x) - g(x)][f(x) + g(x)] \, dx$$

Let R be a region in the xy plane that is bounded on the right by the curve $x = f(y)$, bounded on the left by the curve $x = g(y)$, bounded below by the line $y = a$, and bounded above by the line $y = b$. To find $(\bar{x}, \bar{y})$, the centroid of R, first use the method of Section 7.1 to find A, the measure of the area of R. Then, because the centroid of the horizontal element of area is at the point $(\frac{1}{2}[f(\bar{y}_i) + g(\bar{y}_i)], \bar{y}_i)$, we have

$$A \cdot \bar{x} = \lim_{\|\Delta\| \to 0} \sum_{i=1}^{n} \frac{1}{2} [f(\bar{y}_i) - g(\bar{y}_i)][f(\bar{y}_i) + g(\bar{y}_i)] \, \Delta_i y$$

$$= \frac{1}{2} \int_a^b [f(y) - g(y)][f(y) + \dot{g}(y)] \, dy$$

$$A \cdot \bar{y} = \lim_{\|\Delta\| \to 0} \sum_{i=1}^{n} [f(\bar{y}_i) - g(\bar{y}_i)]\bar{y}_i \, \Delta_i y$$

$$= \int_a^b [f(y) - g(y)]y \, dy$$

Exercises 11.2

4. Find the center of mass of the three particles having masses of 3, 7, and 2 slugs located at the points (2, 3), (−1, 4), and (0, 2), respectively.

SOLUTION: If M slugs is the total mass of the three particles, then

$$M = \sum_{i=1}^{3} m_i = 3 + 7 + 2 = 12$$

Let $(\bar{x}, \bar{y})$ be the coordinates of the center of mass. Then

$$M\bar{x} = \sum_{i=1}^{3} m_i x_i = (3)(2) + (7)(-1) + (2)(0) = -1$$

$$M\bar{y} = \sum_{i=1}^{3} m_i y_i = (3)(3) + (7)(4) + (2)(2) = 41$$

So

$$\bar{x} = \frac{-1}{M} = \frac{-1}{12}$$

$$\bar{y} = \frac{41}{M} = \frac{41}{12}$$

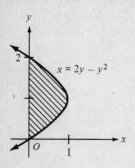

Figure 11. 2. 8

The center of mass is $(-\frac{1}{12}, \frac{41}{12})$.

In Exercises 7–19 find the centroid of the region with the indicated boundaries.

8. The parabola $x = 2y - y^2$ and the y-axis.

SOLUTION: Fig. 11. 2. 8 shows a sketch of the region R. The region is bounded on the right by the curve $x = f(y) = 2y - y^2$, bounded on the left by the line $x = g(y) = 0$, bounded below by the line $y = 0$, and bounded above by the line $y = 2$. Therefore, the area of R is given by

$$A = \lim_{\|\Delta\| \to 0} \sum_{i=1}^{n} [f(\bar{y}_i) - g(\bar{y}_i)] \, \Delta_i y$$

$$= \int_0^2 [f(y) - g(y)] \, dy$$

$$= \int_0^2 (2y - y^2) \, dy$$

$$= y^2 - \frac{1}{3} y^3 \Big]_0^2 = \frac{4}{3}$$

If $(\bar{x}, \bar{y})$ is the centroid of region R, then

$$A \cdot \bar{x} = \lim_{\|\Delta\| \to 0} \sum_{i=1}^{n} \frac{1}{2} [f(\bar{y}_i) - g(\bar{y}_i)][f(\bar{y}_i) + g(\bar{y}_i)] \, \Delta_i y$$

$$= \frac{1}{2} \int_0^2 [f(y) - g(y)][f(y) + g(y)] \, dy$$

$$= \frac{1}{2} \int_0^2 (2y - y^2)^2 \, dy$$

$$= \frac{1}{2} \int_0^2 (4y^2 - 4y^3 + y^4) \, dy$$

$$= \frac{1}{2} \left[\frac{4}{3} y^3 - y^4 + \frac{1}{5} y^5 \right]_0^2 = \frac{8}{15}$$

Because $A = \frac{4}{3}$, we have

$$\tfrac{4}{3} \bar{x} = \tfrac{8}{15}$$

$$\bar{x} = \tfrac{2}{5}$$

Furthermore,

$$A \cdot \bar{y} = \lim_{\|\Delta\| \to 0} \sum_{i=1}^{n} [f(\bar{y}_i) - g(\bar{y}_i)] \, \bar{y}_i \, \Delta_i y$$

$$= \int_0^2 [f(y) - g(y)] \, y \, dy$$

$$= \int_0^2 (2y^2 - y^3) \, dy$$

$$= \frac{2}{3} y^3 - \frac{1}{4} y^4 \Big]_0^2 = \frac{4}{3}$$

Because $A = \frac{4}{3}$, we have

$$\tfrac{4}{3} \bar{y} = \tfrac{4}{3}$$

$$\bar{y} = 1$$

Therefore, the centroid of region R is $(\frac{2}{5}, 1)$.

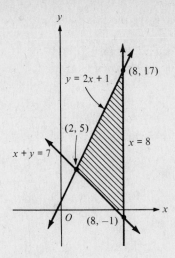

Figure 11. 2. 12

12. The lines $y = 2x + 1$, $x + y = 7$, and $x = 8$.

SOLUTION: Figure 11. 2. 12 shows a sketch of the region R. Because R is bounded above by the line $y = 2x + 1$, we take $f(x) = 2x + 1$. Because R is bounded below by the line $x + y = 7$, or, equivalently, by the line $y = -x + 7$, we take $g(x) = -x + 7$. R is bounded on the left by the line $x = 2$ and bounded on the right by the line $x = 8$. Therefore, the area of R is given by

$$A = \lim_{\|\Delta\| \to 0} \sum_{i=1}^{n} [f(\bar{x}_i) - g(\bar{x}_i)] \Delta_i x$$

$$= \int_2^8 [f(x) - g(x)] \, dx$$

$$= \int_2^8 [(2x + 1) - (-x + 7)] \, dx$$

$$= \int_2^8 (3x - 6) \, dx$$

$$= \frac{3}{2} x^2 - 6x \Big]_2^8 = 54$$

and

$$A \cdot \bar{x} = \lim_{\|\Delta\| \to 0} \sum_{i=1}^{n} [f(\bar{x}_i) - g(\bar{x}_i)] \bar{x}_i \Delta_i x$$

$$= \int_2^8 [f(x) - g(x)] x \, dx$$

$$= \int_2^8 (3x^2 - 6x) \, dx$$

$$= x^3 - 3x^2 \Big]_2^8 = 324$$

Therefore,

$$54\bar{x} = 324$$
$$\bar{x} = 6$$

Furthermore,

$$A \cdot \bar{y} = \lim_{\|\Delta\| \to 0} \sum_{i=1}^{n} \frac{1}{2} [f(\bar{x}_i) - g(\bar{x}_i)][f(\bar{x}_i) + g(\bar{x}_i)] \Delta_i x$$

$$= \frac{1}{2} \int_2^8 [f(x) - g(x)][f(x) + g(x)] \, dx$$

$$= \frac{1}{2} \int_2^8 (3x - 6)(x + 8) \, dx$$

$$= \frac{3}{2} \int_2^8 (x^2 + 6x - 16) \, dx$$

$$= \frac{3}{2} \Big[\frac{1}{3} x^3 + 3x^2 - 16x \Big]_2^8 = 378$$

Therefore

$$54\bar{y} = 378$$
$$\bar{y} = 7$$

The centroid of R is $(6, 7)$.

16. The curves $y = \sin x$ and $y = \cos x$, and the y-axis, in the first quadrant.

SOLUTION: Figure 11. 2. 16 shows a sketch of the region. We take $f(x) = \cos x$ and $g(x) = \sin x$. If A square units is the area of the region, then

$$A = \lim_{\|\Delta\| \to 0} \sum_{i=1}^{n} [f(\bar{x}_i) - g(\bar{x}_i)] \, \Delta_i x$$

$$= \int_{0}^{\pi/4} (\cos x - \sin x) \, dx$$

$$= \sin x + \cos x \Big]_{0}^{\pi/4} = \sqrt{2} - 1$$

If $(\bar{x}, \bar{y})$ is the centroid of the region, then

$$A\bar{x} = \lim_{\|\Delta\| \to 0} \sum_{i=1}^{n} [f(\bar{x}_i) - g(\bar{x}_i)] \bar{x}_i \, \Delta_i x$$

$$= \int_{0}^{\pi/4} (\cos x - \sin x) \, x \, dx$$

We use integration by parts with $u = x$ and $dv = (\cos x - \sin x) \, dx$. Thus, $du = dx$ and $v = \sin x + \cos x$.

$$A\bar{x} = x(\sin x + \cos x) \Big]_{0}^{\pi/4} - \int_{0}^{\pi/4} (\sin x + \cos x) \, dx$$

$$= \frac{1}{4}\sqrt{2}\,\pi - [-\cos x + \sin x \Big]_{0}^{\pi/4}$$

$$= \frac{1}{4}\sqrt{2}\,\pi - 1$$

Because $A = \sqrt{2} - 1$, then

$$(\sqrt{2} - 1)\bar{x} = \tfrac{1}{4}\sqrt{2}\,\pi - 1$$

$$\bar{x} = \frac{\tfrac{1}{4}\sqrt{2}\,\pi - 1}{\sqrt{2} - 1} = 0.267$$

Furthermore,

$$A\bar{y} = \lim_{\|\Delta\| \to 0} \sum_{i=1}^{n} \frac{1}{2}[f(\bar{x}_i) - g(\bar{x}_i)][f(\bar{x}_i) + g(\bar{x}_i)] \, \Delta_i x$$

$$= \frac{1}{2}\int_{0}^{\pi/4} (\cos x - \sin x)(\cos x + \sin x) \, dx$$

$$= \frac{1}{2}\int_{0}^{\pi/4} (\cos^2 x - \sin^2 x) \, dx$$

$$= \frac{1}{2}\int_{0}^{\pi/4} \cos 2x \, dx$$

$$= \frac{1}{4}\sin 2x \Big]_{0}^{\pi/4} = \frac{1}{4}$$

Hence,

$$(\sqrt{2} - 1)\bar{y} = \tfrac{1}{4}$$

$$\bar{y} = \frac{1}{4(\sqrt{2} - 1)} = 0.604$$

The centroid is $(0.267, 0.604)$.

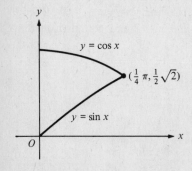

$(\tfrac{1}{4}\pi, \tfrac{1}{2}\sqrt{2})$

$y = \cos x$

$y = \sin x$

Figure 11. 2. 16

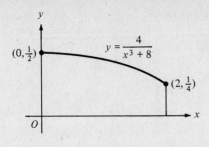

Figure 11. 2. 20

20. Find the abscissa of the centroid of the region bounded by the curve $y = 4/(x^3 + 8)$, the x-axis, the y-axis, and the line $x = 2$.

SOLUTION: Figure 11. 2. 20 shows a sketch of the region. Let $f(x) = 4/(x^3 + 8)$. If A square units is the area of the region, then

$$A = \lim_{\|\Delta\| \to 0} \sum_{i=1}^{n} f(\bar{x}_i) \, \Delta_i x$$

$$= \int_0^2 \frac{4}{x^3 + 8} \, dx \tag{1}$$

Let

$$\frac{4}{x^3 + 8} = \frac{4}{(x + 2)(x^2 - 2x + 4)} = \frac{A}{x + 2} + \frac{B(2x - 2) + C}{x^2 - 2x + 4}$$

Then

$$4 = A(x^2 - 2x + 4) + [B(2x - 2) + C](x + 2)$$

If $x = -2$, then

$$4 = A(12)$$

$$A = \tfrac{1}{3}$$

Thus,

$$4 = \tfrac{1}{3}(x^2 - 2x + 4) + 2Bx^2 - 2Bx + Cx + 4Bx - 4B + 2C$$
$$4 = (\tfrac{1}{3} + 2B)x^2 + (-\tfrac{2}{3} + 2B + C)x + (\tfrac{4}{3} - 4B + 2C)$$

So

$$0 = \tfrac{1}{3} + 2B$$
$$B = -\tfrac{1}{6}$$

And

$$0 = -\tfrac{2}{3} + 2B + C$$
$$0 = -\tfrac{2}{3} + 2(-\tfrac{1}{6}) + C$$
$$C = 1$$

Therefore,

$$\frac{4}{x^3 + 8} = \frac{\tfrac{1}{3}}{x + 2} + \frac{-\tfrac{1}{6}(2x - 2) + 1}{x^2 - 2x + 4}$$

By substituting into Eq. (1), we obtain

$$A = \frac{1}{3} \int_0^2 \frac{dx}{x + 2} - \frac{1}{6} \int_0^2 \frac{(2x - 2) \, dx}{x^2 - 2x + 4} + \int_0^2 \frac{dx}{(x - 1)^2 + 3}$$

$$= \frac{1}{3} \ln|x + 2| \Big]_0^2 - \frac{1}{6} \ln|x^2 - 2x + 4| \Big]_0^2 + \frac{1}{\sqrt{3}} \tan^{-1}\left(\frac{x - 1}{\sqrt{3}}\right) \Big]_0^2$$

$$= \frac{1}{3} (\ln 4 - \ln 2) - \frac{1}{6} (\ln 4 - \ln 4) + \frac{1}{\sqrt{3}} \left[\tan^{-1}\left(\frac{1}{\sqrt{3}}\right) - \tan^{-1}\left(\frac{-1}{\sqrt{3}}\right) \right]$$

$$= \frac{1}{3} \ln 2 + \frac{1}{\sqrt{3}} \left[\frac{\pi}{6} - \left(-\frac{\pi}{6}\right) \right]$$

$$= \frac{1}{3} \ln 2 + \frac{\pi}{3\sqrt{3}}$$

$$= \frac{3 \ln 2 + \sqrt{3}\,\pi}{9} = 0.8356 \tag{2}$$

If $\bar{x}$ is the abscissa of the centroid of the region, then

$$A\bar{x} = \lim_{\|\Delta\| \to 0} \sum_{i=1}^{n} f(\bar{x}_i)\bar{x}_i \, \Delta_i x$$

$$= \int_0^2 \frac{4x}{x^3 + 8} \, dx \tag{3}$$

Let

$$\frac{4x}{x^3 + 8} = \frac{A}{x + 2} + \frac{B(2x - 2) + C}{x^2 - 2x + 4}$$

Then

$$4x = A(x^2 - 2x + 4) + [B(2x - 2) + C](x + 2)$$

If $x = -2$, then

$$-8 = 12A$$
$$A = -\tfrac{2}{3}$$

So

$$4x = -\tfrac{2}{3}(x^2 - 2x + 4) + 2Bx^2 - 2Bx + Cx + 4Bx - 4B + 2C$$
$$4x = (-\tfrac{2}{3} + 2B)x^2 + (\tfrac{4}{3} + 2B + C)x + (-\tfrac{8}{3} - 4B + 2C)$$

Thus

$$0 = -\tfrac{2}{3} + 2B$$
$$B = \tfrac{1}{3}$$

and

$$4 = \tfrac{4}{3} + 2B + C$$
$$4 = \tfrac{4}{3} + 2(\tfrac{1}{3}) + C$$
$$C = 2$$

Hence,

$$\frac{4x}{x^3 + 8} = \frac{-\tfrac{2}{3}}{x + 2} + \frac{\tfrac{1}{3}(2x - 2) + 2}{x^2 - 2x + 4}$$

and from (3) we get

$$A\bar{x} = -\frac{2}{3} \int_0^2 \frac{dx}{x + 2} + \frac{1}{3} \int_0^2 \frac{(2x - 2) \, dx}{x^2 - 2x + 4} + 2 \int_0^2 \frac{dx}{(x - 1)^2 + 3}$$

$$= -\frac{2}{3} \ln|x + 2| \Big]_0^2 + \frac{1}{3} \ln|x^2 - 2x + 4| \Big]_0^2 + \frac{2}{\sqrt{3}} \tan^{-1}\left(\frac{x - 1}{\sqrt{3}}\right) \Big]_0^2$$

$$= -\frac{2}{3}(\ln 4 - \ln 2) + \frac{1}{3}(\ln 4 - \ln 4) + \frac{2}{\sqrt{3}}\left[\frac{\pi}{6} - \left(-\frac{\pi}{6}\right)\right]$$

$$= -\frac{2}{3} \ln 2 + \frac{2\pi}{3\sqrt{3}}$$

$$= \frac{-6 \ln 2 + 2\sqrt{3}\,\pi}{9} = 0.7471$$

By substituting the value of A given in (2), we have

$$\frac{3 \ln 2 + \sqrt{3}\,\pi}{9} \, \bar{x} = \frac{-6 \ln 2 + 2\sqrt{3}\,\pi}{9}$$

$$\bar{x} = \frac{-6 \ln 2 + 2\sqrt{3}\,\pi}{3 \ln 2 + \sqrt{3}\,\pi} = 0.894$$

Thus the abscissa of the centroid of the region is 0.894.

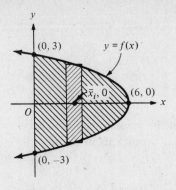

Figure 11. 2. 24

24. Find the center of mass of the lamina bounded by the parabola $2y^2 = 18 - 3x$ and the y-axis if the area density at any point (x, y) is x kg/m^2.

SOLUTION: If ρ is the density function, we are given that $\rho(x) = x$. Figure 11. 2. 24 shows a sketch of the region R. First, we find the mass of the lamina. Because ρ is a function of x, we take elements of area that are perpendicular to the x-axis with width $\Delta_i x$ meters. Solving the given equation for y, we obtain $y = \pm\sqrt{9 - \frac{3}{2}x}$. Thus, we let $f(x) = \sqrt{9 - \frac{3}{2}x}$, and the measure of the element of area is given by $2f(\bar{x}_i) \Delta_i x$. Because the element of mass is the product of the element of area and the element of density the element of mass is given by

$$\Delta_i M = 2\rho(\bar{x}_i)f(\bar{x}_i) \Delta_i x \tag{1}$$

Therefore,

$$M = \lim_{\|\Delta\| \to 0} \sum_{i=1}^{n} 2\rho(\bar{x}_i)f(\bar{x}_i) \Delta_i x$$

Because x is in the interval [0, 6], we have

$$M = \int_0^6 2\rho(x)f(x)\, dx$$

$$= \int_0^6 2x\sqrt{9 - \frac{3}{2}x}\, dx \tag{2}$$

Let $u = \sqrt{9 - \frac{3}{2}x}$. Then $x = \frac{2}{3}(9 - u^2)$; $dx = -\frac{4}{3}u\, du$; $u = 3$ when $x = 0$; and $u = 0$ when $x = 6$. Thus, from (2) we obtain

$$M = \int_3^0 2\left[\frac{2}{3}(9 - u^2)\right]u\left(-\frac{4}{3}u\, du\right)$$

$$= \frac{16}{9}\int_0^3 (9u^2 - u^4)\, du$$

$$= \frac{16}{9}\left[3u^3 - \frac{1}{5}u^5\right]_0^3 = \frac{288}{5} \tag{3}$$

Multiplying on both sides of (1) by $\bar{x}_i$, the abscissa of the centroid of the element of mass, we obtain

$$\Delta_i M \cdot \bar{x} = 2\rho(\bar{x}_i)f(\bar{x}_i)\bar{x}_i \Delta_i x$$

Therefore,

$$M\bar{x} = \lim_{\|\Delta\| \to 0} \sum_{i=1}^{n} 2\rho(\bar{x}_i)f(\bar{x}_i)\bar{x}_i \Delta_i x$$

$$= \int_0^6 2x^2 f(x)\, dx$$

$$= \int_0^6 2x^2\sqrt{9 - \frac{3}{2}x}\, dx \tag{4}$$

With the same choice of u as above, from (4) we obtain

$$M\bar{x} = \int_3^0 2\left[\frac{2}{3}(9 - u^2)\right]^2 u\left(-\frac{4}{3}u\, du\right)$$

$$= \frac{32}{27}\int_0^3 (81u^2 - 18u^4 + u^6)\, du$$

$$= \frac{32}{27}\left[27u^3 - \frac{18}{5}u^5 + \frac{1}{7}u^7\right]_0^3 = \frac{6912}{35} \tag{5}$$

Substituting from (3) into (5), we get

$$\tfrac{288}{5}\bar{x} = \tfrac{6912}{35}$$
$$\bar{x} = \tfrac{24}{7}$$

Because $\bar{y}_i = 0$ for each element of area, $\Delta_i M \, \bar{y}_i = 0$, and thus $\bar{y} = 0$. We conclude that the center of mass of the given lamina is $(\tfrac{24}{7}, 0)$.

28. Use the theorem of Pappus to find the volume of the torus (doughnut-shaped figure) generated by revolving a circle with a radius of r units about a line in its plane at a distance of b units from its center where $b > r$.

SOLUTION: By the theorem of Pappus, if a region R is revolved about a line L in the plane of R that does not cut the region R, then the measure of the volume of the resulting solid of revolution is given by

$$V = 2\pi \bar{r} A$$

where $\bar{r}$ units is the distance between the line L and the centroid of the region R, and A is the measure of the area of R. We have $\bar{r} = b$ and $A = \pi r^2$. Thus

$$V = 2\pi b (\pi r^2) = 2\pi^2 b r^2$$

The volume of the torus is $2\pi^2 b r^2$ cubic units.

32. Let R be the region bounded by the semicircle $y = \sqrt{r^2 - x^2}$ and the x-axis. Use the theorem of Pappus to find the volume of the solid of revolution generated by revolving R about the line $x - y = r$. (*Hint:* Use the result of Exercise 22 in Exercises 4.7.)

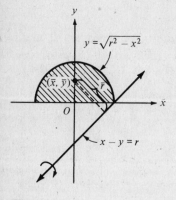

Figure 11. 2. 32

SOLUTION: Figure 11. 2. 32 shows a sketch of the region R and the line $x - y = r$. As in Exercise 28, the volume of the solid of revolution is given by

$$V = 2\pi \bar{r} A \qquad (1)$$

Because the radius of the circle that forms part of the boundary of the region R is r, the area of R is given by

$$A = \tfrac{1}{2}\pi r^2 \qquad (2)$$

Let $(\bar{x}, \bar{y})$ be the centroid of region R. By symmetry, $\bar{x} = 0$. Let $f(x) = \sqrt{r^2 - x^2}$. We have

$$A\bar{y} = \lim_{\|\Delta\| \to 0} \sum_{i=1}^{n} \frac{1}{2} [f(\bar{x}_i)]^2 \, \Delta_i x$$

$$= \int_{-r}^{r} \frac{1}{2} [f(x)]^2 \, dx$$

$$= \frac{1}{2} \int_{-r}^{r} (r^2 - x^2) \, dx$$

$$= \frac{1}{2} \left[r^2 x - \frac{1}{3} x^3 \right]_{-r}^{r} = \frac{2}{3} r^3 \qquad (3)$$

Substituting from (2) into (3), we have

$$\tfrac{1}{2}\pi r^2 \bar{y} = \tfrac{2}{3} r^3$$

$$\bar{y} = \frac{4r}{3\pi}$$

By Exercise 22 in Section 4.7, the distance from the point (x_1, y_1) to the line $Ax + By + C = 0$ is

$$\frac{|Ax_1 + By_1 + C|}{\sqrt{A^2 + B^2}}$$

We use this formula for the given line $x - y - r = 0$ and substitute the coordinates of the centroid $(\bar{x}, \bar{y})$. Thus

$$\bar{r} = \frac{|\bar{x} - \bar{y} - r|}{\sqrt{2}}$$

$$= \frac{\left|0 - \dfrac{4r}{3\pi} - r\right|}{\sqrt{2}}$$

$$= \frac{4r + 3\pi r}{3\sqrt{2}\,\pi} \tag{4}$$

Substituting from (4) and (2) into (1), we obtain

$$V = 2\pi\left(\frac{4r + 3\pi r}{3\sqrt{2}\,\pi}\right)\left(\frac{1}{2}\pi r^2\right)$$

$$= \frac{(4 + 3\pi)\pi\, r^3}{3\sqrt{2}} \tag{5}$$

Thus the volume of the solid of revolution is V cubic units, where V is the number given in Eq. (5).

11.3 CENTROID OF A SOLID OF REVOLUTION

The formulas given in the text for finding the centroid of S, a solid of revolution, apply only to special cases. When these formulas do not apply, we use the following more general method for finding $(\bar{x}, \bar{y}, \bar{z})$, the centroid of S.

1. Use the methods of Sections 7.2 and 7.3 to find V, the measure of the volume of S.

2. Let $\sum\limits_{i=1}^{n} V_i$ be the Riemann sum used in step 1 to find V, and let the centroid (geometric center) of the element of volume used be at the point $(\bar{x}_i, \bar{y}_i, 0)$. Then

$$V\bar{x} = \lim_{\|\Delta\| \to 0} \sum_{i=1}^{n} V_i \bar{x}_i$$

$$V\bar{y} = \lim_{\|\Delta\| \to 0} \sum_{i=1}^{n} V_i \bar{y}_i$$

$$V\bar{z} = 0$$

3. Express each of the above sums as a Riemann sum, find the corresponding definite integral, and evaluate the integral.

If either $\bar{x}_i$ or $\bar{y}_i$ is a constant, then it is not necessary to calculate the corresponding definite integral. If $\bar{x}_i$ is a constant, then $\bar{x} = \bar{x}_i$. And if $\bar{y}_i$ is a constant, then $\bar{y} = \bar{y}_i$.

Exercises 11.3

In Exercises 1–20 find the centroid of the solid of revolution generated by revolving the given region about the indicated line.

4. The region bounded by $x + 2y = 2$, the x-axis, and the y-axis, about the x-axis. Take the rectangular elements parallel to the axis of revolution.

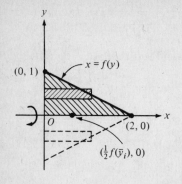

Figure 11. 3. 4

SOLUTION: Figure 11. 3. 4 shows the region R. We take an element of area perpendicular to the y-axis of width $\Delta_i y$ units and revolve the element about the x-axis. This results in an element of volume that is a cylindrical shell with the numbers of units in the thickness given by $\Delta_i r = \Delta_i y$ and the number of units in the mean radius given by $r_i = \bar{y}_i$. Solving the equation of the line for x gives $x = f(y) = -2y + 2$, and thus the measure of the altitude of the shell is given by $h_i = f(\bar{y}_i)$. Therefore,

$$V = \lim_{\|\Delta\| \to 0} \sum_{i=1}^{n} 2\pi r_i h_i \, \Delta_i r$$

$$= \lim_{\|\Delta\| \to 0} \sum_{i=1}^{n} 2\pi \bar{y}_i f(\bar{y}_i) \, \Delta_i y \tag{1}$$

$$V = 2\pi \int_0^1 y f(y) \, dy$$

$$= 2\pi \int_0^1 y(-2y + 2) \, dy$$

$$= 4\pi \int_0^1 (-y^2 + y) \, dy$$

$$= 4\pi \left[-\frac{1}{3} y^3 + \frac{1}{2} y^2 \right]_0^1 = \frac{2}{3} \pi \tag{2}$$

Because the centroid of the cylindrical shell is $(\frac{1}{2} f(\bar{y}_i), 0, 0)$, from Eq. (1) we have

$$V \cdot \bar{x} = \lim_{\|\Delta\| \to 0} \sum_{i=1}^{n} \pi \bar{y}_i [f(\bar{y}_i)]^2 \, \Delta_i y$$

$$= \pi \int_0^1 y [f(y)]^2 \, dy$$

$$= \pi \int_0^1 y(-2y + 2)^2 \, dy$$

$$= 4\pi \int_0^1 (y^3 - 2y^2 + y) \, dy$$

$$= 4\pi \left[\frac{1}{4} y^4 - \frac{2}{3} y^3 + \frac{1}{2} y^2 \right]_0^1 = \frac{1}{3} \pi \tag{3}$$

Substituting from (2) into (3), we get

$$\tfrac{2}{3} \pi \, \bar{x} = \tfrac{1}{3} \pi$$
$$\bar{x} = \tfrac{1}{2}$$

Therefore, the centroid of the solid of revolution is $(\frac{1}{2}, 0, 0)$.

8. The region bounded by $y = x^3$, $x = 2$, and the x-axis, about the line $x = 2$. Take the rectangular elements parallel to the axis of revolution.

SOLUTION: Figure 11. 3. 8 shows a sketch of the region. We take an element of area perpendicular to the x-axis of width $\Delta_i x$ units and revolve the element about the line $x = 2$. This results in an element of volume that is a cylindrical shell with the number of units of thickness given by $\Delta_i r = \Delta_i x$ and the number of units in the mean radius given by $r_i = 2 - \bar{x}_i$. If $f(x) = x^3$ then the measure of the altitude of the shell is given by $h_i = f(\bar{x}_i)$. Therefore,

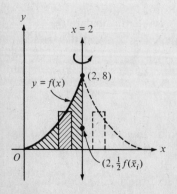

Figure 11. 3. 8

$$V = \lim_{\|\Delta\| \to 0} \sum_{i=1}^{n} 2\pi r_i h_i \, \Delta_i r$$

$$= \lim_{\|\Delta\| \to 0} \sum_{i=1}^{n} 2\pi(2 - \bar{x}_i) f(\bar{x}_i) \, \Delta_i x \tag{1}$$

$$= 2\pi \int_0^2 (2 - x) f(x) \, dx$$

$$= 2\pi \int_0^2 (2x^3 - x^4) \, dx$$

$$= 2\pi \left[\frac{1}{2} x^4 - \frac{1}{5} x^5 \right]_0^2 = \frac{16}{5} \pi \tag{2}$$

Because the centroid of the cylindrical shell is $(2, \frac{1}{2} f(\bar{x}_i), 0)$, from Eq. (1) we obtain

$$V\bar{y} = \lim_{\|\Delta\| \to 0} \sum_{i=1}^{n} \pi(2 - \bar{x}_i)[f(\bar{x}_i)]^2 \, \Delta_i x$$

$$= \pi \int_0^2 (2 - x)[f(x)]^2 \, dx$$

$$= \pi \int_0^2 (2x^6 - x^7) \, dx$$

$$= \pi \left[\frac{2}{7} x^7 - \frac{1}{8} x^8 \right]_0^2 = \frac{32}{7} \pi \tag{3}$$

Substituting from (2) into (3), we obtain

$$\tfrac{16}{5} \pi \bar{y} = \tfrac{32}{7} \pi$$
$$\bar{y} = \tfrac{10}{7}$$

We conclude that the centroid of the solid of revolution is $(2, \frac{10}{7}, 0)$.

12. The region bounded by the portion of the circle $x^2 + y^2 = 4$ in the first quadrant, the portion of the line $2x - y = 4$ in the fourth quadrant, and the y-axis, about the y-axis.

SOLUTION: Figure 11. 3. 12 shows a sketch of the region. We take an element of area perpendicular to the x-axis of width $\Delta_i x$ units and revolve the element about the y-axis. This results in an element of volume that is a cylindrical shell. The measure of the thickness of the shell is given by $\Delta_i r = \Delta_i x$ and the measure of the mean radius of the shell is given by $r_i = \bar{x}_i$. Solving the equation of the circle for y, we have $y = \pm\sqrt{4 - x^2}$, and solving the equation of the line for y we have $y = 2x - 4$. Let $f(x) = \sqrt{4 - x^2}$ and $g(x) = 2x - 4$. Then the measure of the altitude of the shell is given by $h_i = f(\bar{x}_i) - g(\bar{x}_i)$. Therefore,

$$V = \lim_{\|\Delta\| \to 0} \sum_{i=1}^{n} 2\pi \bar{x}_i [f(\bar{x}_i) - g(\bar{x}_i)] \, \Delta_i x \tag{1}$$

$$= 2\pi \int_0^2 x[\sqrt{4 - x^2} - (2x - 4)] \, dx$$

$$= -\pi \int_0^2 (4 - x^2)^{1/2}(-2x \, dx) - 4\pi \int_0^2 (x^2 - 2x) \, dx$$

$$= -\frac{2}{3}\pi(4 - x^2)^{3/2} \Big]_0^2 - 4\pi \left[\frac{1}{3} x^3 - x^2 \right]_0^2$$

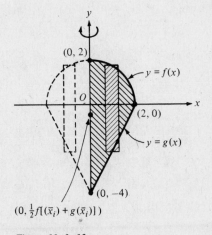

(0, 2)

$y = f(x)$

O

(2, 0)

$y = g(x)$

(0, −4)

$(0, \frac{1}{2} f[(\bar{x}_i) + g(\bar{x}_i)])$

Figure 11. 3. 12

$$= \frac{16}{3}\pi + \frac{16}{3}\pi = \frac{32}{3}\pi \tag{2}$$

Because the centroid of the shell is at the point $(0, \frac{1}{2}[f(\bar{x}_i) + g(\bar{x}_i)])$, from (1) we obtain

$$V \cdot \bar{y} = \lim_{||\Delta|| \to 0} \sum_{i=1}^{n} \pi \bar{x}_i [f(\bar{x}_i) - g(\bar{x}_i)][f(\bar{x}_i) + g(\bar{x}_i)] \, \Delta_i x$$

$$= \pi \int_0^2 x([f(x)]^2 - [g(x)]^2) \, dx$$

$$= \pi \int_0^2 (-5x^3 + 16x^2 - 12x) \, dx$$

$$= \pi \left[-\frac{5}{4}x^4 + \frac{16}{3}x^3 - 6x^2 \right]_0^2 = -\frac{4}{3}\pi \tag{3}$$

Substituting from (2) into (3), we get

$$\tfrac{32}{3}\pi \bar{y} = -\tfrac{4}{3}\pi$$
$$\bar{y} = -\tfrac{1}{8}$$

The centroid of the solid is at the point $(0, -\frac{1}{8}, 0)$.

16. The region bounded by $y = \sqrt{4px}$, the x-axis, and the line $x = p$, about the line $y = 2p$.

SOLUTION: Figure 11. 3. 16 shows a sketch of the region. We take an element of area perpendicular to the x-axis of width $\Delta_i x$ units and revolve the element about the line $y = 2p$. This results in an element of volume that is a circular ring. The measure of the thickness of the ring is given by $\Delta_i h = \Delta_i x$, and the measure of the outer radius is given by $R_i = 2p$. We let $f(x) = \sqrt{4px}$. Then the measure of the inner radius of the ring is given by $r_i = 2p - f(\bar{x}_i)$. Thus,

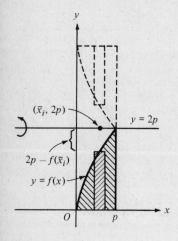

Figure 11. 3. 16

$$V = \lim_{||\Delta|| \to 0} \sum_{i=1}^{n} \pi (R_i^2 - r_i^2) \, \Delta_i h$$

$$= \lim_{||\Delta|| \to 0} \sum_{i=1}^{n} \pi ([2p]^2 - [2p - f(\bar{x}_i)]^2) \, \Delta_i x \tag{1}$$

$$= \pi \int_0^p ([2p]^2 - [2p - f(x)]^2) \, dx$$

$$= \pi \int_0^p (4pf(x) - [f(x)]^2) \, dx$$

$$= \pi \int_0^p [4p\sqrt{4px} - 4px] \, dx$$

$$= 4p\pi \int_0^p (2\sqrt{p}x^{1/2} - x) \, dx$$

$$= 4p\pi \left[\frac{4}{3}\sqrt{p}x^{3/2} - \frac{1}{2}x^2 \right]_0^p = \frac{10}{3}\pi p^3 \tag{2}$$

Because the centroid of the ring is at the point $(\bar{x}_i, 2p)$, from (1) we obtain

$$V \cdot \bar{x} = \lim_{||\Delta|| \to 0} \sum_{i=1}^{n} \pi ([2p]^2 - [2p - f(\bar{x}_i)]^2 \bar{x}_i \, \Delta_i x$$

$$= \int_0^p \pi [4p^2 - (2p - \sqrt{4px})^2] x \, dx$$

$$= \pi \int_0^p (-4px^2 + 8p^{3/2}x^{3/2})\, dx$$

$$= \pi \left[-\frac{4}{3} px^3 + \frac{16}{5} p^{3/2}x^{5/2} \right]_0^p = \frac{28}{15} \pi p^4 \qquad (3)$$

Substituting from (2) into (3), we get

$$\tfrac{10}{3}\pi p^3 \bar{x} = \tfrac{28}{15}\pi p^4$$

$$\bar{x} = \tfrac{14}{25}p$$

We conclude that the centroid of the solid of revolution is at the point $(\frac{14}{25}p,$
$2p, 0)$.

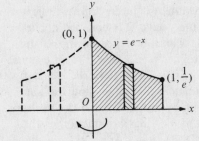

Figure 11. 3. 20

20. The region bounded by $y = e^{-x}$, the coordinate axes, and the line $x = 1$, about the y-axis.

SOLUTION: Figure 11. 3. 20 shows a sketch of the region and a plane section of the solid of revolution. Let $f(x) = e^{-x}$. The element of volume is a cylindrical shell with thickness $\Delta_i r = \Delta_i x$ units, mean radius $r_i = \bar{x}_i$ units and altitude $h_i = f(\bar{x}_i)$ units. If V cubic units is the volume of the solid of revolution, then

$$V = \lim_{\|\Delta\| \to 0} \sum_{i=1}^n 2\pi \bar{x}_i f(\bar{x}_i)\, \Delta_i x$$

$$= 2\pi \int_0^1 xe^{-x}\, dx$$

We integrate by parts with $u = x$ and $dv = e^{-x}\, dx$. Then $du = dx$ and $v = -e^{-x}$. Hence,

$$V = 2\pi \left[-xe^{-x} + \int e^{-x}\, dx \right]_0^1$$

$$= 2\pi \left[-xe^{-x} - e^{-x} \right]_0^1$$

$$= 2\pi[(-e^{-1} - e^{-1}) - (-e^0)]$$

$$= 2\pi(1 - 2e^{-1}) = 1.6603$$

The centroid of the element of volume is at the point $(0, \frac{1}{2}f(\bar{x}_i))$. Therefore, if $(\bar{x}, \bar{y})$ is the centroid of the solid of revolution, then $\bar{x} = 0$, and

$$V\bar{y} = \lim_{\|\Delta\| \to 0} \sum_{i=1}^n [2\pi \bar{x}_i f(\bar{x}_i)] \left[\frac{1}{2}f(\bar{x}_i)\right] \Delta_i x$$

$$= \pi \int_0^1 xe^{-2x}\, dx$$

We integrate by parts with $u = x$ and $dv = e^{-2x}\, dx$. Then $du = dx$ and $v = -\frac{1}{2}e^{-2x}$. Thus,

$$V\bar{y} = \pi \left[-\frac{1}{2} xe^{-2x} + \frac{1}{2}\int e^{-2x}\, dx \right]_0^1$$

$$= \pi \left[-\frac{1}{2} xe^{-2x} - \frac{1}{4} e^{-2x} \right]_0^1$$

$$= \pi[(-\tfrac{1}{2}e^{-2} - \tfrac{1}{4}e^{-2}) - (-\tfrac{1}{4}e^0)]$$

$$= \pi(-\tfrac{3}{4}e^{-2} + \tfrac{1}{4})$$

$$= \tfrac{1}{4}\pi(1 - 3e^{-2}) = 0.4665$$

Substituting the value found for V, we have

$$2\pi(1 - 2e^{-1})\bar{y} = \tfrac{1}{4}\pi(1 - 3e^{-2})$$

$$\bar{y} = \frac{1 - 3e^{-2}}{8(1 - 2e^{-1})} = 0.281$$

The centroid of the solid of revolution is (0, 0.281).

24. Suppose that a cylindrical hole with a radius of r units is bored through a solid wooden hemisphere of radius $2r$ units, so that the axis of the cylinder is the same as the axis of the hemisphere. Find the centroid of the solid remaining.

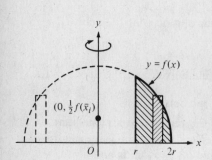

Figure 11. 3. 24

SOLUTION: The solid remaining is a solid of revolution if the region R, indicated by shading in Fig. 11. 3. 24, is revolved about the y-axis. R is bounded above by the circle $x^2 + y^2 = 4r^2$, bounded below by the x-axis, bounded on the left by the line $x = r$, and bounded on the right by the line $x = 2r$. We take an element of area perpendicular to the x-axis of width $\Delta_i x$ units and revolve the element about the y-axis. This results in an element of volume that is a cylindrical shell. The measure of the thickness of the shell is given by $\Delta_i r = \Delta_i x$ and the measure of the mean radius of the shell is given by $r_i = \bar{x}_i$. Solving the equation of the circle for y, we obtain $y = \pm\sqrt{4r^2 - x^2}$. Then let $f(x) = \sqrt{4r^2 - x^2}$, and the measure of the altitude of the shell is given by $h_i = f(\bar{x}_i)$. Therefore,

$$V = \lim_{\|\Delta\| \to 0} \sum_{i=1}^{n} 2\pi\bar{x}_i f(\bar{x}_i)\, \Delta_i x \tag{1}$$

$$= \int_{r}^{2r} 2\pi x\sqrt{4r^2 - x^2}\, dx$$

$$= -\pi \int_{r}^{2r} (4r^2 - x^2)^{1/2}(-2x\, dx)$$

$$= -\frac{2}{3}\,\pi(4r^2 - x^2)^{3/2}\Big]_{r}^{2r} = 2\pi\sqrt{3}\, r^3 \tag{2}$$

Because the centroid of the shell is at the point $(0, \tfrac{1}{2}f(x_i))$, from (1) we obtain

$$V \cdot \bar{y} = \lim_{\|\Delta\| \to 0} \sum_{i=1}^{n} \pi\bar{x}_i[f(\bar{x}_i)]^2\, \Delta_i x$$

$$= \int_{r}^{2r} \pi x(\sqrt{4r^2 - x^2})^2\, dx$$

$$= \pi \int_{r}^{2r} (4r^2 x - x^3)\, dx$$

$$= \pi \left[2r^2 x^2 - \frac{1}{4}x^4 \right]_{r}^{2r} = \frac{9}{4}\,\pi r^4 \tag{3}$$

Substituting from (2) into (3), we get

$$2\pi\sqrt{3}\, r^3\bar{y} = \tfrac{9}{4}\pi r^4$$
$$\bar{y} = \tfrac{3}{8}\sqrt{3}\, r$$

We conclude that the centroid of the solid remaining is $\tfrac{3}{8}\sqrt{3}\, r$ units above the center of the base.

11.4 LIQUID PRESSURE

If a flat surface with area A square units is submerged horizontally at a depth of h units in a liquid with density ρ pounds per cubic unit, then the total force due to liquid pressure on the surface is F pounds, given by

$$F = \rho h A \tag{1}$$

The following definition can sometimes be used to calculate the total force due to liquid pressure. Its use is illustrated in Exercises 4 and 20.

11.4.1 Definition Suppose that a flat plate is submerged vertically in a liquid of weight ρ pounds per cubic unit. The length of the plate at a depth of x units below the surface of the liquid is $f(x)$ units, where f is continuous on the closed interval $[a, b]$ and $f(x) \geq 0$ on $[a, b]$. Then F, the number of pounds of force caused by liquid pressure on the plate, is given by

$$F = \lim_{\|\Delta\|\to 0} \sum_{i=1}^{n} \rho \bar{x}_i f(\bar{x}_i)\, \Delta_i x = \int_a^b \rho x f(x)\, dx$$

When Definition 11.4.1 does not apply, we make a partition of some interval $[a, b]$ on the x-axis and use $\Delta_i A$, h_i, and $\Delta_i F$, the elements of area, depth, and force, respectively, and Formula (1) to obtain

$$\Delta_i F = \rho h_i \Delta_i A$$

from which we have

$$F = \lim_{\|\Delta\|\to 0} \sum_{i=1}^{n} \rho h_i\, \Delta_i A \qquad (2)$$

If the centroid of a vertical plane region is $\bar{x}$ units below the surface of the liquid, then the total force due to the liquid pressure against the vertical region is the same as it would be if the region were horizontal at a depth of $\bar{x}$ units below the surface of the liquid. Therefore, if A square units is the area of such a region, and ρ units is the density of the liquid, then $\rho \bar{x} A$ units is the force due to liquid pressure.

Exercises 11.4

4. A plate in the shape of an isosceles right triangle is submerged vertically in a tank of water, with one leg lying in the surface. The legs are each 6 ft long. Find the force due to liquid pressure on one side of the plate.

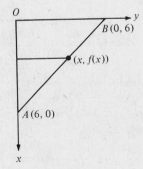

Figure 11. 4. 4

SOLUTION: Figure 11. 4. 4 illustrates the plate AOB. The y-axis is in the surface of the water. Because an equation of line AB is $y = -x + 6$, we let $f(x) = -x + 6$. The length of the plate at a depth of x units is $f(x)$ units with x in $[0, 6]$. Thus, we use Definition 11.4.1 to find the total force F due to liquid pressure on the plate.

$$F = \lim_{\|\Delta\|\to 0} \sum_{i=1}^{n} \rho \bar{x}_i f(\bar{x}_i)\, \Delta_i x$$

$$= \int_a^b \rho x f(x)\, dx$$

$$= \int_0^6 \rho x(-x + 6)\, dx$$

$$= \rho \left[-\frac{1}{3} x^3 + 3x^2 \right]_0^6 = 36\rho$$

The total force due to liquid pressure is 36ρ pounds, where ρ pounds per cubic ft is the density of water.

8. The face of a gate of a dam is vertical and in the shape of an isosceles trapezoid 3 meters wide at the top, 4 meters wide at the bottom, and 3 meters high. If the

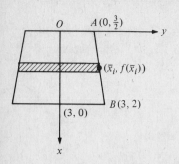

Figure 11. 4. 8

upper base is 20 meters below the surface of the water, find the total force due to liquid pressure on the gate.

SOLUTION: Figure 11. 4. 8 shows the face of the gate. The origin is at the center of the upper base of the trapezoid. Because the surface of the water is not at the origin, x units does not represent the depth of the water, and we cannot use Definition 11.4.1. However, we may partition the interval $[0, 3]$ on the x-axis, and let $\Delta_i A$, h_i, and $\Delta_i F$ be the elements of the area, depth, and force, respectively, for the ith subinterval. Then we have,

$$\Delta_i F = \rho h_i \, \Delta_i A \tag{1}$$

Because an equation of line AB is $y = \frac{1}{6}x + \frac{3}{2}$, we let $f(x) = \frac{1}{6}x + \frac{3}{2}$. Then the element of area is a rectangle with width $\Delta_i x$ meters and length $2 f(\bar{x}_i)$ meters, where $\bar{x}_i$ is some number in the ith subinterval. Thus, the element of area is given by

$$\Delta_i A = 2 f(\bar{x}_i) \, \Delta_i x = (\tfrac{1}{3}\bar{x}_i + 3) \, \Delta_i x \tag{2}$$

Because the upper base is 20 m below the surface of the water, the element of depth is given by

$$h_i = \bar{x}_i + 20 \tag{3}$$

Substituting from (2) and (3) into (1), we obtain

$$\Delta_i F = \rho(\tfrac{1}{3}\bar{x}_i + 3)(\bar{x}_i + 20) \, \Delta_i x$$

Therefore,

$$F = \lim_{\|\Delta\| \to 0} \sum_{i=1}^{n} \rho\left(\frac{1}{3}\bar{x}_i + 3\right)(\bar{x}_i + 20) \, \Delta_i x$$

$$= \int_0^3 \rho\left(\frac{1}{3}x + 3\right)(x + 20) \, dx$$

$$= \rho \int_0^3 \left(\frac{1}{3}x^2 + \frac{29}{3}x + 60\right) dx$$

$$= \rho \left[\frac{1}{9}x^3 + \frac{29}{6}x^2 + 60x\right]_0^3 = 226.5\rho$$

The total force is 226.5ρ newtons, where ρ kg/m³ is the density of water.

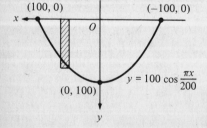

Figure 11. 4. 12

12. The face of a dam is in the shape of one arch of the curve $y = 100 \cos \frac{1}{200}\pi x$ and the surface of the water is at the top of the dam. Find the force due to water pressure on the face of the dam, if distance is measured in feet.

SOLUTION: Figure 11. 4. 12 shows a sketch of the face of the dam. We have taken the positive y-axis downward so that the surface of the water, which is along the x-axis, will be at the top of the dam. Let $f(x) = 100 \cos \frac{1}{200}\pi x$. We take an element of area with width $\Delta_i x$ ft and length $f(\bar{x}_i)$ ft. Thus, the element of area is $\Delta_i A$ sq ft, with

$$\Delta_i A = f(\bar{x}_i) \, \Delta_i x$$

Because the ordinate of the centroid of the element of area is $\frac{1}{2}f(\bar{x}_i)$, the element of depth is h_i ft, where

$$h_i = \tfrac{1}{2}f(\bar{x}_i)$$

Hence, if $\Delta_i F$ pounds is the element of force, then

$$\Delta_i F = \rho h_i \, \Delta_i A$$

$$= \frac{1}{2} \rho [f(\bar{x}_i)]^2 \Delta_i x$$

where ρ pounds per cubic foot is the density of water. Therefore, if F pounds is the force due to water pressure on the face of the dam, then

$$F = \lim_{\|\Delta\| \to 0} \sum_{i=1}^{n} \frac{1}{2} \rho [f(\bar{x}_i)]^2 \Delta_i x$$

$$= \frac{1}{2} \rho \int_{-100}^{100} 100^2 \cos^2 \frac{\pi x}{200} \, dx$$

Because f is an even function, then by symmetry

$$\int_{-a}^{a} f(x) \, dx = 2 \int_{0}^{a} f(x) \, dx$$

Hence,

$$F = 100^2 \rho \int_{0}^{100} \cos^2 \frac{\pi x}{200} \, dx$$

$$= \frac{100^2 \rho}{2} \int_{0}^{100} \left(1 + \cos \frac{\pi x}{100}\right) dx$$

$$= \frac{100^2 \rho}{2} \left[x + \frac{100}{\pi} \sin \frac{\pi x}{100} \right]_{0}^{100}$$

$$= \frac{100^2 \rho}{2} \left[100 + \frac{100}{\pi} \sin \pi \right]$$

$$= \frac{100^3 \rho}{2} = (5.00) 10^5 \rho$$

The force due to water pressure on the face of the dam is $(5.00)10^5 \rho$ pounds.

16. A gate in an irrigation ditch is in the shape of a segment of a circle of radius 4 ft. The top of the gate is horizontal and 3 ft above the lowest point on the gate. If the water level is 2 ft above the top of the gate, find the force on the gate due to water pressure.

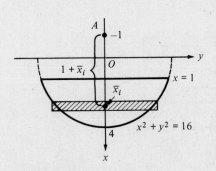

Figure 11. 4. 16

SOLUTION: Refer to Fig. 11. 4. 16. Because the gate is bounded by the lines $x = 1$ and $x = 4$, we partition the interval $[1, 4]$ on the x-axis. Solving the equation of the circle $x^2 + y^2 = 16$ for y, we have $y = \pm\sqrt{16 - x^2}$. Let $f(x) = \sqrt{16 - x^2}$. Then the number of square feet in the element of area is given by $\Delta_i A = 2 f(\bar{x}_i) \Delta_i x$. Because the water level is 2 ft above the top of the gate, the water level is at the point $A = (-1, 0)$ on the x-axis. Thus, the number of feet in the depth is given by $h_i = 1 + \bar{x}_i$. Hence,

$$F = \lim_{\|\Delta\| \to 0} \sum_{i=1}^{n} \rho h_i \Delta_i A$$

$$= \lim_{\|\Delta\| \to 0} \sum_{i=1}^{n} 2\rho (1 + \bar{x}_i) f(\bar{x}_i) \Delta_i x$$

$$= 2\rho \int_{1}^{4} (1 + x) f(x) \, dx$$

$$= 2\rho \int_{1}^{4} (1 + x) \sqrt{16 - x^2} \, dx$$

$$= 2\rho \left[\int_{1}^{4} \sqrt{16 - x^2} \, dx + \int_{1}^{4} x\sqrt{16 - x^2} \, dx \right] \tag{1}$$

We evaluate the integrals on the right-hand side of (1) separately. For the second integral we have

$$\int_1^4 x\sqrt{16 - x^2}\, dx = -\frac{1}{2}\int_1^4 (16 - x^2)^{1/2}(-2x\, dx)$$

$$= -\frac{1}{2}\cdot\frac{2}{3}(16 - x^2)^{3/2}\Big]_1^4$$

$$= -\frac{1}{3}[0 - 15^{3/2}] = 5\sqrt{15} \tag{2}$$

For the first integral on the right-hand side of (1), we use a trigonometric substitution. Thus, we let

$$\theta = \sin^{-1}\tfrac{1}{4}x \tag{3}$$

Then

$$x = 4\sin\theta$$
$$dx = 4\cos\theta\, d\theta$$

and

$$\sqrt{16 - x^2} = \sqrt{16 - 16\sin^2\theta}$$
$$= 4\cos\theta \tag{4}$$

Hence

$$\int\sqrt{16 - x^2}\, dx = \int (4\cos\theta)(4\cos\theta\, d\theta)$$

$$= 8\int (1 + \cos 2\theta)d\theta$$

$$= 8(\theta + \tfrac{1}{2}\sin 2\theta) + C$$
$$= 8(\theta + \sin\theta\cos\theta) + C \tag{5}$$

From (3) we have $\sin\theta = \tfrac{1}{4}x$, and from (4) we have $\cos\theta = \tfrac{1}{4}\sqrt{16 - x^2}$. Substituting in (5) we obtain

$$\int\sqrt{16 - x^2}\, dx = 8\left[\sin^{-1}\left(\frac{1}{4}x\right) + \left(\frac{1}{4}x\right)\left(\frac{1}{4}\sqrt{16 - x^2}\right)\right] + C$$

$$= 8\sin^{-1}\left(\frac{1}{4}x\right) + \frac{1}{2}x\sqrt{16 - x^2} + C$$

Thus,

$$\int_1^4 \sqrt{16 - x^2}\, dx = \left[8\sin^{-1}\left(\frac{1}{4}x\right) + \frac{1}{2}x\sqrt{16 - x^2}\right]_1^4$$

$$= \left[(8\sin^{-1}(1) + 0) - \left(8\sin^{-1}\frac{1}{4} + \frac{1}{2}\sqrt{15}\right)\right]$$

$$= 4\pi - 8\sin^{-1}\left(\frac{1}{4}\right) - \frac{1}{2}\sqrt{15} \tag{6}$$

Substituting from (2) and (6) into (1), we obtain

$$F = 2\rho\left[(4\pi - 8\sin^{-1}(\tfrac{1}{4}) - \tfrac{1}{2}\sqrt{15}) + 5\sqrt{15}\right] \tag{7}$$

The force on the gate due to water pressure is F pounds, where F is given in Eq. (7).

20. If the end of a water tank is in the shape of a rectangle and the tank is full, show that the measure of the force due to water pressure on the end is the product

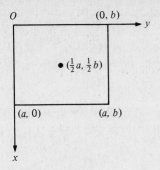

Figure 11. 4. 20

of the measure of the area of the end and the measure of the force at the geometrical center.

SOLUTION: Fig. 11. 4. 20 illustrates the end of the tank. We use Definition 11.4.1 to find the total force due to water pressure. Because the origin is at the surface of the water and the tank is a ft deep, then x units is the distance below the surface of the water for x in the interval $[0, a]$. Because the rectangle is b units long, we let $f(x) = b$. Hence, the total force is given by

$$F = \lim_{\|\Delta\| \to 0} \sum_{i=1}^{n} \rho \bar{x}_i b \, \Delta_i x$$

$$= \int_0^a \rho x b \, dx$$

$$= \frac{1}{2} \rho b x^2 \bigg]_0^a$$

$$= \frac{1}{2} \rho a^2 b \tag{1}$$

Furthermore, the center of the rectangle is at the point $(\frac{1}{2} a, \frac{1}{2} b)$. Thus, $\frac{1}{2} a\rho$ pounds represents the force per square unit of area at this point. Because the area of the rectangle is ab square units and $(\frac{1}{2} a\rho)(ab) = \frac{1}{2}\rho a^2 b$, by comparison with (1), we conclude that the total force due to liquid pressure on the end of the tank is the product of the measure of the area of the end and the measure of the force at the geometrical center.

24. The face of a dam adjacent to the water is vertical and is in the shape of an isosceles trapezoid 90 ft wide at the top, 60 ft wide at the bottom, and 20 ft high. Use Eq. (10) to find the total force due to water pressure of the face of the dam.

SOLUTION: Equation (10) states that if A square units is the area of a vertical region whose centroid is $\bar{x}$ units below the surface of the water with density ρ units, then F units is the total force on the region due to water pressure, with

$$F = \rho A \bar{x}$$

We find $A\bar{x}$ for the trapezoidal region. Figure 11. 4. 24 shows a sketch of the trapezoid. We take the origin at the center of the top of the trapezoid with positive x-axis downward. An equation of the line that contains the points $(0, 45)$ and $(20, 30)$ is $y = -\frac{3}{4} x + 45$. Let $f(x) = -\frac{3}{4} x + 45$. We take an element of the area of width $\Delta_i x$ ft and length $2 f(\bar{x}_i)$ ft. Thus, if $(\bar{x}, \bar{y})$ is the centroid of the region, then

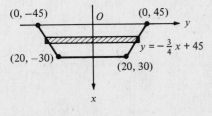

Figure 11. 4. 24

$$A\bar{x} = \lim_{\|\Delta\| \to 0} \sum_{i=1}^{n} 2\bar{x}_i f(\bar{x}_i) \, \Delta_i x$$

$$= \int_0^{20} 2x \left(-\frac{3}{4} x + 45\right) dx$$

$$= \int_0^{20} \left(-\frac{3}{2} x^2 + 90x\right) dx$$

$$= -\frac{1}{2} x^3 + 45x^2 \bigg]_0^{20} = 14{,}000$$

Therefore,

$$F = 14{,}000\rho$$

The total force is $14{,}000\rho$ pounds.

Review Exercises for Chapter 11

4. Three particles, each having the same mass, are located on the x-axis at the points having coordinates −4, 1, and 5, where the distance is measured in meters. Find the coordinates of the center of mass of the system.

SOLUTION: Let m kilograms be the mass of each particle. Then $M = 3m$ kilograms is the total mass of the system. If $(\bar{x}, \bar{y})$ is the center of mass of the system, then

$$M\bar{x} = \sum_{i=1}^{3} m_i x_i = m(-4) + m(1) + m(5) = 2m$$

$$M\bar{y} = \sum_{i=1}^{3} m_i y_i = m(0) + m(0) + m(0) = 0$$

Thus,

$$\bar{x} = \frac{2m}{M} = \frac{2m}{3m} = \frac{2}{3}$$

$$\bar{y} = 0$$

The center of mass of the system is $(\frac{2}{3}, 0)$.

8. Find the center of mass of a rod 4 meters long if the linear density at the point x meters from the left end is $\sqrt{9 + x^2}$ kg/m.

SOLUTION: We are given that $\rho(x) = \sqrt{9 + x^2}$ kg/m is the linear density of the rod x meters from the left end. If M kg is the total mass of the rod, then

$$M = \lim_{\|\Delta\| \to 0} \sum_{i=1}^{n} \rho(\bar{x}_i)\, \Delta_i x$$

$$= \int_0^4 \sqrt{9 + x^2}\, dx$$

Let $\theta = \tan^{-1}(x/3)$. Then $x = 3\tan\theta$, so $dx = 3\sec^2\theta\, d\theta$ and $\sqrt{9 + x^2} = 3\sec\theta$. Thus, the value of the indefinite integral is

$$\int \sqrt{9 + x^2}\, dx = 9 \int \sec^3\theta\, d\theta$$

$$= 9[\tfrac{1}{2}\sec\theta\tan\theta + \tfrac{1}{2}\ln|\sec\theta + \tan\theta|] + C \tag{1}$$

where the value of the integral $\int \sec^3\theta\, d\theta$ is taken from Example 6 of Section 10.2 in the text. Now $\sec\theta = \frac{1}{3}\sqrt{9 + x^2}$ and $\tan\theta = \frac{1}{3}x$. Thus, from (1) we obtain

$$\int \sqrt{9 + x^2}\, dx = \tfrac{9}{2}[\tfrac{1}{3}\sqrt{9 + x^2}(\tfrac{1}{3}x) + \ln|\tfrac{1}{3}\sqrt{9 + x^2} + \tfrac{1}{3}x|] + C$$

$$= \tfrac{1}{2}[x\sqrt{9 + x^2} + 9\ln|\sqrt{9 + x^2} + x|] - \tfrac{9}{2}\ln 3 + C$$

Thus,

$$F(x) = \tfrac{1}{2}[x\sqrt{9 + x^2} + 9\ln|\sqrt{9 + x^2} + x|$$

is a value of the indefinite integral with $C = \frac{9}{2}\ln 3$. Now

$$\int_0^4 \sqrt{9 + x^2}\, dx = F(4) - F(0)$$

$$= \tfrac{1}{2}[4\sqrt{9 + 16} + 9\ln(\sqrt{9 + 16} + 4] - \tfrac{1}{2}[9\ln\sqrt{9}]$$

$$= 10 + \tfrac{9}{2} \ln 9 - \tfrac{9}{2} \ln 3$$

$$= 10 + 9 \ln 3 - \tfrac{9}{2} \ln 3$$

$$= 10 + \tfrac{9}{2} \ln 3$$

Hence,

$$M = 10 + \tfrac{9}{2} \ln 3$$

Furthermore, if the center of mass of the rod is $\bar{x}$ meters from the left end, then

$$M\bar{x} = \lim_{\|\Delta\| \to 0} \sum_{i=1}^{n} \rho(\bar{x}_i)\bar{x}_i \, \Delta_i x$$

$$= \int_0^4 x\sqrt{9 + x^2} \, dx$$

$$= \left(\frac{1}{2}\right)\left(\frac{2}{3}\right)(9 + x^2)^{3/2} \Big]_0^4$$

$$= \tfrac{1}{3}(25^{3/2} - 9^{3/2}) = \tfrac{98}{3}$$

Because $M = 10 + \tfrac{9}{2} \ln 3$, then

$$(10 + \tfrac{9}{2} \ln 3)\bar{x} = \tfrac{98}{3}$$

$$\bar{x} = \frac{98}{3(10 + \tfrac{9}{2} \ln 3)} = 2.19$$

The center of mass of the rod is 2.19 meters from the left end of the rod.

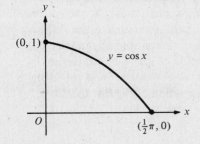

$(0, 1)$

$y = \cos x$

O

$(\tfrac{1}{2}\pi, 0)$

x

Figure 11. 12R

12. Find the centroid of the region in the first quadrant bounded by the coordinate axes and the curve $y = \cos x$.

SOLUTION: Figure 11. 12R shows a sketch of the region. If A square units is the area of the region, then

$$A = \lim_{\|\Delta\| \to 0} \sum_{i=1}^{n} \cos \bar{x}_i \, \Delta_i x$$

$$= \int_0^{\pi/2} \cos x \, dx$$

$$= \sin x \Big]_0^{\pi/2} = 1$$

If $(\bar{x}, \bar{y})$ is the centroid of the region, then

$$A\bar{x} = \lim_{\|\Delta\| \to 0} \sum_{i=1}^{n} (\cos \bar{x}_i)\bar{x}_i \, \Delta_i x$$

$$= \int_0^{\pi/2} x \cos x \, dx$$

$$= x \sin x + \cos x \Big]_0^{\pi/2} = \frac{\pi}{2} - 1$$

where we used integration by parts with $u = x$ and $dv = \cos x \, dx$ to evaluate the integral. Because $A = 1$, we have

$$\bar{x} = \frac{\pi}{2} - 1$$

Furthermore,

$$A\bar{y} = \lim_{||\Delta|| \to 0} \sum_{i=1}^{n} \cos \bar{x}_i \left(\frac{1}{2} \cos \bar{x}_i \right) \Delta_i x$$

$$= \frac{1}{2} \int_0^{\pi/2} \cos^2 x \, dx$$

$$= \frac{1}{4} \int_0^{\pi/2} (1 + \cos 2x) \, dx$$

$$= \frac{1}{4} \left[x + \frac{1}{2} \sin 2x \right]_0^{\pi/2} = \frac{1}{8} \pi$$

Because $A = 1$, then $\bar{y} = \frac{1}{8}\pi$, and the centroid of the region is $(\frac{1}{2}\pi - 1, \frac{1}{8}\pi)$.

16. Give an example to show that the centroid of a region is not necessarily a point within the region.

SOLUTION: Let the region be that which is inside of the circle $x^2 + y^2 = 4$ and outside of the circle $x^2 + y^2 = 1$. The centroid is $(0, 0)$, a point not in the region.

20. Use the theorem of Pappus to find the volume of a right-circular cone with base radius 2 meters and height 3 meters.

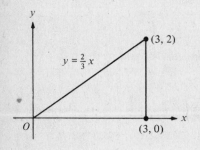

$y = \frac{2}{3} x$

(3, 2)

(3, 0)

Figure 11. 20R

SOLUTION: The cone is the solid generated by revolving the right triangle shown in Fig. 11. 20R about the x-axis. The theorem of Pappus states that the volume of the solid of revolution is $2\pi \bar{y} A$ cubic meters, where A square meters is the area of the triangle and $\bar{y}$ meters is the ordinate of the centroid of the triangular region. Note that $2\pi \bar{y}$ meters is the distance traveled by the centroid during one complete revolution. Because the base of the triangle is 2 meters and the altitude is 3 meters, then $A = \frac{1}{2}(2)(3) = 3$. We find $\bar{y}$. An equation of the line that contains the origin and the point (3, 2) is $y = \frac{2}{3}x = f(x)$. Thus, we have

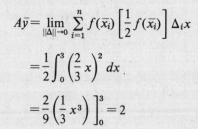

$$A\bar{y} = \lim_{||\Delta|| \to 0} \sum_{i=1}^{n} f(\bar{x}_i) \left[\frac{1}{2} f(\bar{x}_i) \right] \Delta_i x$$

$$= \frac{1}{2} \int_0^3 \left(\frac{2}{3} x \right)^2 dx$$

$$= \frac{2}{9} \left(\frac{1}{3} x^3 \right) \Big]_0^3 = 2$$

Because $A = 3$, we have $\bar{y} = \frac{2}{3}$. Therefore, if V cubic meters is the volume of the cone, then

$$V = 2\pi \bar{y} A$$
$$= 2\pi(\tfrac{2}{3})3 = 4\pi$$

The volume of the cone is 4π cubic meters.

24. The region bounded by the curve $x = \sqrt{y}$, the y-axis, and the line $y = 4$ is revolved about the x-axis. Find the centroid of the solid of revolution.

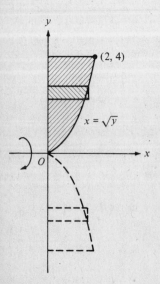

(2, 4)

$x = \sqrt{y}$

O

Figure 11. 24R

SOLUTION: Figure 11. 24R shows a sketch of the region and a plane section of the solid of revolution. The element of volume is a circular shell with thickness $\Delta_i r = \Delta_i y$ units, mean radius $r_i = \bar{y}_i$ units and height $h_i = \sqrt{\bar{y}_i}$ units. If V cubic units is the volume of the solid of revolution, then

$$V = \lim_{||\Delta|| \to 0} \sum_{i=1}^{n} 2\pi r_i h_i \Delta_i r$$

$$= \lim_{\|\Delta\| \to 0} \sum_{i=1}^{n} 2\pi \bar{y}_i \sqrt{\bar{y}_i} \, \Delta_i y$$

$$= 2\pi \int_0^4 y\sqrt{y} \, dy$$

$$= 2\pi \left(\frac{2}{5} y^{5/2} \right) \Big]_0^4 = \frac{128}{5} \pi$$

The centroid of the element of volume is at the point $(\frac{1}{2}\sqrt{\bar{y}_i}, 0)$. Thus, if $(\bar{x}, \bar{y})$ is the centroid of the solid of revolution, then $\bar{y} = 0$, and

$$V\bar{x} = \lim_{\|\Delta\| \to 0} \sum_{i=1}^{n} (2\pi \bar{y}_i \sqrt{\bar{y}_i})(\tfrac{1}{2}\sqrt{\bar{y}_i}) \, \Delta_i y$$

$$= \pi \int_0^4 y^2 \, dy$$

$$= \frac{1}{3} \pi y^3 \Big]_0^4 = \frac{64}{3} \pi$$

Substituting the value for V, we obtain

$$\frac{128\pi}{5} \bar{x} = \frac{64\pi}{3}$$

$$\bar{x} = \frac{5}{6}$$

The centroid of the solid of revolution is $(\frac{5}{6}, 0)$.

28. A semicircular plate with a radius of 3 ft is submerged vertically in a tank of water, with its diameter lying in the surface. Use Eq. (10) of Section 11.4 to find the force due to water pressure on one side of the plate.

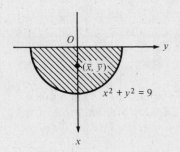

$x^2 + y^2 = 9$

*Figure 11. 28*R

SOLUTION: Figure 11. 28R shows the plate. Eq. (10) of Section 11.4 states that the force due to liquid pressure on one side of the plate is given by

$$F = \rho \bar{x} A \tag{1}$$

where $\bar{x}$ ft is the depth of the centroid of the plate and A square feet is the area of the plate. In Exercise 32 of Exercises 11.2 we found that the centroid of any semicircular region of radius r is on the axis of the region $4r/3\pi$ units from the center of the circle. Because $r = 3$ for the plate, we conclude that

$$\bar{x} = \frac{4}{\pi} \tag{2}$$

Furthermore, the area of the plate is given by the formula $A = \frac{1}{2}\pi r^2$, with $r = 3$. Hence.

$$A = \frac{9}{2} \pi \tag{3}$$

Substituting from (2) and (3) into (1), we obtain

$$F = \rho \left(\frac{4}{\pi} \right)\left(\frac{9}{2} \pi \right) = 18\rho$$

We conclude that the force on the plate is 18ρ pounds.

CHAPTER 12 | Polar Coordinates

12.1 THE POLAR COORDINATE SYSTEM

There is not a unique pair of polar coordinates for a point. If (r, θ) is a polar coordinate representation of point P, then we may add any even multiple of π to θ and obtain another polar coordinate representation of P. That is, for any integer n,

$$(r, \theta) = (r, \theta + 2n\pi)$$

If we add an odd multiple of π to θ and also replace r by $-r$, we obtain another polar coordinate representation of P. That is, for any integer n,

$$(r, \theta) = (-r, \theta + (2n-1)\pi)$$

If (x, y) is the Cartesian coordinate representation of P and (r, θ) is a polar coordinate representation of P, then

$$x = r \cos \theta \qquad r^2 = x^2 + y^2$$

$$y = r \sin \theta \qquad \tan \theta = \frac{y}{x}$$

Exercises 12.1

4. Plot the point having polar coordinates $(3, \frac{3}{2}\pi)$; then find another set of polar coordinates for the same point for which: **(a)** $r < 0$ and $0 \leq \theta < 2\pi$, **(b)** $r > 0$ and $-2\pi < \theta \leq 0$, and **(c)** $r < 0$ and $-2\pi < \theta \leq 0$.

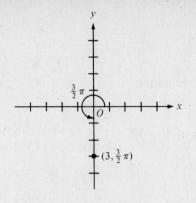

Figure 12. 1. 4

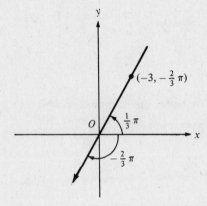

Figure 12. 1. 8

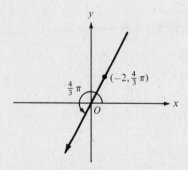

Figure 12. 1. 12

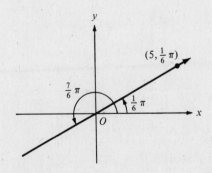

Figure 12. 1. 16

SOLUTION: Figure 12. 1. 4 shows the point.

(a) $(3, \frac{3}{2}\pi) = (-3, \frac{3}{2}\pi - \pi) = (-3, \frac{1}{2}\pi)$

(b) $(3, \frac{3}{2}\pi) = (3, \frac{3}{2}\pi - 2\pi) = (3, -\frac{1}{2}\pi)$

(c) $(3, \frac{3}{2}\pi) = (-3, \frac{3}{2}\pi - 3\pi) = (-3, -\frac{3}{2}\pi)$

8. Plot the point having the polar coordinates $(-3, -\frac{2}{3}\pi)$. Find another set of polar coordinates for this point for which **(a)** $r > 0$ and $0 \leq \theta < 2\pi$; **(b)** $r > 0$ and $-2\pi < \theta \leq 0$; **(c)** $r < 0$ and $2\pi \leq \theta < 4\pi$.

SOLUTION: Figure 12. 1. 8 shows the given point.

(a) $(-3, -\frac{2}{3}\pi) = (3, -\frac{2}{3}\pi + \pi) = (3, \frac{1}{3}\pi)$

(b) $(-3, -\frac{2}{3}\pi) = (3, -\frac{2}{3}\pi - \pi) = (3, -\frac{5}{3}\pi)$

(c) $(-3, -\frac{2}{3}\pi) = (-3, -\frac{2}{3}\pi + 4\pi) = (-3, \frac{10}{3}\pi)$

In Exercises 9–16 plot the point having the given set of polar coordinates; then give two other sets of polar coordinates of the same point, one with the same value of r and one with an r having the opposite sign.

12. $(-2, \frac{4}{3}\pi)$

SOLUTION: We plot the point in Fig. 12. 1. 12.

$(-2, \frac{4}{3}\pi) = (-2, \frac{4}{3}\pi - 2\pi) = (-2, -\frac{2}{3}\pi)$

$(-2, \frac{4}{3}\pi) = (2, \frac{4}{3}\pi - \pi) = (2, \frac{1}{3}\pi)$

16. $(5, \frac{1}{6}\pi)$

SOLUTION: The point is shown in Fig. 12. 1. 16.

$(5, \frac{1}{6}\pi) = (5, \frac{1}{6}\pi - 2\pi) = (5, -\frac{11}{6}\pi)$

$(5, \frac{1}{6}\pi) = (-5, \frac{1}{6}\pi + \pi) = (-5, \frac{1}{6}\pi)$

Other answers are also possible.

20. Find a set of polar coordinates of the points whose rectangular Cartesian coordinates are given. Take $r > 0$ and $0 \leq \theta < 2\pi$: **(a)** $(3, -3)$; **(b)** $(-1, \sqrt{3})$; **(c)** $(0, -2)$; **(d)** $(-2, -2\sqrt{3})$

SOLUTION:

(a) $r = \sqrt{x^2 + y^2} = \sqrt{3^2 + (-3)^2} = 3\sqrt{2}$

$$\tan \theta = \frac{y}{x} = \frac{-3}{3} = -1$$

Because $\tan \frac{1}{4}\pi = 1$, the reference angle is $\frac{1}{4}\pi$. Because $x > 0$ and $y < 0$, the point is in the fourth quadrant. Thus, $\theta = 2\pi - \frac{1}{4}\pi = \frac{7}{4}\pi$. The polar coordinates of the point are $(3\sqrt{2}, \frac{7}{4}\pi)$.

(b) $r = \sqrt{x^2 + y^2} = \sqrt{1 + 3} = 2$

$$\tan \theta = \frac{y}{x} = -\sqrt{3}$$

Because $x < 0$ and $y > 0$, the point is in the second quadrant. Because $\tan \frac{1}{3}\pi = \sqrt{3}$, then $\theta = \pi - \frac{1}{3}\pi = \frac{2}{3}\pi$. Thus, the polar coordinates are $(2, \frac{2}{3}\pi)$.

(c) $r = \sqrt{x^2 + y^2} = \sqrt{4} = 2$. Because $x = 0$, then $\tan \theta$ is not defined. However, because the point is on the negative y-axis, then $\theta = \frac{3}{2}\pi$. Thus, the polar coordinates are $(2, \frac{3}{2}\pi)$.

(d) $r = \sqrt{x^2 + y^2} = \sqrt{4 + 12} = 4$

$$\tan \theta = \frac{y}{x} = \sqrt{3}$$

Because the point is in the third quadrant, and $\tan \frac{1}{3}\pi = \sqrt{3}$, then $\theta = \pi + \frac{1}{3}\pi = \frac{4}{3}\pi$. The polar coordinates are $(4, \frac{4}{3}\pi)$.

In Exercises 21–30 find a polar equation of the graph having the given Cartesian equation.

24. $x^3 = 4y^2$

SOLUTION: Substituting $x = r\cos\theta$ and $y = r\sin\theta$ in the given equation, we have

$$r^3 \cos^3\theta = 4r^2 \sin^2\theta$$
$$r^2(r\cos^3\theta - 4\sin^2\theta) = 0$$

Thus,

$$r^2 = 0 \quad \text{or} \quad r\cos^3\theta - 4\sin^2\theta = 0$$

The graph of $r^2 = 0$ is the pole. Because $(0, 0)$ satisfies the equation $r\cos^3\theta - 4\sin^2\theta = 0$, the graph of this equation also contains the pole. Therefore, a polar equation of the curve is

$$r\cos^3\theta - 4\sin^2\theta = \theta$$

$$r = \frac{4\sin^2\theta}{\cos^3\theta} = 4\tan^2\theta \sec\theta$$

28. $2xy = a^2$

SOLUTION: Because $x = r\cos\theta$ and $y = r\sin\theta$, we have

$$2(r\cos\theta)(r\sin\theta) = a^2$$
$$r^2(2\sin\theta\cos\theta) = a^2$$
$$r^2 \sin 2\theta = a^2$$

In Exercises 31–40, find a Cartesian equation of the graph having the given polar equation.

32. $r^2 \cos 2\theta = 10$

SOLUTION: Because $\cos 2\theta = \cos^2\theta - \sin^2\theta$, the given equation can be written as

$$r^2(\cos^2\theta - \sin^2\theta) = 10$$
$$(r\cos\theta)^2 - (r\sin\theta)^2 = 10$$

Because $r\cos\theta = x$ and $r\sin\theta = y$, we have

$$x^2 - y^2 = 10$$

36. $r = 2\sin 3\theta$

SOLUTION: In Exercise 23 of Exercises 1.7 we proved that

$$\sin 3\theta = 3\sin\theta - 4\sin^3\theta$$

Thus, the given equation may be written

$$r = 6\sin\theta - 8\sin^3\theta$$

Because the graph of this equation contains the pole, we may multiply both sides by r^3 without affecting the graph. Thus,

$$r^4 = 6r^3 \sin \theta - 8r^3 \sin^3 \theta$$
$$(r^2)^2 = 6r^2(r \sin \theta) - 8(r \sin \theta)^3$$

Because $r^2 = x^2 + y^2$ and $r \sin \theta = y$, this becomes

$$(x^2 + y^2)^2 = 6(x^2 + y^2)y - 8y^3$$
$$x^4 + y^4 + 2x^2y^2 - 6x^2y + 2y^3 = 0$$

40. $r = \dfrac{4}{3 - 2 \cos \theta}$

SOLUTION: Eliminating the fraction, we have

$$3r - 2r \cos \theta = 4$$
$$3r = 2(r \cos \theta + 2)$$
$$9r^2 = 4(r \cos \theta + 2)^2$$

Because $r^2 = x^2 + y^2$ and $r \cos \theta = x$, we have

$$9(x^2 + y^2) = 4(x + 2)^2$$
$$5x^2 + 9y^2 - 16x - 16 = 0$$

12.2 GRAPHS OF EQUATIONS IN POLAR COORDINATES

You should be able to identify the graph and make a quick sketch by plotting only a few points for any polar equation of the type listed below. If we replace $\cos n\theta$ by $\sin n\theta$ in the equation, the graph does not change its shape. However, the curve is rotated about the pole through an angle with radian measure $\pi/2n$. The letters a and b are any constants, either positive or negative, but n is a positive integer in each of the following equations.

TYPE OF EQUATION	NAME OF CURVE	EXAMPLE OF GRAPH
1. $r = a$	Circle	Fig. 12. 2. 8
2. $\theta = a$	Line	
3. $r = \theta$	Spiral	Fig. 12. 2. 16
4. $r = \dfrac{a}{\cos \theta}$	Line	Fig. 12. 2. 7 (text)
5. $r = 2a \cos \theta$	Circle	Fig. 12. 2. 4
6. $r^2 = a^2 \cos 2\theta$	Lemniscate	Fig. 12. 2. 32
7. $r = a \cos n\theta$ (n is even)	Rose 2n leaves	Fig. 12. 2. 4 (text)
8. $r = a \cos n\theta$ (n is odd)	Rose n leaves	Fig. 12. 2. 20
9. $r = a(1 + \cos \theta)$	Cardioid	Fig. 12. 2. 24
10. $r = a + b \cos \theta$ ($\lvert a \rvert > \lvert b \rvert$)	Limaçon	Fig. 12. 2. 3 (text)
11. $r = a + b \cos \theta$ ($\lvert b \rvert > \lvert a \rvert$)	Limaçon with loop	Fig. 12. 2. 28

Following are the tests for symmetry for graphs of polar equations.

12.2.1 Theorem RULE 1. If for an equation in polar coordinates an equivalent equation is obtained when (r, θ) is replaced by either $(r, -\theta + 2n\pi)$ or $(-r, \pi - \theta + 2n\pi)$, where n is any integer, the graph of the equation is symmetric with respect to the polar axis.

12.2.2 Theorem RULE 2. If for an equation in polar coordinates an equivalent equation is obtained when (r, θ) is replaced by either $(r, \pi - \theta + 2n\pi)$ or $(-r, -\theta + 2n\pi)$, where n is any integer, the graph of the equation is symmetric with respect to the $\frac{1}{2}\pi$-axis.

12.2.3 Theorem RULE 3. If for an equation in polar coordinates an equivalent equation is obtained when (r, θ) is replaced by either $(-r, \theta + 2n\pi)$ or $(r, \pi + \theta + 2n\pi)$, where n is any integer, the graph of the equation is symmetric with respect to the pole.

Because $\cos(-n\theta) = \cos n\theta$, if $\cos n\theta$ is the only trigonometric expression that occurs in a polar equation, then by Rule 1 the graph of the equation is symmetric with respect to the polar axis. Because $\sin(\pi - \theta) = \sin \theta$, if $\sin \theta$ is the only trigonometric expression that appears in a polar equation, then by Rule 2 the graph of the equation is symmetric with respect to the line $\theta = \frac{1}{2}\pi$. Because $r^2 = (-r)^2$, if r^2 is the only expression involving r that appears in the equation, then by Rule 3 the graph of the equation is symmetric with respect to the pole. Although these conditions are sufficient to prove symmetry, they are not necessary conditions. For example, see Illustration 1 in the text.

To find an equation of a tangent line to the curve at the pole, let $r = 0$ in the equation and solve for θ. If θ_1 is a solution, then the line $\theta = \theta_1$ is a tangent line to the curve at the pole.

Exercises 12.2

In Exercises 1–40 draw a sketch of the graph of the given equation.

4. $r = 2 \sin \theta$

SOLUTION: This is a Type 5 equation (with $\cos \theta$ replaced by $\sin \theta$), and thus the graph is a circle. Furthermore, by Theorem 12.2.2 the graph is symmetric with respect to the line $\theta = \frac{1}{2}\pi$. We plot the points found in Table 4 and use symmetry to make a quick sketch. If $r = 0$, then $\theta = 0$. Thus the line $\theta = 0$, that is, the polar axis, is tangent to the circle. See Fig. 12. 2. 4.

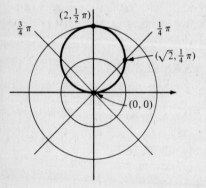

Figure 12. 2. 4

Table 4

θ	0	$\frac{1}{4}\pi$	$\frac{1}{2}\pi$
r	0	1.4	2

8. $r = -4$

SOLUTION: This is a Type 1 equation, and thus the graph is a circle with center at the pole and radius 4. Remember that we may replace r by $-r$ and θ by $\theta + \pi$ in the polar coordinates of a point. Thus, the graph of $r = -4 + 0 \cdot \theta$ is the same as the graph of $-r = -4 + 0 \cdot (\theta + \pi)$. That is, the graph of $r = -4$ is the same as the graph of $r = 4$. See Fig. 12. 2. 8.

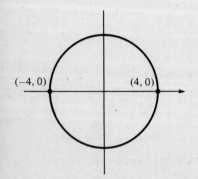

Figure 12. 2. 8

12. $r = -5 \cos \theta$

SOLUTION: This is a Type 5 equation, and thus the graph is a circle. Furthermore, by Theorem 12.2.1 the graph is symmetric with respect to the polar axis. Because $r = 0$ when $\theta = \frac{1}{2}\pi$, the circle is tangent to the line $\theta = \frac{1}{2}\pi$. We plot the points in Table 12 and use symmetry to draw a sketch of the graph, which is shown in Fig. 12. 2. 12.

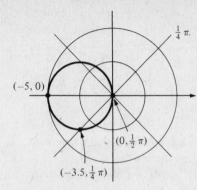

Figure 12. 2. 12

Table 12

θ	0	$\frac{1}{4}\pi$	$\frac{1}{2}\pi$
r	-5	-3.5	0

16. $r = 2\theta$ (spiral of Archimedes)

SOLUTION: This is a Type 3 equation. For $\theta \geq 0$ the graph is similar to Fig. 12. 2. 10 in the text. Table 16 gives several points on the graph with $\theta \geq 0$. Because an equivalent equation is obtained when (r, θ) is replaced by $(-r, -\theta)$, by Theorem 12.2.2 we conclude that the graph is symmetric with respect to the $\frac{1}{2}\pi$-axis. Figure 12. 2. 16 shows a sketch of the graph.

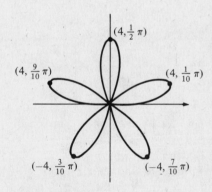

Figure 12. 2. 16

Table 16

θ	0	$\frac{1}{4}\pi$	$\frac{1}{2}\pi$	$\frac{3}{4}\pi$	π	$\frac{3}{2}\pi$	2π
r	0	$\frac{1}{2}\pi$	π	$\frac{3}{2}\pi$	2π	3π	4π

20. $r = 4 \sin 5\theta$ (five-leafed rose)

SOLUTION: This is a Type 8 curve. If $\theta = 0$, then $r = 0$. As θ increases from 0 to $\frac{1}{10}\pi$, r increases from 0 to 4. As θ increases from $\frac{1}{10}\pi$ to $\frac{1}{5}\pi$, r decreases from 4 to 0. Thus, one leaf of the rose occurs when $0 \leq \theta \leq \frac{1}{5}\pi$ and the point $(4, \frac{1}{10}\pi)$ is the end of the leaf. See Fig. 12. 2. 20. When $\frac{1}{5}\pi \leq \theta \leq \frac{2}{5}\pi$, then $r \leq 0$. The largest value of $|r|$ occurs when $\theta = \frac{3}{10}\pi$, so $(-4, \frac{3}{10}\pi)$ is the end of another leaf of the rose. When $\frac{2}{5}\pi \leq \theta \leq \frac{3}{5}\pi$, $r \geq 0$ and the largest value of r occurs when $\theta = \frac{1}{2}\pi$, so $(4, \frac{1}{2}\pi)$ is the end of another leaf. If $\frac{3}{5}\pi \leq \theta \leq \frac{4}{5}\pi$, then $r < 0$ and the largest value of $|r|$ occurs when $\theta = \frac{7}{10}\pi$, so $(-4, \frac{7}{10}\pi)$ is the end of another leaf. When $\frac{4}{5}\pi \leq \theta \leq \pi$, then $r > 0$ and the largest value of r occurs when $\theta = \frac{9}{10}\pi$, so $(4, \frac{9}{10}\pi)$ is the end of a leaf. If $\pi \leq \theta \leq \frac{6}{5}\pi$, then $r < 0$ and the end of a leaf is at the point $(-4, \frac{11}{10}\pi)$. Because $(-4, \frac{11}{10}\pi)$ and $(4, \frac{1}{10}\pi)$ are coordinates of the same point, we conclude that this leaf coincides with the leaf we obtained for $0 \leq \theta \leq \frac{1}{5}\pi$. Thus, the graph is complete.

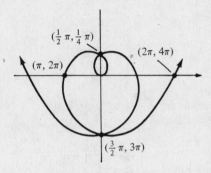

Figure 12. 2. 20

24. $r = 3 + 3 \cos \theta$ (cardioid)

SOLUTION: This is a Type 9 curve. By Theorem 12.2.1 the graph is symmetric with respect to the polar axis. Table 24 gives some points with $0 \leq \theta \leq \pi$. We plot the points in the table and use symmetry to complete the sketch. See Fig. 12. 2. 24.

Table 24

θ	0	$\frac{1}{4}\pi$	$\frac{1}{2}\pi$	$\frac{3}{4}\pi$	π
r	6	5.1	3	0.9	0

28. $r = 3 - 4 \cos \theta$ (limaçon)

SOLUTION: This is a Type 11 curve. The limaçon has a loop. By Theorem 12.2.1 the graph is symmetric with respect to the polar axis. Table 28 gives some points with $0 \leq \theta \leq \pi$. If $r = 0$, then $\cos \theta = \frac{3}{4}$, so $\theta = \cos^{-1}(0.75) = 0.7$. We plot the points from the table and use symmetry to complete the sketch, shown in Fig. 12. 2. 28.

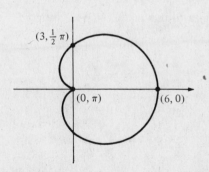

Figure 12. 2. 24

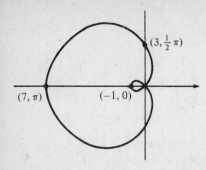

Figure 12. 2. 28

Table 28

θ	0	$\frac{1}{6}\pi$	0.7	$\frac{1}{3}\pi$	$\frac{1}{2}\pi$	$\frac{3}{4}\pi$	π
r	-1	-0.5	0	1	3.0	5.8	7.0

32. $r^2 = -4 \sin 2\theta$ (lemniscate)

SOLUTION: Because $\sin 2\theta > 0$ if $0 < \theta < \frac{1}{2}\pi$, r is not defined when $0 < \theta < \frac{1}{2}\pi$. Table 32 gives some points when $\frac{1}{2}\pi \le \theta \le \pi$. For $\pi < \theta < \frac{3}{2}\pi$ $\sin 2\theta > 0$, so r is not defined. When $\frac{3}{2}\pi \le \theta \le 2\pi$ the same points are obtained as when $\frac{1}{2}\pi \le \theta \le \pi$. A sketch of the graph is shown in Fig. 12. 2. 32.

Table 32

θ	$\frac{1}{2}\pi$	$\frac{7}{12}\pi$	$\frac{2}{3}\pi$	$\frac{3}{4}\pi$	$\frac{5}{6}\pi$	$\frac{11}{12}\pi$	π
r	0	±2	±3.5	±4	±3.5	±2	0

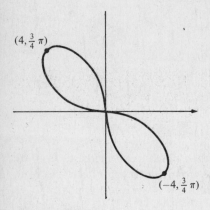

Figure 12. 2. 32

36. $(r-2)^2 = 8\theta$ (parabolic spiral)

SOLUTION: Solving the given equation for r, we obtain $r = 2 \pm \sqrt{8\theta}$. Let

$$r_1 = 2 + \sqrt{8\theta} \quad \text{and} \quad r_2 = 2 - \sqrt{8\theta}$$

In Table 36 we show some points on the curve. Decimal approximations have been calculated for r_1 and r_2. A sketch of the graph is shown in Fig. 12. 2. 36.

Table 36

θ	0	$\frac{1}{8}\pi$	$\frac{1}{4}\pi$	$\frac{3}{8}\pi$	$\frac{1}{2}\pi$	$\frac{3}{4}\pi$	π
r_1	2	3.8	4.5	5.1	5.5	6.3	7.0
r_2	2	0.2	-0.5	-1.1	-1.5	-2.3	-3.0

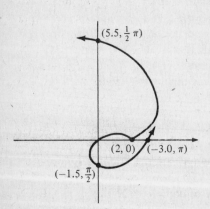

Figure 12. 2. 36

40. $r = 2|\cos\theta|$

SOLUTION: Because $\cos(-\theta) = \cos\theta$, by Theorem 12.2.1 the graph is symmetric with respect to the polar axis. Because $\cos(\pi - \theta) = -\cos\theta$, then $|\cos(\pi - \theta)| = |\cos\theta|$. Thus, by Theorem 12.2.2 the graph is symmetric with respect to the $\frac{1}{2}\pi$ axis. Furthermore, $r = 2\cos\theta$ if $-\frac{1}{2}\pi \le \theta \le \frac{1}{2}\pi$, so the graph is a Type 5 curve if $-\frac{1}{2}\pi \le \theta \le \frac{1}{2}\pi$, and thus the graph is a circle for $-\frac{1}{2}\pi \le \theta \le \pi$. Figure 12. 2. 40 shows a sketch of the graph, which consists of two tangent circles.

44. Find an equation of each of the tangent lines to the given curve at the pole.

$$r^2 = 9 \sin 2\theta$$

SOLUTION: If $r = 0$, then $\sin 2\theta = 0$. Hence,

$$2\theta = n\pi$$
$$\theta = \tfrac{1}{2}n\pi$$

Thus, the lines $\theta = 0$ and $\theta = \frac{1}{2}\pi$ are tangent to the curve at the pole.

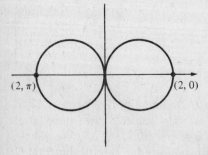

Figure 12. 2. 40

48. Find a polar equation of the line tangent to the curve $r = -6\sin\theta$ at the point $(6, \frac{3}{2}\pi)$.

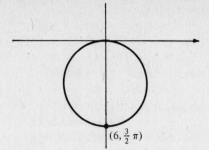

Figure 12. 2. 48

SOLUTION: The graph is the circle shown in Fig. 12. 2. 48. The line tangent to the circle at the point $(6, \frac{3}{2}\pi)$ is a horizontal line whose Cartesian equation is $y = -6$. Because $y = r \sin \theta$, a polar equation of the line is $r \sin \theta = -6$, or, equivalently, $r = -6 \csc \theta$.

12.3 INTERSECTION OF GRAPHS IN POLAR COORDINATES

Often we may find all the points of intersection of two graphs by solving the equations simultaneously. However, we must treat the pole as a special case. We test each equation to see whether there is a replacement for θ (not necessarily the same replacement) that gives r the value 0. If so, then each graph contains the pole, and hence the intersection contains the pole.

Because two or more distinct polar equations may represent the same graph, we must sometimes take additional steps to find all the points of intersection. If the curve C is the graph of the polar equation $r = f(\theta)$, then C is also the graph of any equation of the form $(-1)^n r = f(\theta + n\pi)$, where n is any integer. That is,

$$-r = f(\theta + \pi)$$
$$r = f(\theta + 2\pi)$$
$$-r = f(\theta + 3\pi)$$
$$r = f(\theta + 4\pi)$$

and so on, all represent the curve C. Thus, to find all the points of intersection of a curve C_1, represented by $r = f_1(\theta)$, and a curve C_2, represented by $r = f_2(\theta)$, we do the following:

1. Find all the distinct equations of curve C_1 and of curve C_2.
2. Solve each equation of curve C_1 simultaneously with each equation of curve C_2.
3. Test to see whether the pole is a point of intersection.

Exercises 12.3

In Exercises 1–21, find the points of intersection of the given pair of equations. Draw a sketch of each pair of graphs with the same pole and polar axis.

4. $r = 2 \cos 2\theta$
$r = 2 \sin \theta$

SOLUTION: The graph of $r = 2 \cos 2\theta$ is a four-leafed rose, and the graph of $r = 2 \sin \theta$ is a circle, as shown in Fig. 12. 3. 4. First, we solve the given equations simultaneously. Thus,

$$2 \cos 2\theta = 2 \sin \theta$$
$$2(1 - 2 \sin^2 \theta) = 2 \sin \theta$$
$$2 \sin^2 \theta + \sin \theta - 1 = 0$$
$$(\sin \theta + 1)(2 \sin \theta - 1) = 0$$
$$\sin \theta = -1 \quad \text{or} \quad \sin \theta = \tfrac{1}{2}$$
$$\theta = \tfrac{3}{2}\pi \qquad \theta = \tfrac{1}{6}\pi \qquad \theta = \tfrac{5}{6}\pi$$
$$r = -2 \quad \text{or} \quad r = 1 \quad \text{or} \quad r = 1 \tag{1}$$

Figure 12. 3. 4

Hence, $(-2, \frac{3}{2}\pi)$, $(1, \frac{1}{6}\pi)$, and $(1, \frac{5}{6}\pi)$ are points of intersection.

Next, we test to see whether the pole is a point of intersection. If $r = 0$ in the equation $r = 2 \cos 2\theta$, then $\theta = \frac{1}{4}\pi$. Thus, the rose $r = 2 \cos 2\theta$ contains the pole. If $r = 0$ in the equation $r = 2 \sin \theta$, then $\theta = 0$. Thus, the circle $r = 2 \sin \theta$ also contains the pole, and hence the pole is a point of intersection.

Figure 12. 3. 4 suggests that the four points we have found are the only points of intersection. To prove that this is true, we must find each equation of each curve and solve each pair of equations simultaneously. Replacing r by $-r$ and θ by $\theta + \pi$ in the equation $r = 2 \cos 2\theta$, we obtain $-r = 2 \cos 2(\theta + \pi)$, which is equivalent to $r = -2 \cos 2\theta$. Solving $r = -2 \cos 2\theta$ simultaneously with $r = 2 \sin \theta$, we obtain

$$-2 \cos 2\theta = 2 \sin \theta$$
$$-2(1 - 2 \sin^2 \theta) = 2 \sin \theta$$
$$(\sin \theta - 1)(2 \sin \theta + 1) = 0$$
$$\sin \theta = 1 \quad \text{or} \quad \sin \theta = -\tfrac{1}{2}$$
$$\begin{array}{ccc} \theta = \tfrac{1}{2}\pi & \theta = \tfrac{7}{6}\pi & \theta = \tfrac{11}{6}\pi \\ r = 2 & \text{or} \quad r = -1 & \text{or} \quad r = -1 \end{array} \tag{2}$$

However, $(2, \frac{1}{2}\pi) = (-2, \frac{3}{2}\pi)$; $(-1, \frac{7}{6}\pi) = (1, \frac{1}{6}\pi)$; and $(-1, \frac{11}{6}\pi) = (1, \frac{5}{6}\pi)$. Thus, the three points in (2) are the same as the three points found in (1).

Replacing r by $-r$ and θ by $\theta + \pi$ in the equation $r = 2 \sin \theta$ gives $-r = 2 \sin(\theta + \pi)$, which is equivalent to $r = 2 \sin \theta$. Thus, no distinct equation is found. Similarly, replacing θ by $\theta + 2\pi$ in each of the given equations fails to yield a distinct equation. We conclude that the four points found above are the only points of intersection.

8. $r = 1 - \sin \theta$
$r = \cos 2\theta$

SOLUTION: First, we solve the given pair of equations simultaneously. Thus,

$$1 - \sin \theta = \cos 2\theta$$
$$1 - \sin \theta = 1 - 2 \sin^2 \theta$$
$$\sin \theta(2 \sin \theta - 1) = 0$$
$$\begin{array}{cccc} \theta = 0 & \theta = \pi & \theta = \tfrac{1}{6}\pi & \theta = \tfrac{5}{6}\pi \\ r = 1 & \text{or} \quad r = 1 & \text{or} \quad r = \tfrac{1}{2} & \text{or} \quad r = \tfrac{1}{2} \end{array}$$

Thus, $(1, 0)$, $(1, \pi)$, $(\frac{1}{2}, \frac{1}{6}\pi)$ and $(\frac{1}{2}, \frac{5}{6}\pi)$ are points of intersections. Furthermore, if $r = 0$ in $r = 1 - \sin \theta$, then $\theta = \frac{1}{2}\pi$. Thus, the cardioid $r = 1 - \sin \theta$ contains the pole. If $r = 0$ in the equation $r = \cos 2\theta$, then $\theta = \frac{1}{4}\pi$. Thus, the four-leafed rose $r = \cos 2\theta$ also contains the pole, and hence the pole is a point of intersection.

Next, we replace r by $-r$ and θ by $\theta + \pi$ in $r = 1 - \sin \theta$, resulting in $-r = 1 - \sin(\theta + \pi)$, which is equivalent to $r = -1 - \sin \theta$. Solving this equation simultaneously with $r = \cos 2\theta$, we get

$$\cos 2\theta = -1 - \sin \theta$$
$$1 - 2 \sin^2 \theta = -1 - \sin \theta$$
$$2 \sin^2 \theta - \sin \theta - 2 = 0$$

Using the quadratic formula, we obtain

$$\sin \theta = \frac{1 \pm \sqrt{17}}{4}$$

Because $\frac{1}{4}(1 + \sqrt{17}) > 1$ and $-1 \leq \sin \theta \leq 1$, we disregard the positive square root. However, if

$$\sin \theta = \frac{1 - \sqrt{17}}{4} = -.78$$

then

$$\begin{array}{ccc} \theta = 4.04 & & \theta = 5.38 \\ r = -0.22 & \text{and} & r = -0.22 \end{array}$$

or, equivalently,

$$\begin{array}{ccc} \theta = 0.90 & & \theta = 2.24 \\ r = 0.22 & \text{and} & r = 0.22 \end{array}$$

Thus, the points $(0.22, 0.90)$ and $(0.22, 2.24)$ are also points of intersection. Replacing r by $-r$ and θ by $\theta + \pi$ in the equation $r = \cos 2\theta$ does not lead to any additional points of intersection. As we illustrate in Fig. 12. 3. 8, there are altogether seven points of intersection of the two graphs.

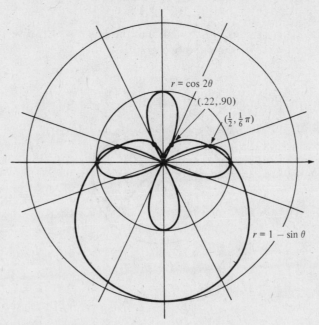

Figure 12. 3. 8

12. $r = 4 \tan \theta \sin \theta$
$r = 4 \cos \theta$

SOLUTION: Solving the given pair of equations simultaneously, we get

$$4 \tan \theta \sin \theta = 4 \cos \theta$$

Dividing both sides by $\cos \theta$, we get

$$\tan^2 \theta = 1$$
$$\tan \theta = \pm 1$$

Thus

$$\begin{array}{cccccccc} \theta = \frac{1}{4}\pi & & \theta = \frac{3}{4}\pi & & \theta = \frac{5}{4}\pi & & \theta = \frac{7}{4}\pi \\ r = 2\sqrt{2} & \text{or} & r = -2\sqrt{2} & \text{or} & r = -2\sqrt{2} & \text{or} & r = 2\sqrt{2} \end{array} \quad (1)$$

There are only two distinct points in (1), namely $(2\sqrt{2}, \frac{1}{4}\pi)$ and $(2\sqrt{2}, \frac{7}{4}\pi)$, which are points of intersection of the graphs. Because $(0, 0)$ satisfies the equation $r = 4 \tan \theta \sin \theta$ and $(0, \frac{1}{2}\pi)$ satisfies the equation $r = 4 \cos \theta$, the pole is also a point of intersection. Replacing r by $(-1)^n r$ and θ by $\theta + n\pi$ in each of the given equations does not result in a new equation. For example, $-r = 4 \tan(\theta + \pi) \sin(\theta + \pi)$ is equivalent to $r = 4 \tan \theta \sin \theta$, because $\tan(\theta + \pi) = \tan \theta$

and $\sin(\theta + \pi) = -\sin \theta$. We conclude that the only points of intersection of the graphs are the three points named above.

The graph of $r = 4 \cos \theta$ is a circle with a diameter that is the line segment joining the points $(4, 0)$ and the pole. Because we do not recognize the polar equation $r = 4 \tan \theta \sin \theta$, we find a Cartesian equation with the same graph. First, we multiply each side by r. Thus,

$$r^2 = 4 \tan \theta (r \sin \theta)$$

Replacing r^2 by $x^2 + y^2$, $\tan \theta$ by y/x, and $r \sin \theta$ by y, we obtain

$$x^2 + y^2 = \frac{4y^2}{x}$$

$$y = \pm x \sqrt{\frac{x}{4 - x}}$$

Thus, $0 \le x < 4$, and the line $x = 4$ is a vertical asymptote. In Fig. 12. 3. 12 we show the two graphs.

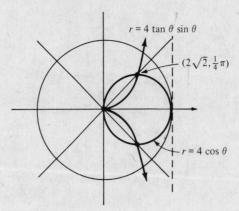

Figure 12. 3. 12

16. $r = 2 \cos \theta$
$\quad\;\; r = 2\sqrt{3} \sin \theta$

SOLUTION: Solving the given pair of equations simultaneously, we obtain

$$2 \cos \theta = 2\sqrt{3} \sin \theta$$

$$\frac{1}{\sqrt{3}} = \frac{\sin \theta}{\cos \theta}$$

$$\frac{1}{\sqrt{3}} = \tan \theta$$

$$\theta = \tfrac{1}{6}\pi \quad \text{or} \quad \theta = \tfrac{7}{6}\pi$$
$$r = \sqrt{3} \qquad\quad\; r = -\sqrt{3}$$

Because $(\sqrt{3}, \tfrac{1}{6}\pi)$ and $(-\sqrt{3}, \tfrac{7}{6}\pi)$ are the same point, we have found only one point of intersection so far. Because $(0, \tfrac{1}{2}\pi)$ satisfies the equation $r = 2 \cos \theta$ and $(0, 0)$ satisfies the equation $r = 2\sqrt{3} \sin \theta$, the pole is also a point of intersection. Figure 12. 3. 16 shows a sketch of the graphs. Each curve is a circle.

20. $r = 4(1 + \sin \theta)$
$\quad\;\; r(1 - \sin \theta) = 3$

SOLUTION: Eliminating r from the system, we obtain

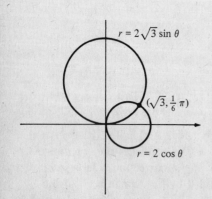

Figure 12. 3. 16

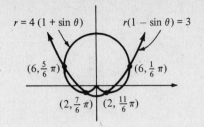

$r = 4(1 + \sin \theta)$ $r(1 - \sin \theta) = 3$

$(6, \frac{5}{6}\pi)$ $(6, \frac{1}{6}\pi)$

$(2, \frac{7}{6}\pi)$ $(2, \frac{11}{6}\pi)$

Figure 12. 3. 20

$$4(1 + \sin \theta)(1 - \sin \theta) = 3$$
$$1 - \sin^2 \theta = \tfrac{3}{4}$$
$$\cos^2 \theta = \tfrac{3}{4}$$
$$\cos \theta = \pm \tfrac{1}{2}\sqrt{3}$$
$$\theta = \tfrac{1}{6}\pi, \tfrac{5}{6}\pi, \tfrac{7}{6}\pi, \tfrac{11}{6}\pi$$

The points of intersection are $(6, \frac{1}{6}\pi)$, $(6, \frac{5}{6}\pi)$, $(2, \frac{7}{6}\pi)$, and $(2, \frac{11}{6}\pi)$. A sketch of the graphs is shown in Fig. 12. 3. 20. The graph of $r = 4(1 + \sin \theta)$ is a cardioid. In Section 13.4 we show that the graph of $r(1 - \sin \theta) = 3$ in a parabola.

24. The graph of the given equation intersects itself. Find the points at which this occurs.

$$r = 1 + 2 \cos 2\theta \tag{1}$$

SOLUTION: Replacing r by $-r$ and θ by $\theta + \pi$ in the given equation, we have

$$-r = 1 + 2 \cos 2(\theta + \pi)$$
$$-r = 1 + 2 \cos 2\theta$$
$$r = -1 - 2 \cos 2\theta \tag{2}$$

We solve (1) and (2) simultaneously. Thus,

$$1 + 2 \cos 2\theta = -1 - 2 \cos 2\theta$$
$$\cos 2\theta = -\tfrac{1}{2}$$

Because $r = 0$ when $\cos 2\theta = -\frac{1}{2}$, we conclude that the curve intersects itself at the pole.

12.4 TANGENT LINES OF POLAR CURVES

Suppose point P is on the graph of the polar equation $r = f(\theta)$, and P is not the pole. Furthermore, suppose χ is an angle (measured counterclockwise as shown in Fig. 12. 4a whose initial side is the radius vector OP and whose terminal side is the line through point P that is tangent to the graph of $r = f(\theta)$. Then

$$\tan \chi = \frac{r}{\dfrac{dr}{d\theta}}$$

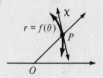

$r = f(\theta)$ χ

P

O

Figure 12. 4a

If two curves intersect at point P and if β is the smaller of the two angles between the tangent lines to the curves at P, then

$$\tan \beta = \left| \frac{\tan \chi_1 - \tan \chi_2}{1 + \tan \chi_1 \tan \chi_2} \right|$$

where χ_1 and χ_2 are the angles between OP and the tangent lines to the curves at P.

Exercises 12.4

In Exercises 1–9 find χ at the point indicated.

4. $r = a \sin \frac{1}{2}\theta$; $(\frac{1}{2}a, \frac{1}{3}\pi)$

SOLUTION:

$$\frac{dr}{d\theta} = \tfrac{1}{2}a \cos \tfrac{1}{2}\theta$$

$$\frac{dr}{d\theta}\bigg]_{\theta = \pi/3} = \tfrac{1}{2}a \cos \tfrac{1}{6}\pi$$

$$= \tfrac{1}{4}a\sqrt{3}$$

We are given that $r = \frac{1}{2}a$ when $\theta = \frac{1}{3}\pi$. Thus

$$\tan \chi = \frac{r}{\dfrac{dr}{d\theta}}$$

$$= \frac{\frac{1}{2}a}{\frac{1}{4}a\sqrt{3}}$$

$$= \frac{2}{\sqrt{3}} = 1.15$$

Hence, $\chi = 49.1°$.

8. $r = a(1 - \sin \theta);\ (a, \pi)$

SOLUTION:

$$\frac{dr}{d\theta} = -a \cos \theta$$

$$\left.\frac{dr}{d\theta}\right]_{\theta=\pi} = a$$

Because $r = a$ when $\theta = \pi$, we have

$$\tan \chi = \frac{a}{a} = 1$$

Thus, $\chi = \frac{1}{4}\pi$.

12. Find a measurement of the smaller angle between the tangent lines to the given pair of curves at the indicated point of intersection.

$$\begin{cases} r = -a \sin \theta \\ r = a \cos 2\theta \end{cases};\quad \text{the pole}$$

SOLUTION: If $r = 0$ in the equation $r = -a \sin \theta$, then $\theta = 0$. Thus, the line $\theta = 0$ is tangent to the circle $r = -a \sin \theta$ at the pole. If $r = 0$ in the equation $r = a \cos 2\theta$, then $\theta = \frac{1}{4}\pi$. Thus, the line $\theta = \frac{1}{4}\pi$ is tangent to the rose $r = a \cos 2\theta$ at the pole. We conclude that $\frac{1}{4}\pi$ radians is a measurement of the angle between the tangent lines at the pole.

16. Find a measurement of the angle between the tangent lines of the given pair of curves at all points of intersection.

$$r = 3 \cos \theta$$
$$r = 1 + \cos \theta$$

SOLUTION: Solving the equations simultaneously, we obtain

$$3 \cos \theta = 1 + \cos \theta$$
$$\cos \theta = \frac{1}{2}$$
$$\theta = \pm \frac{1}{3}\pi$$

Thus, $P_1 = (\frac{3}{2}, \frac{1}{3}\pi)$ and $P_2 = (\frac{3}{2}, -\frac{1}{3}\pi)$ are points of intersection. It is clear from the graphs shown in Fig. 12. 4. 16 that the only other point of intersection is the pole. Let β_1 be the smaller angle between the tangent lines to the curves at point P_1. Then

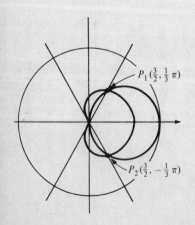

Figure 12. 4. 16

$$\tan \beta_1 = \left| \frac{\tan \chi_1 - \tan \chi_2}{1 + \tan \chi_1 \tan \chi_2} \right| \tag{1}$$

where χ_1 is the angle between OP_1 and the tangent line to the circle $r = 3 \cos \theta$ at the point P_1, and χ_2 is the angle between the line OP_1 and the tangent line to the cardioid $r = 1 + \cos \theta$ at the point P_1. For the curve $r = 3 \cos \theta$, we have

$$\frac{dr}{d\theta} = -3 \sin \theta$$

$$\frac{dr}{d\theta}\bigg]_{\theta=\pi/3} = -\tfrac{3}{2}\sqrt{3}$$

Thus,

$$\tan \chi_1 = \frac{r}{\dfrac{dr}{d\theta}}$$

$$= \frac{\tfrac{3}{2}}{-\tfrac{3}{2}\sqrt{3}} = -\frac{1}{\sqrt{3}} \tag{2}$$

For the curve $r = 1 + \cos \theta$, we have

$$\frac{dr}{d\theta} = -\sin \theta$$

$$\frac{dr}{d\theta}\bigg]_{\theta=\pi/3} = -\tfrac{1}{2}\sqrt{3}$$

Thus,

$$\tan \chi_2 = \frac{\tfrac{3}{2}}{-\tfrac{1}{2}\sqrt{3}}$$

$$= -\sqrt{3} \tag{3}$$

Substituting from (2) and (3) into (1), we obtain

$$\tan \beta_1 = \left| \frac{-\dfrac{1}{\sqrt{3}} + \sqrt{3}}{1 + \left(-\dfrac{1}{\sqrt{3}}\right)(-\sqrt{3})} \right| = \frac{1}{\sqrt{3}}$$

Thus, $\beta_1 = \tfrac{1}{6}\pi$. By symmetry we conclude that $\beta_2 = \tfrac{1}{6}\pi$, where β_2 is the angle between the tangent lines to the curves at point P_2. Because the line $\theta = \tfrac{1}{2}\pi$ is tangent to the circle $r = 3 \cos \theta$ at the pole, and the line $\theta = \pi$ is tangent to the cardioid $r = 1 + \cos \theta$ at the pole, the angle between the tangent lines to the curves at the pole has measure $\tfrac{1}{2}\pi$.

20. For the curve $r \cos \theta = 4$ find $\lim\limits_{\theta \to \pi/2} \chi$.

SOLUTION: We are given that

$$r = \frac{4}{\cos \theta} = 4 \sec \theta$$

Thus,

$$\frac{dr}{d\theta} = 4 \sec \theta \tan \theta$$

Hence,

$$\tan \chi = \frac{r}{\dfrac{dr}{d\theta}}$$

$$= \frac{4 \sec \theta}{4 \sec \theta \tan \theta}$$

$$= \cot \theta$$

$$= \tan(\tfrac{1}{2}\pi - \theta)$$

Thus,

$$\chi = \tfrac{1}{2}\pi - \theta$$

$$\lim_{\theta \to \pi/2} \chi = \lim_{\theta \to \pi/2} (\tfrac{1}{2}\pi - \theta) = 0$$

Figure 12. 4. 20 shows a sketch of the graph.

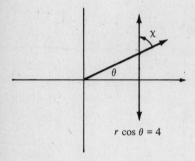

$r \cos \theta = 4$

Figure 12. 4. 20

24. Prove that at the points of intersection of the two curves $r = a \sec^2 \tfrac{1}{2}\theta$ and $r = b \csc^2 \tfrac{1}{2}\theta$ their tangent lines are perpendicular.

SOLUTION: Let P be any point at which the given curves intersect and let β be the angle between the tangent lines to the curves at point P. The tangent lines are perpendicular if and only if $\cot \beta = 0$. Let C_1 be the graph of the equation $r = a \sec^2 \tfrac{1}{2}\theta$ and let C_2 be the graph of the equation $r = b \csc^2 \tfrac{1}{2}\theta$. If χ_1 is the angle between the line OP and the tangent line to the curve C_1, and χ_2 is the angle between the line OP and the tangent line to the curve C_2, then

$$\tan \beta = \left| \frac{\tan \chi_1 - \tan \chi_2}{1 + \tan \chi_1 \tan \chi_2} \right|$$

Thus, $\cot \beta = 0$ if and only if

$$\frac{1 + \tan \chi_1 \tan \chi_2}{\tan \chi_1 - \tan \chi_2} = 0$$

or, equivalently, if and only if

$$\tan \chi_1 \cdot \tan \chi_2 = -1 \tag{1}$$

Suppose that $P = (r_0, \theta_0)$ is a simultaneous solution of the given equations. Then, for the curve $r = a \sec^2 \tfrac{1}{2}\theta$ we have

$$\frac{dr}{d\theta} = a \sec^2 \tfrac{1}{2}\theta \tan \tfrac{1}{2}\theta$$

Thus,

$$\tan \chi_1 = \frac{r}{\dfrac{dr}{d\theta}}\Bigg]_{\theta = \theta_0}$$

$$= \frac{a \sec^2 \tfrac{1}{2}\theta_0}{a \sec^2 \tfrac{1}{2}\theta_0 \tan \tfrac{1}{2}\theta_0}$$

$$= \frac{1}{\tan \tfrac{1}{2}\theta_0} \tag{2}$$

And for the curve $r = b \csc^2 \tfrac{1}{2}\theta$ we have

$$\frac{dr}{d\theta} = -b \csc^2 \tfrac{1}{2}\theta \cot \tfrac{1}{2}\theta$$

Thus,

$$\tan \chi_2 = \frac{b \csc^2 \frac{1}{2} \theta_0}{-b \csc^2 \frac{1}{2} \theta_0 \cot \frac{1}{2} \theta_0}$$

$$= -\tan \tfrac{1}{2} \theta_0 \tag{3}$$

Therefore, by (2) and (3) we have (1); hence the tangent lines are perpendicular. To find another representation of C_1, we replace r by $-r$ and θ by $\theta + \pi$ in the equation $r = a \sec^2 \frac{1}{2}\theta$. This results in

$$-r = a \sec^2 \tfrac{1}{2}(\theta + \pi)$$
$$= a \csc^2 \tfrac{1}{2}\theta \tag{4}$$

If $a = -b$, Eq. (4) and the given equation of curve C_2 are equivalent, which is impossible because C_1 and C_2 are two different curves. If $a \neq -b$, Eq. (4) and the given equation of curve C_2 have no simultaneous solution. Thus, we find no additional points of intersection. Replacing r by $-r$ and θ by $\theta + \pi$ in the equation of curve C_2 leads to the same conclusion. That is, there are no additional points of intersection. Because there are no other representations of C_1 and C_2, we conclude that the tangent lines are perpendicular at every point of intersection.

12.5 AREA OF A REGION IN POLAR COORDINATES

12.5.1 Definition

Let R be the region bounded by the lines $\theta = \alpha$ and $\theta = \beta$ and the curve whose equation is $r = f(\theta)$, where f is continuous and nonnegative on the closed interval $[\alpha, \beta]$. Then if A square units is the area of region R,

$$A = \lim_{\|\Delta\| \to 0} \sum_{i=1}^{n} \frac{1}{2}[f\overline{\theta_i)}]^2 \Delta_i\theta$$

$$= \frac{1}{2} \int_{\alpha}^{\beta} [f(\theta)]^2 \, d\theta$$

If $f(\theta) \geq g(\theta) \geq 0$ for $\alpha \leq \theta \leq \beta$, then the area of the region bounded by the curves $r = f(\theta)$ and $r = g(\theta)$ and the lines $\theta = \alpha$ and $\theta = \beta$ is given by

$$A = \lim_{\|\Delta\| \to 0} \sum_{i=1}^{n} \frac{1}{2}([f(\overline{\theta_i})]^2 - [g(\overline{\theta_i})]^2) \Delta_i\theta$$

$$= \frac{1}{2} \int_{\alpha}^{\beta} ([f(\theta)]^2 - [g(\theta)]^2) \, d\theta$$

Exercises 12.5

Figure 12. 5. 4

4. Find the area of the region enclosed by the graph of $r = 4 \sin^2 \frac{1}{2}\theta$.

SOLUTION: First, we simplify the given equation.

$$r = 4 \sin^2 \tfrac{1}{2}\theta$$
$$= 2(1 - \cos \theta)$$

The graph is the cardioid shown in Fig. 12. 5. 4. Because the region R is symmetric with respect to the polar axis, the area of the region R_1 above the horizontal axis is one-half the area of the entire region R. Because $r = 0$ when $\theta = 0$, we take the line $\theta = 0$ as one of the boundaries of R_1. Thus, R_1 is bounded by the lines $\theta = 0$, $\theta = \pi$, and the curve $r = 2(1 - \cos \theta)$. Hence, the area of R_1 is given by

$$A_1 = \lim_{\|\Delta\| \to 0} \sum_{i=1}^{n} \frac{1}{2}[2(1 - \cos \bar{\theta}_i)]^2 \, \Delta_i\theta$$

$$= \frac{1}{2} \int_0^\pi [2(1 - \cos \theta)]^2 \, d\theta$$

Thus, the area of the entire region R is given by

$$A = 2A_1$$

$$= \int_0^\pi [2(1 - \cos \theta)]^2 \, d\theta$$

$$= 4 \int_0^\pi (1 - 2 \cos \theta + \cos^2\theta) \, d\theta$$

$$= 4 \int_0^\pi \left[1 - 2 \cos \theta + \frac{1}{2}(1 + \cos 2\theta) \right] d\theta$$

$$= 4 \int_0^\pi \left(\frac{3}{2} - 2 \cos \theta + \frac{1}{2} \cos 2\theta \right) d\theta$$

$$= 4 \left[\frac{3}{2}\theta - 2 \sin \theta + \frac{1}{4} \sin 2\theta \right]_0^\pi = 6\pi$$

Thus, the area of the region R is 6π square units.

8. Find the area of the region enclosed by the graph of $r = e^\theta$ and the lines $\theta = 0$ and $\theta = 1$.

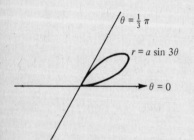

$(e, 1)$

$\theta = 1$

$r = e^\theta$

$(1, 0)$

Figure 12. 5. 8

SOLUTION: The curve is a spiral with endpoints $(1, 0)$ and $(e, 1)$. A sketch of the graph of the region is shown in Fig. 12. 5. 8. Let $f(\theta) = e^\theta$. By Definition 12.5.1, if A square units is the area of the region, then

$$A = \lim_{\|\Delta\| \to 0} \sum_{i=1}^{n} \frac{1}{2}[f(\bar{\theta}_i)]^2 \, \Delta_i\theta$$

$$= \frac{1}{2} \int_0^1 e^{2\theta} \, d\theta$$

$$= \frac{1}{4} e^{2\theta} \Big]_0^1 = \frac{1}{4}(e^2 - 1)$$

The area of the region is $\frac{1}{4}(e^2 - 1)$ square units.

$\theta = \frac{1}{3}\pi$

$r = a \sin 3\theta$

$\theta = 0$

Figure 12. 5. 12

12. Find the area of the region enclosed by one loop of the graph of $r = a \sin 3\theta$.

SOLUTION: The graph is a three-leafed rose. We take the first quadrant loop. Because $r = 0$ when $3\theta = n\pi$, or equivalently, $\theta = \frac{1}{3}n\pi$, we take the lines $\theta = 0$ and $\theta = \frac{1}{3}\pi$ as boundaries. See Fig. 12. 5. 12. Thus, the area of the loop is given by

$$A = \lim_{\|\Delta\| \to 0} \sum_{i=1}^{n} \frac{1}{2}[a \sin 3\bar{\theta}_i]^2 \, \Delta_i\theta$$

$$= \frac{1}{2} \int_0^{\pi/3} [a \sin 3\theta]^2 \, d\theta$$

$$= \frac{1}{2} a^2 \int_0^{\pi/3} \sin^2 3\theta \, d\theta$$

$$= \frac{1}{4} a^2 \int_0^{\pi/3} (1 - \cos 6\theta) \, d\theta$$

$$= \frac{1}{4} a^2 \left[\theta - \frac{1}{6} \sin 6\theta \right]_0^{\pi/3} = \frac{1}{12} \pi a^2$$

Thus, the area of the loop is $\frac{1}{12} \pi a^2$ square units.

16. Find the area of the intersection of the regions enclosed by the graphs of the two given equations.

$$r^2 = 2 \cos 2\theta$$
$$r = 1$$

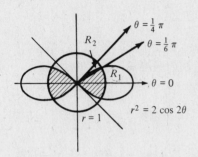

Figure 12. 5. 16

SOLUTION: We want the area of the region that is the intersection of the region bounded by the lemniscate $r^2 = 2 \cos 2\theta$ and the circle $r = 1$, as shown in Fig. 12. 5. 16. Let R be the first quadrant portion of the required region. Because of symmetry, the area of R is one-fourth the area of the entire region. Solving the equations simultaneously, we obtain

$$1 = 2 \cos 2\theta$$
$$\cos 2\theta = \frac{1}{2}$$
$$2\theta = \frac{1}{3} \pi$$
$$\theta = \frac{1}{6} \pi$$

Thus, the line $\theta = \frac{1}{6}\pi$ contains the first-quadrant intersection of the lemniscate and the circle. Let R_1 be the part of region R that is between the lines $\theta = 0$ and $\theta = \frac{1}{6}\pi$. Because the region R_1 is bounded by the circle $r = 1$ and the lines $\theta = 0$ and $\theta = \frac{1}{6}\pi$, we have

$$A_1 = \lim_{\|\Delta\| \to 0} \sum_{i=1}^{n} \frac{1}{2} \Delta_i \theta$$

$$= \frac{1}{2} \int_0^{\pi/6} d\theta$$

$$= \frac{1}{2} \left[\theta \right]_0^{\pi/6} = \frac{1}{12} \pi \tag{1}$$

Let R_2 be the part of region R that is not in region R_1. Setting $r = 0$ in the equation $r^2 = 2 \cos 2\theta$, we get $\theta = \frac{1}{4}\pi$. Thus, R_2 is bounded by the lines $\theta = \frac{1}{6}\pi$ and $\theta = \frac{1}{4}\pi$ and the lemniscate $r^2 = 2 \cos 2\theta$. Hence,

$$A_2 = \lim_{\|\Delta\| \to 0} \sum_{i=1}^{n} \frac{1}{2} [2 \cos 2\bar{\theta}_i] \Delta_i \theta$$

$$= \frac{1}{2} \int_{\pi/6}^{\pi/4} (2 \cos 2\theta) \, d\theta$$

$$= \frac{1}{2} \sin 2\theta \Big]_{\pi/6}^{\pi/4} = \frac{1}{2} \left(1 - \frac{1}{2} \sqrt{3} \right) \tag{2}$$

Because A, the area of the entire region, is given by

$$A = 4(A_1 + A_2)$$

we have from (1) and (2)

$$A = 4(\tfrac{1}{12}\pi + \tfrac{1}{2} - \tfrac{1}{4}\sqrt{3}) = \tfrac{1}{3}\pi + 2 - \sqrt{3}$$

Thus, the area of the region is $\frac{1}{3}\pi + 2 - \sqrt{3}$ square units.

20. Find the area of the region that is inside the graph of the first given equation and outside the graph of the second equation.

$$r = 2a \sin \theta$$
$$r = a$$

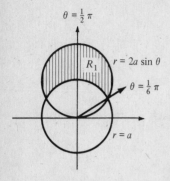

Figure 12. 5. 20

SOLUTION: Figure 12. 5. 20 shows a sketch of the region. Let R_1 be the first quadrant part of the required region. Solving the equations simultaneously, we have

$$2a \sin \theta = a$$
$$\sin \theta = \tfrac{1}{2}$$
$$\theta = \tfrac{1}{6}\pi$$

Thus, the region R_1 is bounded by the lines $\theta = \tfrac{1}{6}\pi$ and $\theta = \tfrac{1}{2}\pi$ and by the curves $r = a$ and $r = 2a \sin \theta$, where $a \le 2a \sin \theta$ when $\tfrac{1}{6}\pi \le \theta \le \tfrac{1}{2}\pi$. Thus, the area of region R_1 is given by

$$A_1 = \lim_{\|\Delta\| \to 0} \sum_{i=1}^{n} \frac{1}{2} [2a \sin \bar{\theta}_i)^2 - a^2] \Delta_i \theta \qquad (1)$$

Hence, the area of the entire region is given by $A = 2A_1$, and from (1) we have

$$A = \int_{\pi/6}^{\pi/2} [(2a \sin \theta)^2 - a^2] \, d\theta$$

$$= a^2 \int_{\pi/6}^{\pi/2} [2(1 - \cos 2\theta) - 1] \, d\theta$$

$$= a^2 \int_{\pi/6}^{\pi/2} (1 - 2 \cos 2\theta) \, d\theta$$

$$= a^2 \left[\theta - \sin 2\theta \right]_{\pi/6}^{\pi/2}$$

$$= a^2 \left[\frac{1}{2}\pi - \left(\frac{1}{6}\pi - \frac{1}{2}\sqrt{3} \right) \right] = a^2 \left(\frac{1}{3}\pi + \frac{1}{2}\sqrt{3} \right)$$

Thus, the area of the region is $a^2(\tfrac{1}{3}\pi + \tfrac{1}{2}\sqrt{3})$ square units.

24. Find the area of the region swept out by the radius vector of the spiral $r = a\theta$ during its second revolution that was not swept out during its first revolution.

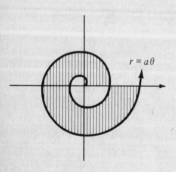

Figure 12. 5. 24

SOLUTION: Let R_1 be the region swept out by the radius vector of the spiral $r = a\theta$ during its first revolution and let R_2 be the region swept out by the radius vector of the spiral $r = a\theta$ during its second revolution. We find $A = A_2 - A_1$ where A_1 is the measure of the area of region R_1 and A_2 is the measure of the area of region R_2. See Fig. 12. 5. 24. We have

$$A_1 = \frac{1}{2} \int_{0}^{2\pi} (a\theta)^2 \, d\theta$$

$$= \frac{1}{6} a^2 \theta^3 \bigg]_{0}^{2\pi}$$

$$= \frac{4}{3} a^2 \pi^3$$

and

$$A_2 = \frac{1}{2} \int_{2\pi}^{4\pi} (a\theta)^2 \, d\theta$$

$$= \frac{1}{6} a^2 \theta^3 \Big]_{2\pi}^{4\pi} = \frac{28}{3} a^2 \pi^3$$

Thus,

$$A = A_2 - A_1 = 8a^2\pi^3$$

The area is $8a^2\pi^3$ square units.

Review Exercises for Chapter 12

4. Find a polar equation of the graph having the given Cartesian equation.
$y^4 = x^2(a^2 - y^2)$

SOLUTION: We let $x = r \cos \theta$ and $y = r \sin \theta$. Thus, a Cartesian equation of the given curve is

$$r^4 \sin^4 \theta = r^2 \cos^2 \theta (a^2 - r^2 \sin^2 \theta)$$

$$r^2 \sin^4 \theta = \cos^2 \theta (a^2 - r^2 \sin^2 \theta)$$

$$r^2 \sin^4 \theta = a^2 \cos^2 \theta - r^2 \sin^2 \theta \cos^2 \theta$$

$$r^2(\sin^4 \theta + \sin^2 \theta \cos^2 \theta) = a^2 \cos^2 \theta$$

$$r^2 = \frac{a^2 \cos^2 \theta}{\sin^2 \theta (\sin^2 \theta + \cos^2 \theta)}$$

$$r^2 = \frac{a^2 \cos^2 \theta}{\sin^2 \theta}$$

$$r^2 = a^2 \cot^2 \theta$$

$$r = \pm a \cot \theta$$

Because $\cot(-\theta) = -\cot \theta$ and $\cot(\pi - \theta) = -\cot \theta$, the graphs of $r = a \cot \theta$ and $r = -a \cot \theta$ are the same. Thus, we may reduce the answer to

$$r = a \cot \theta$$

8. Find a Cartesian equation of the graph having the polar equation $r = a \tan^2 \theta$.

SOLUTION: We replace r by $\pm\sqrt{x^2 + y^2}$ and $\tan \theta$ by y/x. Thus,

$$\pm\sqrt{x^2 + y^2} = a \left(\frac{y}{x}\right)^2$$

$$x^2 + y^2 = \frac{a^2 y^4}{x^4}$$

$$x^4(x^2 + y^2) = a^2 y^4$$

$$x^6 + x^4 y^2 - a^2 y^4 = 0$$

12. Draw a sketch of the graph of $r = \sqrt{|\cos 2\theta|}$.

SOLUTION: Because $\cos[2(-\theta)] = \cos 2\theta$, by Theorem 12.2.1 the graph is symmetric with respect to the polar axis. Because $\cos[2(\pi - \theta)] = \cos 2\theta$, by Theorem 12.2.2 the graph is also symmetric with respect to the line $\theta = \frac{1}{2}\pi$. We plot the points from Table 12 and use symmetry to complete the sketch, which is shown in Fig. 12. 12R.

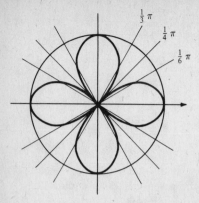

Figure 12. 12R

Table 12

θ	0	$\frac{1}{6}\pi$	$\frac{1}{4}\pi$	$\frac{1}{3}\pi$	$\frac{1}{2}\pi$
r	1	0.7	0	0.7	1

16. Find an equation of each of the tangent lines at the pole to the four-leafed rose $r = 4 \cos 2\theta$.

SOLUTION: We let $r = 0$ and solve for θ.

$$\cos 2\theta = 0$$
$$2\theta = \tfrac{1}{2}\pi + n\pi$$
$$\theta = \tfrac{1}{4}\pi + \tfrac{1}{2}n\pi$$

Replacing n by 0, 1, 2, and 3, we obtain

$$\theta = \tfrac{1}{4}\pi; \quad \theta = \tfrac{3}{4}\pi; \quad \theta = \tfrac{5}{4}\pi; \quad \theta = \tfrac{7}{4}\pi$$

The graphs of $\theta = \tfrac{1}{4}\pi$ and $\theta = \tfrac{5}{4}\pi$ are the same line. Likewise, $\theta = \tfrac{3}{4}\pi$ and $\theta = \tfrac{7}{4}\pi$ have the same graph. We conclude that $\theta = \tfrac{1}{4}\pi$ and $\theta = \tfrac{3}{4}\pi$ are equations of each of the tangent lines at the pole.

20. Find the area of the region inside the graph of the lemniscate $r^2 = 2 \sin 2\theta$ and outside the graph of the circle $r = 1$.

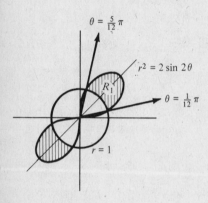

Figure 12. 20R

SOLUTION: Figure 12. 20R shows a sketch of the region. Solving the equations simultaneously, we get

$$2 \sin 2\theta = 1$$
$$\sin 2\theta = \tfrac{1}{2}$$
$$2\theta = \tfrac{1}{6}\pi \quad \text{or} \quad 2\theta = \tfrac{5}{6}\pi$$
$$\theta = \tfrac{1}{12}\pi \quad \text{or} \quad \theta = \tfrac{5}{12}\pi$$

Thus, R_1, the first quadrant part of the region, is bounded by the lines $\theta = \tfrac{1}{12}\pi$ and $\theta = \tfrac{5}{12}\pi$ and by the curves $r = 1$ and $r = \sqrt{2 \sin 2\theta}$, where $\sqrt{2 \sin 2\theta} \geq 1$ for $\tfrac{1}{12}\pi \leq \theta \leq \tfrac{5}{12}\pi$. The number of square units in the area of R_1 is given by

$$A_1 = \lim_{\|\Delta\| \to 0} \sum_{i=1}^{n} \tfrac{1}{2}[2 \sin 2\bar{\theta}_i - 1] \, \Delta_i\theta$$

$$= \tfrac{1}{2} \int_{\pi/12}^{5\pi/12} (2 \sin 2\theta - 1) \, d\theta$$

Because the number of square units in the area of the entire region is given by $A = 2A_1$, we have

$$A = \int_{\pi/12}^{5\pi/12} (2 \sin 2\theta - 1) \, d\theta$$

$$= \Big[-\cos 2\theta - \theta \Big]_{\pi/12}^{5\pi/12}$$

$$= \left(\frac{1}{2}\sqrt{3} - \frac{5}{12}\pi \right) - \left(-\frac{1}{2}\sqrt{3} - \frac{1}{12}\pi \right) = \sqrt{3} - \frac{1}{3}\pi$$

Thus, the area is $\sqrt{3} - \tfrac{1}{3}\pi$ square units.

24. Find all points of intersection of the graphs of the two given equations.

$$r \cos \theta = 1$$
$$r = 1 + 2 \cos \theta$$

SOLUTION: Eliminating r from the equations, we obtain

$$1 + 2 \cos \theta = \frac{1}{\cos \theta}$$

$$2 \cos^2 \theta + \cos \theta - 1 = 0$$

$$(\cos \theta + 1)(2 \cos \theta - 1) = 0$$

$$\cos \theta = -1 \qquad \cos \theta = \tfrac{1}{2}$$

$$\theta = \pi \qquad \theta = \pm \tfrac{1}{3}\pi$$

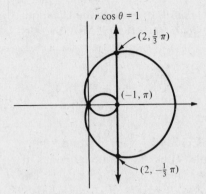

Figure 12. 24R

Thus, $(-1, \pi)$, $(2, \tfrac{1}{3}\pi)$, and $(2, -\tfrac{1}{3}\pi)$ are the points of intersection, as illustrated in Fig. 12. 24R. We note that no additional equations result from replacing r by $(-1)^n r$ and replacing θ by $(\theta + n\pi)$ in the equation $r \cos \theta = 1$. Replacing r by $-r$ and θ by $\theta + \pi$ in the equation $r = 1 + 2 \cos \theta$ results in the equation $-r = 1 - 2 \cos \theta$, but there are no additional points of intersection found by using this equation.

28. Find the slope of the tangent line to the curve $r = 6 \cos \theta - 2$ at the point $(1, \tfrac{5}{3}\pi)$.

SOLUTION: We find dy/dx at the point $(1, \tfrac{5}{3}\pi)$. Because $y = r \sin \theta$ and $r = 6 \cos \theta - 2$, we have

$$
\begin{aligned}
y &= (6 \cos \theta - 2)\sin \theta \\
&= 6 \sin \theta \cos \theta - 2 \sin \theta \\
&= 3 \sin 2\theta - 2 \sin \theta
\end{aligned}
$$

Thus

$$\frac{dy}{d\theta} = 6 \cos 2\theta - 2 \cos \theta$$

$$\frac{dy}{d\theta}\bigg]_{\theta = 5\pi/3} = 6 \cos \tfrac{10}{3}\pi - 2 \cos \tfrac{5}{3}\pi$$

$$= -4 \qquad\qquad (1)$$

Because $x = r \cos \theta$ and $r = 6 \cos \theta - 2$, we have

$$x = 6 \cos^2 \theta - 2 \cos \theta$$

Thus,

$$\frac{dx}{d\theta} = -12 \cos \theta \sin \theta + 2 \sin \theta$$

$$= -6 \sin 2\theta + 2 \sin \theta$$

$$\frac{dx}{d\theta}\bigg]_{\theta = 5\pi/3} = -6 \sin \frac{10}{3}\pi + 2 \sin \frac{5}{3}\pi$$

$$= 2\sqrt{3} \qquad\qquad (2)$$

Because

$$\frac{dy}{dx} = \frac{dy}{d\theta} \div \frac{dx}{d\theta} \qquad\qquad (3)$$

we substitute from (1) and (2) into (3) to obtain

$$\frac{dy}{dx}\bigg]_{\theta = 5\pi/3} = \frac{-4}{2\sqrt{3}} = -\frac{2}{3}\sqrt{3}$$

Thus, the slope of the tangent line is $-\tfrac{2}{3}\sqrt{3}$.

32. Find the points of intersection of the graphs of the equations $r = \tan \theta$ and $r = \cot \theta$.

SOLUTION: If $\tan \theta = \cot \theta$, then $\tan^2 \theta = 1$, so $\tan \theta = \pm 1$, and $\theta = \frac{1}{4}\pi$, $\theta = \frac{3}{4}\pi$, $\theta = \frac{5}{4}\pi$, or $\theta = \frac{7}{4}\pi$. Thus, the points $(1, \frac{1}{4}\pi)$, $(-1, \frac{3}{4}\pi)$, $(1, \frac{5}{4}\pi)$, and $(-1, \frac{7}{4}\pi)$ are points of intersection. Furthermore, because $(0, 0)$ satisfies the equation $r = \tan \theta$ and $(0, \frac{1}{2}\pi)$ satisfies the equation $r = \cot \theta$, then the pole is a point of intersection. If we replace r by $-r$ and θ by $\theta + \pi$ in the equation $r = \tan \theta$ the result is

$$-r = \tan(\theta + \pi)$$
$$-r = \tan \theta$$
$$r = -\tan \theta$$

However, if we attempt to solve the system consisting of $r = -\tan \theta$ and $r = \cot \theta$, we obtain $\tan^2 \theta = -1$, which is impossible. Thus, no additional points of intersection are found. A similar conclusion is reached if we replace r by $-r$ and θ by $\theta + \pi$ in the equation $r = \cot \theta$. Therefore, we have found all the points of intersection.

CHAPTER 13 | The Conic Sections

13.1 THE PARABOLA

13.1.1 Definition A *parabola* is the set of all points in a plane equidistant from a fixed point and a fixed line. The fixed point is called the *focus,* and the fixed line is called the *directrix.*

13.1.2 Theorem An equation of the parabola having its focus at $(p, 0)$ and the line $x = -p$ as its directrix is

$$y^2 = 4px$$

If $p > 0$ in Theorem 13.1.2, the parabola opens to the right; if $p < 0$, the parabola opens to the left. In either case, the vertex is at the origin, and the x-axis is the axis of the parabola.

13.1.3 Theorem An equation of the parabola having its focus at $(0, p)$ and the line $y = -p$ as its directrix is

$$x^2 = 4py$$

If $p > 0$ in Theorem 13.1.3, the parabola opens upward; if $p < 0$, the parabola opens downward. In either case, the vertex is at the origin, and the y-axis is the axis of the parabola.

The chord through the focus perpendicular to the axis of the parabola is called the *latus rectum.* The length of the latus rectum is $|4p|$. This segment is helpful in drawing a sketch of the parabola.

Exercises 13.1

For each of the parabolas in Exercises 1–8, find the coordinates of the focus, an equation of the directrix, and the length of the latus rectum. Draw a sketch of the curve.

4. $x^2 = -16y$

SOLUTION: By Theorem 13.1.3 with $4p = -16$, or $p = -4$, we conclude that the focus is $(0, -4)$; an equation of the directrix is $y = 4$; and the length of the latus rectum is $|-16| = 16$. A sketch of the curve is shown in Fig. 13. 1. 4.

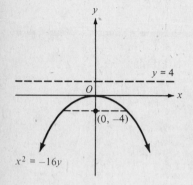

Figure 13. 1. 4

8. $3x^2 + 4y = 0$

SOLUTION: We apply Theorem 13.1.3. The given equation is equivalent to

$$x^2 = -\tfrac{4}{3}y$$

Thus, we have $4p = -\tfrac{4}{3}$, or $p = -\tfrac{1}{3}$. The focus is $(0, -\tfrac{1}{3})$, an equation of the directrix is $y = \tfrac{1}{3}$, and the length of the latus rectum is $|-\tfrac{4}{3}| = \tfrac{4}{3}$. A sketch of the curve is shown in Fig. 13. 1. 8.

In Exercises 9–17 find an equation of the parabola having the given properties.

12. Focus $(-\tfrac{5}{3}, 0)$; directrix $5 - 3x = 0$

SOLUTION: Because the focus is on the x-axis and the directrix is perpendicular to the x-axis, the x-axis is the axis of the parabola. Furthermore, the midpoint of the segment from the focus perpendicular to the directrix is the origin. Thus, we may use Theorem 13.1.2 with $p = -\tfrac{5}{3}$. We have

$$y^2 = 4px$$
$$y^2 = -\tfrac{20}{3}x$$
$$3y^2 = -20x$$

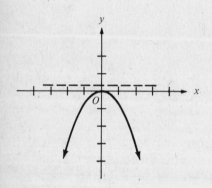

Figure 13. 1. 8

16. Vertex $(0, 0)$; opens upward; length of latus rectum is 3

SOLUTION: Because the vertex is at the origin and the parabola opens upward, we apply Theorem 13.1.3 with $p > 0$. Because the length of the latus rectum is 3, we have $4p = 3$. Thus,

$$x^2 = 4py$$
$$x^2 = 3y$$

20. Prove that the length of the latus rectum of a parabola is $|4p|$.

SOLUTION: Let the vertex of the parabola be at the origin and let the focus be located on the y-axis at the point $F(0, p)$. By Theorem 13.1.3 an equation of the parabola is $x^2 = 4py$. The latus rectum is the chord that is perpendicular to the y-axis at the focus F. Let A and B be the endpoints of the latus rectum. See Fig. 13. 1. 20. Because AB is a horizontal line, the y-coordinates of points A, F, and B are all the same. Thus, both A and B have ordinate p. Replacing y with p in the equation of the parabola, we obtain $x^2 = 4p^2$, or, equivalently, $x = \pm 2|p|$. We conclude that $A = (-2|p|, p)$ and $B = (2|p|, p)$. Thus, the distance between points A and B is $2|p| - (-2|p|) = 4|p| = |4p|$. Hence, the length of the latus rectum is $|4p|$.

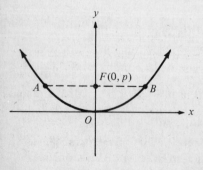

Figure 13. 1. 20

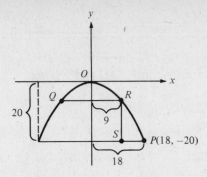

Figure 13. 1. 24

24. A parabolic arch has a height of 20 meters and a width of 36 meters at the base. If the vertex of the parabola is at the top of the arch, at what height above the base is it 18 meters wide?

SOLUTION: Refer to Fig. 13. 1. 24. The vertex of the parabola is at the origin. Because the height of the arch is 20 m and the width of the base is 36 m, the point $P = (18, -20)$ must be on the parabola. By Theorem 13.1.3, an equation of the parabola is of the form

$$x^2 = 4py \tag{1}$$

Because the parabola contains the point $(18, -20)$, we replace x by 18 and y by -20 in Eq. (1). Thus,

$$18^2 = 4p(-20)$$
$$4p = -\tfrac{81}{5}$$

Substituting this value of $4p$ into (1), we obtain

$$x^2 = -\tfrac{81}{5} y \tag{2}$$

Let QR be the chord that is 18 m long and let RS be a segment perpendicular to the base. Because the x coordinate of point R is 9, we let $x = 9$ in Eq. (2). Thus,

$$9^2 = -\tfrac{81}{5} y$$
$$y = -5$$

Because the y-coordinate of point R is -5, we conclude that $|\overline{RS}| = -5 - (-20) = 15$. Thus, the arch is 18 m wide at a height of 15 m above the base.

28. Use Definition 13.1.1 to find an equation of the parabola having as its directrix the line $y = 4$ and as its focus the point $(-3, 8)$.

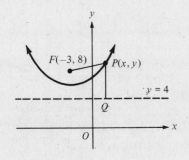

Figure 13. 1. 28

SOLUTION: Refer to Fig. 13. 1. 28. Let $F(-3, 8)$ be the focus; let $P(x, y)$ be any point on the parabola; and let $|\overline{PQ}|$ be the distance from point P to the directrix $y = 4$. By Definition 13.1.1, we have

$$|\overline{FP}| = |\overline{PQ}| \tag{1}$$

From the distance formula, we have

$$|\overline{FP}| = \sqrt{(x+3)^2 + (y-8)^2} \tag{2}$$

Because segment PQ is parallel to the y-axis,

$$|\overline{PQ}| = y - 4 \tag{3}$$

Substituting from (2) and (3) into (1), we obtain

$$\sqrt{(x+3)^2 + (y-8)^2} = y - 4$$
$$x^2 + 6x + 9 + y^2 - 16y + 64 = y^2 - 8y + 16$$
$$x^2 + 6x - 8y + 57 = 0$$

32. In Fig. 13. 1. 7 in the text, prove $\alpha = \beta$. (*Hint:* Choose the coordinate axes so that the parabola has its vertex at the origin, has its axis along the y-axis, and it opens upward. Let Q be the point of intersection of the tangent line PT with the y-axis. Prove $\alpha = \beta$ by showing that triangle QPF is isosceles.)

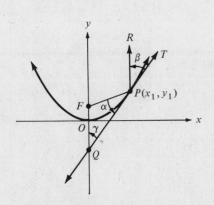

Figure 13. 1. 32

SOLUTION: In Fig. 13. 1. 32 the point $P(x_1, y_1)$ is any point on the parabola. Line PT is tangent to the parabola and line PR is parallel to the y-axis. We find an equation of line PT. An equation of the parabola is $x^2 = 4py$. Thus, $dy/dx = x/2p$, so $m = x_1/2p$ is the slope of line PT, and

$$y - y_1 = \frac{x_1}{2p}(x - x_1) \tag{1}$$

is an equation of the line. If Q is the point where line PT intersects the y-axis, then the x-coordinate of point Q is 0. Replacing x by 0 in Eq. (1) and solving for y, we obtain

$$y - y_1 = \frac{x_1}{2p}(-x_1)$$

$$y = y_1 - \frac{x_1^2}{2p} \tag{2}$$

Because $P = (x_1, y_1)$ is a point on the parabola, then the coordinates of P satisfy the equation of the parabola, so $x_1^2 = 4py_1$. Substituting for x_1 in Eq. (2), we get

$$y = y_1 - \frac{4py_1}{2p} = -y_1$$

Thus, $Q = (0, -y_1)$. If F is the focus of the parabola, then $F = (0, p)$, so the distance between points F and Q is

$$|\overline{FQ}| = p + y_1 \tag{3}$$

Furthermore, the distance between points F and P is

$$\begin{aligned}
|\overline{FP}| &= \sqrt{x_1^2 + (y_1 - p)^2} \\
&= \sqrt{4py_1 + (y_1 - p)^2} \\
&= \sqrt{4py_1 + y_1^2 - 2py_1 + p^2} \\
&= \sqrt{y_1^2 + 2py_1 + p^2} \\
&= \sqrt{(y_1 + p)^2} \\
&= y_1 + p \tag{4}
\end{aligned}$$

From Eqs. (3) and (4) we conclude that triangle QPF is an isosceles triangle. Let α be the measure of angle FPQ, let β be the measure of angle TPR, and let γ be the measure of angle PQF. Because triangle QPF is isosceles, then $\alpha = \gamma$. Because line PR is parallel to the y-axis, then $\beta = \gamma$. Therefore, $\alpha = \beta$.

36. Prove that the distance from the midpoint of a focal chord (a line segment through the focus with endpoints on the parabola) of a parabola to the directrix is half the length of the focal chord.

SOLUTION: Refer to Fig. 13. 1. 36. We take the origin at the vertex of the parabola and position the axes so that the focus is at the point $(0, p)$ with $p > 0$, and the directrix is the line $y = -p$. By Theorem 13.1.3, an equation of parabola is

$$x^2 = 4py \tag{1}$$

Let segment AB be a focal chord with slope m. The slope-intercept form of an equation of line AB is

$$y = mx + p \tag{2}$$

Eliminating y from Eqs. (1) and (2) and solving for x, we obtain

$$\begin{aligned}
x^2 &= 4p(mx + p) \\
x^2 - 4pmx &= 4p^2 \\
x^2 - 4pmx + 4p^2m^2 &= 4p^2 + 4p^2m^2 \\
(x - 2pm)^2 &= 4p^2(1 + m^2) \\
x - 2pm &= \pm 2p\sqrt{1 + m^2} \\
x &= 2p(m \pm \sqrt{1 + m^2})
\end{aligned}$$

Thus, if $A = (x_1, y_1)$ and $B = (x_2, y_2)$ we have

$$x_1 = 2p(m + \sqrt{1 + m^2}) \quad \text{and} \quad x_2 = 2p(m - \sqrt{1 + m^2}) \tag{3}$$

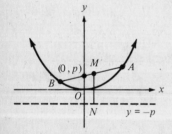

Figure 13. 1. 36

Substituting the values of x into Eq. (2), we obtain

$$y_1 = 2pm(m + \sqrt{1 + m^2}) + p \quad \text{and} \quad y_2 = 2pm(m - \sqrt{1 + m^2}) + p \qquad (4)$$

Using the distance formula with (3) and (4), we obtain

$$\begin{aligned}
|\overline{AB}| &= \sqrt{(x_1 - x_2)^2 + (y_1 - y_2)^2} \\
&= \sqrt{(4p\sqrt{1 + m^2})^2 + (4pm\sqrt{1 + m^2})^2} \\
&= 4p\sqrt{(1 + m^2) + m^2(1 + m^2)} \\
&= 4p(1 + m^2)
\end{aligned} \qquad (5)$$

Let $M(\bar{x}, \bar{y})$ be the midpoint of segment AB. Using the midpoint formula with (4), we obtain

$$\bar{y} = \frac{y_1 + y_2}{2}$$

$$= \frac{2pm(2m) + 2p}{2} = 2pm^2 + p \qquad (6)$$

If $|\overline{MN}|$ is the distance from point M to the directrix, then from (6) we have

$$|\overline{MN}| = \bar{y} + p = 2p(m^2 + 1) \qquad (7)$$

From (5) and (7) we have $|\overline{MN}| = \frac{1}{2}|\overline{AB}|$, and thus the distance from the midpoint of the focal chord to the directrix is half the length of the focal chord.

13.2 TRANSLATION OF AXES

13.2.1 Theorem If (x, y) represents a point P with respect to a given set of axes, and (x', y') is a representation of P after the axes are translated to a new origin having coordinates (h, k) with respect to the given axes, then

$$x = x' + h \quad \text{and} \quad y = y' + k$$

or, equivalently,

$$x' = x - h \quad \text{and} \quad y' = y - k$$

13.2.2 Theorem If p is the directed distance from the vertex to the focus, an equation of the parabola with its vertex at (h, k) and with its axis parallel to the x-axis is

$$(y - k)^2 = 4p(x - h) \qquad (1)$$

A parabola with the same vertex and with its axis parallel to the y-axis has for an equation

$$(x - h)^2 = 4p(y - k) \qquad (2)$$

Exercises 13.2

In Exercises 1–12 find a new equation of the graph of the given equation after a translation of axis to the new origin as indicated. Draw the original and the new axes and a sketch of the graph.

4. $y^2 + 3x - 2y + 7 = 0$; $(-2, 1)$

SOLUTION: By Theorem 13.2.1 we let $x = x' - 2$ and $y = y' + 1$. Thus,

$$(y' + 1)^2 + 3(x' - 2) - 2(y' + 1) + 7 = 0$$
$$(y')^2 + 2y' + 1 + 3x' - 6 - 2y' - 2 + 7 = 0$$
$$(y')^2 = -3x' \qquad (1)$$

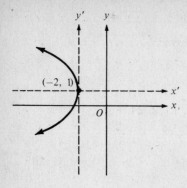

Figure 13. 2. 4

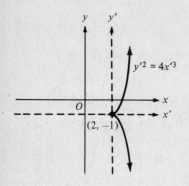

Figure 13. 2. 8

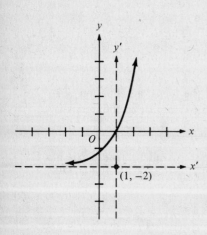

Figure 13. 2. 12

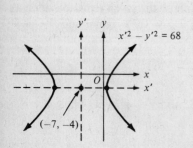

Figure 13. 2. 16

From Eq. (1) we conclude that with respect to the x'- and y'-axes, the graph is a parabola with vertex at the origin, opening to the left with $4p = -3$. With respect to the x- and y-axes, the graph is a parabola with vertex at $(-2, 1)$. A sketch of the parabola is shown in Fig. 13. 2. 4.

8. $(y + 1)^2 = 4(x - 2)^3$; $(2, -1)$

SOLUTION: We let $x = x' + 2$ and $y = y' - 1$. Thus, $y'^2 = 4x'^3$. This is not an equation of a parabola. However, the graph contains the origin of the x'- and y'-axes and is symmetric with respect to the x'-axis. A sketch of the curve is shown in Fig. 13. 2. 8.

12. $2e^{2x-2} = y + 2$; $(1, -2)$

SOLUTION: We let $x = x' + 1$ and $y = y' - 2$. The resulting equation is

$$2e^{x'} = y'$$

The graph is an exponential curve with the negative x'-axis as a horizontal asymptote, and which intersects the y'-axis at $y' = 2$. A sketch of the graph is shown in Fig. 13. 2. 12.

16. Translate the axes so that an equation of the graph with respect to the new axes will contain no first-degree terms. Draw the original and the new axes and a sketch of the graph.

$$x^2 - y^2 + 14x - 8y - 35 = 0$$

SOLUTION: We complete the square on the terms in x and the terms in y.

$$(x^2 + 14x) - (y^2 + 8y) = 35$$
$$(x^2 + 14x + 49) - (y^2 + 8y + 16) = 35 + 49 - 16$$
$$(x + 7)^2 - (y + 4)^2 = 68$$

We let $x' = x + 7$ and $y' = y + 4$. Thus

$$x'^2 - y'^2 = 68$$

We have translated the origin to the point $(-7, -4)$. With respect to the x'- and y'-axes, the graph is a hyperbola with center at the origin, as shown in Fig. 13. 2. 16.

20. Find the vertex, the focus, an equation of the axis, and an equation of the directrix of the given parabola. Draw a sketch of the graph.

$$3y^2 - 8x - 12y - 4 = 0$$

SOLUTION: We complete the square of the terms in y.

$$3y^2 - 12y = 8x + 4$$
$$y^2 - 4y = \tfrac{8}{3}x + \tfrac{4}{3}$$
$$y^2 - 4y + 4 = \tfrac{8}{3}x + \tfrac{4}{3} + 4$$
$$(y - 2)^2 = \tfrac{8}{3}(x + 2)$$

Comparing this equation with Eq. (1) in Theorem 13.2.2, we have $k = 2$, $h = -2$, and $4p = \tfrac{8}{3}$ or $p = \tfrac{2}{3}$. Thus, the vertex is at $V(-2, 2)$. The axis is parallel to the x-axis and contains the vertex. Thus, an equation of the axis is $y = 2$. Because $p > 0$, the parabola opens to the right, as shown in Fig. 13. 2. 20. Because the distance between the vertex and the focus is p, the focus is at $F(-\tfrac{4}{3}, 2)$. Because the distance between the directrix and the vertex is p, an equation of the directrix is $x = -\tfrac{8}{3}$.

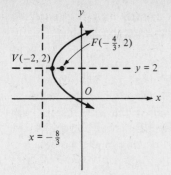

$V(-2, 2)$
$F(-\frac{4}{3}, 2)$
$y = 2$
O
x
$x = -\frac{8}{3}$

Figure 13. 2. 20

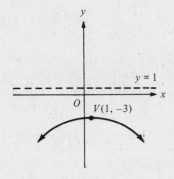

$y = 1$
O $V(1, -3)$
x

Figure 13. 2. 24

In Exercises 23–32 find an equation of the parabola having the given properties. Draw a sketch of the graph.

24. Vertex $(1, -3)$; directrix $y = 1$

SOLUTION: A sketch of the graph is shown in Fig. 13. 2. 24. Because the axis of the parabola is perpendicular to the directrix, the axis is parallel to the y-axis. Thus, by Theorem 13.2.2, an equation of the parabola is of the form

$$(x - h)^2 = 4p(y - k) \tag{1}$$

Because the vertex is $(1, -3)$, we have $h = 1$ and $k = -3$. Because the vertex is below the directrix, $p < 0$, and because the distance between the vertex and the directrix is 4, we conclude that $p = -4$. Thus, from (1) we have

$$(x - 1)^2 = -16(y + 3)$$
$$x^2 - 2x + 16y + 49 = 0$$

which is an equation of the parabola.

28. Axis parallel to the x-axis; through the points $(1, 2)$, $(5, 3)$, and $(11, 4)$

SOLUTION: Because the axis is parallel to the x-axis, by Theorem 13.2.2 an equation of the parabola is of the form

$$(y - k)^2 = 4p(x - h) \tag{1}$$

Because this equation is of the second degree in y and first degree in x, there are constants a, b, and c such that

$$x = ay^2 + by + c \tag{2}$$

and the graph of (2) is the same as the graph of (1). Because the parabola contains the point $(1, 2)$, the coordinates of this point must satisfy Eq. (2). Thus, we let $x = 1$ and $y = 2$ in (2), obtaining

$$1 = 4a + 2b + c \tag{3}$$

Similarly, we substitute $(5, 3)$ and $(11, 4)$ into (2) and obtain

$$5 = 9a + 3b + c \tag{4}$$
$$11 = 16a + 4b + c \tag{5}$$

Solving Eqs. (3), (4), and (5) simultaneously, we get

$$a = 1 \qquad b = -1 \qquad c = -1$$

Substituting these values into (2), we obtain

$$x = y^2 - y - 1$$
$$y^2 - y - x - 1 = 0$$

which is an equation of the parabola. To draw a sketch of the graph, we complete the square of the terms in y. Thus,

$$y^2 - y = x + 1$$
$$y^2 - y + \tfrac{1}{4} = x + 1 + \tfrac{1}{4}$$
$$(y - \tfrac{1}{2})^2 = (x + \tfrac{5}{4})$$

Hence, the vertex is at the point $(-\frac{5}{4}, \frac{1}{2})$, and the parabola opens to the right, as shown in Fig. 13. 2. 28.

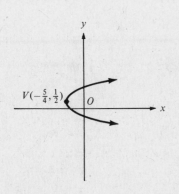

$V(-\frac{5}{4}, \frac{1}{2})$ O
x

Figure 13. 2. 28

32. Endpoints of the latus rectum are $(1, 3)$ and $(7, 3)$.

SOLUTION: Because the latus rectum is parallel to the *x*-axis, the axis of the parabola is parallel to the *y*-axis. Thus, an equation of the parabola is of the form

$$(x - h)^2 = 4p(y - k) \tag{1}$$

Because the length of the latus rectum is $|4p|$, we have $|4p| = 6$. Thus $4p = \pm 6$, and there are two parabolas possible. Refer to Fig. 13. 2. 32, which shows the case where $4p = 6$. Because the focus is at the midpoint of the latus rectum, we have $F = (4, 3)$. Because $p = \frac{3}{2}$ and the distance between the focus and the vertex is p, we have $V = (4, \frac{3}{2})$. Thus, $h = 4$ and $k = \frac{3}{2}$. Substituting in (1), we obtain

$$(x - 4)^2 = 6(y - \tfrac{3}{2})$$
$$x^2 - 8x - 6y + 25 = 0$$

which is an equation of the parabola for which $4p = 6$. If $4p = -6$, then the vertex is at the point $(4, \frac{9}{2})$ and

$$(x - 4)^2 = -6(y - \tfrac{9}{2})$$
$$x^2 - 8x + 6y - 11 = 0$$

which is an equation of the other parabola possible.

36. Given the equation $4x^3 - 12x^2 + 12x - 3y - 10 = 0$, translate the axes so that the equation of the graph with respect to the new axes will contain no second-degree term and no constant term. Draw a sketch of the graph and the two sets of axes. (*Hint:* Let $x = x' + h$ and $y = y' + k$ in the given equation.)

SOLUTION: If $x = x' + h$ and $y = y' + k$ in the given equation, we have

$$4(x' + h)^3 - 12(x' + h)^2 + 12(x' + h) - 3(y' + k) - 10 = 0$$
$$4(x')^3 + 12(x')^2 h + 12x' h^2 + 4h^3 - 12(x')^2 - 24x' h - 12h^2 + 12x'$$
$$+ 12h - 3y' - 3k - 10 = 0 \tag{1}$$

Setting the coefficient of x'^2 in Eq. (1) equal to zero, we have

$$12h - 12 = 0$$
$$h = 1$$

Setting the constant term in Eq. (1) equal to zero, we get

$$4h^3 - 12h^2 + 12h - 3k - 10 = 0$$

Because $h = 1$, this gives $k = -2$. Thus, we translate the origin to the point $(1, -2)$. With $h = 1$ and $k = -2$ in Eq. (1), we get

$$4(x')^3 + 12(x')^2 + 12x' + 4 - 12(x')^2 - 24x'$$
$$- 12 + 12x' + 12 - 3y' + 6 - 10 = 0$$

which reduces to

$$3y' = 4x'^3$$

In Fig. 13. 2. 36 we show a sketch of the graph.

40. Show that after a translation of axes to the new origin $(-\frac{1}{4}\pi, 1)$ the equation $y = \frac{1}{2}\sqrt{2}(\sin x + \cos x) + 1$ becomes $y' = \sin x'$.

SOLUTION: We let $x = x' - \frac{1}{4}\pi$ and $y = y' + 1$. Thus,

$$y' + 1 = \tfrac{1}{2}\sqrt{2}[\sin(x' - \tfrac{1}{4}\pi) + \cos(x' - \tfrac{1}{4}\pi)] + 1$$
$$y' = \tfrac{1}{2}\sqrt{2}[\sin x' \cos \tfrac{1}{4}\pi - \cos x' \sin \tfrac{1}{4}\pi + \cos x' \cos \tfrac{1}{4}\pi + \sin x' \sin \tfrac{1}{4}\pi]$$
$$= \tfrac{1}{2}\sqrt{2}[\tfrac{1}{2}\sqrt{2} \sin x' - \tfrac{1}{2}\sqrt{2} \cos x' + \tfrac{1}{2}\sqrt{2} \cos x' + \tfrac{1}{2}\sqrt{2} \sin x']$$
$$= \tfrac{1}{2}\sqrt{2}(\sqrt{2} \sin x')$$
$$= \sin x'$$

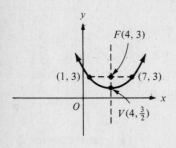

Figure 13. 2. 32

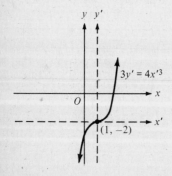

Figure 13. 2. 36

13.3 SOME PROPERTIES OF CONICS

13.3.1 Definition

A *conic* is the set of all points P in a plane such that the undirected distance of P from a fixed point is in a constant ratio to the undirected distance of P from a fixed line that does not contain the fixed point.

Let F be a fixed point and let D be a fixed line that does not contain the point F. For each point P in the plane that contains line D and point F, let d_1 be the undirected distance between point P and point F, and let d_2 be the undirected distance between point P and line D, as illustrated in Fig. 13. 3a. The conic C is the set of all points P for which

$$d_1 = ed_2$$

where e is a nonnegative constant.

(i) If $0 < e < 1$, then the conic C is an ellipse.
(ii) If $e = 1$, then the conic C is a parabola.
(iii) If $e > 1$, then the conic C is a hyperbola.

The fixed point F is called a *focus* of the conic. The fixed line D is called a *directrix* of the conic. The constant e is called the *eccentricity* of the conic. The line through focus F perpendicular to the directrix D is called the *principal axis* of the conic. Any point V of intersection of the principal axis with the conic is called a *vertex* of the conic. The theorems in this section prove that a parabola has one focus, one directrix, and one vertex. Both the ellipse and the hyperbola have two foci, two directrices, and two vertices. Each of the conics has only one principal axis, and the conic is symmetric with respect to its principal axis. the ellipse also has a minor axis, defined in Section 13.5, and the ellipse is symmetric with respect to its minor axis. The hyperbola has a conjugate axis, defined in Section 13.5, and the hyperbola is symmetric with respect to its conjugate axis.

In order to use Definition 13.3.1 to find an equation of a conic, we need the formula of Exercise 22 of Exercises 4.7, which follows.

The undirected distance between the line $Ax + By + C = 0$ and the point (x_1, y_1) is given by

$$d = \frac{|Ax_1 + By_1 + C|}{\sqrt{A^2 + B^2}} \tag{1}$$

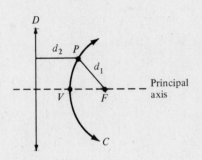

Figure 13. 3a

Exercises 13.3

In Exercises 1–12 use Definition 13.3.1 and the formula of Exercise 22 of Exercises 4.7 to find an equation of the conic having the given focus, corresponding directrix, and eccentricity. Identify the conic as a parabola, an ellipse, or a hyperbola.

4. Focus at $(\tfrac{3}{2}, 0)$; directrix $x = 0$; $e = \tfrac{4}{3}$

SOLUTION: Let $P(x, y)$ be any point on the conic. If d_1 is the distance between P and the focus $F(\tfrac{3}{2}, 0)$, then

$$d_1 = \sqrt{(x - \tfrac{3}{2})^2 + y^2}$$

If d_2 is the distance between P and the directrix $x = 0$, then

$$d_2 = |x|$$

Because $d_1 = ed_2$, and $e = \tfrac{4}{3}$, we have

$$\sqrt{(x - \tfrac{3}{2})^2 + y^2} = \tfrac{4}{3}|x|$$
$$x^2 - 3x + \tfrac{9}{4} + y^2 = \tfrac{16}{9}x^2$$

$$36x^2 - 108x + 81 + 36y^2 = 64x^2$$
$$-28x^2 + 36y^2 - 108x + 81 = 0$$

Because $e = \frac{4}{3} > 1$, the conic is a hyperbola.

8. Focus at $(2, -5)$; directrix $x = -\frac{5}{3}$; $e = 1$

SOLUTION: If d_1 is the distance between any point $P(x, y)$ on the conic and the focus $F(2, -5)$, then

$$d_1 = \sqrt{(x - 2)^2 + (y + 5)^2}$$

If d_2 is the distance between P and the directrix $x = -\frac{5}{3}$, then

$$d_2 = |x + \tfrac{5}{3}|$$

Because $d_1 = ed_2$ and $e = 1$, we have

$$\sqrt{(x - 2)^2 + (y + 5)^2} = |x + \tfrac{5}{3}|$$
$$x^2 - 4x + 4 + y^2 + 10y + 25 = x^2 + \tfrac{10}{3}x + \tfrac{25}{9}$$
$$9x^2 - 36x + 36 + 9y^2 + 90y + 225 = 9x^2 + 30x + 25$$
$$9y^2 - 66x + 90y + 236 = 0$$

Because $e = 1$, the conic is a parabola.

12. Focus at $(-1, 4)$; directrix $2x - y + 3 = 0$; $e = 2$

SOLUTION: Let $P(x, y)$ be any point on the conic. If d_1 is the distance between P and $(-1, 4)$, we have

$$d_1 = \sqrt{(x + 1)^2 + (y - 4)^2}$$

If d_2 is the distance between P and the line $2x - y + 3 = 0$, then by the distance formula (1) with $A = 2$, $B = -1$, and $C = 3$, we have

$$d_2 = \frac{|2x - y + 3|}{\sqrt{5}}$$

Because $d_1 = ed_2$, and $e = 2$, we obtain

$$\sqrt{(x + 1)^2 + (y - 4)^2} = \frac{2}{\sqrt{5}}|2x - y + 3|$$

$$x^2 + 2x + 1 + y^2 - 8y + 16 = \tfrac{4}{5}(4x^2 + y^2 + 9 - 4xy + 12x - 6y)$$

$$11x^2 - 16xy - y^2 + 38x + 16y - 49 = 0$$

Because $e = 2 > 1$, the conic is a hyperbola.

16. Find the eccentricity of the conic having focus at $(2, -3)$ and corresponding directrix $2x - 2y - 1 = 0$ if the point $P = (1, -2)$ is on the conic.

SOLUTION: If d_1 is the distance between the given focus and the given point P, then

$$d_1 = \sqrt{(2 - 1)^2 + (-3 + 2)^2} = \sqrt{2}$$

Let d_2 be the distance between the point P and the given directrix. Applying the distance formula

$$d = \frac{|Ax_1 + By_1 + C|}{\sqrt{A^2 + B^2}}$$

with $A = 2$, $B = -2$, $C = -1$, and $(x_1, y_1) = (1, -2)$, we obtain

$$d_2 = \frac{|2(1) + (-2)(-2) - 1|}{\sqrt{2^2 + (-2)^2}}$$

$$= \frac{5}{2\sqrt{2}}$$

$$= \tfrac{5}{4}\sqrt{2}$$

Therefore,

$$e = \frac{d_1}{d_2}$$

$$= \frac{\sqrt{2}}{\tfrac{5}{4}\sqrt{2}} = \tfrac{4}{5}$$

20. Find an equation whose graph consists of all points P in a plane such that the undirected distance of P from the point $(-ae, 0)$ is in a constant ratio e to the undirected distance of P from the line $x = -a/e$. Let $a^2(1 - e^2) = \pm b^2$ and consider the three cases: $e = 1$, $e < 1$, and $e > 1$.

SOLUTION: If d_1 is the distance between point $P(x, y)$ and point $(-ae, 0)$, then

$$d_1 = \sqrt{(x + ae)^2 + y^2}$$

If d_2 is the distance between point $P(x, y)$ and line $x = -a/e$, then

$$d_2 = \left| x + \frac{a}{e} \right|$$

If $d_1 = ed_2$, with $e > 0$, we have

$$\sqrt{(x + ae)^2 + y^2} = e \left| x + \frac{a}{e} \right|$$

$$= |ex + a|$$

$$x^2 + 2aex + a^2e^2 + y^2 = e^2x^2 + 2aex + a^2$$

$$x^2(1 - e^2) + y^2 = a^2(1 - e^2) \tag{1}$$

Case 1: $e = 1$
Then (1) reduces to $y = 0$.

Case 2: $0 < e < 1$
Then $1 - e^2 > 0$. We let $a^2(1 - e^2) = b^2$. Dividing both sides of (1) by $a^2(1 - e^2)$, we obtain

$$\frac{x^2}{a^2} + \frac{y^2}{a^2(1 - e^2)} = 1 \tag{2}$$

Because $a^2(1 - e^2) = b^2$, Eq. (2) is equivalent to

$$\frac{x^2}{a^2} + \frac{y^2}{b^2} = 1$$

which is an equation of an ellipse.

Case 3: $e > 1$
Then $1 - e^2 < 0$. We let $a^2(1 - e^2) = -b^2$, and Eq. (2) becomes

$$\frac{x^2}{a^2} - \frac{y^2}{b^2} = 1$$

which is an equation of a hyperbola.

13.4 POLAR EQUATIONS OF THE CONICS

Let e be the eccentricity of a conic and let d be the undirected distance from a focus to the corresponding directrix. If the focus is at the pole and if the axis of the conic is either horizontal or vertical (we assume the polar axis is horizontal), then there is a polar equation of the conic that is one of the following four standard forms, depending on whether the vertex nearest the pole is to the right or to the left of the pole, or whether it is above or below the pole.

Standard Form of Equation of Conic	Equation of Directrix Nearest the Pole	Position of Vertex Nearest the Pole
1. $r = \dfrac{ed}{1 + e\cos\theta}$	$r\cos\theta = d$	To the right of the pole
2. $r = \dfrac{ed}{1 - e\cos\theta}$	$r\cos\theta = -d$	To the left of the pole
3. $r = \dfrac{ed}{1 + e\sin\theta}$	$r\sin\theta = d$	Above the pole
4. $r = \dfrac{ed}{1 - e\sin\theta}$	$r\sin\theta = -d$	Below the pole

Exercises 13.4

In Exercises 3–14 the equation is that of a conic having a focus at the pole. In each exercise

    (a) find the eccentricity
    (b) identify the conic
    (c) write an equation of the directrix that corresponds to the focus at the pole
    (d) draw a sketch of the curve.

4. $r = \dfrac{4}{1 + \cos\theta}$

SOLUTION:

    (a) By comparing the given equation with the standard form (1)

$$r = \frac{ed}{1 + e\cos\theta}$$

    we conclude that $e = 1$ and $d = 4$. The eccentricity is 1.
    (b) Because $e = 1$, the conic is a parabola.
    (c) Because $d = 4$, an equation of the directrix nearest the pole is
    $r\cos\theta = 4$.
    (d) Table 4 gives some points on the curve. Figure 13. 4. 4 shows a sketch of the parabola.

Figure 13. 4. 4

Table 4

θ	0	$\frac{1}{2}\pi$	π	$\frac{3}{2}\pi$
r	2	4	not defined	4

8. $r = \dfrac{1}{2 + \sin \theta}$

SOLUTION: For each of the standard forms, the constant term in the denominator of the fraction is 1. Thus, we divide the numerator and denominator by 2. We have

$$r = \frac{\frac{1}{2}}{1 + \frac{1}{2} \sin \theta} \tag{1}$$

(a) Comparing (1) with standard form 3, we conclude that $e = \frac{1}{2}$.
(b) Because $e < 1$, the conic is an ellipse.
(c) Comparing (1) with standard form 3, we conclude that $ed = \frac{1}{2}$. Because $e = \frac{1}{2}$, then $d = 1$. Thus, an equation of the directrix nearest the pole is

$$r \sin \theta = 1$$

(d) We plot the points in Table 8 and draw a sketch of the graph shown in Fig. 13. 4. 8.

Table 8

θ	0	$\frac{1}{2}\pi$	π	$\frac{3}{2}\pi$
r	$\frac{1}{2}$	$\frac{1}{3}$	$\frac{1}{2}$	1

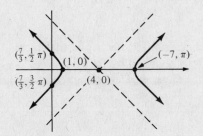

Figure 13. 4. 8

12. $r = \dfrac{7}{3 + 4 \cos \theta}$

SOLUTION: First, we write the given equation in standard form. Dividing the numerator and denominator by 3, we obtain

$$r = \frac{\frac{7}{3}}{1 + \frac{4}{3} \cos \theta}$$

(a) Comparing with standard form 1, we have $e = \frac{4}{3}$.
(b) Because $e > 1$, the conic is a hyperbola.
(c) Because $ed = \frac{7}{3}$ and $e = \frac{4}{3}$, we have $d = \frac{7}{4}$. Thus, an equation of the directrix is

$$r \cos \theta = \tfrac{7}{4}$$

(d) We plot the points from Table 12. Because the points $(1, 0)$ and $(-7, \pi)$ are the vertices, the center of the hyperbola is at $(4, 0)$. Furthermore, $3 + 4 \cos \theta = 0$ when $\cos \theta = -\frac{3}{4}$ or equivalently, when $\theta = \pm 139°$. We conclude that the asymptotes of the hyperbola intersect at the point $(4, 0)$ and form angles of $\pm 41°$ with the polar axis. A sketch of the graph is shown in Fig. 13. 4. 12.

Figure 13. 4. 12

Table 12

θ	0	$\frac{1}{2}\pi$	π	$\frac{3}{2}\pi$
r	1	$\frac{7}{3}$	-7	$\frac{7}{3}$

16. Find a polar equation of the ellipse with $e = \frac{1}{2}$, focus at the pole, and a vertex at $(4, \pi)$.

SOLUTION: There are two solutions possible. If the given vertex is the vertex nearest the pole, we use standard form 2 with $e = \frac{1}{2}$. Thus, an equation of the ellipse is of the form

$$r = \frac{\frac{1}{2} d}{1 - \frac{1}{2} \cos \theta} = \frac{d}{2 - \cos \pi} \qquad (1)$$

Because the ellipse has vertex at $(4, \pi)$, we let $r = 4$ and $\theta = \pi$ in (1). Thus,

$$4 = \frac{d}{2 - \cos \pi}$$

$$d = 12$$

Hence, from (1), with $d = 12$, we get

$$r = \frac{12}{2 - \cos \theta}$$

which is an equation of the ellipse.

And if the given vertex is not the vertex nearest the pole, we use standard form 1 with $e = \frac{1}{2}$. Thus,

$$r = \frac{\frac{1}{2} d}{1 + \frac{1}{2} \cos \theta} = \frac{d}{2 + \cos \theta} \qquad (2)$$

with $r = 4$ and $\theta = \pi$ in Eq. (2), we get $d = 4$. Thus, an equation of the ellipse is

$$r = \frac{4}{2 + \cos \theta}$$

20. Find a polar equation of the parabola having a focus at the pole and a vertex at the point $(6, \frac{1}{2}\pi)$.

SOLUTION: Because the vertex is at the point $(6, \frac{1}{2}\pi)$, we use standard form 3 with $d = 6$. For any parabola we have $e = 1$. Thus, an equation of the parabola is

$$r = \frac{ed}{1 + e \sin \theta} = \frac{6}{1 + \sin \theta}$$

24. Find the area of the region inside the ellipse $r = 6/(2 - \sin \theta)$ and below the parabola $r = 3/(1 + \sin \theta)$.

SOLUTION: Figure 13. 4. 24 shows a sketch of the region R. Let R_1 be that part of the region R in the fourth quadrant. Then R_1 is bounded by the ellipse $r = 6/(2 - \sin \theta)$ and the lines $\theta = -\frac{1}{2}\pi$ and $\theta = 0$. Thus,

$$A_1 = \lim_{\|\Delta\| \to 0} \sum_{i=1}^{n} \frac{1}{2} \left[\frac{6}{2 - \sin \bar\theta_i} \right]^2 \Delta_i \theta$$

$$= \frac{1}{2} \int_{-\pi/2}^{0} \frac{36 \, d\theta}{(2 - \sin \theta)^2}$$

We let $z = \tan \frac{1}{2}\theta$ as in Section 10.7. Then

$$\sin \theta = \frac{2z}{1 + z^2} \qquad d\theta = \frac{2 \, dz}{1 + z^2}$$

Thus,

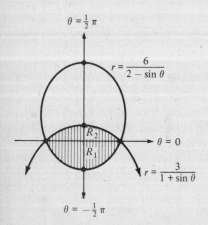

Figure 13. 4. 24

$$A_1 = 18 \int_{-1}^{0} \frac{\dfrac{2\,dz}{1 + z^2}}{\left(2 - \dfrac{2z}{1 + z^2}\right)^2}$$

$$= 9 \int_{-1}^{0} \frac{(z^2 + 1)\,dz}{(z^2 - z + 1)^2} \tag{1}$$

We use partial fractions. If

$$\frac{z^2 + 1}{(z^2 - z + 1)^2} = \frac{Az + B}{(z^2 - z + 1)^2} + \frac{Cz + D}{z^2 - z + 1}$$

then $A = 1$, $B = 0$, $C = 0$, and $D = 1$. Thus,

$$\frac{z^2 + 1}{(z^2 - z + 1)^2} = \frac{z}{(z^2 - z + 1)^2} + \frac{1}{z^2 - z + 1}$$

From (1) we have

$$A_1 = 9 \int_{-1}^{0} \frac{z\,dz}{(z^2 - z + 1)^2} + 9 \int_{-1}^{0} \frac{dz}{z^2 - z + 1}$$

$$= \frac{9}{2} \int_{-1}^{0} \frac{(2z - 1)\,dz}{(z^2 - z + 1)^2} + \frac{9}{2} \int_{-1}^{0} \frac{dz}{(z^2 - z + 1)^2} + 9 \int_{-1}^{0} \frac{dz}{z^2 - z + 1} \tag{2}$$

For the first integral on the right of (2), we have

$$\frac{9}{2} \int_{-1}^{0} \frac{(2z - 1)\,dz}{(z^2 - z + 1)^2} = -\frac{9}{2}(z^2 - z + 1)^{-1}\Big]_{-1}^{0} = -3 \tag{3}$$

For the second integral on the right of (2), we complete the square.

$$\frac{9}{2} \int_{-1}^{0} \frac{dz}{(z^2 - z + 1)^2} = \frac{9}{2} \int_{-1}^{0} \frac{dz}{[(z - \frac{1}{2})^2 + \frac{3}{4}]^2} \tag{4}$$

Now we let

$$t = \tan^{-1}\left(\frac{z - \frac{1}{2}}{\frac{1}{2}\sqrt{3}}\right)$$

Thus, when $z = -1$, $t = -\frac{1}{3}\pi$; when $z = 0$, $t = -\frac{1}{6}\pi$.

$$z - \tfrac{1}{2} = \tfrac{1}{2}\sqrt{3}\,\tan t$$
$$dz = \tfrac{1}{2}\sqrt{3}\,\sec^2 t\,dt$$
$$[(z - \tfrac{1}{2})^2 + \tfrac{3}{4}]^2 = [\tfrac{3}{4}\tan^2 t + \tfrac{3}{4}]^2$$
$$= \tfrac{9}{16}\sec^4 t$$

Substituting in the right side of (4), we obtain

$$\frac{9}{2} \int_{-1}^{0} \frac{dz}{(z^2 - z + 1)^2} = \frac{9}{2} \int_{-\pi/3}^{-\pi/6} \frac{\frac{1}{2}\sqrt{3}\,\sec^2 t\,dt}{\frac{9}{16}\sec^4 t}$$

$$= 4\sqrt{3} \int_{-\pi/3}^{-\pi/6} \cos^2 t\,dt$$

$$= 2\sqrt{3} \int_{-\pi/3}^{-\pi/6} (1 + \cos 2t)\,dt$$

$$= 2\sqrt{3}\left[\theta + \frac{1}{2}\sin 2t\right]_{-\pi/3}^{-\pi/6}$$

$$= \frac{1}{3}\sqrt{3}\,\pi \tag{5}$$

For the third integral on the right side of (2), we have

$$9 \int_{-1}^{0} \frac{dz}{z^2 - z + 1} = 9 \int_{-1}^{0} \frac{dz}{(z - \frac{1}{2})^2 + \frac{3}{4}}$$

$$= 9 \left(\frac{2}{\sqrt{3}} \right) \tan^{-1} \frac{z - \frac{1}{2}}{\frac{1}{2}\sqrt{3}} \Big]_{-1}^{0}$$

$$= \sqrt{3}\, \pi \tag{6}$$

Substituting from (3), (5), and (6) into (2), we have

$$A_1 = -3 + \tfrac{1}{3}\sqrt{3}\,\pi + \sqrt{3}\,\pi = -3 + \tfrac{4}{3}\sqrt{3}\,\pi \tag{7}$$

Next, let R_2 be the part of the region R in first quadrant. Then R_2 is bounded by the parabola $r = 3/(1 + \sin\theta)$ and the lines $\theta = 0$ and $\theta = \frac{1}{2}\pi$. Thus,

$$A_2 = \frac{1}{2} \int_0^{\pi/2} \frac{9\, d\theta}{(1 + \sin\theta)^2} \tag{8}$$

We let $\theta = \frac{1}{2}\pi - t$. Then

$$(1 + \sin\theta)^2 = [1 + \sin(\tfrac{1}{2}\pi - t)]^2$$
$$= (1 + \cos t)^2$$
$$= (2\cos^2 \tfrac{1}{2} t)^2$$
$$= 4\cos^4 \tfrac{1}{2} t$$

Furthermore, $d\theta = -dt$; $t = \frac{1}{2}\pi$ when $\theta = 0$; and $t = 0$ when $\theta = \frac{1}{2}\pi$. Thus, from (8) we have

$$A_2 = \frac{1}{2} \int_{\pi/2}^{0} \frac{-9\, dt}{4\cos^4 \tfrac{1}{2} t}$$

$$= \frac{9}{8} \int_0^{\pi/2} \sec^4 \tfrac{1}{2} t\, dt$$

$$= \frac{9}{4} \int_0^{\pi/2} \left(1 + \tan^2 \tfrac{1}{2} t \right) \sec^2 \tfrac{1}{2} t \left(\tfrac{1}{2}\, dt \right)$$

$$= \frac{9}{4} \left[\tan \tfrac{1}{2} t + \tfrac{1}{3} \tan^3 \tfrac{1}{2} t \right]_0^{\pi/2} = 3 \tag{9}$$

Because the region R is symmetric with respect to the line $\theta = \frac{1}{2}\pi$, then the area of R is given by $A = 2(A_1 + A_2)$. Substituting from (7) and (9), we get

$$A = 2[(-3 + \tfrac{4}{3}\sqrt{3}\,\pi) + 3] = \tfrac{8}{3}\sqrt{3}\,\pi$$

Thus, the area of region R is $\frac{8}{3}\sqrt{3}\,\pi$ square units.

28. Show that the equation $r = k \csc^2 \frac{1}{2}\theta$, where k is a constant, is a polar equation of a parabola.

SOLUTION: We use the identity $\sin^2 t = \frac{1}{2}(1 - \cos 2t)$. Thus,

$$r = k \csc^2 \tfrac{1}{2}\theta$$

$$= \frac{k}{\sin^2 \tfrac{1}{2}\theta}$$

$$= \frac{k}{\tfrac{1}{2}(1 - \cos\theta)} = \frac{2k}{1 - \cos\theta} \tag{1}$$

Because Eq. (1) is standard form 2, we conclude that the graph of (1) is a conic. Furthermore, comparing (1) with standard form 2, we have $e = 1$. Thus, (1) is an equation of a parabola.

32. Show that the tangent lines at the points of intersection of the parabolas $r = a/(1 + \cos \theta)$ and $r = b/(1 - \cos \theta)$ are perpendicular.

SOLUTION: Because $\cos^2 t = \frac{1}{2}(1 + \cos 2t)$, $1 + \cos \theta = 2 \cos^2 \frac{1}{2}\theta$. Thus the parabola

$$r = \frac{a}{1 + \cos \theta}$$

has equation

$$r = \frac{a}{2 \cos^2 \frac{1}{2}\theta} = \frac{1}{2} a \sec^2 \frac{1}{2}\theta \qquad (1)$$

By using the identity $\sin^2 t = \frac{1}{2}(1 - \cos 2t)$, we show that the parabola

$$r = \frac{b}{1 - \cos \theta}$$

has equation

$$r = \frac{1}{2} b \csc^2 \frac{1}{2}\theta \qquad (2)$$

In Exercise 24 of Exercises 12.4 we showed that for the curves $r = a \sec^2 \frac{1}{2}\theta$ and $r = b \csc^2 \frac{1}{2}\theta$ the tangent lines at the points of intersection are perpendicular, where a and b are any constants. From (1) and (2) we conclude that for the given parabolas the tangent lines at the points of intersection are perpendicular.

13.5 CARTESIAN EQUATIONS OF THE CONICS

Following are the standard forms for Cartesian equations of the ellipse and hyperbola with center at the origin, principal axis horizontal, and eccentricity e.

Ellipse	Hyperbola
$\dfrac{x^2}{a^2} + \dfrac{y^2}{b^2} = 1$	$\dfrac{x^2}{a^2} - \dfrac{y^2}{b^2} = 1$
if $a > b > 0$ and $b = a\sqrt{1 - e^2}$	if $a > 0$, $b > 0$ and $b = a\sqrt{e^2 - 1}$

For each of the above central conics, we have the following facts:

Vertices are at $(a, 0)$ and $(-a, 0)$.
Foci are at $(ae, 0)$ and $(-ae, 0)$.
Equations of directrices are $x = a/e$ and $x = -a/e$.

For the ellipse, the segment joining the vertices $(a, 0)$ and $(-a, 0)$ is called the *major axis;* for the hyperbola, this segment is called the *transverse axis.* For the ellipse, the segment joining the points $(0, b)$ and $(0, -b)$ is called the *minor axis;* for the hyperbola, this segment is called the *conjugate axis.* The ends of the minor axis of the ellipse are points on the curve, but the ends of the conjugate axis of the hyperbola do not lie on the hyperbola.

In Sections 13.6 and 13.7 we consider ellipses and hyperbolas that do not have their center at the origin and principal axis horizontal.

Exercises 13.5

4. Find the vertices, foci, directrices, eccentricity, and ends of the minor axis of the given ellipse. Draw a sketch of the curve and show the foci and directrices.

$$3x^2 + 4y^2 = 9$$

SOLUTION: First we write the given equation in the standard form

$$\frac{x^2}{a^2} + \frac{y^2}{b^2} = 1$$

Dividing both sides of the given equation by 9, we have

$$\frac{x^2}{3} + \frac{y^2}{\frac{9}{4}} = 1$$

Thus, $a = \sqrt{3}$ and $b = \frac{3}{2}$. The vertices are at $(\sqrt{3}, 0)$ and $(-\sqrt{3}, 0)$, and the ends of the minor axes are at $(0, \frac{3}{2})$. Because

$$b = a\sqrt{1 - e^2}$$

we have

$$\frac{3}{2} = \sqrt{3}\sqrt{1 - e^2}$$

from which we get $e = \frac{1}{2}$. Thus, $ae = \frac{1}{2}\sqrt{3}$, and hence the foci are at $(\frac{1}{2}\sqrt{3}, 0)$ and $(-\frac{1}{2}\sqrt{3}, 0)$. Because $a/e = 2\sqrt{3}$, the directrices are the lines $x = 2\sqrt{3}$ and $x = -2\sqrt{3}$. In Fig. 13. 5. 4 we show a sketch of the ellipse, the foci, and the directrices.

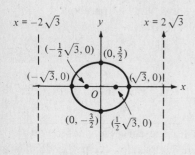

Figure 13. 5. 4

8. Find the vertices, foci, directrices, eccentricity, and lengths of the transverse and conjugate axes of the given hyperbola. Draw a sketch of the curve and show the foci and the directrices.

$$25x^2 - 25y^2 = 1$$

SOLUTION: Because the given equation may be written as

$$\frac{x^2}{\frac{1}{25}} - \frac{y^2}{\frac{1}{25}} = 1$$

we have $a = \frac{1}{5}$ and $b = \frac{1}{5}$. Thus, the vertices are at the points $(\frac{1}{5}, 0)$ and $(-\frac{1}{5}, 0)$, and the transverse axis has length $\frac{2}{5}$. The ends of the conjugate axis are at $(0, \frac{1}{5})$ and $(0, -\frac{1}{5})$, and the length of the conjugate axis is $\frac{2}{5}$. Because

$$b = a\sqrt{e^2 - 1}$$

we have

$$\frac{1}{5} = \frac{1}{5}\sqrt{e^2 - 1}$$

from which we get $e = \sqrt{2}$. Because $ae = \frac{1}{5}\sqrt{2}$, the foci are at $(\frac{1}{5}\sqrt{2}, 0)$ and $(-\frac{1}{5}\sqrt{2}, 0)$. Because $a/e = \frac{1}{10}\sqrt{2}$, the directrices are the lines $x = \frac{1}{10}\sqrt{2}$ and $x = -\frac{1}{10}\sqrt{2}$. In Fig. 13. 5. 8 we show a sketch of the hyperbola, the foci, and the directrices.

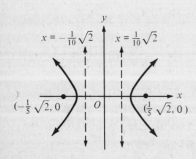

Figure 13. 5. 8

12. Find an equation of the hyperbola having the ends of its conjugate axis at $(0, -3)$ and $(0, 3)$ and $e = 2$.

SOLUTION: Because the midpoint of the conjugate axis is at $(0, 0)$, the hyperbola has its center at the origin. Because the endpoints of the conjugate axis are on the y-axis, the principal axis of the hyperbola is the x-axis. Thus, an equation of the hyperbola is of the form

$$\frac{x^2}{a^2} - \frac{y^2}{b^2} = 1 \tag{1}$$

Because the ends of the conjugate axis are at $(0, 3)$ and $(0, -3)$, we are given that $b = 3$. Because $e = 2$ and

$$b = a\sqrt{e^2 - 1}$$

we have

$$3 = a\sqrt{4-1}$$

Thus, $a = \sqrt{3}$. From (1) with $a = \sqrt{3}$ and $b = 3$, we obtain

$$\frac{x^2}{3} - \frac{y^2}{9} = 1$$

$$3x^2 - y^2 = 9$$

which is an equation of the hyperbola.

16. Find an equation of the normal line to the hyperbola $4x^2 - 3y^2 = 24$ at the point $(3, 2)$.

SOLUTION: We differentiate implicitly with respect to x on both sides of the given equation.

$$8x - 6y\, D_x y = 0$$

$$D_x y = \frac{4x}{3y}$$

At the point $(3, 2)$ the slope of the tangent line to the curve is

$$D_x y \Big]_{\substack{x=3 \\ y=2}} = \frac{4(3)}{3(2)} = 2$$

Thus, the slope of the normal line is $-\frac{1}{2}$ and an equation of the normal line is

$$y - 2 = -\tfrac{1}{2}(x - 3)$$
$$x + 2y - 7 = 0$$

20. Find an equation of the ellipse whose foci are the vertices of the hyperbola $11x^2 - 7y^2 = 77$ and whose vertices are the foci of this hyperbola.

SOLUTION: Let

$$\frac{x^2}{a^2} - \frac{y^2}{b^2} = 1 \tag{1}$$

be the standard form of the equation of the hyperbola, with eccentricity e. Let

$$\frac{x^2}{\bar{a}^2} + \frac{y^2}{\bar{b}^2} = 1 \tag{2}$$

be the standard form of the equation of the ellipse with eccentricity $\bar{e}$. Because the foci of the ellipse are the vertices of the hyperbola, we are given that

$$\bar{a}\bar{e} = a \tag{3}$$

Because the vertices of the ellipse are the foci of the hyperbola, we are given that

$$\bar{a} = ae \tag{4}$$

We write the given equation of the hyperbola in the form (1). Thus,

$$\frac{x^2}{7} - \frac{y^2}{11} = 1$$

from which we have $a = \sqrt{7}$ and $b = \sqrt{11}$.
  Because $b = a\sqrt{e^2 - 1}$, we obtain

$$\sqrt{11} = \sqrt{7}\sqrt{e^2 - 1}$$
$$e = 3\sqrt{\tfrac{2}{7}}$$

Hence, $ae = 3\sqrt{2}$, and from (4) we get

$$\bar{a} = 3\sqrt{2} \qquad\qquad (5)$$

Furthermore, from (3) we get

$$3\sqrt{2}\bar{e} = \sqrt{7}$$
$$\bar{e} = \tfrac{1}{6}\sqrt{14}$$

And because $\bar{b} = \bar{a}\sqrt{1 - \bar{e}^2}$

$$\bar{b} = 3\sqrt{2}\sqrt{1 - \tfrac{7}{18}} = \sqrt{11} \qquad\qquad (6)$$

Substituting from (5) and (6) into (2), we obtain

$$\frac{x^2}{18} + \frac{y^2}{11} = 1$$

$$11x^2 + 18y^2 = 198$$

which is an equation of the ellipse.

24. The orbit of the earth around the sun is elliptical in shape with the sun at one focus, a semimajor axis of length 92.9 million miles, and an eccentricity of 0.017. Find **(a)** how close the earth gets to the sun and **(b)** the greatest possible distance between the earth and the sun.

SOLUTION:

(a) Let one unit be 1 million miles and take the origin at the center of the orbit of the earth. Let the points $A = (a, 0)$ and $A' = (-a, 0)$ be the vertices of the ellipse. Let the sun be at the focus $F(ae, 0)$. The point on an ellipse that is closest to a focus is at the vertex that is nearest the focus, and the point on an ellipse that is farthest from the focus is at the vertex that is farthest from the focus. Thus, the distance between the earth and the sun when the earth is closest to the sun is given by

$$|\overline{AF}| = a - ae \qquad\qquad (1)$$

and the distance between the earth and the sun when the earth is farthest from the sun is given by

$$|\overline{FA'}| = ae + a \qquad\qquad (2)$$

Because the length of the semimajor axis is 92.9 million miles, we are given that $a = 92.9$. Furthermore, we are given that $e = 0.017$. Thus,

$$a - ae = a(1 - e)$$
$$= (92.9)(1 - 0.017) = 91.3$$

From (1) we conclude that the closest the earth gets to the sun is 91.3 million miles.

(b) From (2),

$$a + ae = a(1 + e)$$
$$= (92.9)(1 + 0.017) = 94.5$$

and thus we conclude that the greatest distance between the earth and the sun is 94.5 million miles.

28. Find the centroid of the solid of revolution generated by revolving the region bounded by the hyperbola $x^2/a^2 - y^2/b^2 = 1$ and the line $x = 2a$ about the x-axis.

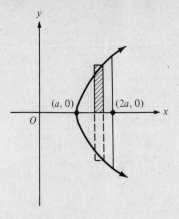

Figure 13. 5. 28

SOLUTION: In Fig. 13. 5. 28 we show the region and a plane section of the solid of revolution. Solving the given equation of the hyperbola for y, we obtain

$$y = \pm \frac{b}{a} \sqrt{x^2 - a^2}$$

We let f be the function defined by

$$f(x) = \frac{b}{a} \sqrt{x^2 - a^2}$$

Then the element of volume is a circular disk with thickness $\Delta_i x$ units and radius $f(\overline{x}_i)$ units. Thus,

$$V = \lim_{\|\Delta\| \to 0} \sum_{i=1}^{n} \pi [f(\overline{x}_i)]^2 \, \Delta_i x \qquad (1)$$

$$= \pi \int_{a}^{2a} [f(x)]^2 \, dx$$

$$= \frac{\pi b^2}{a^2} \int_{a}^{2a} (x^2 - a^2) \, dx$$

$$= \frac{\pi b^2}{a^2} \left[\frac{1}{3} x^3 - a^2 x \right]_{a}^{2a}$$

$$= \frac{4}{3} \pi a b^2 \qquad (2)$$

Because the centroid of the element of volume is at the point $(\overline{x}_i, 0, 0)$, from (1) we have

$$V \cdot \overline{x} = \lim_{\|\Delta\| \to 0} \sum_{i=1}^{n} \pi [f(\overline{x}_i)]^2 \overline{x}_i \, \Delta_i x$$

$$= \pi \int_{a}^{2a} [f(x)]^2 x \, dx$$

$$= \frac{\pi b^2}{a^2} \int_{a}^{2a} (x^3 - a^2 x) \, dx$$

$$= \frac{\pi b^2}{a^2} \left[\frac{1}{4} x^4 - \frac{1}{2} a^2 x^2 \right]_{a}^{2a}$$

$$= \frac{9}{4} \pi a^2 b^2 \qquad (3)$$

Substituting from (2) into (3), we obtain

$$\tfrac{4}{3} \pi a b^2 \overline{x} = \tfrac{9}{4} \pi a^2 b^2$$
$$\overline{x} = \tfrac{27}{16} a$$

We conclude that the centroid of the solid of revolution is at the point $(\tfrac{27}{16} a, 0, 0)$.

13.6 THE ELLIPSE   If the center of an ellipse is at the point (h, k), if the major axis is parallel to a coordinate axis, and if $b = a\sqrt{1 - e^2}$, where e is the eccentricity of the ellipse, and thus $a > b > 0$, then we have the following facts:

	Major Axis Is Horizontal	Major Axis Is Vertical
Equation of ellipse	$\dfrac{(x-h)^2}{a^2}+\dfrac{(y-k)^2}{b^2}=1$	$\dfrac{(x-h)^2}{b^2}+\dfrac{(y-k)^2}{a^2}=1$
Vertices	$(h\pm a,\ k)$	$(h,\ k\pm a)$
Foci	$(h\pm ae,\ k)$	$(h,\ k\pm ae)$
Equation of directrices	$x=h\pm\dfrac{a}{e}$	$y=k\pm\dfrac{a}{e}$

13.6.1 Theorem   If in the general second-degree equation

$$Ax^2 + Bxy + Cy^2 + Dx + Ey + F = 0$$

$B = 0$ and $AC > 0$, then the graph is either an ellipse, a point-ellipse, or the empty set. In addition, if $A = C$, then the graph is either a circle, a point-circle, or the empty set.

13.6.2 Theorem   An ellipse can be defined as the set of points such that the sum of the distances from any point of the set to two given points (the foci) is a constant.

Exercises 13.6

In Exercises 1–8 find the eccentricity, center, foci, and directrices of the given ellipse and draw a sketch of the graph.

4. $2x^2 + 2y^2 - 2x + 18y + 33 = 0$

SOLUTION: We complete the square on the terms in x and y and write the equation in a standard form.

$$2(x^2 - x) + 2(y^2 + 9y) = -33$$

$$2(x^2 - x + \tfrac{1}{4}) + 2(y^2 + 9y + \tfrac{81}{4}) = -33 + \tfrac{1}{2} + \tfrac{81}{2}$$

$$2(x - \tfrac{1}{2})^2 + 2(y + \tfrac{9}{2})^2 = 8$$

$$\frac{(x - \tfrac{1}{2})^2}{4} + \frac{(y + \tfrac{9}{2})^2}{4} = 1$$

$$(x - \tfrac{1}{2})^2 + (y + \tfrac{9}{2})^2 = 4$$

The graph is a circle with center at the point $(\tfrac{1}{2}, -\tfrac{9}{2})$ and radius 2. Because $a = b = 2$, and $b = a\sqrt{1 - e^2}$, then $e = 0$. The foci and directrices do not exist. A sketch of the graph is shown in Fig. 13. 6. 4.

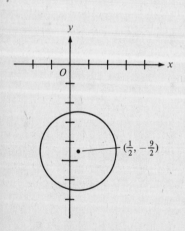

$(\tfrac{1}{2}, -\tfrac{9}{2})$

Figure 13. 6. 4

8. $2x^2 + 3y^2 - 4x + 12y + 2 = 0$

SOLUTION: We complete the squares and find a standard form of the given equation.

$$2(x^2 - 2x + 1) + 3(y^2 + 4y + 4) = -2 + 2 + 12$$

$$\frac{(x - 1)^2}{6} + \frac{(y + 2)^2}{4} = 1 \tag{1}$$

Because $6 > 4$, we take $a^2 = 6$ and $b^2 = 4$. Thus, $a = \sqrt{6}$ and $b = 2$. Because $b = a\sqrt{1 - e^2}$, then

$$2 = \sqrt{6}\sqrt{1 - e^2}$$
$$4 = 6 - 6e^2$$
$$e^2 = \tfrac{1}{3}$$
$$e = \tfrac{1}{3}\sqrt{3}$$

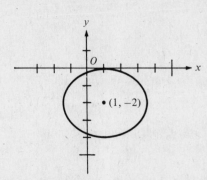

Figure 13. 6. 8

Comparing Eq. (1) with the standard form for an ellipse with major axis horizontal, we conclude that $h = 1$, $k = -2$, so the center is at the point $(1, -2)$. Because $ae = \sqrt{2}$, the foci are $(1 + \sqrt{2}, -2)$ and $(1 - \sqrt{2}, -2)$. Because $a/e = 3\sqrt{2}$, equations of the directrices are $x = 1 + 3\sqrt{2}$ and $x = 1 - 3\sqrt{2}$. A sketch of the graph is shown in Fig. 13. 6. 8.

In Exercises 11–18 find an equation of the ellipse satisfying the given conditions and draw a sketch of the graph.

12. Vertices at $(0, 5)$ and $(0, -5)$ and through the point $(2, -\tfrac{5}{3}\sqrt{5})$.

SOLUTION: Figure 13. 6. 12 shows a sketch of the ellipse. Because the origin is the midpoint of the segment joining the vertices, the center of the ellipse is at the origin. Because the major axis is vertical, an equation of the ellipse is of the form

$$\frac{x^2}{b^2} + \frac{y^2}{a^2} = 1$$

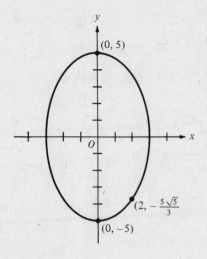

Figure 13. 6. 12

Because the vertices are at $(0, \pm 5)$, then $a = 5$. Because the ellipse contains the point $(2, -\tfrac{5}{3}\sqrt{5})$, then the equation is satisfied when $x = 2$ and $y = -\tfrac{5}{3}\sqrt{5}$. With these replacements, we have

$$\frac{4}{b^2} + \frac{\frac{125}{9}}{25} = 1$$

$$\frac{4}{b^2} + \frac{5}{9} = 1$$

$$\frac{4}{b^2} = \frac{4}{9}$$

$$b^2 = 9$$

Therefore, an equation of the ellipse is

$$\frac{x^2}{9} + \frac{y^2}{25} = 1$$

16. Foci at $(2, 3)$ and $(2, -7)$ and eccentricity of $\tfrac{2}{3}$.

SOLUTION: Because the foci are in a vertical line, and because the midpoint of the segment joining the foci is $(2, -2)$, we use the standard form of the equation of an ellipse with major axis vertical with $h = 2$ and $k = -2$. Thus, an equation of the ellipse is of the form.

$$\frac{(x - 2)^2}{b^2} + \frac{(y + 2)^2}{a^2} = 1 \tag{1}$$

Because the distance between the foci is 10, we are given that $2ae = 10$. Because $e = \tfrac{2}{3}$, we have $a = \tfrac{15}{2}$. Because $b = a\sqrt{1 - e^2}$, we have

$$b = \tfrac{15}{2}\sqrt{1 - \tfrac{4}{9}}$$
$$= \tfrac{5}{2}\sqrt{5}$$

Thus, from (1) we obtain

$$\frac{(x-2)^2}{\frac{125}{2}} + \frac{(y+2)^2}{\frac{225}{4}} = 1$$

$$\frac{4(x-2)^2}{125} + \frac{4(y+2)^2}{225} = 1$$

which is an equation of the ellipse. A sketch of the graph is shown in Fig. 13. 6. 16.

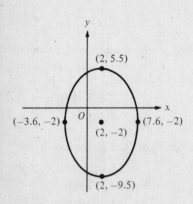

Figure 13. 6. 16

20. Use Theorem 13.6.2 to find an equation of the ellipse for which the sum of the distances from any point on the ellipse to $(4, -1)$ and $(4, 7)$ is equal to 12.

SOLUTION: Let $P(x, y)$ be any point on the ellipse and let d_1 be the distance between P and $(4, -1)$. Thus,

$$d_1 = \sqrt{(x-4)^2 + (y+1)^2}$$

Let d_2 be the distance between P and $(4, 7)$. Thus,

$$d_2 = \sqrt{(x-4)^2 + (y-7)^2}$$

Because $d_1 + d_2 = 12$, we have

$$\sqrt{(x-4)^2 + (y+1)^2} + \sqrt{(x-4)^2 + (y-7)^2} = 12$$
$$\sqrt{(x-4)^2 + (y+1)^2} = 12 - \sqrt{(x-4)^2 + (y-7)^2}$$

Squaring on both sides we obtain

$$(x-4)^2 + y^2 + 2y + 1 = 144 - 24\sqrt{(x-4)^2 + (y-7)^2} + (x-4)^2 + y^2$$
$$- 14y + 49$$
$$16y - 192 = -24\sqrt{(x-4)^2 + (y-7)^2}$$
$$2y - 24 = -3\sqrt{(x-4)^2 + (y-7)^2}$$
$$4y^2 - 96y + 576 = 9(x-4)^2 + 9(y^2 - 14y + 49)$$

Thus,

$$9(x-4)^2 + 5y^2 - 30y - 135 = 0$$

$$9(x-4)^2 + 5(y^2 - 6y + 9) = 135 + 45$$

$$9(x-4)^2 + 5(y-3)^2 = 180$$

$$\frac{(x-4)^2}{20} + \frac{(y-3)^2}{36} = 1$$

24. A plate is in the shape of the region bounded by the ellipse having a semimajor axis of length 3 ft and a semiminor axis of length 2 ft. If the plate is lowered vertically in a tank of water until the minor axis lies in the surface of the water, find the force due to water pressure on one side of the submerged portion of the plate.

SOLUTION: Refer to Fig. 13. 6. 24. We take the center of the ellipse at the origin with the major axis along the x-axis, which is directed downward, and the y-axis in the surface of the water. An equation of the ellipse is

$$\frac{x^2}{9} + \frac{y^2}{4} = 1$$

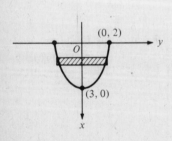

Figure 13. 6. 24

Solving for y we obtain $y = \pm\frac{2}{3}\sqrt{9 - x^2}$. We let f be the function defined by $f(x) = \frac{2}{3}\sqrt{9 - x^2}$. The number of square feet in the element of area is given by $\Delta_i A = 2f(\bar{x}_i) \Delta_i x$. The element of depth is $\bar{x}_i$ ft. Thus, the number of pounds in the force is given by

$$F = \lim_{\|\Delta\| \to 0} \sum_{i=1}^{n} 2\rho \bar{x}_i f(\bar{x}_i) \, \Delta_i x$$

$$= 2\rho \int_0^3 x \left(\frac{2}{3}\right) \sqrt{9 - x^2} \, dx$$

$$= -\frac{2}{3} \rho \int_0^3 9 - x^2)^{1/2}(-2x \, dx)$$

$$= -\frac{4}{9} \rho (9 - x^2)^{3/2} \Big]_0^3 = 12\rho$$

The total force is 12ρ pounds.

13.7 THE HYPERBOLA   If the center of a hyperbola is at the point (h, k), if the transverse axis is parallel to a coordinate axis, and if $b = a\sqrt{e^2 - 1}$, where e is the eccentricity of the hyperbola, with $a > 0$ and $b > 0$, we have the following facts:

	Transverse Axis Is Horizontal	Transverse Axis Is Vertical
Equation of hyperbola	$\dfrac{(x - h)^2}{a^2} - \dfrac{(y - k)^2}{b^2} = 1$	$\dfrac{(y - k)^2}{a^2} - \dfrac{(x - h)^2}{b^2} = 1$
Vertices	$(h \pm a, k)$	$(h, k \pm a)$
Foci	$(h \pm ae, k)$	$(h, k \pm ae)$
Equations of directrices	$x = h \pm \dfrac{a}{e}$	$y = k \pm \dfrac{a}{e}$
Equations of asymptotes	$\dfrac{x - h}{a} = \pm \dfrac{y - k}{b}$	$\dfrac{y - k}{a} = \pm \dfrac{x - h}{b}$

13.7.3 Theorem   If in the general second-degree equation

$$Ax^2 + Bxy + Cy^2 + Dx + Ey + F = 0$$

$B = 0$ and $AC < 0$, then the graph is either a hyperbola or two intersecting straight lines.

13.7.4 Theorem   A hyperbola can be defined as the set of points such that the absolute value of the difference of the distances from any point of the set to two given points (the foci) is a constant.

13.7.5 Theorem   If in the general second-degree equation

$$Ax^2 + Bxy + Cy^2 + Dx + Ey + F = 0$$

$B = 0$ and either $A = 0$ and $C \neq 0$ or $C = 0$ and $A \neq 0$, then the graph is one of the following: a parabola, two parallel lines, one line, or the empty set.

13.7.6 Theorem   The graph of the equation $Ax^2 + Cy^2 + Dx + Ey + F = 0$, where A and C are not both zero, is either a conic or a degenerate of a conic; if it is a conic, then the graph is

   (i) a parabola if either $A = 0$ or $C = 0$, that is, if $AC = 0$;
   (ii) an ellipse if A and C have the same sign, that is, if $AC > 0$;
   (iii) a hyperbola if A and C have opposite signs, that is, if $AC < 0$.

Exercises 13.7

In Exercises 1–8 find the eccentricity, center, foci, directrices, and equations of the asymptotes of the given hyperbola and draw a sketch of the graph.

4. $x^2 - y^2 + 6x + 10y - 4 = 0$

SOLUTION: We complete the squares on the terms in x and y and write the given equation in a standard form.

$$(x^2 + 6x) - (y^2 - 10y) = 4$$

$$(x^2 + 6x + 9) - (y^2 - 10y + 25) = 4 + 9 - 25$$

$$(x + 3)^2 - (y - 5)^2 = -12$$

$$\frac{(y - 5)^2}{12} - \frac{(x + 3)^2}{12} = 1$$

This is the standard form for a hyperbola with transverse axis vertical. We have $a^2 = b^2 = 12$, so $a = b = 2\sqrt{3}$. Because $b = a\sqrt{e^2 - 1}$, then $2\sqrt{3} = 2\sqrt{3}\sqrt{e^2 - 1}$, or $1 = \sqrt{e^2 - 1}$, so $e = \sqrt{2}$. Because $h = -3$ and $k = 5$, the center is at the point $(-3, 5)$. Because $a = 2\sqrt{3}$, the vertices are at the points $(-3, 5 + 2\sqrt{3})$ and $(-3, 5 - 2\sqrt{3})$. Because $ae = 2\sqrt{6}$, the foci are at the points $(-3, 5 + 2\sqrt{6})$ and $(-3, 5 - 2\sqrt{6})$. Because $a/e = \sqrt{6}$, equations of the directrices are $y = 5 + \sqrt{6}$ and $y = 5 - \sqrt{6}$. Equations of the asymptotes may be found by replacing the constant 1 by 0 in the standard form. This results in

$$\frac{(y - 5)^2}{12} - \frac{(x + 3)^2}{12} = 0$$

$$(y - 5)^2 = (x + 3)^2$$

$$y - 5 = \pm(x + 3)$$

Hence,

$$x - y + 8 = 0$$

and

$$x + y - 2 = 0$$

are equations of the asymptotes. A sketch of the graph is shown in Fig. 13. 7. 4.

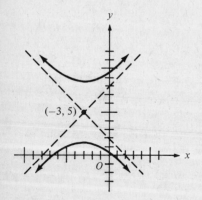

Figure 13. 7. 4

8. $y^2 - x^2 + 2y - 2x - 1 = 0$

SOLUTION: We complete the squares on the terms in x and y and write the equation in a standard form.

$$(y^2 + 2y) - (x^2 + 2x) = 1$$
$$(y^2 + 2y + 1) - (x^2 + 2x + 1) = 1 + 1 - 1$$
$$(y + 1)^2 - (x + 1)^2 = 1 \tag{1}$$

Comparing this equation with the standard form of the equation of a hyperbola with vertical transverse axis, we have $a = 1$ and $b = 1$. Because $b = a\sqrt{e^2 - 1}$, we have $1 = \sqrt{e^2 - 1}$. Thus, $e = \sqrt{2}$. From (1) we see that $k = -1$ and $h = -1$. Thus, the center of the hyperbola is at the point $(-1, -1)$. Because $a = 1$, the vertices are at $(-1, 0)$ and $(-1, -2)$. Because $ae = \sqrt{2}$, the foci are at $(-1, -1 + \sqrt{2})$ and $(-1, -1 - \sqrt{2})$. Because $a/e = \frac{1}{2}\sqrt{2}$, the equations of the directrices are $y = -1 + \frac{1}{2}\sqrt{2}$ and $y = -1 - \frac{1}{2}\sqrt{2}$. To find the equations of the asymptotes, we replace the constant term by 0. Thus,

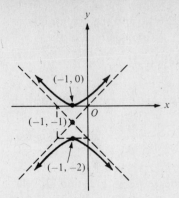

$$(y+1)^2 - (x+1)^2 = 0$$
$$y+1 = \pm(x+1)$$

Hence, the equations of the asymptotes are $y = x$ and $x = -x - 2$. A sketch of the graph is shown in Fig. 13. 7. 8. We use the auxiliary rectangle and the asymptotes to aid in making the sketch.

In Exercises 9–15 find an equation of the hyperbola satisfying the given conditions and draw a sketch of the graph.

12. Foci at (3, 6) and (3, 0) and passing through the point $(5, 3 + 6/\sqrt{5})$

SOLUTION: Because the foci lie in a vertical line, an equation of the hyperbola is of the form

$$\frac{(y-k)^2}{a^2} - \frac{(x-h)^2}{b^2} = 1 \tag{1}$$

Because the midpoint of the segment joining the foci (3, 6) and (3, 0) is the point (3, 3), then (3, 3) is the center of the hyperbola, and we have $h = 3$ and $k = 3$. Thus, from (1) we have

$$\frac{(y-3)^2}{a^2} - \frac{(x-3)^2}{b^2} = 1 \tag{2}$$

Because the hyperbola contains the point $(5, 3 + 6/\sqrt{5})$, we may let $x = 5$ and $y = 3 + 6/\sqrt{5}$ in Eq. (2). Thus,

$$\frac{36}{5a^2} - \frac{4}{b^2} = 1 \tag{3}$$

Because the distance between the center (3, 3) and the focus (3, 6) is 3, we have $ae = 3$. Because $b = a\sqrt{e^2 - 1}$, we have

$$b^2 = a^2(e^2 - 1)$$
$$a^2 + b^2 = a^2 e^2$$

and since $ae = 3$, we have

$$a^2 + b^2 = 9 \tag{4}$$

Eliminating a from (3) and (4), we obtain

$$\frac{36}{5(9 - b^2)} - \frac{4}{b^2} = 1$$

$$36b^2 - 180 + 20b^2 = 45b^2 - 5b^4$$

$$5b^4 + 11b^2 - 180 = 0$$

$$(b^2 - 5)(5b^2 + 36) = 0$$

$$b^2 - 5 = 0$$

$$b^2 = 5$$

Thus, $a^2 = 9 - b^2 = 4$, and substituting the values for a^2 and b^2 into Eq. (2), we obtain

$$\frac{(y-3)^2}{4} - \frac{(x-3)^2}{5} = 1$$

A sketch of the graph is shown in Fig. 13. 7. 12.

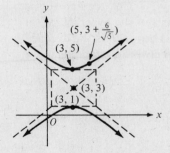

Figure 13. 7. 12

16. The vertices of a hyperbola are at (2, 7) and (2, −1), and the eccentricity is $\sqrt{3}$. Find **(a)** an equation of the hyperbola, and **(b)** equations of the asymptotes.

SOLUTION:

(a) Because the vertices are on a vertical line, a standard form of the hyperbola is

$$\frac{(y-k)^2}{a^2} - \frac{(x-h)^2}{b^2} = 1 \tag{1}$$

The midpoint of the line segment joining the vertices (2, 7) and (2, −1) is the point (2, 3). Thus, (2, 3) is the center and we have $h = 2$ and $k = 3$. The distance from the center to each vertex is 4, so $a = 4$. Because $b = a\sqrt{e^2 - 1}$ and we are given that $e = \sqrt{3}$, then $b = 4\sqrt{2}$. Substituting for h, k, a, and b in the standard form (1), we have

$$\frac{(y-3)^2}{16} - \frac{(x-2)^2}{32} = 1$$

which is an equation of the hyperbola.

(b) Equations of the asymptotes are found by replacing the constant 1 by 0, which results in

$$\frac{(y-3)^2}{16} = \frac{(x-2)^2}{32}$$

$$2(y-3)^2 = (x-2)^2$$

$$\sqrt{2}(y-3) = \pm(x-2)$$

Thus, equations of the asymptotes are

$$\sqrt{2}(y-3) = x - 2$$

$$x - \sqrt{2}y - 2 + 3\sqrt{2} = 0$$

and

$$\sqrt{2}(y-3) = -x + 2$$

$$x + \sqrt{2}y - 2 - 3\sqrt{2} = 0$$

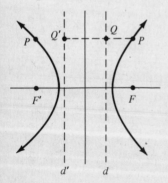

Figure 13. 7. 20

20. Prove that any hyperbola is the set of points such that the absolute value of the difference of the distances from any point of the set to two given points (the foci) is a constant.

SOLUTION: Refer to Fig. 13. 7. 20. Let the foci be at points F and F' with the corresponding directrices lines d and d'. Let P be any point on the hyperbola, and let segment PQQ' be parallel to the transverse axis of the hyperbola with point Q on directrix d and point Q' on directrix d'. Then $|\overline{PQ}|$ is the distance between P and directrix d, and $|\overline{PQ'}|$ is the distance between P and directrix d'. By Definition 13.3.1 we have

$$|\overline{PF}| = e|\overline{PQ}| \quad \text{and} \quad |\overline{PF'}| = e|\overline{PQ'}|$$

Subtracting each member of the first equation from the corresponding member of the second equation, we obtain

$$|\overline{PF'}| - |\overline{PF}| = e(|\overline{PQ'}| - |\overline{PQ}|) \tag{1}$$

Because the distance between the directrices is $2a/e$, we have $|\overline{QQ'}| = 2a/e$. Thus, if P is on the branch of the hyperbola with focus F, we have

$$|\overline{PQ'}| = |\overline{QQ'}| + |\overline{PQ}|$$

$$= \frac{2a}{e} + |\overline{PQ}| \qquad (2)$$

Substituting from (2) into (1), we obtain

$$|\overline{PF'}| - |\overline{PF}| = 2a \qquad (3)$$

If P is on the branch of the hyperbola with focus F', we have

$$|\overline{PQ'}| + |\overline{Q'Q}| = |\overline{PQ}|$$

$$|\overline{PQ'}| = |\overline{PQ}| - \frac{2a}{e} \qquad (4)$$

Substituting from (4) into (1), we obtain

$$|\overline{PF'}| - |\overline{PF}| = -2a \qquad (5)$$

From (3) and (5) we conclude that the absolute value of the difference of the distances from any point of the hyperbola to the foci is the constant $2a$.

24. Prove that the eccentricity of an equilateral hyperbola is equal to $\sqrt{2}$.

SOLUTION: An equilateral hyperbola is one in which $a = b$. In every hyperbola $b = a\sqrt{e^2 - 1}$. Because $a = b$, we have

$$a = a\sqrt{e^2 - 1}$$
$$1 = \sqrt{e^2 - 1}$$
$$1 = e^2 - 1$$
$$2 = e^2$$
$$e = \sqrt{2}$$

13.8 ROTATION OF AXES The following steps may be followed to eliminate the xy term in the general second-degree equation

$$Ax^2 + Bxy + Cy^2 + Dx + Ey + F = 0 \qquad (1)$$

where $B \neq 0$:

1. Let $\cot 2\alpha = \dfrac{A - C}{B}$

2. If $A = C$, then $\alpha = \frac{1}{4}\pi$ and

$$\sin \alpha = \frac{1}{2}\sqrt{2} \quad \text{and} \quad \cos \alpha = \frac{1}{2}\sqrt{2}$$

3. If $A \neq C$, then find $\sin \alpha$ and $\cos \alpha$ by using the following identities in the order indicated:

a. $\tan 2\alpha = \dfrac{1}{\cot 2\alpha}$

b. $\cos^2 2\alpha = \dfrac{1}{1 + \tan^2 2\alpha}$

c. $\cos 2\alpha = \begin{cases} \sqrt{\cos^2 2\alpha} & \text{if } \cot 2\alpha > 0 \\ -\sqrt{\cos^2 2\alpha} & \text{if } \cot 2\alpha < 0 \end{cases}$

d. $\sin \alpha = \sqrt{\dfrac{1 - \cos 2\alpha}{2}}$

e. $\cos \alpha = \sqrt{\dfrac{1 + \cos 2\alpha}{2}}$

4. Make the following substitutions in Eq. (1):

$$x = \bar{x} \cos \alpha - \bar{y} \sin \alpha$$
$$y = \bar{x} \sin \alpha + \bar{y} \cos \alpha$$

13.8.2 Theorem   The graph of the equation

$$Ax^2 + Bxy + Cy^2 + Dx + Ey + F = 0$$

is either a conic or a degenerate conic. If it is a conic, then it is

(i) a *parabola* if $B^2 - 4AC = 0$
(ii) an *ellipse* if $B^2 - 4AC < 0$
(iii) a *hyperbola* if $B^2 - 4AC > 0$.

Exercises 13.8

In Exercises 4–10 remove the xy term from the given equation by a rotation of axes. Draw a sketch of the graph and show both sets of axes.

4. $x^2 + xy + y^2 = 3$

SOLUTION: Because $A = 1$, $B = 1$, and $C = 1$, then

$$\cot 2\alpha = \frac{A - C}{B} = 0$$

Therefore, $\alpha = \frac{1}{4}\pi$ and

$$\sin \alpha = \tfrac{1}{2}\sqrt{2} \quad \cos \alpha = \tfrac{1}{2}\sqrt{2}$$

Thus,

$$x = \bar{x} \cos \alpha - \bar{y} \sin \alpha \qquad y = \bar{x} \sin \alpha + \bar{y} \cos \alpha$$
$$x = \tfrac{1}{2}\sqrt{2}\,(\bar{x} - \bar{y}) \qquad y = \tfrac{1}{2}\sqrt{2}(\bar{x} + \bar{y}) \tag{1}$$

And hence,

$$x^2 = \tfrac{1}{2}(\bar{x}^2 - 2\bar{x}\bar{y} + \bar{y}^2) \tag{2}$$
$$xy = \tfrac{1}{2}(\bar{x}^2 - \bar{y}^2) \tag{3}$$
$$y^2 = \tfrac{1}{2}(\bar{x}^2 + 2\bar{x}\bar{y} + \bar{y}^2) \tag{4}$$

Substituting from (2), (3), and (4), into the given equation, we obtain

$$\tfrac{1}{2}(\bar{x}^2 - 2\bar{x}\bar{y} + \bar{y}^2) + \tfrac{1}{2}(\bar{x}^2 - \bar{y}^2) + \tfrac{1}{2}(\bar{x}^2 + 2\bar{x}\bar{y} + \bar{y}^2) = 3$$

$$\tfrac{3}{2}\bar{x}^2 + \tfrac{1}{2}\bar{y}^2 = 3$$

$$\frac{\bar{x}^2}{2} + \frac{\bar{y}^2}{6} = 1 \tag{5}$$

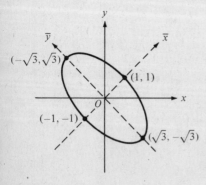

Figure 13. 8. 4

From (5) we conclude that relative to the $\bar{x}\bar{y}$-axes the graph is an ellipse with semimajor axis $\sqrt{6}$ and semiminor axis $\sqrt{2}$. Thus, the vertices are at the points whose $\bar{x}\bar{y}$ coordinates are $(0, \pm\sqrt{6})$. Substituting these coordinates into (1), we learn that the xy coordinates of the vertices are $(-\sqrt{3}, \sqrt{3})$ and $(\sqrt{3}, -\sqrt{3})$. Because the $\bar{x}\bar{y}$ coordinates of the ends of the minor axis are $(\pm\sqrt{2}, 0)$, from (1) we obtain the xy coordinates of the ends of the minor axis, namely $(1, 1)$ and $(-1, -1)$. A sketch of the graph is shown in Fig. 13. 8. 4.

8. $6x^2 + 20\sqrt{3}xy + 26y^2 = 324$

SOLUTION:

$$\cot 2\alpha = \frac{6-26}{20\sqrt{3}} = -\frac{1}{\sqrt{3}}$$

Thus, $\tan 2\alpha = -\sqrt{3}$

$$\cos^2 2\alpha = \frac{1}{\sec^2 2\alpha}$$

$$= \frac{1}{1 + \tan^2 2\alpha}$$

$$= \frac{1}{1+3} = \frac{1}{4}$$

Because $\cot 2\alpha < 0$, then $\cos 2\alpha < 0$. Thus

$$\cos 2\alpha = -\sqrt{\tfrac{1}{4}} = -\tfrac{1}{2}$$

Then

$$\sin \alpha = \sqrt{\frac{1-\cos 2\alpha}{2}} \qquad \cos \alpha = \sqrt{\frac{1+\cos 2\alpha}{2}}$$

$$= \sqrt{\frac{1+\tfrac{1}{2}}{2}} \qquad\qquad = \sqrt{\frac{1-\tfrac{1}{2}}{2}}$$

$$= \tfrac{1}{2}\sqrt{3} \qquad\qquad\quad = \tfrac{1}{2}$$

Hence,

$$x = \bar{x}\cos\alpha - \bar{y}\sin\alpha \qquad y = \bar{x}\sin\alpha + \bar{y}\cos\alpha$$
$$x = \tfrac{1}{2}(\bar{x} - \sqrt{3}\,\bar{y}) \qquad\quad y = \tfrac{1}{2}(\sqrt{3}\,\bar{x} + \bar{y}) \tag{1}$$

from which we have

$$x^2 = \tfrac{1}{4}(\bar{x}^2 - 2\sqrt{3}\bar{x}\bar{y} + 3\bar{y}^2) \tag{2}$$
$$xy = \tfrac{1}{4}(\sqrt{3}\bar{x}^2 - 2\bar{x}\bar{y} - \sqrt{3}\bar{y}^2) \tag{3}$$
$$y^2 = \tfrac{1}{4}(3\bar{x}^2 + 2\sqrt{3}\bar{x}\bar{y} + \bar{y}^2) \tag{4}$$

Substituting from (2), (3), and (4) into the given equation, we obtain

$$\tfrac{3}{2}(\bar{x}^2 - 2\sqrt{3}\bar{x}\bar{y} + 3\bar{y}^2) + 5\sqrt{3}(\sqrt{3}\bar{x}^2 - 2\bar{x}\bar{y} - \sqrt{3}\bar{y}^2) + \tfrac{13}{2}(3\bar{x}^2 + 2\sqrt{3}\bar{x}\bar{y} + \bar{y}^2) = 324$$

$$36\bar{x}^2 - 4\bar{y}^2 = 324$$

$$\frac{\bar{x}^2}{9} - \frac{\bar{y}^2}{81} = 1$$

Therefore, relative to the $\bar{x}\bar{y}$ axes, the graph is a hyperbola with transverse axis along the x-axis. The $\bar{x}\bar{y}$ coordinates of the vertices are $(\pm 3, 0)$. From (1) we obtain the xy coordinates of the vertices, $(\tfrac{3}{2}, \tfrac{3}{2}\sqrt{3})$ and $(-\tfrac{3}{2}, -\tfrac{3}{2}\sqrt{3})$. The $\bar{x}\bar{y}$ coordinates of the ends of the conjugate axis are $(0, \pm 9)$, and from (1) we obtain the xy coordinates, namely $(-\tfrac{9}{2}\sqrt{3}, \tfrac{9}{2})$ and $(\tfrac{9}{2}\sqrt{3}, -\tfrac{9}{2})$. A sketch of the graph is shown in Fig. 13. 8. 8.

In Exercises 11–18 simplify the given equation by a rotation and translation of axes. Draw a sketch of the graph and show the three sets of axes.

12. $x^2 - 10xy + y^2 + x + y + 1 = 0$

SOLUTION: $A = 1$, $B = -10$, and $C = 1$. Thus, $\cot 2\alpha = 0$, and $\alpha = \tfrac{1}{4}\pi$, so $\sin \alpha = \tfrac{1}{2}\sqrt{2}$ and $\cos \alpha = \tfrac{1}{2}\sqrt{2}$. We let

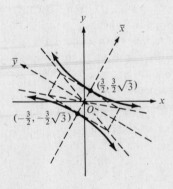

Figure 13. 8. 8

$$x = \bar{x} \cos \alpha - \bar{y} \sin \alpha \qquad y = \bar{x} \sin \alpha + \bar{y} \cos \alpha$$

$$x = \frac{\bar{x} - \bar{y}}{\sqrt{2}} \qquad\qquad y = \frac{\bar{x} + \bar{y}}{\sqrt{2}} \tag{1}$$

Then

$$x^2 = \frac{\bar{x}^2 - 2\bar{x}\bar{y} + \bar{y}^2}{2}$$

$$xy = \frac{\bar{x}^2 - \bar{y}^2}{2}$$

$$y^2 = \frac{\bar{x}^2 + 2\bar{x}\bar{y} + \bar{y}^2}{2}$$

Substituting into the given equation, we have

$$\frac{\bar{x}^2 - 2\bar{x}\bar{y} + \bar{y}^2}{2} - \frac{10(\bar{x}^2 - \bar{y}^2)}{2} + \frac{\bar{x}^2 + 2\bar{x}\bar{y} + \bar{y}^2}{2} + \frac{\bar{x} - \bar{y}}{\sqrt{2}} + \frac{\bar{x} + \bar{y}}{\sqrt{2}} + 1 = 0$$

$$\frac{-8\bar{x}^2 + 12\bar{y}^2}{2} + \frac{2\bar{x}}{\sqrt{2}} + 1 = 0$$

$$-4\bar{x}^2 + 6\bar{y}^2 + \sqrt{2}\,\bar{x} + 1 = 0$$

We complete the square on the terms in $\bar{x}$ and write the equation in a standard form.

$$-4(\bar{x}^2 - \tfrac{1}{4}\sqrt{2}\,\bar{x} + \tfrac{1}{32}) + 6\bar{y}^2 = -1 - \tfrac{1}{8}$$

$$-4(\bar{x} - \tfrac{1}{8}\sqrt{2})^2 + 6\bar{y}^2 = -\tfrac{9}{8}$$

$$\frac{(\bar{x} - \tfrac{1}{8}\sqrt{2})^2}{\tfrac{9}{32}} - \frac{\bar{y}^2}{\tfrac{3}{16}} = 1 \tag{2}$$

We translate the origin to the point with $\bar{x}\bar{y}$ coordinates $(\tfrac{1}{8}\sqrt{2}, 0)$. Thus,

$$\bar{x} = \bar{x}' + \tfrac{1}{8}\sqrt{2} \quad \text{and} \quad \bar{y} = \bar{y}' \tag{3}$$

Substituting into Eq. (2), we get

$$\frac{(\bar{x}')^2}{\tfrac{9}{32}} - \frac{(\bar{y}')^2}{\tfrac{3}{16}} = 1 \tag{4}$$

The graph of Eq. (4) is a hyperbola with its center at the origin of the $\bar{x}'\,\bar{y}'$ system and its transverse axis the $\bar{x}'$-axis. Thus, the center is at the point whose $\bar{x}\bar{y}$ coordinates are $\bar{x} = \tfrac{1}{8}\sqrt{2}$ and $\bar{y} = 0$. By substitution into Eqs. (1) we determine that the center is at the point whose xy-coordinates are $(\tfrac{1}{8}, \tfrac{1}{8})$. Furthermore, because $a^2 = \tfrac{9}{32}$, then $a = \tfrac{3}{8}\sqrt{2}$. Thus, the vertices are at the points where $\bar{x}' = \pm\tfrac{3}{8}\sqrt{2}$ and $\bar{y}' = 0$. By substitution into Eqs. (3) we determine that the vertices have $\bar{x}\bar{y}$ coordinates $\bar{x} = \tfrac{1}{2}\sqrt{2}$, $\bar{y} = 0$ and $\bar{x} = -\tfrac{1}{4}\sqrt{2}$, $\bar{y} = 0$. By substitution into Eqs. (1) we obtain the xy-coordinates of the vertices $(\tfrac{1}{2}, \tfrac{1}{2})$ and $(-\tfrac{1}{4}, \tfrac{1}{4})$. Moreover, because $b^2 = \tfrac{3}{16}$ then $b = \tfrac{1}{4}\sqrt{3}$, so the ends of the conjugate axis are at the points where $\bar{x}' = 0$ and $\bar{y}' = \pm\tfrac{1}{4}\sqrt{3}$. Figure 13. 8. 12 shows a sketch of the graph.

16. $3x^2 - 4xy + 8x - 1 = 0$

SOLUTION: Because $A = 3$, $B = -4$, and $C = 0$, then

$$\cot 2\alpha = -\tfrac{3}{4}$$

Hence,

Figure 13. 8. 12

$$\tan 2\alpha = -\tfrac{4}{3}$$

$$\cos^2 2\alpha = \frac{1}{1 + \tan^2 2\alpha}$$

$$= \frac{1}{1 + \tfrac{16}{9}} = \tfrac{9}{25}$$

Because $\cot 2\alpha < 0$, then $\cos 2\alpha < 0$. Thus,

$$\cos 2\alpha = -\sqrt{\tfrac{9}{25}} = -\tfrac{3}{5}$$

we have

$$\sin \alpha = \sqrt{\frac{1 - \cos 2\alpha}{2}} \qquad \cos \alpha = \sqrt{\frac{1 + \cos 2\alpha}{2}}$$

$$= \sqrt{\frac{1 + \tfrac{3}{5}}{2}} \qquad\qquad = \sqrt{\frac{1 - \tfrac{3}{5}}{2}}$$

$$= \tfrac{2}{5}\sqrt{5} \qquad\qquad\quad = \tfrac{1}{5}\sqrt{5}$$

Thus,

$$\begin{aligned}
x &= \bar{x}\cos\alpha - \bar{y}\sin\alpha & y &= \bar{x}\sin\alpha + \bar{y}\cos\alpha \\
x &= \tfrac{1}{5}\sqrt{5}(\bar{x} - 2\bar{y}) & y &= \tfrac{1}{5}\sqrt{5}(2\bar{x} + \bar{y})
\end{aligned} \qquad (1)$$

from which we have

$$x^2 = \tfrac{1}{5}(\bar{x}^2 - 4\bar{x}\bar{y} + 4\bar{y}^2) \qquad (2)$$
$$xy = \tfrac{1}{5}(2\bar{x}^2 - 3\bar{x}\bar{y} - 2\bar{y}^2) \qquad (3)$$

Substituting from (1), (2), and (3) into the given equation, we obtain

$$\tfrac{3}{5}(\bar{x}^2 - 4\bar{x}\bar{y} + 4\bar{y}^2) - \tfrac{4}{5}(2\bar{x}^2 - 3\bar{x}\bar{y} - 2\bar{y}^2) + \tfrac{8}{5}\sqrt{5}(\bar{x} - 2\bar{y}) - 1 = 0$$
$$-\bar{x}^2 + 4\bar{y}^2 + \tfrac{8}{5}\sqrt{5}\,\bar{x} - \tfrac{16}{5}\sqrt{5}\,\bar{y} - 1 = 0$$

Next, we complete the squares on the terms in $\bar{x}$ and $\bar{y}$.

$$-(\bar{x}^2 - \tfrac{8}{5}\sqrt{5}\,\bar{x} + \tfrac{16}{5}) + 4(\bar{y}^2 - \tfrac{4}{5}\sqrt{5}\,\bar{y} + \tfrac{4}{5}) = 1 - \tfrac{16}{5} + \tfrac{16}{5}$$
$$-(\bar{x} - \tfrac{4}{5}\sqrt{5})^2 + 4(\bar{y} - \tfrac{2}{5}\sqrt{5})^2 = 1 \qquad (4)$$

We translate the origin to the point whose $\bar{x}\bar{y}$-coordinates are $(\tfrac{4}{5}\sqrt{5}, \tfrac{2}{5}\sqrt{5})$ by the equations

$$\bar{x} = \bar{x}' + \tfrac{4}{5}\sqrt{5} \quad \text{and} \quad \bar{y} = \bar{y}' + \tfrac{2}{5}\sqrt{5} \qquad (5)$$

Thus, Eq. (4) becomes

$$-(\bar{x}')^2 + 4(\bar{y}')^2 = 1 \qquad (6)$$

The graph of Eq. (6) is a hyperbola with center at the origin of the $\bar{x}'\bar{y}'$ system and transverse axis the $\bar{y}'$-axis. The vertices are at the points where $\bar{x}' = 0$ and $\bar{y}' = \pm\tfrac{1}{2}$, and the ends of the conjugate axis are at the points where $\bar{x}' = \pm 1$ and $\bar{y}' = 0$. Because the center of the hyperbola is at the point where $\bar{x}' = 0$ and $\bar{y}' = 0$, from (5) the center is at the point where $\bar{x} = \tfrac{4}{5}\sqrt{5}$ and $\bar{y} = \tfrac{2}{5}\sqrt{5}$, and from (1) the center is at the point where $x = 0$ and $y = 2$. In Fig. 13. 8. 16 we show a sketch of the graph and the three sets of axes. We use the fact that $\tan \alpha = \sin \alpha/\cos \alpha = 2$ to construct the angle of rotation.

20. Given the equation $(a^2 + b^2)xy = 1$, where $a > 0$ and $b > 0$, find an equation of the graph with respect to the $\bar{x}$- and $\bar{y}$-axes after a rotation of the axes through an angle of radian measure $\tan^{-1}(b/a)$.

SOLUTION: If $\alpha = \tan^{-1}(b/a)$ with $a > 0$ and $b > 0$, then by Fig. 13. 8. 20 we have

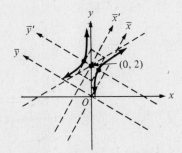

Figure 13. 8. 16

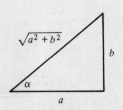

Figure 13. 8. 20

$$\sin \alpha = \frac{b}{\sqrt{a^2 + b^2}} \quad \text{and} \quad \cos \alpha = \frac{a}{\sqrt{a^2 + b^2}}$$

Thus,

$$x = \bar{x} \cos \alpha - \bar{y} \sin \alpha \qquad y = \bar{x} \sin \alpha + \bar{y} \cos \alpha$$

$$x = \frac{a\bar{x} - b\bar{y}}{\sqrt{a^2 + b^2}} \qquad y = \frac{b\bar{x} + a\bar{y}}{\sqrt{a^2 + b^2}}$$

Hence,

$$xy = \frac{(a\bar{x} - b\bar{y})(b\bar{x} + a\bar{y})}{a^2 + b^2}$$

Substituting for xy into the given equation, we obtain

$$(a\bar{x} - b\bar{y})(b\bar{x} + a\bar{y}) = 1$$
$$ab\bar{x}^2 + (a^2 - b^2)\bar{x}\bar{y} - ab\bar{y}^2 = 1$$

Review Exercises for Chapter 13

4. Find an equation of the parabola having its vertex at $(-3, 5)$ and its focus at $(-3, -1)$.

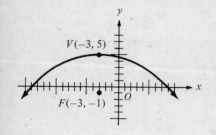

Figure 13. 4R

SOLUTION: Figure 13. 4R shows a sketch of the parabola. By Theorem 13.2.2 an equation of the parabola is of the form

$$(x - h)^2 = 4p(y - k) \tag{1}$$

Because the vertex is at $(-3, 5)$, we have $h = -3$ and $k = 5$. Because the directed distance from the vertex $(-3, 5)$ to the focus $(-3, -1)$ is -6, we have $p = -6$. Substituting into Eq. (1) we obtain

$$(x + 3)^2 = -24(y - 5)$$

which is an equation of the parabola.

8. Find the focus of the parabola having the equation $y = ax^2 + bx$.

SOLUTION: We complete the square on terms in x and write the equation in a standard form given by Theorem 13.2.2.

$$x^2 + \frac{bx}{a} = \frac{y}{a}$$

$$x^2 + \frac{bx}{a} + \frac{b^2}{4a^2} = \frac{y}{a} + \frac{b^2}{4a^2}$$

$$\left(x + \frac{b}{2a}\right)^2 = \frac{1}{a}\left(y + \frac{b^2}{4a}\right)$$

Thus, $h = -b/2a$ and $k = -b^2/4a$, so the vertex is at the point $(-b/2a, -b^2/4a)$ and $4p = 1/a$, or $p = 1/4a$. Because the axis is parallel to the y-axis, then the coordinates of the focus are (x, y), where

$$x = h \qquad \text{and} \quad y = k + p$$

$$x = -\frac{b}{2a} \qquad y = -\frac{b^2}{4a} + \frac{1}{4a} = \frac{-b^2 + 1}{4a}$$

12. Find a Cartesian equation of the conic with foci at $(-5, 1)$ and $(1, 1)$ and one vertex at $(-4, 1)$, and draw a sketch of the graph.

SOLUTION: Because the vertex is between the foci, the conic is a hyperbola. Because the foci are in a horizontal line, an equation of the hyperbola is of the form

$$\frac{(x-h)^2}{a^2} - \frac{(y-k)^2}{b^2} = 1 \tag{1}$$

The center of the hyperbola is at the midpoint of the segment joining the foci. Thus, the center is at $(-2, 1)$, and we have $h = -2$, $k = 1$. Because the distance between the center $(-2, 1)$ and the vertex $(-4, 1)$ is 2, we have $a = 2$. Because the distance between the center $(-2, 1)$ and the focus $(1, 1)$ is 3, we have $ae = 3$. Thus, $e = \frac{3}{2}$. Because $b = a\sqrt{e^2 - 1}$, we have $b = 2\sqrt{(\frac{3}{2})^2 - 1} = \sqrt{5}$. Substituting the values of h, k, a, and b into (1), we obtain

$$\frac{(x+2)^2}{4} - \frac{(y-1)^2}{5} = 1$$

A sketch of the graph is shown in Fig. 13. 12R.

16. Find a polar equation of the conic with a focus at the pole, a vertex at $(6, \frac{1}{2}\pi)$, and $e = \frac{3}{4}$, and draw a sketch of the graph.

SOLUTION: There are two conics possible. Because the focus is at the pole and a vertex is above the pole, a polar equation of the conic is either of the form

$$r = \frac{ed}{1 + e \sin \theta} \tag{1}$$

or of the form

$$r = \frac{ed}{1 - e \sin \theta} \tag{2}$$

We let $e = \frac{3}{4}$ and $(r, \theta) = (6, \frac{1}{2}\pi)$ in Eq. (1). Thus,

$$6 = \frac{\frac{3}{4}d}{1 + \frac{3}{4}}$$

from which we get $d = 14$. Thus, with $e = \frac{3}{4}$ and $d = 14$, from (1) we have

$$r = \frac{\frac{3}{4} \cdot 14}{1 + \frac{3}{4} \sin \theta} = \frac{42}{4 + 3 \sin \theta} \tag{3}$$

which is a polar equation of one of the conics. A sketch of the graph of (3) is shown in Fig. 13. 16aR. Next, we let $e = \frac{3}{4}$ and $(r, \theta) = (6, \frac{1}{2}\pi)$ in Eq. (2). Thus,

$$6 = \frac{\frac{3}{4}d}{1 - \frac{3}{4}}$$

from which we get $d = 2$. With $e = \frac{3}{4}$ and $d = 2$ in Eq. (2), we have

$$r = \frac{6}{4 - 3 \sin \theta} \tag{4}$$

which is a polar equation of the other conic possible. A sketch of the graph of (4) is shown in Fig. 13. 16bR.

20. The equation is that of a conic having a focus at the pole.
    (a) Find the eccentricity.
    (b) Identify the conic.
    (c) Write an equation of the directrix that corresponds to the focus at the pole.

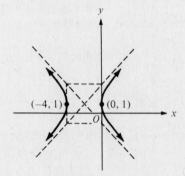

Figure 13. 12R

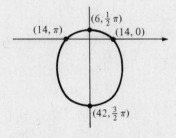

Figure 13. 16aR

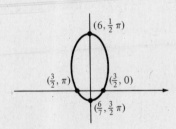

Figure 13. 16bR

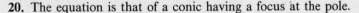

(d) Draw a sketch of the curve.

$$r = \frac{5}{3 + 3\sin\theta}$$

SOLUTION: Dividing the numerator and denominator by 3, we obtain

$$r = \frac{\frac{5}{3}}{1 + \sin\theta}$$

(a) Comparing this with standard form 3 of Section 13.4, we see that $e = 1$.

(b) Because $e = 1$, the conic is a parabola.

(c) Because $ed = \frac{5}{3}$, we have $d = \frac{5}{3}$. Thus, an equation of the directrix is

$$r\sin\theta = \frac{5}{3}$$

(d) A sketch of the parabola is shown in Fig. 13. 20R.

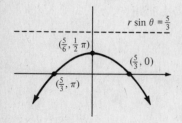

Figure 13. 20R

24. The equation is that of either an ellipse or a hyperbola. Find the eccentricity, center, foci, and directrices, and draw a sketch of the graph. If it is a hyperbola, also find the equations of the asymptotes.

$$3x^2 - 2y^2 + 6x - 8y + 11 = 0$$

SOLUTION: We complete the square on the terms in x and y.

$$3(x^2 + 2x + 1) - 2(y^2 + 4y + 4) = -11 + 3 - 8$$

$$3(x + 1)^2 - 2(y + 2)^2 = -16$$

$$\frac{(y + 2)^2}{8} - \frac{(x + 1)^2}{\frac{16}{3}} = 1 \qquad (1)$$

Thus, the graph is a hyperbola with center at $(-1, -2)$ and vertical transverse axis. We have $a = 2\sqrt{2}$ and $b = \frac{4}{3}\sqrt{3}$. Because $b = a\sqrt{e^2 - 1}$, we have $e = \frac{1}{3}\sqrt{15}$. Because $ae = \frac{2}{3}\sqrt{30}$, the foci are at $(-1, -2 \pm \frac{2}{3}\sqrt{30})$. Because $a/e = \frac{2}{5}\sqrt{30}$, the directrices are the lines $y = -2 \pm \frac{2}{5}\sqrt{30}$. To find equations of the asymptotes, we replace the constant term by 0 in Eq. (1). Thus,

$$2(y + 2)^2 = 3(x + 1)^2$$
$$\sqrt{2}(y + 2) = \pm\sqrt{3}(x + 1)$$
$$\sqrt{3}x - \sqrt{2}y + \sqrt{3} - 2\sqrt{2} = 0 \quad \text{and} \quad \sqrt{3}x + \sqrt{2}y + 2\sqrt{2} + \sqrt{3} = 0$$

are equations of the asymptotes. A sketch of the graph is shown in Fig. 13. 24R.

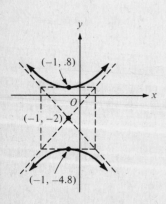

Figure 13. 24R

28. Simplify the given equation by a rotation and translation of axes. Draw a sketch of the graph and show the three sets of axes.

$$4x^2 + 3xy + y^2 - 6x + 12y = 0$$

SOLUTION: Let α be the angle of rotation. Then $\cot 2\alpha = (A - C)/B = 1$, so $2\alpha = \frac{1}{4}\pi$ and $\alpha = \frac{1}{8}\pi$. Hence, $\cos 2\alpha = \frac{1}{2}\sqrt{2}$ and

$$\sin\alpha = \sqrt{\frac{1 - \frac{1}{2}\sqrt{2}}{2}} = \frac{\sqrt{2 - \sqrt{2}}}{2}$$

$$\cos\alpha = \sqrt{\frac{1 + \frac{1}{2}\sqrt{2}}{2}} = \frac{\sqrt{2 + \sqrt{2}}}{2}$$

Thus,

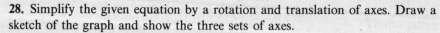

$$x = \frac{\sqrt{2 + \sqrt{2}}\,\bar{x} - \sqrt{2 - \sqrt{2}}\,\bar{y}}{2} \qquad y = \frac{\sqrt{2 - \sqrt{2}}\,\bar{x} + \sqrt{2 + \sqrt{2}}\,\bar{y}}{2}$$

and

$$x^2 = \frac{(2+\sqrt{2})\bar{x}^2 - 2\sqrt{2}\,\bar{x}\bar{y} + (2-\sqrt{2})\,\bar{y}^2}{4}$$

$$xy = \frac{\sqrt{2}\,\bar{x}^2 + 2\sqrt{2}\,\bar{x}\bar{y} - \sqrt{2}\,\bar{y}^2}{4}$$

$$y^2 = \frac{(2-\sqrt{2})\bar{x}^2 + 2\sqrt{2}\,\bar{x}\bar{y} + (2+\sqrt{2})\,\bar{y}^2}{4}$$

Substituting into the given equation and combining similar terms, we obtain

$$\frac{(5+3\sqrt{2})\bar{x}^2}{2} + \frac{(5-3\sqrt{2})\bar{y}^2}{2} + \frac{(-6\sqrt{2}+\sqrt{2}+12\sqrt{2-\sqrt{2}})\bar{x}}{2} + \frac{(6\sqrt{2-\sqrt{2}}+12\sqrt{2+\sqrt{2}})\bar{y}}{2} = 0$$

We use decimal approximations for the coefficients.

$$4.62\bar{x}^2 + 0.379\bar{y}^2 - 0.951\bar{x} + 13.383\bar{y} = 0$$

We complete the squares and write the equation in a standard form.

$$4.62(\bar{x}^2 - 0.206\bar{x} + 0.011) + 0.379(\bar{y}^2 + 35.3\bar{y} + 311.7) = 118.6$$

$$\frac{(\bar{x} - 0.103)^2}{25.7} + \frac{(\bar{y} + 17.6)^2}{313} = 1$$

Translating the origin to the point where $\bar{x} = 0.103$ and $\bar{y} = -17.6$, we have

$$\bar{x}' = \bar{x} + 0.103 \quad \text{and} \quad \bar{y}' = \bar{y} - 17.6$$

and the equation becomes

$$\frac{(\bar{x}')^2}{25.7} + \frac{(\bar{y}')^2}{313} = 1$$

The graph is an ellipse with center at the point where $\bar{x} = 0.103$ and $\bar{y} = -17.6$. A sketch is shown in Fig. 13. 28R.

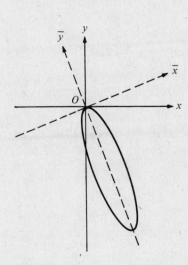

Figure 13. 28R

32. Show that the hyperbola $x^2 - y^2 = 4$ has the same foci as the ellipse $x^2 + 9y^2 = 9$.

SOLUTION: A standard form of the equation of the hyperbola is

$$\frac{x^2}{4} - \frac{y^2}{4} = 1$$

We have $a = b = 2$ and therefore the equilateral hyperbola has eccentricity $e = \sqrt{2}$, because $b = a\sqrt{e^2 - 1}$. Thus, $ae = 2\sqrt{2}$. Because the center of the hyperbola is at the origin and the transverse axis is along the x-axis, we conclude that the foci are at the points $(2\sqrt{2}, 0)$ and $(-2\sqrt{2}, 0)$. For the equation of the ellipse a standard form is

$$\frac{x^2}{9} + \frac{y^2}{1} = 1$$

Then $a = 3$ and $b = 1$. Because $b = a\sqrt{1 - e^2}$, then $1 = 3\sqrt{1 - e^2}$, or $1 - e^2 = \frac{1}{9}$, so $e = \frac{2}{3}\sqrt{2}$. Hence, $ae = 2\sqrt{2}$. Because the center of the ellipse is at the origin and the major axis is along the x axis, we conclude that the foci are at the points $(2\sqrt{2}, 0)$ and $(-2\sqrt{2}, 0)$. Thus, the hyperbola has the same foci as the ellipse.

36. The directrix of the parabola $y^2 = 4px$ is tangent to a circle having the focus of the parabola as its center. Find an equation of the circle and the points of intersection of the two curves.

SOLUTION: The focus of the parabola is at the point $(p, 0)$ and the directrix is the line $x = -p$. Therefore, the radius of the circle is $2p$, the distance between the focus and the directrix. An equation of the circle is

$$(x - p)^2 + y^2 = (2p)^2 \qquad (1)$$

We find the intersection of the circle and the parabola whose equation is given.

$$y^2 = 4px \qquad (2)$$

Eliminating y from the system of Eqs. (1) and (2), we obtain

$$(x - p)^2 + 4px = (2p)^2$$
$$x^2 - 2px + p^2 + 4px = 4p^2$$
$$x^2 + 2px - 3p^2 = 0$$
$$(x + 3p)(x - p) = 0$$

$$x = -3p \qquad \text{or} \qquad x = p$$

Because $y^2 \geq 0$, from Eq. (2) we see that $px \geq 0$. Thus, $x = -3p$ is impossible, so $x = p$ is the only solution possible. If $x = p$ in Eq. (2), the result is $y = \pm 2p$. Therefore, $(p, 2p)$ and $(p, -2p)$ are the points of intersection of the parabola and the circle.

40. the arch of a bridge is in the shape of a semiellipse having a horizontal span of 40 m and a height of 16 m at its center. How high is the arch 9 m to the right or left of the center?

SOLUTION: Take the origin at the center of the ellipse with x-axis along the horizontal base of the arch, as shown in Fig. 13. 40R. An equation of the ellipse is of the form

$$\frac{x^2}{a^2} + \frac{y^2}{b^2} = 1 \qquad (1)$$

Because the length of the major axis is 40 m, we have $a = 20$. Because the length of the semiminor axis is 16 m, we have $b = 16$. Substituting these values of a and b into (1), we obtain

$$\frac{x^2}{400} + \frac{y^2}{256} = 1 \qquad (2)$$

To find the height of the arch 9 m to the right or left of the center, we let $x = 9$ in Eq. (2) and solve for y. Thus,

$$\frac{9^2}{400} + \frac{y^2}{256} = 1$$

$$y = \frac{4}{5}\sqrt{319} = 14.3$$

Hence, the height of the arch is 14.3 m at a point that is 9 m to the left or right of the center.

44. A focal chord of a conic is divided into two segments by the focus. Prove that the sum of the reciprocals of the measures of the lengths of the two segments is the same, regardless of what chord is taken. (*Hint:* Use polar coordinates.)

SOLUTION: Refer to Fig. 13. 44R. Take the pole at a focus of the conic with the corresponding directrix the line $r \cos \theta = d$. An equation of the conic is

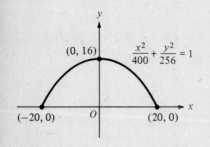

Figure 13. 40R

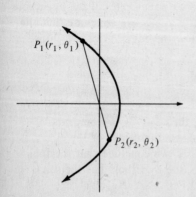

Figure 13. 44R

$$r = \frac{ed}{1 + e \cos \theta} \tag{1}$$

Let $P_1 P_2$ be any focal chord through the pole and chose polar coordinates $P_1 = (r_1, \theta_1)$ and $P_2 = (r_2, \theta_2)$ so that $r_1 > 0$ and $r_2 > 0$. Because point P_1 lies on the conic, its coordinates must satisfy (1). Thus

$$r_1 = \frac{ed}{1 + e \cos \theta_1}$$

and

$$\frac{1}{r_1} = \frac{1 + e \cos \theta_1}{ed} \tag{2}$$

Because point P_2 lies on the conic, its coordinates must satisfy (1). Moreover, because segment $P_1 P_2$ contains the pole, then $\theta_2 = \theta_1 - \pi$. Thus,

$$r_2 = \frac{ed}{1 + e \cos \theta_2}$$

$$= \frac{ed}{1 + e \cos(\theta_1 - \pi)}$$

$$= \frac{ed}{1 - e \cos \theta_1}$$

and

$$\frac{1}{r_2} = \frac{1 - e \cos \theta_1}{ed} \tag{3}$$

By adding the corresponding members of (2) and (3), we obtain

$$\frac{1}{r_1} + \frac{1}{r_2} = \frac{2}{ed}$$

Thus, the sum of the reciprocals of the segments of the focal chord is always $2/ed$, a constant.

48. The graph of the equation $(1 - e^2)x^2 + y^2 - 2px + p^2 = 0$ is a central conic having eccentricity e and a focus at $(p, 0)$. Simplify the equation by a translation of axes. Find the new origin with respect to the x- and y-axes, and determine the type of conic.

SOLUTION: If $e = 1$, the given equation reduces to

$$y^2 - 2px + p^2 = 0$$
$$y^2 = 2p(x - \tfrac{1}{2}p)$$

which is an equation of the parabola with vertex at the point $(\tfrac{1}{2}p, 0)$ and axis along the x-axis. A translation of the origin to the point $(\tfrac{1}{2}p, 0)$ reduces the equation to

$$\bar{y}^2 = 2p\bar{x}.$$

If $e \neq 1$, we complete the square on terms in x and find a standard form of the equation.

$$(1 - e^2)x^2 - 2px + y^2 = -p^2$$

$$(1 - e^2)\left[x^2 - \frac{2px}{1 - e^2} + \left(\frac{p}{1 - e^2} \right)^2 \right] + y^2 = -p^2 + \frac{p^2}{1 - e^2}$$

$$(1 - e^2)\left(x - \frac{p}{1 - e^2} \right)^2 + y^2 = \frac{p^2 e^2}{1 - e^2}$$

$$\frac{\left(x - \dfrac{p}{1-e^2}\right)^2}{\left(\dfrac{pe}{1-e^2}\right)^2} + \frac{y^2}{\dfrac{(pe)^2}{1-e^2}} = 1$$

Translating the origin to the point $(p/(1 - e^2),\ 0)$, we obtain

$$\frac{\bar{x}^2}{\left(\dfrac{pe}{1-e^2}\right)^2} + \frac{\bar{y}^2}{\dfrac{(pe)^2}{1-e^2}} = 1$$

If $|e| > 1$, then the coefficients of $\bar{x}^2$ and $\bar{y}^2$ are opposite in sign, so the graph is a hyperbola. If $|e| < 1$ and $e \neq 0$, the coefficients of $\bar{x}^2$ and $\bar{y}^2$ are both positive, so the graph is an ellipse.

CHAPTER 14 | Indeterminate Forms, Improper Integrals, and Taylor's Formula

14.1 THE INDETERMINATE FORM 0/0

By Limit Theorem 9, if

$$\lim_{x \to a} f(x) = 0 \quad \text{and} \quad \lim_{x \to a} g(x) = c \neq 0$$

then

$$\lim_{x \to a} \frac{f(x)}{g(x)} = 0$$

By Limit Theorem 12, if

$$\lim_{x \to a} f(x) = c \neq 0 \quad \text{and} \quad \lim_{x \to a} g(x) = 0$$

then

$$\lim_{x \to a} \frac{f(x)}{g(x)} = \pm \infty$$

14.1.1 Definition   If f and g are two functions such that $\lim_{x \to a} f(x) = 0$ and $\lim_{x \to a} g(x) = 0$, then the function f/g has the *indeterminate form* 0/0 *at* a.

We now have two forms of L'Hôpital's rule that can sometimes be used to find the limit of $f(x)/g(x)$ if f/g has the indeterminate form 0/0.

14.1.2 Theorem   (L'Hôpital's rule) Let f and g be functions that are differentiable on an open interval I, except possibly at the number a in I. Suppose that for all $x \neq a$ in I, $g'(x) \neq 0$. Then if $\lim_{x \to a} f(x) = 0$ and $\lim_{x \to a} g(x) = 0$, and if

$$\lim_{x \to a} \frac{f'(x)}{g'(x)} = L$$

it follows that

$$\lim_{x \to a} \frac{f(x)}{g(x)} = L$$

The theorem is valid if all the limits are right-hand limits or all the limits are left-hand limits.

14.1.4 Theorem (L'Hôpital's rule) Let f and g be functions that are differentiable for all $x > N$, where N is a positive constant, and suppose that for all $x > N$, $g'(x) \neq 0$. Then if $\lim\limits_{x \to +\infty} f(x) = 0$ and $\lim\limits_{x \to +\infty} g(x) = 0$ and if

$$\lim_{x \to +\infty} \frac{f'(x)}{g'(x)} = L$$

it follows that

$$\lim_{x \to +\infty} \frac{f(x)}{g(x)} = L$$

The theorem is also valid if "$x \to +\infty$" is replaced by "$x \to -\infty$."

The proof of L'Hôpital's rule depends on the following theorem.

14.1.3 Theorem (Cauchy's mean-value theorem) If f and g are two functions such that

  (i) f and g are continuous on the closed interval $[a, b]$
  (ii) f and g are differentiable on the open interval (a, b)
  (iii) for all x in the open interval (a, b), $g'(x) \neq 0$

then there exists a number z in the open interval (a, b) such that

$$\frac{f(b) - f(a)}{g(b) - g(a)} = \frac{f'(z)}{g'(z)}$$

Exercises 14.1

4. Find all values of z in the given interval (a, b) satisfying the conclusion of Theorem 14.1.3 for the given pair of functions.

$$f(x) = \cos 2x, \ g(x) = \sin x; \ (a, b) = (0, \tfrac{1}{2}\pi)$$

SOLUTION:

$$\frac{f(b) - f(a)}{g(b) - g(a)} = \frac{\cos \pi - \cos 0}{\sin \tfrac{1}{2}\pi - \sin 0} = -2$$

$$\frac{f'(x)}{g'(x)} = \frac{-2 \sin 2x}{\cos x}$$

Thus, we want to find z in $(0, \tfrac{1}{2}\pi)$ such that

$$\frac{-2 \sin 2z}{\cos z} = -2$$

$$\frac{2 \sin z \cos z}{\cos z} = 1$$

$$\sin z = \tfrac{1}{2}$$

Hence, $z = \frac{1}{6}\pi$.

In Exercises 6–30 evaluate the limit, if it exists.

8. $\displaystyle\lim_{x\to 0}\frac{\tan x - x}{x - \sin x}$

SOLUTION: Because $\displaystyle\lim_{x\to 0}(\tan x - x) = 0$ and $\displaystyle\lim_{x\to 0}(x - \sin x) = 0$, we apply L'Hôpital's rule. Thus,

$$\lim_{x\to 0}\frac{\tan x - x}{x - \sin x} = \lim_{x\to 0}\frac{\sec^2 x - 1}{1 - \cos x} \tag{1}$$

Because $\displaystyle\lim_{x\to 0}(\sec^2 x - 1) = 0$ and $\displaystyle\lim_{x\to 0}(1 - \cos x) = 0$, we apply L'Hôpital's rule to the right side of (1). Thus,

$$\lim_{x\to 0}\frac{\tan x - x}{x - \sin x} = \lim_{x\to 0}\frac{2\sec^2 x \tan x}{\sin x}$$

$$= \lim_{x\to 0}\frac{\dfrac{2}{\cos^2 x}\cdot\dfrac{\sin x}{\cos x}}{\sin x}$$

$$= \lim_{x\to 0}\frac{2}{\cos^3 x} = 2$$

12. $\displaystyle\lim_{x\to 0}\frac{e^x - \cos x}{x\sin x}$

SOLUTION: Because $\displaystyle\lim_{x\to 0}(e^x - \cos x) = 0$ and $\displaystyle\lim_{x\to 0} x\sin x = 0$, we apply L'Hôpital's rule. Thus,

$$\lim_{x\to 0}\frac{e^x - \cos x}{x\sin x} = \lim_{x\to 0}\frac{e^x + \sin x}{x\cos x + \sin x} \tag{1}$$

Because $\displaystyle\lim_{x\to 0}(e^x + \sin x) = 1$ and $\displaystyle\lim_{x\to 0}(x\cos x + \sin x) = 0$, we conclude that the limit in (1) is either $+\infty$ or $-\infty$. Because $x\cos x + \sin x$ approaches zero through positive values as $x\to 0^+$, we have

$$\lim_{x\to 0^+}\frac{e^x - \cos x}{x\sin x} = +\infty$$

Because $x\cos x + \sin x$ approaches zero through negative values as $x\to 0^-$, we have

$$\lim_{x\to 0^-}\frac{e^x - \cos x}{x\sin x} = -\infty$$

16. $\displaystyle\lim_{t\to 2}\frac{t^n - 2^n}{t - 2}$

SOLUTION: Because the hypothesis of L'Hôpital's rule is satisfied, we have

$$\lim_{t\to 2}\frac{t^n - 2^n}{t - 2} = \lim_{t\to 2}\frac{nt^{n-1}}{1}$$

$$= n\cdot 2^{n-1}$$

20. $\lim\limits_{y \to 0} \dfrac{y^2}{1 - \cosh y}$

SOLUTION: The hypothesis of L'Hôpital's rule is satisfied. Thus, two applications of the rule yield

$$\lim_{y \to 0} \frac{y^2}{1 - \cosh y} = \lim_{y \to 0} \frac{2y}{-\sinh y}$$

$$= \lim_{y \to 0} \frac{2}{-\cosh y} = -2$$

24. $\lim\limits_{x \to 0} \dfrac{\tanh 2x}{\tanh x}$

SOLUTION: Applying L'Hôpital's rule, we obtain

$$\lim_{x \to 0} \frac{\tanh 2x}{\tanh x} = \lim_{x \to 0} \frac{2 \operatorname{sech}^2 2x}{\operatorname{sech} x} = 2$$

28. $\lim\limits_{x \to +\infty} \dfrac{\dfrac{1}{x^2} - 2 \tan^{-1} \dfrac{1}{x}}{\dfrac{1}{x}}$

SOLUTION: Because

$$\lim_{x \to +\infty} \left(\frac{1}{x^2} - 2 \tan^{-1} \frac{1}{x} \right) = 0 \quad \text{and} \quad \lim_{x \to +\infty} \frac{1}{x} = 0$$

We apply L'Hôpital's rule. Thus,

$$\lim_{x \to +\infty} \frac{\dfrac{1}{x^2} - 2 \tan^{-1} \dfrac{1}{x}}{\dfrac{1}{x}} = \lim_{x \to +\infty} \frac{-\dfrac{2}{x^3} + \dfrac{\dfrac{2}{x^2}}{1 + \dfrac{1}{x^2}}}{-\dfrac{1}{x^2}}$$

$$= \lim_{x \to +\infty} \left[\frac{2}{x} - \frac{2}{1 + \dfrac{1}{x^2}} \right] = -2$$

32. In a geometric progression, if a is the first term, r is the common ratio of two successive terms, and S is the sum of the first n terms, then if $r \neq 1$,

$$S = \frac{a(r^n - 1)}{r - 1}$$

Find $\lim\limits_{r \to 1} S$. Is the result consistent with the sum of the first n terms if $r = 1$?

SOLUTION: Because the hypothesis of L'Hôpital's rule is satisfied, we have

$$\lim_{r \to 1} S = \lim_{r \to 1} \frac{a(r^n - 1)}{r - 1}$$

$$= \lim_{r \to 1} \frac{anr^{n-1}}{1}$$

$$= an \tag{1}$$

If $r = 1$, then each term is equal to the first term a. Hence, the sum of the first n terms is

$$a + a + a + \cdots + a = na$$

Thus, $\lim\limits_{r \to 1} S$ is consistent with the sum of the first n terms if $r = 1$.

36. Suppose that f is a function defined for all $x > N$, where N is a positive constant. If $t = 1/x$ and $F(t) = f(1/t)$, where $t \neq 0$, prove that the statements $\lim\limits_{x \to +\infty} f(x) = M$ and $\lim\limits_{t \to 0^+} F(t) = M$ have the same meaning.

SOLUTION: Because $F(t) = f(1/t)$ when $t \neq 0$, we have

$$\lim_{t \to 0^+} F(t) = \lim_{t \to 0^+} f\left(\frac{1}{t}\right) \tag{1}$$

If $t = 1/x$, then $x = 1/t$. Thus, $t \to 0^+$ is equivalent to $x \to +\infty$. Hence,

$$\lim_{t \to 0^+} f\left(\frac{1}{t}\right) = \lim_{x \to +\infty} f(x) \tag{2}$$

From (1) and (2) we conclude that

$$\lim_{t \to 0^+} F(t) = \lim_{x \to +\infty} f(x)$$

Thus, the given equations have the same meaning.

14.2 OTHER INDETERMINATE FORMS

We may also use L'Hôpital's rule when both the numerator and denominator of a fraction have limits $\pm\infty$.

14.2.1 Theorem (L'Hôpital's rule) Let f and g be functions that are differentiable on an open interval I, except possibly at the number a in I, and suppose that for all $x \neq a$ in I, $g'(x) \neq 0$. Then if $\lim\limits_{x \to a} f(x) = +\infty$ or $-\infty$, and $\lim\limits_{x \to a} g(x) = +\infty$ or $-\infty$, and if

$$\lim_{x \to a} \frac{f'(x)}{g'(x)} = L$$

it follows that

$$\lim_{x \to a} \frac{f(x)}{g(x)} = L$$

The theorem is valid if all the limits are right-hand limits or if all the limits are left-hand limits.

14.2.2 Theorem (L'Hôpital's rule) Let f and g be functions that are differentiable for all $x > N$, where N is a positive constant, and suppose that for all $x > N$, $g'(x) \neq 0$. Then if $\lim\limits_{x \to +\infty} f(x) = +\infty$ or $-\infty$, and $\lim\limits_{x \to +\infty} g(x) = +\infty$ or $-\infty$, and if

$$\lim_{x \to +\infty} \frac{f'(x)}{g'(x)} = L$$

it follows that

$$\lim_{x \to +\infty} \frac{f(x)}{g(x)} = L$$

The theorem is also valid if "$x \to +\infty$" is replaced by "$x \to -\infty$."

For the indeterminate forms $0 \cdot (\pm\infty)$ and $+\infty - (+\infty)$, we must first write the given expression in the indeterminate form $0/0$ or $\pm\infty/\pm\infty$. This is illustrated in Exercises 8 and 16. For the indeterminate forms 0^0, $(\pm\infty)^0$, or $1^{\pm\infty}$, we introduce the ln function in order to obtain the indeterminate form $0/0$ or $\pm\infty/\pm\infty$. This is illustrated in Exercises 12, 20, 24, 28, and 32.

Exercises 14.2

In Exercises 1–34 evaluate the limit, if it exists.

4. $\lim_{x \to +\infty} \dfrac{(\ln x)^3}{x}$

SOLUTION: Because $\lim\limits_{x \to +\infty} (\ln x)^3 = +\infty$ and $\lim\limits_{x \to +\infty} x = +\infty$, we apply L'Hôpital's rule. Thus,

$$\lim_{x \to +\infty} \frac{(\ln x)^3}{x} = \lim_{x \to +\infty} \frac{3(\ln x)^2 \cdot \dfrac{1}{x}}{1}$$

$$= \lim_{x \to +\infty} \frac{3(\ln x)^2}{x} \tag{1}$$

Because the hypothesis of L'Hôpital's rule is satisfied by the function defined on the right side of (1), we apply the rule again to obtain

$$\lim_{x \to +\infty} \frac{(\ln x)^3}{x} = \lim_{x \to +\infty} \frac{6 \ln x}{x} \tag{2}$$

Applying L'Hôpital's rule to the right-hand side of (2), we obtain

$$\lim_{x \to +\infty} \frac{(\ln x)^3}{x} = \lim_{x \to +\infty} \frac{6}{x}$$

$$= 0$$

8. $\lim\limits_{x \to 0^+} \tan^{-1} x \cot x$

SOLUTION: Because $\lim\limits_{x \to 0^+} \tan^{-1} x = 0$ and $\lim\limits_{x \to 0^+} \cot x = +\infty$, we cannot apply L'Hôpital's rule in the present form. However, we may apply L'Hôpital's rule as follows.

$$\lim_{x \to 0^+} \tan^{-1} x \cot x = \lim_{x \to 0^+} \frac{\tan^{-1} x}{\tan x}$$

$$= \lim_{x \to 0^+} \frac{\dfrac{1}{1 + x^2}}{\sec^2 x} = 1$$

12. $\lim\limits_{x \to 0^+} (\sinh x)^{\tan x}$

SOLUTION: Because $\lim\limits_{x\to0^+}\sinh x=0$ and $\lim\limits_{x\to0^+}\tan x=0$, we have the indeterminate form 0^0. We note that if $x>0$, then $\sinh x>0$, so $\ln(\sinh x)$ is defined when $x>0$ and we may let

$$y=(\sinh x)^{\tan x}$$

and

$$\ln y=\ln(\sinh x)^{\tan x}$$

$$=\tan x\ln(\sinh x)$$

$$=\frac{\ln(\sinh x)}{\cot x}$$

Because $\lim\limits_{x\to0^+}\ln(\sinh x)=-\infty$ and $\lim\limits_{x\to0^+}\cot x=+\infty$, we may now apply L'Hôpital's rule as follows.

$$\lim_{x\to0^+}\ln y=\lim_{x\to0^+}\frac{\ln(\sinh x)}{\cot x}$$

$$=\lim_{x\to0^+}\frac{\dfrac{\cosh x}{\sinh x}}{-\csc^2 x}$$

$$=\lim_{x\to0^+}\frac{-\sin^2 x}{\tanh x} \tag{1}$$

Because $\lim\limits_{x\to0^+}\sin^2 x=0$ and $\lim\limits_{x\to0^+}\tanh x=0$, we may apply L'Hôpital's rule again. Thus,

$$\lim_{x\to0^+}\frac{-\sin^2 x}{\tanh x}=\lim_{x\to0^+}\frac{-2\sin x\cos x}{\operatorname{sech}^2 x}=0 \tag{2}$$

Substituting from (2) into (1), we have

$$\lim_{x\to0^+}\ln y=0$$

or, equivalently,

$$\ln(\lim_{x\to0^+}y)=0$$

$$\lim_{x\to0^+}y=e^0=1$$

Therefore,

$$\lim_{x\to0^+}(\sinh x)^{\tan x}=1$$

16. $\lim\limits_{x\to2}\left(\dfrac{5}{x^2+x-6}-\dfrac{1}{x-2}\right)$

SOLUTION:

$$\lim_{x\to2}\frac{5}{x^2+x-6}=\lim_{x\to2}\frac{5}{(x+3)(x-2)}=\pm\infty$$

$$\lim_{x\to2}\frac{1}{x-2}=\pm\infty$$

Thus, we have an indeterminate form for which L'Hôpital's rule does not apply. However,

$$\lim_{x \to 2} \left(\frac{5}{x^2 + x - 6} - \frac{1}{x - 2} \right) = \lim_{x \to 2} \frac{-x + 2}{(x + 3)(x - 2)}$$

$$= \lim_{x \to 2} \frac{-1}{x + 3} = -\frac{1}{5}$$

20. $\displaystyle\lim_{x \to 0^+} x^{1/\ln x}$

SOLUTION: Because $\displaystyle\lim_{x \to 0^+} \ln x = -\infty$, then

$$\lim_{x \to 0^+} \frac{1}{\ln x} = 0$$

Thus, we have the indeterminate form 0^0. Let

$$y = x^{1/\ln x}$$

Then

$$\ln y = \ln x^{1/\ln x}$$

$$= \frac{1}{\ln x} \cdot \ln x = 1$$

Hence,

$$y = e$$

Thus,

$$x^{1/\ln x} = e$$

and

$$\lim_{x \to 0^+} x^{1/\ln x} = e$$

24. $\displaystyle\lim_{x \to +\infty} \left(1 + \frac{1}{2x} \right)^{x^2}$

SOLUTION: Because we have the indeterminate form $1^{+\infty}$, we let

$$y = \left(1 + \frac{1}{2x} \right)^{x^2}$$

Then

$$\ln y = \ln \left(1 + \frac{1}{2x} \right)^{x^2}$$

$$= x^2 \ln \left(1 + \frac{1}{2x} \right)$$

$$= \frac{\ln \left(1 + \dfrac{1}{2x} \right)}{x^{-2}} \tag{1}$$

Because

$$\lim_{x \to +\infty} \ln \left(1 + \frac{1}{2x} \right) = 0 \quad \text{and} \quad \lim_{x \to +\infty} x^{-2} = 0$$

we use L'Hôpital's rule on the right side of (1). Thus,

$$\lim_{x \to +\infty} \ln y = \lim_{x \to +\infty} \frac{\ln\left(1 + \dfrac{1}{2x}\right)}{x^{-2}}$$

$$= \lim_{x \to +\infty} \frac{\dfrac{-\frac{1}{2}x^{-2}}{1 + \frac{1}{2}x^{-1}}}{-2x^{-3}}$$

$$= \lim_{x \to +\infty} \frac{\frac{1}{4}x}{1 + \frac{1}{2}x^{-1}}$$

$$= \lim_{x \to +\infty} \frac{\frac{1}{4}}{x^{-1} + \frac{1}{2}x^{-2}} = +\infty$$

Thus,

$$\lim_{x \to +\infty} y = +\infty$$

or, equivalently,

$$\lim_{x \to +\infty} \left(1 + \frac{1}{2x}\right)^{x^2} = +\infty$$

28. $\displaystyle\lim_{x \to 0} (\cos x)^{1/x^2}$

SOLUTION: Because

$$\lim_{x \to 0} \cos x = 1 \quad \text{and} \quad \lim_{x \to 0} \frac{1}{x^2} = +\infty$$

we have the indeterminate form $1^{+\infty}$. Thus, we let

$$y = (\cos x)^{1/x^2}$$

$$\ln y = \ln(\cos x)^{1/x^2} = \frac{\ln(\cos x)}{x^2} \tag{1}$$

Because

$$\lim_{x \to 0} \ln(\cos x) = 0 \quad \text{and} \quad \lim_{x \to 0} x^2 = 0,$$

we use L'Hôpital's rule on the right-hand side of (1). Thus,

$$\lim_{x \to 0} \ln y = \lim_{x \to 0} \frac{\ln(\cos x)}{x^2} = \lim_{x \to 0} \frac{-\sin x}{2x \cos x} \tag{2}$$

We use L'Hôpital's rule again on the right-hand side of (2). Thus,

$$\lim_{x \to 0} \ln y = \lim_{x \to 0} \frac{-\cos x}{-2x \sin x + 2 \cos x} = -\frac{1}{2}$$

Thus,

$$\ln(\lim_{x \to 0} y) = -\frac{1}{2}$$

$$\lim_{x \to 0} y = e^{-1/2}$$

or, equivalently,

$$\lim_{x \to 0} (\cos x)^{1/x^2} = e^{-1/2}$$

32. $\lim\limits_{x \to 0^+} x^{x^x}$

SOLUTION: First we find $\lim\limits_{x \to 0^+} x^x$. Let

$$y = x^x$$

$$\ln y = x \ln x$$

$$\lim\limits_{x \to 0^+} \ln y = \lim\limits_{x \to 0^+} x \ln x$$

$$= \lim\limits_{x \to 0^+} \frac{\ln x}{\dfrac{1}{x}} \tag{1}$$

Because

$$\lim\limits_{x \to 0^+} \ln x = -\infty \quad \text{and} \quad \lim\limits_{x \to 0^+} \frac{1}{x} = +\infty$$

we apply L'Hôpital's rule on the right-hand side of (1). Thus,

$$\lim\limits_{x \to 0^+} \ln y = \lim\limits_{x \to 0^+} \frac{x^{-1}}{-x^{-2}}$$

$$= \lim\limits_{x \to 0^+} -x = 0$$

Thus,

$$\ln \lim\limits_{x \to 0^+} y = 0$$

$$\lim\limits_{x \to 0^+} y = e^0$$

or, equivalently,

$$\lim\limits_{x \to 0^+} x^x = 1 \tag{2}$$

Now let

$$z = x^{x^x}$$

Then

$$\ln z = \ln x^{x^x}$$

$$= x^x \cdot \ln x$$

$$\lim\limits_{x \to 0^+} \ln z = \lim\limits_{x \to 0^+} (x^x \cdot \ln x) \tag{3}$$

Because $\lim\limits_{x \to 0^+} \ln x = -\infty$ and because of (2), we conclude from (3) that

$$\lim\limits_{x \to 0^+} \ln z = -\infty$$

$$\ln \lim\limits_{x \to 0^+} z = -\infty$$

Thus,

$$\lim\limits_{x \to 0^+} z = 0$$

or, equivalently,

$$\lim_{x \to 0^+} x^{x^x} = 0$$

36. If $\displaystyle \lim_{x \to +\infty} \left(\frac{nx+1}{nx-1} \right)^x = 9$, find n.

SOLUTION: Let

$$y = \left(\frac{nx+1}{nx-1} \right)^x$$

Then

$$\ln y = x \ln \frac{nx+1}{nx-1}$$

$$= \frac{\ln \dfrac{nx+1}{nx-1}}{\dfrac{1}{x}} \tag{1}$$

Because

$$\lim_{x \to +\infty} \frac{nx+1}{nx-1} = \lim_{x \to +\infty} \frac{n + \dfrac{1}{x}}{n - \dfrac{1}{x}} = 1$$

then

$$\lim_{x \to +\infty} \ln \left(\frac{nx+1}{nx-1} \right) = 0$$

Furthermore, $\displaystyle \lim_{x \to +\infty} 1/x = 0$. Thus we may apply L'Hôpital's rule to the right-hand side of (1). Therefore,

$$\lim_{x \to +\infty} \ln y = \lim_{x \to +\infty} \frac{\ln \left(\dfrac{nx+1}{nx-1} \right)}{\dfrac{1}{x}}$$

$$= \lim_{x \to +\infty} \frac{\ln(nx+1) - \ln(nx-1)}{x^{-1}}$$

$$= \lim_{x \to +\infty} \frac{\dfrac{n}{nx+1} - \dfrac{n}{nx-1}}{-x^{-2}}$$

$$= \lim_{x \to +\infty} \frac{2n}{n^2 - \dfrac{1}{x^2}} = \frac{2}{n}$$

or, equivalently,

$$\lim_{x \to +\infty} \ln \left(\frac{nx+1}{nx-1} \right)^x = \frac{2}{n}$$

$$\ln \lim_{x \to +\infty} \left(\frac{nx+1}{nx-1} \right)^x = \frac{2}{n} \tag{2}$$

We are given that

$$\lim_{x \to +\infty} \left(\frac{nx+1}{nx-1} \right)^x = 9 \tag{3}$$

Substituting from (3) into (2), we get

$$\ln 9 = \frac{2}{n}$$

$$n = \frac{2}{\ln 9} = \frac{1}{\ln 3}$$

14.3 IMPROPER INTEGRALS WITH INFINITE LIMITS OF INTEGRATION

14.3.1 Definition If f is continuous for all $x \geq a$, then

$$\int_a^{+\infty} f(x) \, dx = \lim_{b \to +\infty} \int_a^b f(x) \, dx$$

if this limit exists.

14.3.2 Definition If f is continuous for all $x \leq b$, then

$$\int_{-\infty}^b f(x) \, dx = \lim_{a \to -\infty} \int_a^b f(x) \, dx$$

if this limit exists.

14.3.3 Definition If f is continuous for all values of x, then

$$\int_{-\infty}^{+\infty} f(x) \, dx = \lim_{a \to -\infty} \int_a^0 f(x) \, dx + \lim_{b \to +\infty} \int_0^b f(x) \, dx$$

if these limits exist.

The following limit theorems are helpful when applying the definitions of this section.

$$\lim_{x \to +\infty} \ln x = +\infty$$

$$\lim_{x \to +\infty} a^x = +\infty \qquad \text{if } a > 0$$

$$\lim_{x \to -\infty} a^x = 0 \qquad \text{if } a > 0$$

$$\lim_{x \to +\infty} \tan^{-1} x = \tfrac{1}{2}\pi$$

$$\lim_{x \to -\infty} \tan^{-1} x = -\tfrac{1}{2}\pi$$

Exercises 14.3

In Exercises 1–18 determine whether the improper integral is convergent or divergent. If it is convergent, evaluate it.

4. $\displaystyle\int_1^{+\infty} 2^{-x} \, dx$

SOLUTION: We use Definition 14.3.1. Thus,

$$\int_1^{+\infty} 2^{-x}\,dx = \lim_{b \to +\infty} \int_1^b 2^{-x}\,dx$$

$$= \lim_{b \to +\infty} \left[\frac{-2^{-x}}{\ln 2} \right]_1^b$$

$$= \lim_{b \to +\infty} \left[\frac{-2^{-b}}{\ln 2} + \frac{1}{2 \ln 2} \right]$$

$$= \frac{1}{2 \ln 2}$$

Thus, the integral is convergent with value $1/(2 \ln 2)$.

8. $\displaystyle \int_{-\infty}^0 x^2 e^x\,dx$

SOLUTION: We use Definition 14.3.2. Thus,

$$\int_{-\infty}^0 x^2 e^x\,dx = \lim_{a \to -\infty} \int_a^0 x^2 e^x\,dx \tag{1}$$

We use integration by parts. Let

$$u = x^2 \qquad dv = e^x\,dx$$
$$du = 2x\,dx \qquad v = e^x$$

Thus,

$$\int x^2 e^x\,dx = x^2 e^x - \int 2x e^x\,dx \tag{2}$$

We integrate by parts again. Let

$$\bar{u} = 2x \qquad d\bar{v} = e^x\,dx$$
$$d\bar{u} = 2\,dx \qquad \bar{v} = e^x$$

Then from (2) we obtain

$$\int x^2 e^x\,dx = x^2 e^x - 2x e^x + \int 2 e^x\,dx$$

$$= x^2 e^x - 2x e^x + 2 e^x + C \tag{3}$$

Substituting from (3) into (1), we obtain

$$\int_{-\infty}^0 x^2 e^x\,dx = \lim_{a \to -\infty} \left[e^x(x^2 - 2x + 2) \right]_a^0$$

$$= \lim_{a \to -\infty} [2 - e^a(a^2 - 2a + 2)]$$

$$= 2 - \lim_{a \to -\infty} \frac{(a-1)^2 + 1}{e^{-a}} \tag{4}$$

Because $\lim\limits_{a \to -\infty} [(a-1)^2 + 1] = +\infty$ and $\lim\limits_{a \to -\infty} e^{-a} = +\infty$, we may use L'Hôpital's rule for the limit in (4). Thus,

$$\lim_{a \to -\infty} \frac{(a-1)^2 + 1}{e^{-a}} = \lim_{a \to -\infty} \frac{2(a-1)}{-e^{-a}} \tag{5}$$

Because $\lim\limits_{a \to -\infty} 2(a-1) = -\infty$ and $\lim\limits_{a \to -\infty} -e^{-a} = -\infty$, we may use L'Hôpital's rule again on the right-hand side of (5). Thus,

$$\lim_{a \to -\infty} \frac{(a-1)^2 + 1}{e^{-a}} = \lim_{a \to -\infty} \frac{2}{e^{-a}}$$

$$= \lim_{a \to -\infty} 2e^{a}$$

$$= 0 \tag{6}$$

Substituting from (6) into (4), we obtain

$$\int_{-\infty}^{0} x^2 e^x \, dx = 2$$

Thus, the integral is convergent and has value 2.

12. $\displaystyle \int_{e}^{+\infty} \frac{dx}{x \ln x}$

By Definition 14.3.1, we have

$$\int_{e}^{+\infty} \frac{dx}{x \ln x} = \lim_{b \to +\infty} \int_{e}^{b} \frac{dx}{x \ln x}$$

$$= \lim_{b \to +\infty} \ln(\ln x) \Big]_{e}^{b}$$

$$= \lim_{b \to +\infty} [\ln(\ln b) - \ln 1] = +\infty$$

Thus, the integral is divergent.

16. $\displaystyle \int_{-\infty}^{+\infty} \frac{dx}{16 + x^2}$

SOLUTION: We use Definition 14.3.3. Thus,

$$\int_{-\infty}^{+\infty} \frac{dx}{16 + x^2} = \lim_{a \to -\infty} \int_{a}^{0} \frac{dx}{16 + x^2} + \lim_{b \to +\infty} \int_{0}^{b} \frac{dx}{16 + x^2}$$

$$= \lim_{a \to -\infty} \left[\frac{1}{4} \tan^{-1} \frac{x}{4} \right]_{a}^{0} + \lim_{b \to +\infty} \left[\frac{1}{4} \tan^{-1} \frac{x}{4} \right]_{0}^{b}$$

$$= \lim_{a \to -\infty} \left[-\frac{1}{4} \tan^{-1} \frac{a}{4} \right] + \lim_{b \to +\infty} \left[\frac{1}{4} \tan^{-1} \frac{b}{4} \right]$$

$$= -\frac{1}{4} \left(-\frac{1}{2} \pi \right) + \frac{1}{4} \left(\frac{1}{2} \pi \right) = \frac{1}{4} \pi$$

Thus, the integral is convergent and has value $\frac{1}{4}\pi$.

20. Prove that if $\displaystyle \int_{-\infty}^{b} f(x) \, dx$ is convergent, then $\displaystyle \int_{-b}^{+\infty} f(-x) \, dx$ is also convergent and has the same value.

SOLUTION: We are given that

$$\int_{-\infty}^{b} f(x) \, dx = \lim_{a \to -\infty} \int_{a}^{b} f(x) \, dx \tag{1}$$

exists. Furthermore,

$$\int_{-b}^{+\infty} f(-x)\,dx = \lim_{c \to +\infty} \int_{-b}^{c} f(-x)\,dx \qquad (2)$$

We let $z = -x$. Then $dz = -dx$; when $x = -b$, $z = b$; and when $x = c$, $z = -c$. Substituting in the right-hand side of (2), we obtain

$$\int_{-b}^{+\infty} f(-x)\,dx = \lim_{c \to +\infty} \int_{b}^{-c} f(z)(-dz)$$

$$= \lim_{c \to +\infty} \int_{-c}^{b} f(z)\,dz$$

$$= \lim_{c \to -\infty} \int_{c}^{b} f(x)\,dx \qquad (3)$$

Comparing (1) with (3), we conclude that $\int_{-b}^{+\infty} f(-x)\,dx$ is convergent and has the same value as $\int_{-\infty}^{b} f(x)\,dx$.

24. Determine whether it is possible to assign a finite number to represent the measure of the area of the region bounded by the x-axis, the line $x = 2$, and the curve whose equation is $y = 1/(x^2 - 1)$. If a finite number can be assigned, find it.

SOLUTION: Let f be the function defined by

$$f(x) = \frac{1}{x^2 - 1}$$

We find the area of the region R, bounded above by the curve $y = f(x)$, bounded below by the x-axis, bounded on the left by the line $x = 2$, and bounded on the right by the line $x = b$, where $b > 2$ as illustrated in Fig. 14. 3. 24. The number of square units in the area of R is given by

$$A = \lim_{\|\Delta\| \to 0} \sum_{i=1}^{n} f(\bar{x}_i)\,\Delta_i x$$

$$= \int_{2}^{b} \frac{dx}{x^2 - 1}$$

We use partial fractions for this integral. Thus,

$$A = \frac{1}{2} \int_{2}^{b} \frac{dx}{x - 1} - \frac{1}{2} \int_{2}^{b} \frac{dx}{x + 1}$$

$$= \frac{1}{2} \ln \frac{x - 1}{x + 1} \Big]_{2}^{b}$$

$$= \frac{1}{2} \left[\ln \frac{b - 1}{b + 1} - \ln \frac{1}{3} \right]$$

Then,

$$\lim_{b \to +\infty} A = \lim_{b \to +\infty} \frac{1}{2} \left[\ln \frac{b - 1}{b + 1} + \ln 3 \right]$$

$$= \frac{1}{2} [\ln 1 + \ln 3] = \frac{1}{2} \ln 3$$

Because the above limit exists, the area is $\frac{1}{2} \ln 3$ square units.

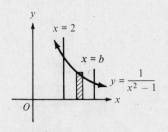

Figure 14. 3. 24

28. The continuous flow of profit for a company is increasing with time, and at t years the number of dollars in the profit per year is proportional to t. Show that the present value of the company is inversely proportional to i^2, where $100i$ percent is the interest rate compounded continuously.

SOLUTION: Let f be the function giving the number of dollars in the profit per year. We are given that $f(t) = kt$, with $k > 0$, because the profit per year is proportional to t. Thus, if V dollars is the present value of the company, then

$$V = \int_0^{+\infty} f(t)e^{-it}\,dt$$

$$= \int_0^{+\infty} kte^{-it}\,dt$$

$$= k \lim_{b \to +\infty} \int_0^b te^{-it}\,dt \qquad (1)$$

We integrate by parts with

$$u = t \qquad dv = e^{-it}\,dt$$

$$du = dt \qquad v = -\frac{1}{i}e^{-it}$$

Thus,

$$\int te^{-it}\,dt = -\frac{t}{i}e^{-it} + \frac{1}{i}\int e^{-it}\,dt$$

$$= -\frac{t}{i}e^{-it} - \frac{1}{i^2}e^{-it}$$

$$= -\frac{it+1}{i^2 e^{it}} \qquad (2)$$

Substituting the value of the indefinite integral given in (2) into (1), we obtain

$$V = k \lim_{b \to +\infty} \left(-\frac{it+1}{i^2 e^{it}}\right)\Bigg]_0^b$$

$$= -\frac{k}{i^2}\left[\lim_{b \to +\infty}\frac{ib+1}{e^{ib}} - 1\right] \qquad (3)$$

Because $\lim_{b \to +\infty}(ib+1) = +\infty$ and $\lim_{b \to +\infty} e^{ib} = +\infty$, we apply L'Hôpital's rule to find the limit in (3). We obtain

$$\lim_{b \to +\infty}\frac{ib+1}{e^{ib}} = \lim_{b \to +\infty}\frac{i}{ie^{ib}}$$

$$= 0 \qquad (4)$$

Substituting from (4) into (3), we get

$$V = -\frac{k}{i^2}[0 - 1] = \frac{k}{i^2}$$

Thus, the present value of the company is inversely proportional to i^2.

32. Determine a value of n for which the improper integral is convergent and evaluate the integral for this value of n.

$$\int_1^{+\infty}\left(\frac{nx^2}{x^3+1} - \frac{1}{3x+1}\right)dx$$

SOLUTION:

$$\int_1^{+\infty} \left(\frac{nx^2}{x^3+1} - \frac{1}{3x+1} \right) dx = \lim_{b\to+\infty} \int_1^b \left(\frac{nx^2}{x^3+1} - \frac{1}{3x+1} \right) dx$$

$$= \lim_{b\to+\infty} \left[\frac{n}{3} \ln(x^3+1) - \frac{1}{3} \ln(3x+1) \right]_1^b$$

$$= \frac{1}{3} \lim_{b\to+\infty} \left(\ln \frac{(x^3+1)^n}{3x+1} \right) \Big]_1^b$$

$$= \frac{1}{3} \lim_{b\to+\infty} \left[\ln \frac{(b^3+1)^n}{3b+1} - \ln \left(\frac{2^n}{4} \right) \right]$$

$$= \frac{1}{3} \ln \left[\lim_{b\to+\infty} \frac{(b^3+1)^n}{3b+1} \right] - \frac{1}{3} \ln \left(\frac{2^n}{4} \right) \qquad (1)$$

If $n = \frac{1}{3}$, then

$$\lim_{b\to+\infty} \frac{(b^3+1)^n}{3b+1} = \lim_{b\to+\infty} \frac{\sqrt[3]{b^3+1}}{3b+1}$$

$$= \lim_{b\to+\infty} \frac{\sqrt[3]{1 + \dfrac{1}{b^3}}}{3 + \dfrac{1}{b}} = \frac{1}{3} \qquad (2)$$

Substituting from (2) into (1), we have if $n = \frac{1}{3}$, then

$$\int_1^{+\infty} \left(\frac{nx^2}{x^3+1} - \frac{1}{3x+1} \right) dx = \frac{1}{3} \ln \frac{1}{3} - \frac{1}{3} \ln \frac{\sqrt[3]{2}}{4}$$

$$= \frac{1}{3} \ln \frac{4}{3\sqrt[3]{2}}$$

$$= \frac{1}{9} \ln \left(\frac{4}{3\sqrt[3]{2}} \right)^3$$

$$= \frac{1}{9} \ln \frac{32}{27}$$

14.4 OTHER IMPROPER INTEGRALS The definite integral $\int_a^b f(x)\, dx$ is improper if f is discontinuous on the closed interval $[a, b]$. We consider the three cases covered by the following definitions.

14.4.1 Definition If f is continuous at all x in the interval half open on the left $(a, b]$, and if $\lim_{x\to a^+} f(x) = \pm\infty$, then

$$\int_a^b f(x)\, dx = \lim_{\epsilon\to 0^+} \int_{a+\epsilon}^b f(x)\, dx$$

if this limit exists.

14.4.2 Definition If f is continuous at all x in the interval half open on the right $[a, b)$ and if $\lim_{x\to b^-} f(x) = \pm\infty$, then

$$\int_a^b f(x)\, dx = \lim_{\epsilon\to 0^+} \int_a^{b-\epsilon} f(x)\, dx$$

if this limit exists.

14.4.3 Definition If f is continuous at all x in the interval $[a, b]$, except c when $a < c < b$, and if $\lim_{x\to c} |f(x)| = +\infty$, then

$$\int_a^b f(x)\,dx = \lim_{\epsilon \to 0^+} \int_a^{c-\epsilon} f(x)\,dx + \lim_{\delta \to 0^+} \int_{c-\delta}^b f(x)\,dx$$

if these limits exist.

If $\int_a^b f(x)\,dx$ is an improper integral, it is convergent if the corresponding limit exists; otherwise, it is divergent.

If there is more than one number in $[a,\ b]$ at which the function f has limit $\pm\infty$, we must express the given integral as a sum of two or more integrals of the types covered by 14.4.1 and 14.4.2. We illustrate this in Exercise 16.

Exercises 14.4

In Exercises 1–25 determine whether the improper integral is convergent or divergent. If it is convergent, evaluate it.

4. $\displaystyle\int_0^4 \frac{x\,dx}{\sqrt{16-x^2}}$

SOLUTION: Because there is an infinite discontinuity at 4, we use Definition 14.4.2. Thus,

$$\int_0^4 \frac{x\,dx}{\sqrt{16-x^2}} = \lim_{\epsilon \to 0^+} \int_0^{4-\epsilon} \frac{x\,dx}{\sqrt{16-x^2}}$$

$$= \lim_{b \to 4^-} \int_0^b \frac{x\,dx}{\sqrt{16-x^2}}$$

$$= \lim_{b \to 4^-} \left[-\sqrt{16-x^2}\ \right]_0^b$$

$$= \lim_{b \to 4^-} [-\sqrt{16-b^2} + 4] = 4$$

8. $\displaystyle\int_{-2}^0 \frac{dx}{\sqrt{4-x^2}}$

SOLUTION: Because there is an infinite discontinuity at -2, we apply Definition 14.4.1.

$$\int_{-2}^0 \frac{dx}{\sqrt{4-x^2}} = \lim_{\epsilon \to 0^+} \int_{-2+\epsilon}^0 \frac{dx}{\sqrt{4-x^2}}$$

$$= \lim_{\epsilon \to 0^+} \sin^{-1} \frac{x}{2} \bigg]_{-2+\epsilon}^0$$

$$= \lim_{\epsilon \to 0^+} \left[\sin^{-1} 0 - \sin^{-1} \left(\frac{-2+\epsilon}{2} \right) \right]$$

$$= -\sin^{-1}(-1) = \frac{1}{2}\pi$$

12. $\displaystyle\int_0^2 \frac{dx}{(x-1)^{2/3}}$

SOLUTION: Because there is an infinite discontinuity at 1, we use Definition 14.4.3. Thus,

$$\int_0^2 \frac{dx}{(x-1)^{2/3}} = \lim_{\epsilon \to 0^+} \int_0^{1-\epsilon} (x-1)^{-2/3}\, dx + \lim_{\delta \to 0^+} \int_{1+\delta}^2 (x-1)^{-2/3}\, dx$$

$$= \lim_{\epsilon \to 0^+} \left[3(x-1)^{1/3} \right]_0^{1-\epsilon} + \lim_{\delta \to 0^+} \left[3(x-1)^{1/3} \right]_{1+\delta}^2$$

$$= \lim_{\epsilon \to 0^+} [3(-\epsilon)^{1/3} + 3] + \lim_{\delta \to 0^+} [3 - 3\delta^{1/3}]$$

$$= 3 + 3 = 6$$

16. $\displaystyle \int_0^2 \frac{dx}{\sqrt{2x - x^2}}$

SOLUTION: There are infinite discontinuities at both 0 and 2. Thus, we express the given integral as a sum of two integrals each with an infinite discontinuity at one endpoint.

$$\int_0^2 \frac{dx}{\sqrt{2x - x^2}} = \int_0^1 \frac{dx}{\sqrt{2x - x^2}} + \int_1^2 \frac{dx}{\sqrt{2x - x^2}}$$

$$= \lim_{a \to 0^+} \int_a^1 \frac{dx}{\sqrt{1 - (x-1)^2}} + \lim_{b \to 2^-} \int_1^b \frac{dx}{\sqrt{1 - (x-1)^2}}$$

$$= \lim_{a \to 0^+} \left[\sin^{-1}(x-1) \right]_a^1 + \lim_{b \to 2^-} \left[\sin^{-1}(x-1) \right]_1^b$$

$$= \lim_{a \to 0^+} [-\sin^{-1}(a-1)] + \lim_{b \to 2^-} [\sin^{-1}(b-1)]$$

$$= \frac{1}{2}\pi + \frac{1}{2}\pi = \pi$$

20. $\displaystyle \int_0^{+\infty} \frac{e^{-\sqrt{x}}}{\sqrt{x}}\, dx$

SOLUTION: There is an infinite discontinuity at 0 and an infinite upper limit. Thus, we express the given integral as a sum of two improper integrals.

$$\int_0^{+\infty} \frac{e^{-\sqrt{x}}}{\sqrt{x}}\, dx = \int_0^1 \frac{e^{-\sqrt{x}}}{\sqrt{x}}\, dx + \int_1^{+\infty} \frac{e^{-\sqrt{x}}}{\sqrt{x}}\, dx$$

$$= \lim_{a \to 0^+} \int_a^1 \frac{e^{-\sqrt{x}}}{\sqrt{x}}\, dx + \lim_{b \to +\infty} \int_1^b \frac{e^{-\sqrt{x}}}{\sqrt{x}}\, dx \qquad (1)$$

To find the integral, we let $u = -\sqrt{x}$. Then $du = -\frac{1}{2}x^{-1/2}\, dx$. Thus,

$$\int \frac{e^{-\sqrt{x}}}{\sqrt{x}}\, dx = \int e^u(-2\, du)$$

$$= -2e^u + C$$

$$= -2e^{-\sqrt{x}} + C \qquad (2)$$

Substituting from (2) into (1), we obtain

$$\int_0^{+\infty} \frac{e^{-\sqrt{x}}}{\sqrt{x}}\, dx = \lim_{a \to 0^+} \left[-2e^{-\sqrt{x}} \right]_a^1 + \lim_{b \to +\infty} \left[-2e^{-\sqrt{x}} \right]_1^b$$

$$= \lim_{a \to 0^+} [-2e^{-1} + 2e^{-\sqrt{a}}] + \lim_{b \to +\infty} [-2e^{-\sqrt{b}} + 2e^{-1}]$$

$$= [-2e^{-1} + 2] + 2e^{-1} = 2$$

24. $\displaystyle\int_0^1 \frac{dx}{x\sqrt{4-x^2}}$

SOLUTION: There is an infinite discontinuity at 0. Thus

$$\int_0^1 \frac{dx}{x\sqrt{4-x^2}} = \lim_{a\to0^+} \int_a^1 \frac{dx}{x\sqrt{4-x^2}} \tag{1}$$

To find the integral we let $u = \sqrt{4-x^2}$. Then $u^2 = 4 - x^2$, and $x\,dx = -u\,du$. Thus,

$$\int \frac{dx}{x\sqrt{4-x^2}} = \int \frac{x\,dx}{x^2\sqrt{4-x^2}}$$

$$= \int \frac{-u\,du}{(4-u^2)u}$$

$$= \int \frac{du}{u^2-4}$$

$$= \frac{1}{4}\ln\left|\frac{u-2}{u+2}\right| + C$$

$$= \frac{1}{4}\ln\left|\frac{\sqrt{4-x^2}-2}{\sqrt{4-x^2}+2}\right| + C \tag{2}$$

Substituting from (2) into (1), we obtain

$$\int_0^1 \frac{dx}{x\sqrt{4-x^2}} = \lim_{a\to0^+}\left[\frac{1}{4}\ln\left|\frac{\sqrt{4-x^2}-2}{\sqrt{4-x^2}+2}\right|\right]_a^1$$

$$= \lim_{a\to0^+}\left[\frac{1}{4}\ln\left|\frac{\sqrt{3}-2}{\sqrt{3}+2}\right| - \frac{1}{4}\ln\left|\frac{\sqrt{4-a^2}-2}{\sqrt{4-a^2}+2}\right|\right] \tag{3}$$

Because

$$\lim_{a\to0^+}\frac{\sqrt{4-a^2}-2}{\sqrt{4-a^2}+2} = 0$$

then

$$\lim_{a\to0^+}\ln\left|\frac{\sqrt{4-a^2}-2}{\sqrt{4-a^2}+2}\right| = -\infty$$

Therefore, the limit in (3) does not exist. Thus, the integral is divergent.

28. Find the values of n for which the improper integral converges and evaluate the integral for these values of n.

$$\int_0^1 x^n \ln x\, dx$$

SOLUTION: First, we find the indefinite integral. If $n \neq -1$, we may integrate by parts. Let

$$u = \ln x \qquad dv = x^n\, dx$$

$$du = \frac{dx}{x} \qquad v = \frac{x^{n+1}}{n+1}$$

Thus, if $n \neq -1$

$$\int x^n \ln x \, dx = \frac{x^{n+1} \ln x}{n+1} - \int \frac{x^n \, dx}{n+1}$$

$$= \frac{x^{n+1} \ln x}{n+1} - \frac{x^{n+1}}{(n+1)^2} + C \qquad \text{if } n \neq -1 \tag{1}$$

If $n = -1$, then

$$\int x^n \ln x \, dx = \int (\ln x) x^{-1} \, dx$$

$$= \frac{1}{2} (\ln x)^2 + C \qquad \text{if } n = -1 \tag{2}$$

We consider three cases: $n > -1$, $n < -1$, and $n = -1$. If $n \neq -1$, we use the antiderivative in (1), and we have

$$\int_0^1 x^n \ln x \, dx = \lim_{a \to 0^+} \int_a^1 x^n \ln x \, dx$$

$$= \lim_{a \to 0^+} \left[\frac{x^{n+1} \ln x}{n+1} - \frac{x^{n+1}}{(n+1)^2} \right]_a^1$$

$$= \lim_{a \to 0^+} \left[\frac{\ln 1}{n+1} - \frac{1}{(n+1)^2} - \frac{a^{n+1} \ln a}{n+1} + \frac{a^{n+1}}{(n+1)^2} \right]$$

$$= -\frac{1}{(n+1)^2} - \frac{1}{n+1} \lim_{a \to 0^+} a^{n+1} \ln a \tag{3}$$

Case 1: $n > -1$
We use L'Hôpital's rule to find the limit in (3). Because $n + 1 > 0$

$$\lim_{a \to 0^+} a^{n+1} \ln a = \lim_{a \to 0^+} \frac{\ln a}{\dfrac{1}{a^{n+1}}}$$

$$= \lim_{a \to 0^+} \frac{\dfrac{1}{a}}{\dfrac{-(n+1)}{a^{n+2}}}$$

$$= \lim_{a \to 0^+} \frac{-a^{n+1}}{n+1} = 0 \tag{4}$$

Substituting from (4) into (3), we have

$$\int_0^1 x^n \ln x \, dx = -\frac{1}{(n+1)^2} \qquad \text{if } n > -1$$

Case 2: $n < -1$
Then $n + 1 < 0$ and because

$$\lim_{a \to 0^+} a^{n+1} = +\infty \quad \text{and} \quad \lim_{a \to 0^+} \ln a = -\infty$$

the limit in (3) does not exist. Thus, the integral is divergent if $n < -1$.

Case 3: $n = -1$
Using the antiderivative in (2), we have

$$\int_0^1 x^n \ln x \, dx = \lim_{a \to 0^+} \int_a^1 x^{-1} \ln x \, dx$$

$$= \lim_{a \to 0^+} \left[\frac{1}{2} (\ln x)^2 \right]_a^1$$

$$= \lim_{a \to 0^+} \left[-\frac{1}{2} (\ln a)^2 \right] = -\infty$$

Thus, the integral is divergent if $n = -1$.

32. Determine whether the given improper integral is convergent or divergent in each case: **(a)** $0 < n < 1$;   **(b)** $n = 1$;   **(c)** $n > 1$. If the integral is convergent, evaluate it.

$$\int_a^b \frac{dx}{(x-a)^n} \qquad \text{where } b > a$$

SOLUTION: In each case there is an infinite discontinuity at $x = a$.
  (a) If $0 < n < 1$, then $1 - n > 0$. Thus,

$$\int_a^b \frac{dx}{(x-a)^n} = \lim_{\epsilon \to 0^+} \int_{a+\epsilon}^b (x-a)^{-n} \, dx$$

$$= \lim_{\epsilon \to 0^+} \frac{(x-a)^{-n+1}}{-n+1} \Big]_{a+\epsilon}^b$$

$$= \frac{1}{1-n} \lim_{\epsilon \to 0^+} [(b-a)^{1-n} - \epsilon^{1-n}]$$

$$= \frac{(b-a)^{1-n}}{1-n}$$

Thus, the improper integral is convergent.
  (b) If $n = 1$, then

$$\int_a^b \frac{dx}{(x-a)^n} = \lim_{\epsilon \to 0^+} \int_{a+\epsilon}^b \frac{dx}{x-a}$$

$$= \lim_{\epsilon \to 0^+} \ln(x-a) \Big]_{a+\epsilon}^b$$

$$= \lim_{\epsilon \to 0^+} [\ln(b-a) - \ln \epsilon] = +\infty$$

Thus, the improper integral is divergent.
  (c) If $n > 1$, then $n - 1 > 0$. Hence,

$$\int_a^b \frac{dx}{(x-a)^n} = \lim_{\epsilon \to 0^+} \int_{a+\epsilon}^b (x-a)^{-n} \, dx$$

$$= \lim_{\epsilon \to 0^+} \frac{(x-a)^{-n+1}}{-n+1} \Big]_{a+\epsilon}^b$$

$$= \frac{-1}{n-1} \lim_{\epsilon \to 0^+} \left[\frac{1}{(b-a)^{n-1}} - \frac{1}{\epsilon^{n-1}} \right] = +\infty$$

Thus, the improper integral is divergent.

14.5 TAYLOR'S FORMULA

14.5.1 Theorem Let f be a function such that f and its first n derivatives are continuous on the closed interval $[a, b]$. Furthermore, let $f^{(n+1)}(x)$ exist for all x in the open interval (a, b). Then there is a number ξ in the open interval (a, b) such that

$$f(b) = f(a) + \frac{f'(a)}{1!}(b-a) + \frac{f''(a)}{2!}(b-a)^2 + \cdots + \frac{f^{(n)}(a)}{n!}(b-a)^n$$

$$+ \frac{f^{(n+1)}(\xi)}{(n+1)!}(b-a)^{n+1} \tag{1}$$

If in (1) b is replaced by x, *Taylor's formula* if obtained. It is

$$f(x) = f(a) + \frac{f'(a)}{1!}(x-a) + \frac{f''(a)}{2!}(x-a)^2 + \cdots + \frac{f^{(n)}(a)}{n!}(x-a)^n$$

$$+ \frac{f^{(n+1)}(\xi)}{(n+1)!}(x-a)^{n+1} \tag{2}$$

where ξ is between a and x.

If $f(a)$ exists and if the first n derivatives of f exist at the number a, we define the nth degree *Taylor polynomial* of the function f at the number a as follows.

$$P_n(x) = f(a) + \frac{f'(a)(x-a)}{1!} + \frac{f''(a)(x-a)^2}{2!} + \cdots + \frac{f^{(n)}(a)(x-a)^n}{n!} \tag{3}$$

If the Taylor polynomial P_n is used to approximate the function f, then the error, $f(x) - P_n(x)$, denoted by $R_n(x)$, is given by

$$R_n(x) = \frac{f^{(n+1)}(\xi)(x-a)^{n+1}}{(n+1)!} \qquad \text{where } \xi \text{ is between } a \text{ and } x, \text{ provided } f^{(n+1)} \text{ exists} \tag{4}$$

The term $R_n(x)$ as given in (4) is called the *Lagrange form* of the remainder. Using (3) and (4), Taylor's formula (2) can be written as

$$f(x) = P_n(x) + R_n(x)$$

Sometimes $|R_n(x)|$ is small, in which case the Taylor polynomial $P_n(x)$ is a good approximation of $f(x)$. However, sometimes $|R_n(x)|$ is large, in which case $P_n(x)$ is of no use for approximating $f(x)$. Whether $|R_n(x)|$ is small or large depends on the function f, the value of $(x - a)$, and the value of n. In Chapter 15 we will learn tests that can be used to determine whether or not $|R_n(x)|$ can be made as small as we like by choosing n to be large. In Section 15.8 we provide a flowchart that can be used to program a computer to do the calculations required to evaluate a Taylor polynomial with many terms.

Exercises 14.5

In Exercises 1–14 find the Taylor polynomial of degree n with the Lagrange form of the remainder at the number a for the function defined by the given equation.

4. $f(x) = \tan x$; $a = 0$; $n = 3$

SOLUTION: The third-degree Taylor polynomial at 0 for the function f is defined by

$$P_3(x) = f(0) + \frac{f'(0)x}{1!} + \frac{f''(0)x^2}{2!} + \frac{f'''(0)x^3}{3!} \tag{1}$$

and the Lagrange form of the remainder is given by

$$R_3(x) = \frac{f^{(iv)}(\xi)x^4}{4!} \qquad \text{where } \xi \text{ is between 0 and } x \tag{2}$$

Thus, we find the first four derivatives of f.

$$f'(x) = \sec^2 x$$
$$f''(x) = 2 \sec^2 x \tan x = 2 \tan^3 x + 2 \tan x$$
$$f'''(x) = 6 \tan^2 x \sec^2 x + 2 \sec^2 x = 6 \tan^4 x + 8 \tan^2 x + 2$$
$$f^{(iv)}(x) = 24 \tan^3 x \sec^2 x + 16 \tan x \sec^2 x$$
$$= 24 \tan^5 x + 40 \tan^3 x + 16 \tan x$$

from which we obtain

$$f(0) = \tan 0 = 0$$
$$f'(0) = \sec^2 0 = 1$$
$$f''(0) = 2 \tan^3 0 + 2 \tan 0 = 0$$
$$f'''(0) = 6 \tan^4 0 + 8 \tan^2 0 + 2 = 2$$

Substituting in (1), we obtain

$$P_3(x) = 0 + 1 \cdot x + \frac{0 \cdot x^2}{2!} + \frac{2x^3}{3!}$$

$$= x + \tfrac{1}{3} x$$

Substituting in (2), we obtain

$$R_3(x) = \frac{(24 \tan^5 \xi + 40 \tan^3 \xi + 16 \tan \xi)x^4}{4!}$$

$$= \tfrac{1}{3}(3 \tan^5 \xi + 5 \tan^3 \xi + 2 \tan \xi)x^4 \qquad \text{where } \xi \text{ is between 0 and } x$$

8. $f(x) = \sqrt{x}$; $a = 4$; $n = 4$

SOLUTION: The fourth-degree Taylor polynomial at 4 for the function f is defined by

$$P_4(x) = f(4) + \frac{f'(4)(x-4)}{1!} + \frac{f''(4)(x-4)^2}{2!} + \frac{f'''(4)(x-4)^3}{3!}$$

$$+ \frac{f^{(iv)}(4)(x-4)^4}{4!} \tag{1}$$

and the Lagrange form of the remainder is given by

$$R_4(x) = \frac{f^{(v)}(\xi)(x-4)^5}{5!} \qquad \text{where } \xi \text{ is between 4 and } x \tag{2}$$

We have $f(4) = 2$, and

$$f'(x) = \tfrac{1}{2} x^{-1/2} \qquad \text{thus, } f'(4) = \tfrac{1}{4}$$
$$f''(x) = -\tfrac{1}{4} x^{-3/2} \qquad \text{thus, } f''(4) = -\tfrac{1}{32}$$
$$f'''(x) = \tfrac{3}{8} x^{-5/2} \qquad \text{thus } f'''(4) = \tfrac{3}{256}$$
$$f^{(iv)}(x) = -\tfrac{15}{16} x^{-7/2} \qquad \text{thus, } f^{(iv)}(4) = -\tfrac{15}{2048}$$

Substituting in (1), we obtain

$$P_4(x) = 2 + \tfrac{1}{4}(x-4) - \frac{\tfrac{1}{32}(x-4)^2}{2!} + \frac{\tfrac{3}{256}(x-4)^3}{3!} - \frac{\tfrac{15}{2048}(x-4)^4}{4!}$$

$$= 2 + \tfrac{1}{4}(x-4) - \tfrac{1}{64}(x-4)^2 + \tfrac{1}{512}(x-4)^3 - \tfrac{5}{16384}(x-4)^4$$

Because

$$f^{(v)}(x) = \tfrac{105}{32} x^{-9/2}$$

from (2) we obtain

$$R_4(x) = \frac{\tfrac{105}{32}\xi^{-9/2}(x-4)^5}{5!}$$

$$= \tfrac{7}{256}\xi^{-9/2}(x-4)^5 \qquad \text{where } \xi \text{ is between 4 and } x$$

12. $f(x) = e^{-x^2};\ a = 0;\ n = 3$

SOLUTION: As in Exercise 4 we must find

$$P_3(x) = f(0) + \frac{f'(0)x}{1!} + \frac{f''(0)x^2}{2!} + \frac{f'''(0)x^3}{3!} \qquad (1)$$

and

$$R_3(x) = \frac{f^{(iv)}(\xi)x^4}{4!} \qquad \text{where } \xi \text{ is between 0 and } x \qquad (2)$$

We have

$$
\begin{aligned}
f(x) &= e^{-x^2} \\
f'(x) &= -2xe^{-x^2} \\
f''(x) &= (-2x)(-2x)e^{-x^2} + (-2)e^{-x^2} \\
&= 2e^{-x^2}(2x^2 - 1) \\
f'''(x) &= 2e^{-x^2}(4x) + (2x^2 - 1)(-4xe^{-x^2}) \\
&= -4e^{-x^2}(2x^3 - 3x) \\
f^{(iv)}(x) &= -4e^{-x^2}(6x^2 - 3) + (2x^3 - 3x)(8xe^{-x^2}) \\
&= 4e^{-x^2}[-(6x^2 - 3) + 2x(2x^3 - 3x)] \\
&= 4e^{-x^2}(4x^4 - 12x^2 + 3) \qquad (3)
\end{aligned}
$$

Thus,

$$
\begin{aligned}
f(0) &= 1 \\
f'(0) &= 0 \\
f''(0) &= -2 \\
f'''(0) &= 0
\end{aligned}
$$

Substituting into Eq. (1), we obtain

$$P_3(x) = 1 - x^2$$

which is the required Taylor polynomial. Substituting from (3) into (2), we obtain

$$R_3(x) = \frac{4e^{-\xi^2}(4\xi^4 - 12\xi^2 + 3)x^4}{4!}$$

$$= \frac{(4\xi^4 - 12\xi^2 + 3)x^4}{6e^{\xi^2}} \qquad \text{where } \xi \text{ is between 0 and } x$$

$R_3(x)$ is the Lagrange form of the remainder.

16. Use the Taylor polynomial in Exercise 8 to compute $\sqrt{5}$ accurate to as many places as is justified when R_4 is neglected.

SOLUTION: Because $f(x) = \sqrt{x}$, then $f(5) = \sqrt{5}$. In Exercise 8 we showed that

$$R_4(x) = \frac{7}{256}\xi^{-9/2}(x-4)^5 \qquad \text{where } \xi \text{ is between 4 and } x$$

Thus,

$$R_4(5) = \frac{7}{256} \xi^{-9/2} \qquad \text{where } 4 < \xi < 5 \tag{1}$$

From (1) we conclude that

$$\frac{7}{256} \cdot 5^{-9/2} < R_4(5) < \frac{7}{256} \cdot 4^{-9/2}$$

Thus,

$$R_4(5) < \frac{7}{256} \cdot \frac{1}{512} = 0.00005$$

Because $R_4(5) < 10^{-4}$ we use $P_4(5)$ to find a four-place decimal approximation for $f(5)$. From Exercise 8 we have

$$P_4(x) = 2 + \frac{1}{4}(x-4) - \frac{1}{64}(x-4)^2 + \frac{1}{512}(x-4)^3 - \frac{5}{16,384}(x-4)^4$$

Thus,

$$P_4(5) = 2 + \frac{1}{4} - \frac{1}{64} + \frac{1}{512} - \frac{5}{16,384}$$

$$= 2.00000 + 0.25000 - 0.01563 + 0.00195 - 0.00031 = 2.2360$$

Therefore $\sqrt{5} \approx 2.2360$, and the fourth digit to the right of the decimal differs from the actual value by at most 1.

20. Use a Taylor polynomial at 0 for the function defined by $f(x) = \ln(1 + x)$ to compute the value of $\ln 1.2$, accurate to four decimal places.

SOLUTION: $f(0) = \ln 1 = 0$

$$
\begin{array}{ll}
f'(x) = (1+x)^{-1} & f'(0) = 1 \\
f''(x) = -(1+x)^{-2} & f''(0) = -1 \\
f'''(x) = 2(1+x)^{-3} & f'''(0) = 2 \\
f^{(iv)}(x) = -6(1+x)^{-4} & f^{(iv)}(0) = -6 \\
f^{(v)}(x) = 24(1+x)^{-5} & f^{(v)}(0) = 24
\end{array}
$$

Thus, the fifth-degree Taylor polynomial is

$$P_5(x) = 0 + \frac{1 \cdot x}{1!} - \frac{1 \cdot x^2}{2!} + \frac{2x^3}{3!} - \frac{6x^4}{4!} + \frac{24x^5}{5!}$$

$$= x - \frac{1}{2}x^2 + \frac{1}{3}x^3 - \frac{1}{4}x^4 + \frac{1}{5}x^5 \tag{1}$$

Because $f(0.2) = \ln(1 + 0.2) = \ln 1.2$, we replace x by 0.2 in (1). Thus,

$$P_5(0.2) = 0.2 - \frac{1}{2}(0.2)^2 + \frac{1}{3}(0.2)^3 - \frac{1}{4}(0.2)^4 + \frac{1}{5}(0.2)^5$$

$$= 0.20000 - 0.02000 + 0.002667 - 0.00040 + 0.00006$$
$$= 0.18233 \tag{2}$$

We find $|R_5(0.2)|$. Because $f^{(vi)}(x) = -120(1 + x)^{-6}$, then

$$R_5(x) = \frac{-120(1 + \xi)^{-6}x^6}{6!}$$

$$= -\frac{x^6}{6(1 + \xi)^6} \qquad \text{where } \xi \text{ is between 0 and } x$$

Thus,

$$|R_5(0.2)| = \frac{(0.2)^6}{6(1 + \xi)^6} \quad \text{where } 0 < \xi < 0.2$$

Because $0 < \xi < 0.2$, then

$$\frac{1}{(1 + \xi)^6} < 1$$

Thus,

$$|R_5(0.2)| < \frac{(0.2)^6}{6} = 0.000011$$

Because the error is less than 10^{-4} when $P_5(0.2)$ is used to approximate $f(0.2)$, we conclude that (2) gives an approximation for $f(0.2)$ that is accurate to four decimal places. Rounding off the result in (2) we have $\ln 1.2 = 0.1823$.

24. Show that the formula

$$(1 + x)^{3/2} \approx 1 + \tfrac{3}{2}x$$

is accurate to three decimal places if $-0.03 \le x \le 0$.

SOLUTION: Let f be the function defined by

$$f(x) = (1 + x)^{3/2}$$

We show that $1 + \tfrac{3}{2}x$ is the first-degree Taylor polynomial about the number 0 for the function f. We have $f(0) = 1$. Because $f'(x) = \tfrac{3}{2}(1 + x)^{1/2}$, then $f'(0) = \tfrac{3}{2}$. Hence,

$$P_1(x) = f(0) + f'(0) \cdot x$$
$$= 1 + \tfrac{3}{2}x$$

The Lagrange form of the remainder for $P_1(x)$ is given by

$$R_1(x) = \frac{f''(\xi)x^2}{2!} \quad \text{where } \xi \text{ is between 0 and } x$$

Because $f''(x) = \tfrac{3}{4}(1 + x)^{-1/2}$, then we have

$$R_1(x) = \frac{\tfrac{3}{4}(1 + \xi)^{-1/2}x^2}{2!}$$

$$= \frac{3x^2}{8(1 + \xi)^{1/2}} \quad \text{where } \xi \text{ is between 0 and } x \tag{1}$$

We find an upper bound for $R_1(x)$. Because $-0.03 \le x \le 0$, then

$$x^2 \le 0.0009 \tag{2}$$

Because $x < \xi < 0$, then

$$-0.03 < \xi < 0$$
$$0.97 < 1 + \xi < 1$$
$$\sqrt{0.97} < \sqrt{1 + \xi} < 1$$
$$1 < \frac{1}{\sqrt{1 + \xi}} < \frac{1}{\sqrt{0.97}}$$
$$\frac{1}{\sqrt{1 + \xi}} < 1.02 \tag{3}$$

Multiplying the members of (2) and (3), we have

$$\frac{x^2}{\sqrt{1+\xi}} < 0.00092$$

Multiplying both sides by $\frac{3}{8}$ we get

$$\frac{3x^2}{8\sqrt{1+\xi}} < 0.00034 \qquad (4)$$

By (1) and (4) we have

$$R_1(x) < 10^{-3} \qquad \text{if } -0.03 \leq x \leq 0$$

Thus, the first-degree Taylor polynomial $1 + \frac{3}{2}x$ is accurate to three decimal places if used to approximate $(1+x)^{3/2}$ when $-0.03 \leq x \leq 0$.

28. (a) Use the first-degree Taylor polynomial at 0 to approximate e^k if $0 < k < 0.01$.

(b) Estimate the error in terms of k.

SOLUTION:

(a) Let $f(x) = e^x$. Because $f'(x) = e^x$, we have $f(0) = 1$ and $f'(0) = 1$.

Thus, the first-degree Taylor polynomial at 0 for the function f is given by

$$P_1(x) = 1 + x$$

Because $f(k) = e^k$ and $f(k) \approx P_1(k)$, then

$$e^k \approx 1 + k$$

(b) The Lagrange form of the remainder for $P_1(x)$ is given by

$$R_1(x) = \frac{f''(\xi)x^2}{2!} \qquad \text{where } \xi \text{ is between 0 and } x$$

Because $f''(x) = e^x$, then

$$R_1(k) = \frac{e^\xi k^2}{2} \qquad \text{where } \xi \text{ is between 0 and } k$$

If $\xi < k$, then $e^\xi < e^k$, so

$$R_1(k) < \frac{e^k k^2}{2} \qquad (1)$$

which estimates the error in terms of k. We are given that $0 < k < 0.01$. Thus,

$$e^k < e^{0.01} < 1.01 \qquad (2)$$

and

$$k^2 < 10^{-4} \qquad (3)$$

From (1), (2), and (3), we conclude that

$$R_1(k) < \frac{(1.01)(10^{-4})}{2}$$

$$\leq (5.1)10^{-5}$$

Therefore, for all k where $0 < k < 0.01$, the error is less than $(5.1)10^{-5}$ when the Taylor polynomial $1 + k$ is used to approximate e^k.

Review Exercises for Chapter 14

In Exercises 1–18 evaluate the limit, if it exists.

4. $\lim\limits_{x \to 0} \dfrac{e - (1 + x)^{1/x}}{x}$

SOLUTION: From Section 8.5 we have the limit

$$\lim_{x \to 0} (1 + x)^{1/x} = e \tag{1}$$

Therefore, $\lim\limits_{x \to 0} [e - (1 + x)^{1/x}] = 0$ and $\lim\limits_{x \to 0} x = 0$. Thus, the hypothesis of L'Hôpital's rule is satisfied for the given function. Because $D_x e = 0$, and $D_x x = 1$, we have

$$\lim_{x \to 0} \frac{e - (1 + x)^{1/x}}{x} = \lim_{x \to 0} [-D_x (1 + x)^{1/x}] \tag{2}$$

Let

$$y = (1 + x)^{1/x}$$
$$|y| = |1 + x|^{1/x}$$

Then

$$\ln|y| = \frac{\ln|1 + x|}{x}$$

$$\frac{1}{y} \cdot D_x y = \frac{\dfrac{x}{1 + x} - \ln|1 + x|}{x^2}$$

$$D_x y = y \cdot \frac{\dfrac{x}{1 + x} - \ln|1 + x|}{x^2}$$

$$D_x (1 + x)^{1/x} = (1 + x)^{1/x} \frac{\dfrac{x}{1 + x} - \ln|1 + x|}{x^2} \tag{3}$$

Substituting from (3) into (2), we obtain

$$\lim_{x \to 0} \frac{e - (1 + x)^{1/x}}{x} = \lim_{x \to 0} (1 + x)^{1/x} \cdot \lim_{x \to 0} \frac{\ln|1 + x| - \dfrac{x}{1 + x}}{x^2} \tag{4}$$

By L'Hôpital's rule

$$\lim_{x \to 0} \frac{\ln|1 + x| - \dfrac{x}{1 + x}}{x^2} = \lim_{x \to 0} \frac{\dfrac{1}{1 + x} - \dfrac{1}{(1 + x)^2}}{2x}$$

$$= \lim_{x \to 0} \frac{1}{2(1 + x)^2} = \frac{1}{2} \tag{5}$$

Substituting from (1) and (5) into (4), we get

$$\lim_{x \to 0} \frac{e - (1 + x)^{1/x}}{x} = \frac{1}{2} e$$

8. $\lim\limits_{t \to +\infty} \dfrac{t^{100}}{e^t}$

SOLUTION: If n is any positive integer, then

$$\lim_{t \to +\infty} t^n = +\infty$$

Furthermore, $\lim\limits_{t \to +\infty} e^t = +\infty$. Therefore, we may apply L'Hôpital's rule to obtain

$$\lim_{t \to +\infty} \frac{t^{100}}{e^t} = \lim_{t \to +\infty} \frac{100\, t^{99}}{e^t}$$

$$= \lim_{t \to +\infty} \frac{100 \cdot 99\, t^{98}}{e^t}$$

$$= \lim_{t \to +\infty} \frac{100 \cdot 99 \cdot 98\, t^{97}}{e^t}$$

and so forth. After 100 applications of L'Hôpital's rule we obtain

$$\lim_{t \to +\infty} \frac{t^{100}}{e^t} = \lim_{t \to +\infty} \frac{100!}{e^t} = 0$$

12. $\lim\limits_{y \to +\infty} (1 + e^{2y})^{-2/y}$

SOLUTION: Let

$$z = (1 + e^{2y})^{-2/y}$$

Then

$$\ln z = \ln(1 + e^{2y})^{-2/y}$$

$$= \frac{-2\ln(1 + e^{2y})}{y}$$

and

$$\lim_{y \to +\infty} (\ln z) = \lim_{y \to +\infty} \frac{-2\ln(1 + e^{2y})}{y}$$

Because $\lim\limits_{y \to +\infty} -2\ln(1 + e^{2y}) = -\infty$ and $\lim\limits_{y \to +\infty} y = +\infty$, we apply L'Hôpital's rule. Thus

$$\lim_{y \to +\infty} (\ln z) = \lim_{y \to +\infty} \frac{-4\, e^{2y}}{1 + e^{2y}}$$

$$= \lim_{y \to +\infty} \frac{-4}{e^{-2y} + 1} = -4$$

Thus,

$$\ln(\lim_{y \to +\infty} z) = -4$$

$$\lim_{y \to +\infty} z = e^{-4}$$

Hence,

$$\lim_{y \to +\infty} (1 + e^{2y})^{-2/y} = e^{-4}$$

16. $\lim\limits_{x \to 0} \left(\dfrac{\sin x}{x} \right)^{1/x}$

SOLUTION: By (10.2.1),

$$\lim\limits_{x \to 0} \frac{\sin x}{x} = 1$$

Thus, we have the indeterminate form 1^∞. We let

$$y = \left(\frac{\sin x}{x} \right)^{1/x}$$

$$\ln y = \frac{\ln \left(\dfrac{\sin x}{x} \right)}{x}$$

By L'Hôpital's rule

$$\lim\limits_{x \to 0} \ln y = \lim\limits_{x \to 0} \frac{\ln \left(\dfrac{\sin x}{x} \right)}{x}$$

$$= \lim\limits_{x \to 0} \frac{\ln(\sin x) - \ln x}{x}$$

$$= \lim\limits_{x \to 0} \frac{\dfrac{\cos x}{\sin x} - \dfrac{1}{x}}{1}$$

$$= \lim\limits_{x \to 0} \frac{x \cos x - \sin x}{x \sin x} \tag{1}$$

We apply L'Hôpital's rule two times on the right-hand side of (1). Thus,

$$\lim\limits_{x \to 0} \ln y = \lim\limits_{x \to 0} \frac{-x \sin x + \cos x - \cos x}{x \cos x + \sin x}$$

$$= \lim\limits_{x \to 0} \frac{-x \cos x - \sin x}{-x \sin x + \cos x + \cos x} = 0$$

Therefore,

$$\ln \lim\limits_{x \to 0} y = 0$$

$$\lim\limits_{x \to 0} y = e^0$$

or, equivalently,

$$\lim\limits_{x \to 0} \left(\frac{\sin x}{x} \right)^{1/x} = 1$$

In Exercises 19–32 determine whether the improper integral is convergent or divergent. If it is convergent, evaluate it.

20. $\displaystyle\int_0^{+\infty} \frac{dx}{\sqrt{e^x}}$

SOLUTION:

$$\int_0^{+\infty} \frac{dx}{\sqrt{e^x}} = \lim_{b \to +\infty} \int_0^b e^{-x/2} \, dx$$

$$= \lim_{b \to +\infty} \left[-2 \, e^{-x/2} \right]_0^b$$

$$= \lim_{b \to +\infty} \left[-2 e^{-b/2} + 2 e^0 \right] = 2$$

24. $\displaystyle \int_0^{+\infty} \frac{dt}{t^4 + t^2}$

SOLUTION:

$$\int_0^{+\infty} \frac{dt}{t^4 + t^2} = \lim_{b \to +\infty} \int_0^b \frac{dt}{t^2(t^2 + 1)}$$

We use partial fractions.

$$\frac{1}{t^2(t^2 + 1)} = \frac{A}{t^2} + \frac{B}{t} + \frac{Ct + D}{t^2 + 1}$$

$$1 = A(t^2 + 1) + Bt(t^2 + 1) + (Ct + D)t^2$$

from which we get $A = 1$, $B = 0$, $C = 0$, and $D = -1$. Thus

$$\frac{1}{t^2(t^2 + 1)} = \frac{1}{t^2} - \frac{1}{t^2 + 1}$$

Substituting in (1), we obtain

$$\int_0^{+\infty} \frac{dt}{t^4 + t^2} = \lim_{b \to +\infty} \int_1^b \frac{dt}{t^2} - \lim_{b \to +\infty} \int_1^b \frac{dt}{t^2 + 1}$$

$$= \lim_{b \to +\infty} \left[-\frac{1}{t} \right]_1^b - \lim_{b \to +\infty} \left[\tan^{-1} t \right]_1^b$$

$$= \lim_{b \to +\infty} \left[-\frac{1}{b} + 1 \right] - \lim_{b \to +\infty} \left[\tan^{-1} b - \frac{1}{4}\pi \right]$$

$$= 1 - \left[\frac{1}{2}\pi - \frac{1}{4}\pi \right] = 1 - \frac{1}{4}\pi$$

28. $\displaystyle \int_0^{+\infty} \frac{3 - \sqrt{x}}{\sqrt{x}} \, dx$

SOLUTION: There is an infinite discontinuity at $x = 0$. We express the given integral as the sum of two improper intergrals.

$$\int_0^{+\infty} \frac{3 - \sqrt{x}}{\sqrt{x}} \, dx = \int_0^1 \frac{3 - \sqrt{x}}{\sqrt{x}} \, dx + \int_1^{+\infty} \frac{3 - \sqrt{x}}{\sqrt{x}} \, dx$$

$$= \lim_{a \to 0^+} \int_a^1 \frac{3 - \sqrt{x}}{\sqrt{x}} \, dx + \lim_{b \to +\infty} \int_1^b \frac{3 - \sqrt{x}}{\sqrt{x}} \, dx \qquad (1)$$

We find the value of the indefinite integral. Let

$$u = 3^{-\sqrt{x}}$$
$$du = 3^{-\sqrt{x}}(-\tfrac{1}{2}x^{-1/2})(\ln 3) \, dx$$

Thus,

$$\frac{3^{-\sqrt{x}}\,dx}{\sqrt{x}} = \frac{-2\,du}{\ln 3}$$

and

$$\int \frac{3^{-\sqrt{x}}\,dx}{\sqrt{x}} = \int \frac{-2\,du}{\ln 3}$$

$$= \frac{-2\,u}{\ln 3} + C$$

$$= \frac{-2(3^{-\sqrt{x}})}{\ln 3} + C$$

$$= \frac{-2}{3^{\sqrt{x}}\ln 3} + C \tag{2}$$

Substituting the value of the indefinite integral (2) into (1), we obtain

$$\int_0^{+\infty} \frac{3^{-\sqrt{x}}}{\sqrt{x}}\,dx = \lim_{a\to 0^+}\left[\frac{-2}{3^{\sqrt{x}}\ln 3}\right]_a^1 + \lim_{b\to +\infty}\left[\frac{-2}{3^{\sqrt{x}}\ln 3}\right]_1^b$$

$$= \frac{-2}{\ln 3}\left[\lim_{a\to 0^+}\left(\frac{1}{3} - \frac{1}{3^{\sqrt{a}}}\right) + \lim_{b\to +\infty}\left(\frac{1}{3^{\sqrt{b}}} - \frac{1}{3}\right)\right]$$

$$= \frac{-2}{\ln 3}\left[\frac{1}{3} - \frac{1}{3^0} + 0 - \frac{1}{3}\right] = \frac{2}{\ln 3}$$

32. $\displaystyle\int_{-3}^0 \frac{dx}{\sqrt{3 - 2x - x^2}}$

SOLUTION: There is an infinite discontinuity at -3. Thus,

$$\int_{-3}^0 \frac{dx}{\sqrt{3 - 2x - x^2}} = \lim_{a\to -3^+}\int_a^0 \frac{dx}{\sqrt{4 - (x+1)^2}}$$

$$= \lim_{a\to -3^+}\left[\sin^{-1}\frac{x+1}{2}\right]_a^0$$

$$= \lim_{a\to -3^+}\left[\sin^{-1}\frac{1}{2} - \sin^{-1}\frac{a+1}{2}\right]$$

$$= \frac{1}{6}\pi - \left(-\frac{1}{2}\pi\right) = \frac{2}{3}\pi$$

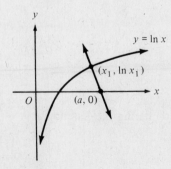

Figure 14. 36R

36. If the normal line to the curve $y = \ln x$ at the point $(x_1, \ln x_1)$ intersects the x-axis at the point having an abscissa of a, prove that $\displaystyle\lim_{x_1\to +\infty}(a - x_1) = 0$.

SOLUTION: Refer to Fig. 14. 36R. Because $D_x y = 1/x$, the slope of the normal line to the curve $y = \ln x$ at the point where $x = x_1$ is $-x_1$. Thus, the point-slope form of the equation of the normal line is

$$y - \ln x_1 = -x_1(x - x_1) \tag{1}$$

Because the normal line intersects the x-axis at $(a, 0)$, we let $x = a$ and $y = 0$ in Eq. (1). Thus,

$$-\ln x_1 = -x_1(a - x_1)$$

$$a - x_1 = \frac{\ln x_1}{x_1}$$

Hence,

$$\lim_{x_1 \to +\infty} (a - x_1) = \lim_{x_1 \to +\infty} \frac{\ln x_1}{x_1} \qquad (2)$$

We use L'Hôpital's rule to find the limit on the right side of (2). Thus,

$$\lim_{x_1 \to +\infty} (a - x_1) = \lim_{x_1 \to +\infty} \frac{1}{x_1} = 0$$

40. Find the Taylor polynomial of degree n with the Lagrange form of the remainder at the number a for the function defined by the given equation.

$$f(x) = (1 + x^2)^{-1}; \quad a = 1; \quad n = 3$$

SOLUTION: We have

$$P_3(x) = f(1) + f'(1)(x - 1) + \frac{f''(1)(x-1)^2}{2!} + \frac{f'''(1)(x-1)^3}{3!} \qquad (1)$$

and

$$R_3(x) = \frac{f^{(iv)}(\xi)(x-1)^4}{4!} \qquad \text{where } \xi \text{ is between 1 and } x \qquad (2)$$

Differentiating the given function, we obtain

$$f'(x) = -2x(1 + x^2)^{-2}$$
$$f''(x) = 2(3x^2 - 1)(1 + x^2)^{-3}$$
$$f'''(x) = -24(x^3 - x)(1 + x^2)^{-4}$$
$$f^{(iv)}(x) = 24(5x^4 - 10x^2 + 1)(1 + x^2)^{-5}$$

Thus,

$$f(1) = \tfrac{1}{2} \qquad f'(1) = -\tfrac{1}{2} \qquad f''(1) = \tfrac{1}{2} \qquad f'''(1) = 0$$

Substituting in (1), we obtain

$$P_3(x) = \tfrac{1}{2} - \tfrac{1}{2}(x - 1) + \tfrac{1}{4}(x - 1)^2$$

Substituting in (2), we get

$$R_3(x) = \frac{24(5\xi^4 - 10\xi^2 + 1)(1 + \xi^2)^{-5}(x - 1)^4}{4!}$$

$$= \frac{(5\xi^4 - 10\xi^2 + 1)(x - 1)^4}{(1 + \xi^2)^5} \qquad \text{where } \xi \text{ is between 1 and } x$$

44. Apply Taylor's formula to express the polynomial

$$P(x) = 4x^3 + 5x^2 - 2x + 1$$

as a polynomial in powers of $(x + 2)$.

SOLUTION: We let $a = -2$ in Taylor's formula. Differentiating the function P, we have

$$P'(x) = 12x^2 + 10x - 2$$
$$P''(x) = 24x + 10$$
$$P'''(x) = 24$$

Thus,

$$P(-2) = -7 \qquad P'(-2) = 26 \qquad P''(-2) = -38 \qquad P'''(-2) = 24$$

Now the third-degree Taylor polynomial about -2 for P is

$$P_3(x) = P(-2) + P'(-2)(x+2) + \frac{P''(-2)(x+2)^2}{2!} + \frac{P'''(-2)(x+2)^3}{3!}$$

$$= -7 + 26(x+2) - 19(x+2)^2 + 4(x+2)^3 \tag{1}$$

Furthermore, because $P^{(iv)}(x) = 0$, then the remainder $R_4(x) = 0$. Thus, $P_3(x) = P(x)$, and the polynomial in (1) is exactly $P(x)$. [Multiply out the terms in (1) and simplify. The result is $P(x)$.]

48. Find the values of n for which the improper integral

$$\int_1^{+\infty} \frac{\ln x}{x^n} \, dx$$

converges and evaluate the integral for those values of n.

SOLUTION:

$$\int_1^{+\infty} \frac{\ln x}{x^n} \, dx = \lim_{b \to +\infty} \int_1^b \frac{\ln x}{x^n} \, dx \tag{1}$$

To find the indefinite integral, we apply Eqs. (1) and (2) of Exercise 28 in Exercises 14.4 with n replaced by $-n$. Thus,

$$\int \frac{\ln x}{x^n} \, dx = \begin{cases} \dfrac{x^{1-n} \ln x}{1-n} - \dfrac{x^{1-n}}{(1-n)^2} + C & \text{if } n \neq 1 \\[2ex] \dfrac{1}{2} (\ln x)^2 + C & \text{if } n = 1 \end{cases} \tag{2}$$

We consider three cases: $n > 1$, $n < 1$, and $n = 1$. If $n \neq 1$, we substitute from (2) into (1). We have

$$\int_1^{+\infty} \frac{\ln x}{x^n} \, dx = \lim_{b \to +\infty} \left[\frac{x^{1-n} \ln x}{1-n} - \frac{x^{1-n}}{(1-n)^2} \right]_1^b$$

$$= \lim_{b \to +\infty} \left[\frac{b^{1-n} \ln b}{1-n} - \frac{b^{1-n}}{(1-n)^2} - \frac{\ln 1}{1-n} + \frac{1}{(1-n)^2} \right]$$

$$= \frac{1}{1-n} \lim_{b \to +\infty} \left[\frac{\ln b}{b^{n-1}} \right] - \frac{1}{(1-n)^2} \lim_{b \to +\infty} \left[\frac{1}{b^{n-1}} \right] + \frac{1}{(1-n)^2} \tag{3}$$

Case 1: $n > 1$

Thus, $n - 1 > 0$. By L'Hôpital's rule,

$$\lim_{b \to +\infty} \frac{\ln b}{b^{n-1}} = \lim_{b \to +\infty} \frac{\dfrac{1}{b}}{(n-1)b^{n-2}}$$

$$= \frac{1}{n-1} \lim_{b \to +\infty} \frac{1}{b^{n-1}} = 0$$

Thus, both of the limits on the right-hand side of (3) are 0, and (3) reduces to

$$\int_1^{+\infty} \frac{\ln x}{x^n} \, dx = \frac{1}{(1-n)^2} \qquad \text{if } n > 1$$

Case 2: $n < 1$

Then $1 - n > 0$. Because

$$\lim_{n \to +\infty} b^{1-n} = +\infty$$

the limits on the right-hand side of Eq. (3) do not exist. Thus, the integral is divergent if $n < 1$.

Case 3: $n = 1$

Substituting from (2) into (1), we have

$$\int_1^{+\infty} \frac{\ln x}{x^n} \, dx = \lim_{b \to +\infty} \left[\frac{1}{2} (\ln x)^2 \right]_1^b$$

$$= \lim_{b \to +\infty} \left[\frac{1}{2} (\ln b)^2 - \frac{1}{2} (\ln 1)^2 \right] = +\infty$$

Thus, the integral is divergent if $n = 1$.

CHAPTER 15 | Infinite Series

15.1 SEQUENCES

15.1.1 Definition A *sequence* is a function whose domain is the set of positive integers.

15.1.2 Definition A sequence $\{a_n\}$ has the limit L if for any $\epsilon > 0$ there exists a number $N > 0$ such that $|a_n - L| < \epsilon$ for every integer $n > N$; and we write

$$\lim_{n \to +\infty} a_n = L$$

15.1.3 Theorem If $\lim_{x \to +\infty} f(x) = L$ and f is defined for every positive integer, then also $\lim_{n \to +\infty} f(n) = L$ when n is any positive integer.

By Theorem 15.1.3 we may use the limit theorems of Chapter 2 and L'Hôpital's rule to find the limit of a sequence function.

15.1.4 Definition If a sequence $\{a_n\}$ has a limit, the sequence is said to be *convergent,* and a_n *converges* to that limit. If the sequence is not convergent, it is *divergent.*

Thus, if $\lim_{n \to +\infty} a_n = \pm\infty$ or if $\lim_{n \to +\infty} a_n$ does not exist, the sequence $\{a_n\}$ is divergent.

15.1.5 Theorem If $\{a_n\}$ and $\{b_n\}$ are convergent sequences and c is a constant, then

    (i) the constant sequence $\{c\}$ has c as its limit;

    (ii) $\lim\limits_{n \to +\infty} ca_n = c \lim\limits_{n \to +\infty} a_n;$

    (iii) $\lim\limits_{n \to +\infty} (a_n \pm b_n) = \lim\limits_{n \to +\infty} a_n \pm \lim\limits_{n \to +\infty} b_n;$

(iv) $\lim\limits_{n \to +\infty} a_n b_n = (\lim\limits_{n \to +\infty} a_n)(\lim\limits_{n \to +\infty} b_n);$

(v) $\lim\limits_{n \to +\infty} \dfrac{a_n}{b_n} = \dfrac{\lim\limits_{n \to +\infty} a_n}{\lim\limits_{n \to +\infty} b_n}$ if $\lim\limits_{n \to +\infty} b_n \neq 0.$

The result of Example 3 is often stated as a theorem.

Theorem   If $|r| < 1$, the sequence $\{r^n\}$ is convergent and

$$\lim_{n \to +\infty} r^n = 0$$

Exercises 15.1

In Exercises 1–19 write the first four terms of the sequence and determine whether the sequence is convergent or divergent. If the sequence converges, find its limit.

4. $\left\{\dfrac{3n^3 + 1}{2n^2 + n}\right\}$

SOLUTION: Let

$$a_n = \frac{3n^3 + 1}{2n^2 + n}$$

The first four terms of the sequence are found by replacing n by 1, 2, 3, and 4.

Thus,

$$a_1 = \frac{3(1^3) + 1}{2(1^2) + 1} = \frac{4}{3}$$

$$a_2 = \frac{3(2^3) + 1}{2(2^2) + 2} = \frac{5}{2}$$

$$a_3 = \frac{3(3^3) + 1}{2(3^2) + 3} = \frac{82}{21}$$

$$a_4 = \frac{3(4^3) + 1}{2(4^2) + 4} = \frac{193}{36}$$

We apply Theorem 15.1.3 to determine whether the sequence is convergent or divergent. Let

$$f(x) = \frac{3x^3 + 1}{2x^2 + x}$$

Then $f(n) = a_n$ for every positive integer n. Furthermore,

$$\lim_{x \to +\infty} f(x) = \lim_{x \to +\infty} \frac{3x^3 + 1}{2x^2 + x}$$

$$= \lim_{x \to +\infty} \frac{3 + \dfrac{1}{x^3}}{\dfrac{2}{x} + \dfrac{1}{x^2}} = +\infty$$

Therefore, the sequence $\{a_n\}$ is divergent.

8. $\left\{\dfrac{\log_b n}{n}\right\}; \ b > 1$

SOLUTION: Let

$$a_n = \frac{\log_b n}{n} = \frac{\ln n}{n \ln b}$$

The first four terms of the sequence are

$$a_1 = 0$$

$$a_2 = \frac{\ln 2}{2 \ln b}$$

$$a_3 = \frac{\ln 3}{3 \ln b}$$

$$a_4 = \frac{\ln 4}{4 \ln b}$$

We apply Theorem 15.1.3 to determine whether the sequence is convergent or divergent. Let

$$f(x) = \frac{\ln x}{x \ln b}$$

Then $f(n) = a_n$ for every positive integer n. Furthermore, by L'Hôpital's rule

$$\lim_{x \to +\infty} f(x) = \lim_{x \to +\infty} \frac{\ln x}{x \ln b}$$

$$= \frac{1}{\ln b} \lim_{x \to +\infty} \frac{1}{x} = 0$$

Therefore, the sequence $\{a_n\}$ is convergent and has limit 0.

12. $\left\{ \dfrac{\sinh n}{\sin n} \right\}$

SOLUTION: If $a_n = \dfrac{\sinh n}{\sin n}$, then

$$a_1 = \frac{\sinh 1}{\sin 1} = 1.4$$

$$a_2 = \frac{\sinh 2}{\sin 2} = 4.0$$

$$a_3 = \frac{\sinh 3}{\sin 3} = 71.5$$

$$a_4 = \frac{\sinh 4}{\sin 4} = -36.1$$

Because $\lim\limits_{n \to +\infty} \sin n$ does not exist, and $\lim\limits_{n \to +\infty} \sinh n = +\infty$, we conclude that

$\lim\limits_{n \to +\infty} a_n$ does not exist. Hence, the sequence $\{a_n\}$ is divergent.

16. $\left\{ \left(1 + \dfrac{2}{n} \right)^n \right\}$

SOLUTION: Let

$$a_n = \left(1 + \frac{2}{n} \right)^n$$

Then the first four terms are

$$a_1 = 3$$
$$a_2 = 2^2 = 4$$
$$a_3 = (\tfrac{5}{3})^3 = \tfrac{125}{27}$$
$$a_4 = (\tfrac{3}{2})^4 = \tfrac{81}{16}$$

If

$$f(x) = \left(1 + \frac{2}{x}\right)^x$$

then

$$\lim_{x \to +\infty} f(x) = \lim_{x \to +\infty} \left(1 + \frac{2}{x}\right)^x \tag{1}$$

Let $z = 2/x$. Then $x \to +\infty$ is equivalent to $z \to 0^+$. Substituting in the right-hand side of (1), we obtain

$$\lim_{x \to +\infty} f(x) = \lim_{z \to 0^+} (1 + z)^{2/z}$$
$$= \lim_{z \to 0^+} [(1 + z)^{1/z}]^2$$
$$= [\lim_{z \to 0^+} (1 + z)^{1/z}]^2 \tag{2}$$

Furthermore,

$$\lim_{z \to 0^+} (1 + z)^{1/z} = e$$

Substituting in (2) yields

$$\lim_{x \to +\infty} f(x) = e^2$$

Thus, the sequence $\{a_n\}$ is convergent and has limit e^2.

In Exercises 20–25 use Definition 15.1.2 to prove that the given sequence has the limit L.

20. $\left\{\dfrac{4}{2n-1}\right\}; \ L = 0$

SOLUTION: For every $\epsilon > 0$ we must find a number $N > 0$ such that

$$\left|\frac{4}{2n-1}\right| < \epsilon \qquad \text{for every integer } n > N \tag{1}$$

Now for every positive integer n, we have

$$\frac{4}{2n-1} < \epsilon$$

if and only if

$$4 < 2n\epsilon - \epsilon$$

$$\frac{4 + \epsilon}{2\epsilon} < n \tag{2}$$

Thus, we let $N = (4 + \epsilon)/2\epsilon$. Then (2) holds for every integer n, if $n > N$. Thus statement (1) follows. Therefore, by Definition 15.1.2,

$$\lim_{n \to +\infty} \frac{4}{2n - 1} = 0$$

24. $\left\{\dfrac{5 - n}{2 + 3n}\right\}; \; L = -\dfrac{1}{3}$

SOLUTION: For every $\epsilon > 0$ we must find some number $N > 0$ such that

$$\left| \frac{5 - n}{2 + 3n} + \frac{1}{3} \right| < \epsilon \quad \text{for every integer } n > N \tag{1}$$

For every positive integer n,

$$\left| \frac{5 - n}{2 + 3n} + \frac{1}{3} \right| = \frac{\frac{17}{3}}{2 + 3n}$$

Thus, we must find $N > 0$ such that

$$\frac{\frac{17}{3}}{2 + 3n} < \epsilon \quad \text{for every integer } n > N \tag{2}$$

Now (2) is true if and only if

$$2 + 3n > \frac{17}{3\epsilon}$$

$$n > \frac{17 - 6\epsilon}{9\epsilon} \tag{3}$$

Thus, we let N be either $(17 - 6\epsilon)/9\epsilon$ or 1, whichever is larger. We have $N > 0$. Furthermore, whenever n is an integer with $n > N$, it follows that

$$n > \frac{17 - 6\epsilon}{9\epsilon}$$

and, since this is (3), we may reverse the steps above and obtain (1). Thus,

$$\lim_{n \to +\infty} \frac{5 - n}{2 + 3n} = -\frac{1}{3}$$

28. For the given sequence a and b are constants and $b \neq 0$. Determine whether the sequence is convergent or divergent. If the sequence converges, find its limit.

$$\left\{ \frac{1 - \left(1 - \dfrac{1}{n}\right)^a}{1 - \left(1 - \dfrac{1}{n}\right)^b} \right\}$$

SOLUTION: Let

$$a_n = \frac{1 - \left(1 - \dfrac{1}{n}\right)^a}{1 - \left(1 - \dfrac{1}{n}\right)^b} \quad \text{with } b \neq 0$$

and

$$f(x) = \frac{1 - (1 - x^{-1})^a}{1 - (1 - x^{-1})^b} \quad \text{with } b \neq 0$$

We note that $f(n) = a_n$ for every positive integer n. Because

$$\lim_{x \to +\infty} [1 - (1 - x^{-1})^q] = 0$$

and

$$\lim_{x \to +\infty} [1 - (1 - x^{-1})^b] = 0$$

we may apply L'Hôpital's rule to find the limit as follows.

$$\lim_{x \to +\infty} f(x) = \lim_{x \to +\infty} \frac{1 - (1 - x^{-1})^a}{1 - (1 - x^{-1})^b}$$

$$= \lim_{x \to +\infty} \frac{-a(1 - x^{-1})^{a-1}(x^{-2})}{-b(1 - x^{-1})^{b-1}(x^{-2})}$$

$$= \frac{a}{b} \lim_{x \to +\infty} \frac{(1 - x^{-1})^{a-1}}{(1 - x^{-1})^{b-1}}$$

$$= \frac{a}{b} \cdot \frac{1^{a-1}}{1^{b-1}} = \frac{a}{b}$$

Thus, the sequence $\{a_n\}$ is convergent, and its limit is a/b.

32. Prove that if the sequence $\{a_n\}$ is convergent and $\lim\limits_{n \to +\infty} a_n = L$, then the sequence $\{|a_n|\}$ is also convergent and $\lim\limits_{n \to +\infty} |a_n| = |L|$.

SOLUTION: We use Definition 15.1.2. For any $\epsilon > 0$ we must show that there is a number $N > 0$ such that

$$\Big| |a_n| - |L| \Big| < \epsilon \qquad \text{for every integer } n > N \tag{1}$$

Because $\lim\limits_{n \to +\infty} a_n = L$, by 15.1.2 for any $\epsilon > 0$ there is some number $N > 0$ such that

$$|a_n - L| < \epsilon \qquad \text{for every integer } n > N \tag{2}$$

We show that (2) implies (1). Because $|a| - |b| \le |a - b|$, we have

$$|a_n| - |L| \le |a_n - L| \tag{3}$$

and

$$|L| - |a_n| \le |L - a_n| \tag{4}$$

Because $|L - a_n| = |a_n - L|$, we may multiply both sides of (4) by -1 and obtain

$$|a_n| - |L| \ge -|a_n - L| \tag{5}$$

From (3) and (5) we have

$$\Big| |a_n| - |L| \Big| \le |a_n - L|$$

Therefore, (2) implies that (1) is true. Hence,

$$\lim_{n \to +\infty} |a_n| = |L|$$

15.2 MONOTONIC AND BOUNDED SEQUENCES

Sometimes we want to know whether or not a sequence is convergent without actually finding the limit. A monotonic sequence is convergent if and only if it is bounded. Formal statements of the definitions and theorems follow.

15.2.1 Definition   A sequence $\{a_n\}$ is said to be

(i) *increasing* if $a_n \leq a_{n+1}$ for all n
(ii) *decreasing* if $a_n \geq a_{n+1}$ for all n.

If a sequence is increasing or if it is decreasing, it is called *monotonic*.

15.2.2 Definition   The number C is called a *lower bound* of the sequence $\{a_n\}$ if $C \leq a_n$ for all positive integers n, and the number D is called an *upper bound* of the sequence $\{a_n\}$ if $a_n \leq D$ for all positive integers n.

15.2.4 Definition   A sequence $\{a_n\}$ is said to be *bounded* if and only if it has an upper bound and a lower bound.

15.2.6 Theorem   A bounded monotonic sequence is convergent.

15.2.9 Theorem   A convergent monotonic sequence is bounded.

We may use either Definition 15.2.1 or the following theorem to show that the sequence $\{a_n\}$ *is monotonic*.

Theorem   If f is a function such that $f(n) = a_n$ for each positive integer n, and

(i) if $f'(x) > 0$ for all $x > 0$, then $\{a_n\}$ is increasing
(ii) if $f'(x) < 0$ for all $x > 0$, then $\{a_n\}$ is decreasing.

If the sequence $\{a_n\}$ is increasing, then $a_n \geq a_1$ for all n. Thus, a_1 is a lower bound for $\{a_n\}$, and $\{a_n\}$ is convergent if and only if $\{a_n\}$ has an upper bound. Similarly, if $\{a_n\}$ is decreasing, then a_1 is an upper bound and $\{a_n\}$ is convergent if and only if $\{a_n\}$ has a lower bound. We state this formally in the following theorems.

15.2.7 Theorem   Let $\{a_n\}$ be an increasing sequence, and suppose that D is an upper bound of this sequence. Then $\{a_n\}$ is convergent and

$$\lim_{n \to +\infty} a_n \leq D$$

15.2.8 Theorem   Let $\{a_n\}$ be a decreasing sequence, and suppose that C is a lower bound of this sequence. Then $\{a_n\}$ is convergent and

$$\lim_{n \to +\infty} a_n \geq C$$

To complete the discussion in this section, we have the following definition and axiom.

15.2.3 Definition   If A is a lower bound of a sequence $\{a_n\}$ and if A has the property that for every lower bound C of $\{a_n\}$, $C \leq A$, then A is called the *greatest lower bound* of the sequence. Similarly, if B is an upper bound of a sequence $\{a_n\}$ and if B has the property that for every upper bound D of $\{a_n\}$, $B \leq D$, then B is called the *least upper bound* of the sequence.

15.2.5 The Axiom of Completeness   Every nonempty set of real numbers that has a lower bound has a greatest lower bound. Also, every nonempty set of real numbers that has an upper bound has a least upper bound.

Exercises 15.2

In Exercises 1–16 determine if the given sequence is increasing, decreasing, or not monotonic.

4. $\{\sin n\pi\}$

SOLUTION: Let $a_n = \sin n\pi$. For each positive integer n we have $\sin n\pi = 0$. Because $a_n = a_{n+1}$ for each positive integer n, then $a_n \leq a_{n+1}$ for each positive integer n, and by Definition 15.2.1(i) the sequence $\{a_n\}$ is increasing. On the other hand, it is also true that $a_n \geq a_{n+1}$ for each positive integer n. Therefore, by Definition 15.2.1(ii) the sequence $\{a_n\}$ is also decreasing.

8. $\left\{\dfrac{2^n}{1+2^n}\right\}$

SOLUTION: Let

$$a_n = \frac{2^n}{1+2^n}$$

We have $a_1 = \frac{2}{3}$, $a_2 = \frac{4}{5}$, $a_3 = \frac{8}{9}$. Because $a_1 < a_2 < a_3$, we think that the sequence may be increasing. By Definition 15.2.1 we must show for all n

$$a_n \leq a_{n+1} \tag{1}$$

Because

$$a_{n+1} = \frac{2^{n+1}}{1+2^{n+1}}$$

then (1) means that we must show that for all n

$$\frac{2^n}{1+2^n} \leq \frac{2^{n+1}}{1+2^{n+1}} \tag{2}$$

Eliminating the fractions in (2), we obtain

$$2^n + 2^{2n+1} \leq 2^{n+1} + 2^{2n+1}$$
$$2^n \leq 2^{n+1} \tag{3}$$

Dividing on both sides of (3) by 2^n, we get

$$1 \leq 2 \tag{4}$$

Because (4) is true and (4) is equivalent to (1), we conclude that $\{a_n\}$ is increasing.

12. $\left\{\dfrac{n}{2^n}\right\}$

SOLUTION: Let

$$a_n = \frac{n}{2^n}$$

We have $a_1 = \frac{1}{2}$, $a_2 = \frac{1}{2}$, $a_3 = \frac{3}{8}$, and $a_4 = \frac{1}{4}$. Because $a_1 \geq a_2 > a_3 > a_4$, the sequence may be decreasing. We show that for all n,

$$a_n \geq a_{n+1} \tag{1}$$

which is equivalent to

$$\frac{n}{2^n} \geq \frac{n+1}{2^{n+1}}$$

$$n \cdot 2^{n+1} \geq (n+1)2^n$$
$$2n \geq n+1$$
$$n \geq 1 \tag{2}$$

Because (2) is true and (2) is equivalent to (1), we conclude that the sequence $\{a_n\}$ is decreasing.

16. $\left\{ \dfrac{1 \cdot 3 \cdot 5 \cdot \cdots \cdot (2n-1)}{2^n \cdot n!} \right\}$

SOLUTION: Let

$$a_n = \frac{1 \cdot 3 \cdot 5 \cdot \cdots \cdot (2n-1)}{2^n \cdot n!}$$

We have

$$a_1 = \frac{1}{2} = 0.50$$

$$a_2 = \frac{1 \cdot 3}{2^2 \cdot 2!} = \frac{3}{8} = 0.38$$

$$a_3 = \frac{1 \cdot 3 \cdot 5}{2^3 \cdot 3!} = \frac{5}{16} = 0.31$$

$$a_4 = \frac{1 \cdot 3 \cdot 5 \cdot 7}{2^4 \cdot 4!} = \frac{35}{128} = 0.27$$

Because $a_1 > a_2 > a_3 > a_4$, the sequence $\{a_n\}$ may be decreasing. We show that for all n

$$a_n \geq a_{n+1} \tag{1}$$

Replacing n by $n + 1$ in a_n, we get

$$a_{n+1} = \frac{1 \cdot 3 \cdot 5 \cdot \cdots \cdot (2n+1)}{2^{n+1} \cdot (n+1)!}$$

Thus, (1) is equivalent to

$$\frac{1 \cdot 3 \cdot 5 \cdot \cdots \cdot (2n-1)}{2^n \cdot n!} \geq \frac{1 \cdot 3 \cdot 5 \cdot \cdots \cdot (2n+1)}{2^{n+1} \cdot (n+1)!} \tag{2}$$

Because

$$\frac{2^{n+1}}{2^n} = 2 \quad \text{and} \quad \frac{(n+1)!}{n!} = n+1$$

if we multiply on both sides of (2) by $2^{n+1}(n+1)!$, we get

$$2(n+1)[1 \cdot 3 \cdot 5 \cdot \cdots \cdot (2n-1)] \geq 1 \cdot 3 \cdot 5 \cdot \cdots \cdot (2n+1) \tag{3}$$

Because

$$1 \cdot 3 \cdot 5 \cdot \cdots \cdot (2n+1) = 1 \cdot 3 \cdot 5 \cdot \cdots \cdot (2n-1)(2n+1)$$

we may divide on both sides of (3) by $1 \cdot 3 \cdot 5 \cdot \cdots \cdot (2n-1)$ and obtain

$$2(n+1) \geq 2n+1$$
$$2 \geq 1 \tag{4}$$

Because (4) is true and (4) is equivalent to (1), we conclude that the sequence $\{a_n\}$ is decreasing.

20. Determine whether the sequence $\{3 - (-1)^{n-1}\}$ is bounded.

SOLUTION: Let $a_n = 3 - (-1)^{n-1}$. If n is an odd integer, then $n - 1$ is even, so $(-1)^{n-1} = 1$ and $a_n = 2$. If n is an even integer, then $n - 1$ is odd, so

$(-1)^{n-1} = -1$ and $a_n = 4$. Thus, $2 \leq a_n \leq 4$ for all positive integers n, and thus the sequence $\{a_n\}$ is bounded.

In Exercises 21–30 prove that the given sequence is convergent by using Theorem 15.2.6.

24. The sequence of Exercise 8.

SOLUTION: The sequence $\{a_n\}$ is defined by

$$a_n = \frac{2^n}{1 + 2^n}$$

In Exercise 8 we showed that $\{a_n\}$ is increasing. Thus, $\{a_n\}$ is monotonic. We show that $\{a_n\}$ is bounded. Because $\{a_n\}$ is increasing, then $a_n \geq a_1$ for all n. Thus, a_1 is a lower bound for $\{a_n\}$. Because $2^n < 1 + 2^n$, then

$$\frac{2^n}{1 + 2^n} < \frac{1 + 2^n}{1 + 2^n} = 1$$

Thus, $a_n < 1$, and hence, 1 is an upper bound for $\{a_n\}$. Because $\{a_n\}$ is monotonic and bounded, by Theorem 15.2.6 it is convergent.

28. The sequence of Exercise 16.

SOLUTION: The sequence $\{a_n\}$ is defined by

$$a_n = \frac{1 \cdot 3 \cdot 5 \cdot \cdots \cdot (2n - 1)}{2^n \cdot n!}$$

In Exercise 16 we showed that $\{a_n\}$ is decreasing. Thus, $a_n \leq a_1$ for all n, and hence a_1 is an upper bound for $\{a_n\}$. Because $a_n > 0$ for all n, then 0 is a lower bound for $\{a_n\}$. Because $\{a_n\}$ is bounded and monotonic, by Theorem 15.2.6 $\{a_n\}$ is convergent.

32. Given the sequence $\{a_n\}$, where $a_n > 0$ for all n, and $a_{n+1} < ka_n$ with $0 < k < 1$. Prove that $\{a_n\}$ is convergent.

SOLUTION: We apply Theorem 15.2.6. Because $0 < k < 1$, and $a_n > 0$ for all n, then $ka_n < a_n$ for all n. We are given that $a_{n+1} < ka_n$. Therefore, $a_{n+1} < a_n$ and thus the sequence $\{a_n\}$ is decreasing. Furthermore, because $a_n \leq a_1$ for all n, then a_1 is an upper bound for the sequence. And because $a_n > 0$ for all n, then 0 is a lower bound for the sequence. Therefore, $\{a_n\}$ is a bounded monotonic sequence and by Theorem 15.2.6 the sequence is convergent.

15.3 INFINITE SERIES OF CONSTANT TERMS

Let $\{u_n\}$ be a sequence and define $\{s_n\}$, the *sequence of partial sums,* as follows:

$s_1 = u_1$
$s_2 = u_1 + u_2$
$s_3 = u_1 + u_2 + u_3$

and so on, where

$$s_n = u_1 + u_2 + \cdots + u_n$$

The sequence $\{s_n\}$ is called an *infinite series.* If $\{s_n\}$ converges, then we define

$$\sum_{n=1}^{+\infty} u_n = \lim_{n \to +\infty} s_n$$

and we say that the infinite series is *convergent.* The formal definitions are as follows.

15.3.1 Definition If $\{u_n\}$ is a sequence and

$$s_n = \sum_{i=1}^{n} u_i = u_1 + u_2 + u_3 + \cdots + u_n$$

then the sequence $\{s_n\}$ is called an *infinite series*.

15.3.2 Definition Let $\sum\limits_{n=1}^{+\infty} u_n$ be a given infinite series, and let $\{s_n\}$ be the sequence of partial sums defining this infinite series. Then if $\lim\limits_{n\to +\infty} s_n$ exists and is equal to S, we say that the given series is *convergent* and that S is the *sum* of the given infinite series. If $\lim\limits_{n\to +\infty} s_n$ does not exist, the series is said to be *divergent* and the series does not have a sum.

We will not usually be able to use Definition 15.3.2 to tell whether or not an infinite series is convergent. The following theorems allow us to prove that an infinite series is *divergent*.

15.3.3 Theorem If the infinite series $\sum\limits_{n=1}^{+\infty} u_n$ is convergent, then $\lim\limits_{n\to +\infty} u_n = 0$.

From Theorem 15.3.3 it follows that if $\lim\limits_{n\to +\infty} u_n \neq 0$, then the series $\sum\limits_{n=1}^{+\infty} u_n$ is divergent.

The converse of Theorem 15.3.3 is false. That is, if $\lim\limits_{n\to +\infty} u_n = 0$, it does not follow that $\sum\limits_{n=1}^{+\infty} u_n$ is convergent. For example, we have the *harmonic series*, defined by

$$\sum_{n=1}^{+\infty} \frac{1}{n} = 1 + \frac{1}{2} + \frac{1}{3} + \cdots + \frac{1}{n} + \cdots$$

which is divergent.

If the ratio of successive terms in a series is constant, the series is called a *geometric series*. That is, a geometric series is a series of the form

$$\sum_{n=1}^{+\infty} ar^{n-1} = a + ar + ar^2 + \cdots + ar^{n-1} + \cdots$$

where a and r are any constants. We have the following very important theorem.

15.3.5 Theorem The geometric series converges to the sum $a/(1-r)$ if $|r| < 1$, and the geometric series diverges if $|r| \geq 1$. Thus,

$$\sum_{n=1}^{+\infty} ar^{n-1} = \frac{a}{1-r} \qquad \text{if } |r| < 1$$

When testing an infinite series to determine whether or not it is convergent, we may disregard the first N terms, where N is any particular positive integer. Furthermore, it does not affect the convergence or divergence of a series if we multiply each term by the same nonzero constant. If we add or subtract the corresponding terms of two convergent series, the resulting series is also convergent. If we add the terms of a convergent series to the corresponding terms of a divergent series, the resulting series is divergent. However, if we add the corresponding terms of two divergent series, the resulting series may be divergent or it may be convergent. These facts are stated formally in the following theorems.

15.3.6 Theorem If $\sum\limits_{n=1}^{+\infty} a_n$ and $\sum\limits_{n=1}^{+\infty} b_n$ are two infinite series, differing only in their first m terms (i.e., $a_k = b_k$ if $k > m$), then either both series converge or both series diverge.

15.3.7 Theorem   Let c be any nonzero constant.

(i) If the series $\sum\limits_{n=1}^{+\infty} u_n$ is convergent and its sum is S, then the series $\sum\limits_{n=1}^{+\infty} cu_n$ is also convergent and its sum is $c \cdot S$.

(ii) If the series $\sum\limits_{n=1}^{+\infty} u_n$ is divergent, then the series $\sum\limits_{n=1}^{+\infty} cu_n$ is also divergent.

15.3.8 Theorem   If $\sum\limits_{n=1}^{+\infty} a_n$ and $\sum\limits_{n=1}^{+\infty} b_n$ are convergent infinite series whose sums are S and R, respectively, then

(i) $\sum\limits_{n=1}^{+\infty} (a_n + b_n)$ is a convergent series and its sum is $S + R$;

(ii) $\sum\limits_{n=1}^{+\infty} (a_n - b_n)$ is a convergent series and its sum is $S - R$.

15.3.9 Theorem   If the series $\sum\limits_{n=1}^{+\infty} a_n$ is convergent and the series $\sum\limits_{n=1}^{+\infty} b_n$ is divergent, then the series $\sum\limits_{n=1}^{+\infty} (a_n + b_n)$ is divergent.

As we progress through this chapter, we will develop techniques for determining whether or not a series is convergent. In this section we have the following methods.

To show that a series is *convergent*, either

1. find the partial sum s_n and show $\lim\limits_{n \to +\infty} s_n$ exists, or
2. show that the series is a geometric series with $|r| < 1$.

To show that a series is *divergent*, either

1. find s_n, and show that $\lim\limits_{n \to +\infty} s_n$ does not exist
2. show that the $\lim\limits_{n \to +\infty} u_n \neq 0$
3. show that the series is a geometric series with $|r| \geq 1$, or
4. show that the series can be obtained from the harmonic series by multiplying each term of the harmonic series by the same nonzero constant.

Exercises 15.3

In Exercises 1–8 find the first four elements of the sequence of partial sums $\{s_n\}$ and find a formula for s_n in terms of n. Also, determine if the infinite series is convergent or divergent, and if it is convergent, find its sum.

4. $\sum\limits_{n=1}^{+\infty} \dfrac{2}{(4n - 3)(4n + 1)}$

SOLUTION: We have

$$u_n = \frac{2}{(4n - 3)(4n + 1)} \tag{1}$$

Thus, by making natural number replacements for n in (1), we have

$$u_1 = \tfrac{2}{5} \qquad u_2 = \tfrac{2}{45} \qquad u_3 = \tfrac{2}{117} \qquad u_4 = \tfrac{2}{221}$$

from which we obtain the first four terms of the sequence of partial sums.

$$s_1 = u_1 = \tfrac{2}{5}$$
$$s_2 = u_1 + u_2 = \tfrac{2}{5} + \tfrac{2}{45} = \tfrac{4}{9}$$
$$s_3 = u_1 + u_2 + u_3 = s_2 + u_3 = \tfrac{4}{9} + \tfrac{2}{117} = \tfrac{6}{13}$$
$$s_4 = u_1 + u_2 + u_3 + u_4 = s_3 + u_4 = \tfrac{6}{13} + \tfrac{2}{221} = \tfrac{8}{17}$$

To find a formula for s_n, we use partial fractions. Let

$$u_n = \frac{2}{(4n-3)(4n+1)} = \frac{A}{4n-3} + \frac{B}{4n+1}$$

We find $A = \tfrac{1}{2}$ and $B = -\tfrac{1}{2}$. Thus

$$u_n = \frac{2}{(4n-3)(4n+1)} = \frac{\tfrac{1}{2}}{4n-3} - \frac{\tfrac{1}{2}}{4n+1} \tag{2}$$

From (2) we obtain

$$s_n = \sum_{i=1}^{n} u_i$$

$$= \sum_{i=1}^{n} \left[\frac{\tfrac{1}{2}}{4i-3} - \frac{\tfrac{1}{2}}{4i+1} \right]$$

$$= -\frac{1}{2} \sum_{i=1}^{n} \left[\frac{1}{4i+1} - \frac{1}{4i-3} \right] \tag{3}$$

We use

$$\sum_{i=1}^{n} [F(i) - F(i-1)] = F(n) - F(0)$$

with

$$F(i) = \frac{1}{4i+1}$$

to obtain the sum in (3). Thus,

$$s_n = -\frac{1}{2}\left[\frac{1}{4n+1} - 1 \right] = \frac{2n}{4n+1}$$

Because

$$\lim_{n \to +\infty} s_n = \lim_{n \to +\infty} \frac{2n}{4n+1}$$

$$= \lim_{n \to +\infty} \frac{2}{4 + \dfrac{1}{n}} = \frac{1}{2}$$

we conclude that the infinite series is convergent and

$$\sum_{n=1}^{+\infty} \frac{2}{(4n-3)(4n+1)} = \frac{1}{2}$$

8. $\displaystyle\sum_{n=1}^{+\infty} \frac{2^{n-1}}{3^n}$

SOLUTION: we have

$$u_n = \frac{2^{n-1}}{3^n}$$

Thus, $u_1 = \tfrac{1}{3}$, $u_2 = \tfrac{2}{9}$, $u_3 = \tfrac{4}{27}$, and $u_4 = \tfrac{8}{81}$, from which we obtain the first four elements of the sequence of partial sums.

$$s_1 = u_1 = \tfrac{1}{3}$$
$$s_2 = u_1 + u_2 = \tfrac{1}{3} + \tfrac{2}{9} = \tfrac{5}{9}$$
$$s_3 = s_2 + u_3 = \tfrac{5}{9} + \tfrac{4}{27} = \tfrac{19}{27}$$
$$s_4 = s_3 + u_4 = \tfrac{19}{27} + \tfrac{8}{81} = \tfrac{65}{81}$$

Because $u_2 \div u_1 = \tfrac{2}{3}$, $u_3 \div u_2 = \tfrac{2}{3}$, and $u_4 \div u_3 = \tfrac{2}{3}$, the series may be a geometric series. In fact

$$s_n = \sum_{i=1}^{n} u_i$$

$$= \sum_{i=1}^{n} \frac{2^{i-1}}{3^i}$$

$$= \frac{1}{3} \sum_{i=1}^{n} \frac{2^{i-1}}{3^{i-1}} = \frac{1}{3} \sum_{i=1}^{n} \left(\frac{2}{3}\right)^{i-1} \tag{1}$$

Because the sum in (1) is of the form for a geometric series with $a = 1$ and $r = \tfrac{2}{3}$, we may use Formula (10) in the text to obtain the sum. Thus, from (1) we have

$$s_n = \frac{1}{3} \frac{a(1 - r^n)}{1 - r}$$

$$= \frac{1}{3} \cdot \frac{1 - (\tfrac{2}{3})^n}{1 - \tfrac{2}{3}}$$

$$= 1 - (\tfrac{2}{3})^n$$

Because

$$\lim_{n \to +\infty} s_n = \lim_{n \to +\infty} [1 - (\tfrac{2}{3})^n] = 1$$

we conclude that the infinite series is convergent, and

$$\sum_{n=1}^{+\infty} \frac{2^{n-1}}{3^n} = 1$$

We may also use Theorem 15.3.5 to find the sum, as follows:

$$\sum_{n=1}^{+\infty} \frac{2^{n-1}}{3^n} = \frac{1}{3} \sum_{n=1}^{+\infty} \left(\frac{2}{3}\right)^{n-1}$$

$$= \frac{1}{3} \cdot \frac{1}{1 - \tfrac{2}{3}} = 1$$

12. Find the infinite series that is the given sequence of partial sums. Also determine if the infinite series is convergent or divergent, and if it is convergent, find its sum.

$$\{s_n\} = \{3^n\}$$

SOLUTION: Let the infinite series be represented by

$$\sum_{n=1}^{+\infty} u_n$$

Because $s_n = s_{n-1} + u_n$, and we are given that $s_n = 3^n$, then

$$u_n = s_n - s_{n-1}$$
$$= 3^n - 3^{n-1}$$
$$= 3^{n-1}(3 - 1) = 2 \cdot 3^{n-1}$$

Thus, the infinite series is

$$\sum_{n=1}^{+\infty} 2 \cdot 3^{n-1}$$

Because the series is a geometric series with $r = 3$, and thus $|r| > 1$, the series is divergent.

In Exercises 14–29 write the first four terms of the given infinite series and determine if the series is convergent or divergent. If the series is convergent, find its sum.

16. $\displaystyle\sum_{n=1}^{+\infty} \frac{2}{3n}$

SOLUTION:

$$\sum_{n=1}^{+\infty} \frac{2}{3n} = \frac{2}{3} + \frac{1}{3} + \frac{2}{9} + \frac{1}{6} + \cdots \tag{1}$$

Because

$$\sum_{n=1}^{+\infty} \frac{2}{3n} = \frac{2}{3} \sum_{n=1}^{+\infty} \frac{1}{n}$$

each term in (1) is $\frac{2}{3}$ times the corresponding term in the harmonic series. Thus, series (1) is divergent.

20. $\displaystyle\sum_{n=1}^{+\infty} \frac{2}{3^{n-1}}$

SOLUTION:

$$\sum_{n=1}^{+\infty} \frac{2}{3^{n-1}} = 2 + \frac{2}{3} + \frac{2}{9} + \frac{2}{27} + \cdots$$

Because

$$\sum_{n=1}^{+\infty} \frac{2}{3^{n-1}} = \sum_{n=1}^{+\infty} 2 \left(\frac{1}{3}\right)^{n-1}$$

the series is a geometric series with $a = 2$ and $r = \frac{1}{3}$. By Theorem 15.3.5 we conclude that the series is convergent and

$$\sum_{n=1}^{+\infty} \frac{2}{3^{n-1}} = \frac{2}{1 - \frac{1}{3}} = 3$$

24. $\displaystyle\sum_{n=1}^{+\infty} \frac{\sinh n}{n}$

SOLUTION:

$$\sum_{n=1}^{+\infty} \frac{\sinh n}{n} = \sinh 1 + \frac{1}{2} \sinh 2 + \frac{1}{3} \sinh 3 + \frac{1}{4} \sinh 4 + \cdots$$

$$= 1.18 + 1.83 + 3.34 + 6.82 + \cdots$$

Because $u_1 < u_2 < u_3 < u_4$, the sequence $\{u_n\}$ may not have limit 0. By L'Hôpital's rule

$$\lim_{x \to +\infty} \frac{\sinh x}{x} = \lim_{x \to +\infty} \frac{\cosh x}{1} = +\infty$$

Thus, by Theorem 15.3.3 the series is divergent.

28. $\sum\limits_{n=1}^{+\infty} [1 + (-1)^n]$

SOLUTION:

$$\sum\limits_{n=1}^{+\infty} [1 + (-1)^n] = 0 + 2 + 0 + 2 + \cdots$$

Let $a_n = 1 + (-1)^n$. Because $a_{2n} = 2$, then $\lim\limits_{n \to +\infty} a_n \neq 0$. Therefore, by Theorem 15.3.3 the given infinite series is divergent.

32. Express the given nonterminating repeating decimal as a common fraction.

2.045 45 45 . . .

SOLUTION:

$$2.045\ 45\ 45\ \ldots = 2.0 + [0.045 + 0.00045 + 0.0000045 + \cdots] \tag{1}$$

The expression in brackets in (1) is a geometric series with $a = 0.045$ and $r = 0.01$ since $0.00045 \div 0.045 = 0.01$, $0.0000045 \div 0.00045 = 0.01$, and so on. Thus, by Theorem 16.3.6, we have

$$0.045 + 0.00045 + 0.0000045 + \cdots = \frac{0.045}{1 - 0.01} = \frac{1}{22} \tag{2}$$

Substituting from (2) into (1), we obtain

$$2.045\ 45\ 45\ \ldots = 2.0 + \tfrac{1}{22} = \tfrac{45}{22}$$

36. A ball is dropped from a height of 12 meters. Each time it strikes the ground, it bounces back to a height of three-fourths the distance from which it fell. Find the total distance traveled by the ball before it comes to rest.

SOLUTION: Let u_n be the number of meters traveled by the ball during the nth bounce. Because the ball is dropped from a height of 12 m and rebounds to a height of $\tfrac{3}{4} \cdot 12$ m, on the first bounce the ball travels a total distance of 21 m. That is,

$$u_1 = 21$$

Because the ball rebounds to a height of three-fourths the distance from which it fell on each bounce, we have for each n

$$u_n = \tfrac{3}{4} u_{n-1}$$

Thus,

$$u_2 = \tfrac{3}{4} u_1 = (\tfrac{3}{4})21$$
$$u_3 = \tfrac{3}{4} u_2 = (\tfrac{3}{4})^2 21$$
$$u_4 = \tfrac{3}{4} u_3 = (\tfrac{3}{4})^3 21$$
$$\cdot$$
$$\cdot$$
$$\cdot$$
$$u_n = (\tfrac{3}{4})^{n-1} 21$$

Let S be the number of feet in the total distance traveled by the ball before coming to rest. Then

$$S = \sum\limits_{n=1}^{+\infty} u_n$$

$$= \sum\limits_{n=1}^{+\infty} 21 \left(\frac{3}{4}\right)^{n-1} \tag{1}$$

Because (1) is a geometric series with $a = 21$ and $r = \frac{3}{4}$, we have

$$S = \frac{21}{1 - \frac{3}{4}} = 84$$

Thus, the ball travels 84 meters before coming to rest.

15.4 INFINITE SERIES OF POSITIVE TERMS

If every term in a series is positive, we may sometimes determine whether or not the series is convergent by comparing it with a series of positive terms that we know to be convergent or that we know to be divergent.

15.4.2 Theorem (Comparison Test) Let the series $\sum\limits_{n=1}^{+\infty} u_n$ be a series of positive terms.

(i) If $\sum\limits_{n=1}^{+\infty} v_n$ is a series of positive terms that is known to be convergent, and $u_n \leq v_n$ for all positive integers n, then $\sum\limits_{n=1}^{+\infty} u_n$ is convergent.

(ii) If $\sum\limits_{n=1}^{+\infty} v_n$ is a series of positive terms that is known to be divergent, and $u_n \geq v_n$ for all positive integers n, then $\sum\limits_{n=1}^{+\infty} u_n$ is divergent.

The comparison test fails for the two cases not covered. That is, if $0 < u_n \leq v_n$ and $\sum\limits_{n=1}^{+\infty} v_n$ is divergent, we do not know whether $\sum\limits_{n=1}^{+\infty} u_n$ is convergent or divergent. And if $0 < v_n \leq u_n$ and $\sum\limits_{n=1}^{+\infty} v_n$ is convergent, we do not know whether $\sum\limits_{n=1}^{+\infty} u_n$ is convergent or divergent. Sometimes, when Theorem 15.4.2 is difficult to apply, we may use the following test to determine whether or not the series is convergent.

15.4.3 Theorem (Limit Comparison Test) Let $\sum\limits_{n=1}^{+\infty} u_n$ and $\sum\limits_{n=1}^{+\infty} v_n$ be two series of positive terms.

(i) If $\lim\limits_{n \to +\infty} (u_n/v_n) = c > 0$, then the two series either both converge or both diverge.

(ii) If $\lim\limits_{n \to +\infty} (u_n/v_n) = 0$, and if $\sum\limits_{n=1}^{+\infty} v_n$ converges, then $\sum\limits_{n=1}^{+\infty} u_n$ converges.

(iii) If $\lim\limits_{n \to +\infty} (u_n/v_n) = +\infty$, and if $\sum\limits_{n=1}^{+\infty} v_n$ diverges, then $\sum\limits_{n=1}^{+\infty} u_n$ diverges.

The limit comparison test fails in part (ii) if $\sum\limits_{n=1}^{+\infty} v_n$ diverges, and the test fails in part (iii) if $\sum\limits_{n=1}^{+\infty} v_n$ converges.

For the known series $\sum\limits_{n=1}^{+\infty} v_n$ in the comparison tests, we may use a geometric series, the harmonic series, or the *hyperharmonic* or *p-series,* defined as follows.

$$\sum_{n=1}^{+\infty} \frac{1}{n^p} = \frac{1}{1^p} + \frac{1}{2^p} + \frac{1}{3^p} + \cdots + \frac{1}{n^p} + \cdots$$

where p is a constant. The p-series converges if $p > 1$ and diverges if $p \leq 1$.

We may also use the result of Example 1 in the text. That is,

$$\sum_{n=1}^{+\infty} \frac{1}{n!}$$

is convergent.

Be careful to distinguish a geometric series from a p-series. Note that n is the exponent in a geometric series, whereas n is the base in a p-series. We summarize this statement in the following table.

Geometric Series	p-Series
$\displaystyle\sum_{n=1}^{+\infty} r^n$	$\displaystyle\sum_{n=1}^{+\infty} \frac{1}{n^p}$
converges if $\|r\| < 1$ diverges if $\|r\| \geq 1$	converges if $p > 1$ diverges if $p \leq 1$

The following theorems complete the study of series in this section.

15.4.1 Theorem   An infinite series of positive terms is convergent if and only if its sequence of partial sums has an upper bound.

15.4.4 Theorem   If $\displaystyle\sum_{n=1}^{+\infty} u_n$ is a given convergent series of positive terms, its terms can be grouped in any manner, and the resulting series also will be convergent and will have the same sum as the given series.

15.4.5 Theorem   If $\displaystyle\sum_{n=1}^{+\infty} u_n$ is a given convergent series of positive terms, the order of the terms can be rearranged and the resulting series also will be convergent and will have the same sum as the given series.

Exercises 15.4

In Exercises 1–26 determine whether the given series is convergent or divergent.

4. $\displaystyle\sum_{n=1}^{+\infty} \frac{n^2}{4n^3 + 1}$

SOLUTION: For large n, the expression $4n^3 + 1$ is approximately equal to $4n^3$. Thus, for large n

$$\frac{n^2}{4n^3 + 1} \approx \frac{n^2}{4n^3} = \frac{1}{4n}$$

We make a limit comparison test with

$$u_n = \frac{n^2}{4n^3 + 1} \quad \text{and} \quad v_n = \frac{1}{4n}$$

Thus,

$$\lim_{n \to +\infty} \frac{u_n}{v_n} = \lim_{n \to +\infty} \frac{4n^3}{4n^3 + 1}$$

$$= \lim_{n \to +\infty} \frac{4}{4 + \dfrac{1}{n^3}} = 1$$

Because the harmonic series is divergent, and

$$\sum_{n=1}^{+\infty} v_n = \sum_{n=1}^{+\infty} \frac{1}{4n} = \frac{1}{4} \sum_{n=1}^{+\infty} \frac{1}{n}$$

the series $\displaystyle\sum_{n=1}^{+\infty} v_n$ is divergent. Thus, by Theorem 15.4.3(i), $\displaystyle\sum_{n=1}^{+\infty} u_n$ is divergent. That is, the given series is divergent.

8. $\displaystyle\sum_{n=1}^{+\infty} \frac{1}{\ln(n+1)}$

SOLUTION: For all positive integers n we have $0 < \ln(n+1) < n+1$, and thus

$$\frac{1}{\ln(n+1)} > \frac{1}{n+1}$$

We use the comparison test with

$$u_n = \frac{1}{\ln(n+1)} \quad \text{and} \quad v_n = \frac{1}{n+1}$$

Because

$$\sum_{n=1}^{+\infty} v_n = \sum_{n=1}^{+\infty} \frac{1}{n+1} = \tfrac{1}{2} + \tfrac{1}{3} + \tfrac{1}{4} + \cdots$$

then $\displaystyle\sum_{n=1}^{+\infty} v_n$ is the harmonic series with its first term omitted, and thus $\displaystyle\sum_{n=1}^{+\infty} v_n$ is divergent. Because $u_n > v_n$, by Theorem 15.4.2(ii) we conclude that $\displaystyle\sum_{n=1}^{+\infty} u_n$ is divergent. That is, the given series is divergent.

12. $\displaystyle\sum_{n=1}^{+\infty} \frac{1}{\sqrt{n^3+1}}$

SOLUTION: Because

$$\frac{1}{\sqrt{n^3+1}} < \frac{1}{\sqrt{n^3}}$$

we let

$$u_n = \frac{1}{\sqrt{n^3+1}} \quad \text{and} \quad v_n = \frac{1}{\sqrt{n^3}} = \frac{1}{n^{3/2}}$$

Because

$$\sum_{n=1}^{+\infty} v_n = \sum_{n=1}^{+\infty} \frac{1}{n^{3/2}}$$

then $\displaystyle\sum_{n=1}^{+\infty} v_n$ is a p-series with $p = \tfrac{3}{2}$. Because $p > 1$, the p-series converges. Thus, by Theorem 15.4.2(i) the given series $\displaystyle\sum_{n=1}^{+\infty} u_n$ also converges.

16. $\displaystyle\sum_{n=1}^{+\infty} \sin\frac{1}{n}$

SOLUTION: We use the limit comparison test. Let

$$u_n = \sin\frac{1}{n} \quad \text{and} \quad v_n = \frac{1}{n}$$

$$\lim_{n \to +\infty} \frac{u_n}{v_n} = \lim_{n \to +\infty} \frac{\sin\dfrac{1}{n}}{\dfrac{1}{n}}$$

(1)

To find the limit in (1), we let $x = 1/n$. Then $n \to +\infty$ is equivalent to $x \to 0^+$, and we have

$$\lim_{n \to +\infty} \frac{\sin \dfrac{1}{n}}{\dfrac{1}{n}} = \lim_{x \to 0^+} \frac{\sin x}{x} = 1$$

Thus, $\lim\limits_{n \to +\infty} u_n/v_n = 1$. Furthermore, $\sum\limits_{n=1}^{+\infty} v_n$ is divergent because the series is the harmonic series. By Theorem 15.4.3(i), we conclude that the given series is also divergent.

20. $\sum\limits_{n=1}^{+\infty} \dfrac{2^n}{n!}$

SOLUTION: As in Example 6 of the text, the comparison test fails if we attempt to compare the given series with $\sum\limits_{n=1}^{+\infty} 1/n!$. However,

$$\sum_{n=1}^{+\infty} \frac{2^n}{n!} = \frac{2}{1} + \frac{2 \cdot 2}{2 \cdot 1} + \frac{2 \cdot 2 \cdot 2}{3 \cdot 2 \cdot 1} + \frac{2 \cdot 2 \cdot 2 \cdot 2}{4 \cdot 3 \cdot 2 \cdot 1} + \cdots \tag{1}$$

Beginning with the third term in (1), the ratio of successive terms is less than $\frac{2}{3}$. That is, if

$$a_n = \frac{2^n}{n!}$$

Then

$$\frac{a_{n+1}}{a_n} = \frac{2^{n+1}}{(n+1)!} \div \frac{2^n}{n!}$$

$$= \frac{2}{n+1}$$

$$\leq \frac{2}{3} \qquad \text{if } n \geq 2$$

Therefore, we disregard the first two terms of the given series and make a comparison test with a geometric series with ratio $\frac{2}{3}$ and first term a_3. That is, we let

$$u_n = a_{n+2} = \frac{2^{n+2}}{(n+2)!} \tag{1}$$

and

$$v_n = a_3(\tfrac{2}{3})^{n-1}$$
$$= \tfrac{4}{3}(\tfrac{2}{3})^{n-1}$$
$$= 2(\tfrac{2}{3})^n \tag{2}$$

We show that $u_n \leq v_n$. Because

$$(n+2)! = 1 \cdot 2 \cdot 3 \cdot \cdots \cdot (n+2)$$
$$= 2[3 \cdot 4 \cdot \cdots \cdot (n+2)]$$
$$> 2[3 \cdot 3 \cdot 3 \cdot \cdots \cdot 3] = 2 \cdot 3^n$$

then

$$\frac{1}{(n+2)!} < \frac{1}{2 \cdot 3^n}$$

$$\frac{2^{n+2}}{(n+2)!} < \frac{2^{n+2}}{2 \cdot 3^n}$$

$$= 2(\tfrac{2}{3})^n \qquad (3)$$

Substituting from (1) and (2) into (3), we have $u_n \le v_n$. Furthermore, $\sum_{n=1}^{+\infty} v_n$ converges because it is a geometric series with $|r| < 1$. Thus, by Theorems 15.4.2(i) and 15.3.6, the given series is convergent.

24. $\displaystyle\sum_{n=1}^{+\infty} \frac{1}{3^n - \cos n}$

SOLUTION: Because $-1 \le -\cos n \le 1$, then

$$3^n - 1 \le 3^n - \cos n \le 3^n + 1$$

$$\frac{1}{3^n + 1} \le \frac{1}{3^n - \cos n} \le \frac{1}{3^n - 1} \qquad (1)$$

we show that

$$\sum_{n=1}^{+\infty} \frac{1}{3^n - 1} \qquad (2)$$

converges, and by (1) and the comparison test, it follows that the given series is convergent. To show that (2) is convergent, we use a limit comparison test. Let

$$u_n = \frac{1}{3^n - 1} \quad \text{and} \quad v_n = \frac{1}{3^n}$$

Thus,

$$\lim_{n \to +\infty} \frac{u_n}{v_n} = \lim_{n \to +\infty} \frac{3^n}{3^n - 1}$$

$$= \lim_{n \to +\infty} \frac{1}{1 - 3^{-n}} = 1$$

Furthermore, $\sum_{n=1}^{+\infty} v_n$ converges because it is a geometric series with $r = \tfrac{1}{3}$. Thus, by Theorem 15.4.3(i) series (2) converges, and hence the given series is convergent.

28. Suppose f is a function such that $f(n) > 0$ for n any positive integer. Furthermore, suppose that if p is any positive number, $\lim_{n \to +\infty} n^p f(n)$ exists and is positive. Prove that the series $\sum_{n=1}^{+\infty} f(n)$ is convergent if $p > 1$ and divergent if $0 < p \le 1$.

SOLUTION: We use a limit comparison test with a p-series. Let

$$u_n = f(n) \quad \text{and} \quad v_n = \frac{1}{n^p}$$

By hypothesis, $u_n > 0$ for all positive n. Moreover,

$$\lim_{n \to +\infty} \frac{u_n}{v_n} = \lim_{n \to +\infty} n^p f(n) \qquad (1)$$

By hypothesis, the limit in (1) exists and is positive if $p > 0$. Thus, by Theorem 15.4.3(i) the series $\sum\limits_{n=1}^{+\infty} f(n)$ is convergent if $\sum\limits_{n=1}^{+\infty} v_n$ converges, and the series $\sum\limits_{n=1}^{+\infty} f(n)$ is divergent if $\sum\limits_{n=1}^{+\infty} v_n$ diverges. Because $\sum\limits_{n=1}^{+\infty} v_n$ is a p-series, it converges if $p > 1$ and diverges if $0 < p \leq 1$. Thus, the given series is convergent if $p > 1$ and divergent if $0 < p \leq 1$.

15.5 THE INTEGRAL TEST

15.5.1 Theorem (Integral Test) Let f be a function that is continuous, decreasing, and positive valued for all $x \geq 1$. Then the infinite series

$$\sum_{n=1}^{+\infty} f(n) = f(1) + f(2) + f(3) + \cdots + f(n) + \cdots$$

is convergent if the improper integral

$$\int_1^{+\infty} f(x) \, dx$$

exists, and it is divergent if $\lim\limits_{b \to +\infty} \int_1^b f(x) \, dx = +\infty$.

We may also use the integral test for an infinite series that does not begin with $n = 1$. Thus, if f is continuous, decreasing, and positive valued for all $x \geq a$, where a is a positive integer, then $\sum\limits_{n=a}^{+\infty} f(n)$ converges if and only if $\int_a^{+\infty} f(x) \, dx$ is convergent.

Exercises 15.5

In Exercises 1–8 use the integral test to determine whether the given series is convergent or divergent.

4. $\sum\limits_{n=2}^{+\infty} \dfrac{n}{n^2 - 2}$

SOLUTION: Let

$$f(x) = \frac{x}{x^2 - 2}$$

We show that the hypothesis of the integral test is satisfied. First, we note that f is continuous and positive valued for all $x \geq 2$. We show that f is also decreasing for $x \geq 2$.

$$f'(x) = \frac{(x^2 - 2)(1) - x(2x)}{(x^2 - 2)^2}$$

$$= -\frac{x^2 + 2}{(x^2 - 2)^2}$$

Because $f'(x) < 0$ if $x \geq 2$, then f is decreasing on $[2, +\infty)$. Thus, the hypothesis of Theorem 15.5.1 is satisfied.

$$\int_2^{+\infty} f(x) \, dx = \int_2^{+\infty} \frac{x \, dx}{x^2 - 2}$$

$$= \lim_{b \to +\infty} \int_2^b \frac{x \, dx}{x^2 - 2}$$

$$= \lim_{b \to +\infty} \frac{1}{2} \ln|x^2 - 2| \Big]_2^b$$

$$= \frac{1}{2} \lim_{b \to +\infty} [\ln(b^2 - 2) - \ln 2] = +\infty$$

Therefore, the given series is divergent.

8. $\displaystyle\sum_{n=1}^{+\infty} \frac{2n}{n^4 + 1}$

SOLUTION: Let

$$f(x) = \frac{2x}{x^4 + 1}$$

Then

$$f'(x) = \frac{(x^4 + 1)(2) - 2x(4x^3)}{(x^4 + 1)^2}$$

$$= \frac{-6x^4 + 2}{(x^4 + 1)^2}$$

Because $f'(x) < 0$ if $x \geq 1$, then f is decreasing on $[1, +\infty)$. Furthermore, f is continuous and positive valued for all $x \geq 1$. Thus, the hypothesis of the integral test (15.5.1) is satisfied.

$$\int_1^{+\infty} f(x)\, dx = \int_1^{+\infty} \frac{2x\, dx}{x^4 + 1}$$

$$= \lim_{b \to +\infty} \int_1^b \frac{2x\, dx}{1 + (x^2)^2}$$

$$= \lim_{b \to +\infty} \tan^{-1} x^2 \Big]_1^b$$

$$= \lim_{b \to +\infty} [\tan^{-1} b^2 - \tan^{-1} 1]$$

$$= \frac{\pi}{2} - \frac{\pi}{4} = \frac{\pi}{4}$$

Therefore, the given series is convergent.

In Exercises 9–22 determine whether the given series is convergent or divergent.

12. $\displaystyle\sum_{n=1}^{+\infty} ne^{-n^2}$

SOLUTION: We use the integral test with $f(x) = xe^{-x^2}$. It is clear that f is continuous and positive valued. To show that f is decreasing for $x > 1$, we find the derivative of f. Thus,

$$f'(x) = e^{-x^2}(1 - 2x^2)$$

Because $f'(x) < 0$ if $x > 1$, then f is decreasing on $[1, +\infty)$. Thus, the hypothesis of the integral test is satisfied. Furthermore,

$$\int_1^{+\infty} f(x)\, dx = \lim_{b \to +\infty} \int_1^b xe^{-x^2}\, dx$$

$$= \lim_{b \to +\infty} \left[-\frac{1}{2} e^{-x^2} \right]_1^b$$

$$= \lim_{b \to +\infty} \left[-\frac{1}{2} e^{-b^2} + \frac{1}{2} e^{-1} \right] = \frac{1}{2e}$$

Thus, the given series is convergent.

16. $\displaystyle\sum_{n=1}^{+\infty} \cot^{-1} n$

SOLUTION: We use a limit comparison test with

$$u_n = \cot^{-1} n \quad \text{and} \quad v_n = \frac{1}{n}$$

Because

$$\lim_{x \to +\infty} \cot^{-1} x = 0 \quad \text{and} \quad \lim_{x \to +\infty} \frac{1}{x} = 0$$

we may apply L'Hôpital's rule to determine that

$$\lim_{n \to +\infty} \frac{u_n}{v_n} = \lim_{x \to +\infty} \frac{\cot^{-1} x}{\dfrac{1}{x}}$$

$$= \lim_{x \to +\infty} \frac{\dfrac{-1}{1 + x^2}}{\dfrac{-1}{x^2}}$$

$$= \lim_{x \to +\infty} \frac{x^2}{1 + x^2}$$

$$= \lim_{x \to +\infty} \frac{1}{\dfrac{1}{x^2} + 1} = 1$$

Because $\displaystyle\sum_{n=1}^{+\infty} v_n$ is the harmonic series that is divergent, by Theorem 15.4.3(i) we conclude that $\displaystyle\sum_{n=1}^{+\infty} u_n$ is also divergent. That is, the given series is divergent.

20. $\displaystyle\sum_{n=1}^{+\infty} \operatorname{sech}^2 n$

SOLUTION: We use the integral test with $f(x) = \operatorname{sech}^2 x$. Because $f'(x) = -2 \operatorname{sech}^2 x \tanh x$, then $f'(x) < 0$ if $x \geq 1$. Thus, f is decreasing on $[1, +\infty)$. Furthermore, f is continuous and positive valued for $x \geq 1$. Thus, the hypothesis of the integral test is satisfied. We have

$$\int_1^{+\infty} f(x)\, dx = \lim_{b \to +\infty} \int_1^b \operatorname{sech}^2 x\, dx$$

$$= \lim_{b \to +\infty} \left[\tanh x \right]_1^b$$

$$= \lim_{b \to +\infty} [\tanh b + \tanh 1]$$

$$= \lim_{b \to +\infty} \tanh b + \tanh 1 \qquad (1)$$

We use L'Hôpital's rule to find the limit in (1). Thus,

$$\lim_{b \to +\infty} \tanh b = \lim_{b \to +\infty} \frac{e^b - e^{-b}}{e^b + e^{-b}} \cdot \frac{e^b}{e^b}$$

$$= \lim_{b \to +\infty} \frac{e^{2b} - 1}{e^{2b} + 1}$$

$$= \lim_{b \to +\infty} \frac{2e^{2b}}{2e^{2b}} = 1 \qquad (2)$$

Substituting from (2) into (1), we obtain

$$\int_1^{+\infty} f(x)\, dx = 1 + \tanh 1$$

Hence, the given series is convergent.

24. Prove that the given series is convergent if and only if $p > 1$.

$$\sum_{n=3}^{+\infty} \frac{1}{n(\ln n)[\ln(\ln n)]^p}$$

SOLUTION: We apply the integral test. Let

$$f(x) = \frac{1}{x(\ln x)[\ln(\ln x)]^p}$$

If $x \geq 3$, then $\ln x > 1$, and $\ln(\ln x) > 0$, and thus $f(x) > 0$. Hence f is continuous, decreasing, and positive valued for all $x \geq 3$. Furthermore,

$$\int_3^{+\infty} f(x)\, dx = \int_3^{+\infty} \frac{dx}{x(\ln x)[\ln(\ln x)]^p}$$

$$= \lim_{b \to +\infty} \int_3^b \frac{dx}{x(\ln x)[\ln(\ln x)]^p} \qquad (1)$$

To evaluate the indefinite integral we let

$$u = \ln(\ln x) \quad \text{and} \quad du = \frac{dx}{x(\ln x)}$$

Thus,

$$\int \frac{dx}{x(\ln x)[\ln(\ln x)]^p} = \int \frac{du}{u^p}$$

$$= \begin{cases} \ln|u| + C & \text{if } p = 1 \\ \dfrac{u^{-p+1}}{-p+1} + C & \text{if } p \neq 1 \end{cases} \qquad (2)$$

We consider three cases.

Case 1: $p = 1$
Substituting from (2) into (1), we have

$$\int_3^{+\infty} f(x)\, dx = \lim_{b \to +\infty} \left[\ln|u| \right]_3^b$$

$$= \lim_{b \to +\infty} [\ln b - \ln 3] = +\infty$$

Thus, the given series is divergent if $p = 1$.

Case 2: $p < 1$

Then $1 - p > 0$, and substituting from (2) into (1) we have

$$\int_3^{+\infty} f(x)\, dx = \lim_{b \to +\infty} \left[\frac{u^{-p+1}}{-p+1} \right]_3^b$$

$$= \frac{1}{1-p} \lim_{b \to +\infty} [b^{1-p} - 3^{1-p}] = +\infty$$

Thus, the given series is divergent if $p < 1$.

Case 3: $p > 1$

Then $p - 1 > 0$, and

$$\int_3^{+\infty} f(x)\, dx = \lim_{b \to +\infty} \left[\frac{u^{-p+1}}{-p+1} \right]_3^b$$

$$= \frac{-1}{p-1} \lim_{b \to +\infty} \left[\frac{1}{u^{p-1}} \right]_3^b$$

$$= \frac{-1}{p-1} \lim_{b \to +\infty} \left[\frac{1}{b^{p-1}} - \frac{1}{3^{p-1}} \right]$$

$$= \frac{1}{(p-1)3^{p-1}}$$

Thus, the given series is convergent if $p > 1$. From Cases 1, 2, and 3 we conclude that the series is convergent if and only if $p > 1$.

15.6 INFINITE SERIES OF POSITIVE AND NEGATIVE TERMS

An alternating series is a series whose terms are alternately positive and negative. If the terms in an alternating series are decreasing in absolute value and have limit zero, the alternating series is convergent. Furthermore, if the sum of such an alternating series is approximated by the first k terms, then the absolute value of the error is less than the absolute value of the $k + 1$st term. We state these facts formally in the following definitions and theorems.

15.6.1 Definition   If $a_n > 0$ for all positive integers n, then the series

$$\sum_{n=1}^{+\infty} (-1)^{n+1} a_n = a_1 - a_2 + a_3 - a_4 + \cdots + (-1)^{n+1} a_n + \cdots$$

and the series

$$\sum_{n=1}^{+\infty} (-1)^n a_n = -a_1 + a_2 - a_3 + a_4 - \cdots + (-1)^n a_n + \cdots$$

are called *alternating series.*

15.6.2 Theorem   (Alternating Series Test) If the numbers u_1, u_2, u_3, . . . , u_n, . . . are alternately positive and negative, $|u_{n+1}| < |u_n|$ for all positive integers n, and $\lim_{n \to +\infty} u_n = 0$, then the alternating series $\sum_{n=1}^{+\infty} u_n$ is convergent.

15.6.3 Definition   If an infinite series $\sum_{n=1}^{+\infty} u_n$ is convergent and its sum is S, then the *remainder* obtained by approximating the sum of the series by the kth partial sum s_k is denoted by R_k and

$$R_k = S - s_k$$

15.6.4 Theorem   Suppose $\sum_{n=1}^{+\infty} u_n$ is an alternating series, $|u_{n+1}| < |u_n|$, and $\lim_{n \to +\infty} u_n = 0$. Then, if

R_k is the remainder obtained by approximating the sum of the series by the sum of the first k terms, $|R_k| < |u_{k+1}|$.

If we replace each term in a convergent series by its absolute value, the resulting series may be either convergent or divergent. If convergent, we say that the original series is *absolutely convergent;* if divergent, we say that the original series is *conditionally convergent.* The formal definitions and theorems follow.

15.6.5 Definition The infinite series $\sum\limits_{n=1}^{+\infty} u_n$ is said to be *absolutely convergent* if the series $\sum\limits_{n=1}^{+\infty} |u_n|$ is convergent.

15.6.6 Definition A series that is convergent, but not absolutely convergent, is said to be *conditionally convergent.*

15.6.7 Theorem If the infinite series $\sum\limits_{n=1}^{+\infty} u_n$ is absolutely convergent, it is convergent and

$$\left| \sum_{n=1}^{+\infty} u_n \right| \leq \sum_{n=1}^{+\infty} |u_n|$$

The last and most powerful of all the tests for convergence is the ratio test, which follows. The ratio test is particularly helpful for those series in which the nth term contains either $n!$ or a^n.

15.6.8 Theorem (Ratio Test) Let $\sum\limits_{n=1}^{+\infty} u_n$ be a given infinite series for which every u_n is nonzero. Then

(i) if $\lim\limits_{n \to +\infty} |u_{n+1}/u_n| = L < 1$, the given series is absolutely convergent;

(ii) if $\lim\limits_{n \to +\infty} |u_{n+1}/u_n| = L > 1$ or if $\lim\limits_{n \to +\infty} |u_{n+1}/u_n| = +\infty$, the series is divergent;

(iii) if $\lim\limits_{n \to +\infty} |u_{n+1}/u_n| = 1$, no conclusion regarding convergence may be made from this text.

The following steps may often be used to determine whether the infinite series $\sum\limits_{n=1}^{+\infty} u_n$ is convergent or divergent.

---

1. If $u_n = ar^{n-1}$, with $a \neq 0$, then the series is a geometric series.
   a. If $|r| < 1$, the series is convergent and has limit $a/(1 - r)$.
   b. If $|r| \geq 1$, the series is divergent.
2. If $u_n = a/n^p$, with $a \neq 0$, the series is a constant multiple of a p-series.
   a. If $p > 1$, the series is convergent.
   b. If $p \leq 1$, the series is divergent.
3. Find $\lim\limits_{n \to +\infty} u_n = L$.
   a. If $L \neq 0$, the series is divergent.
   b. If $L = 0$, and the series is an alternating series, and if $|u_{n+1}| < |u_n|$, the series is convergent.
   c. If $L = 0$ and the alternating series test does not apply, no conclusion can be made.

4. Find $\lim\limits_{n \to +\infty} \left| \dfrac{u_{n+1}}{u_n} \right| = L$ (ratio test)

   a. If $L < 1$, the series is absolutely convergent.
   b. If $L > 1$ or $L = +\infty$, the series is divergent.
   c. If $L = 1$ or L does not exist, no conclusion can be made.
5. If $u_n > 0$ for all n, try the following:
   a. Use a comparison test or a limit comparison test with some known series.
   b. Use the integral test.

Exercises 15.6

In Exercises 1–10 determine whether the given alternating series is convergent or divergent.

4. $\displaystyle\sum_{n=1}^{+\infty} (-1)^{n+1} \frac{\ln n}{n}$

SOLUTION: We use the alternating series test (15.6.2). Let

$$u_n = (-1)^{n+1} \frac{\ln n}{n}$$

By L'Hôpital's rule,

$$\lim_{n \to +\infty} |u_n| = \lim_{n \to +\infty} \frac{\ln n}{n}$$

$$= \lim_{n \to +\infty} \frac{1}{n} = 0$$

To show that $|u_{n+1}| < |u_n|$, we let f be the function defined by $f(n) = |u_n|$. Thus,

$$f(x) = \frac{\ln x}{x}$$

$$f'(x) = \frac{1 - \ln x}{x^2}$$

Because $f'(x) < 0$ for all $x > e$, then f is decreasing on $[e, +\infty)$. Thus, $|u_{n+1}| < |u_n|$ if $n \geq 3$. By the alternating series test, we conclude that the series is convergent.

8. $\displaystyle\sum_{n=1}^{+\infty} (-1)^n \frac{\sqrt{n}}{3n - 1}$

SOLUTION: Let

$$u_n = (-1)^n \frac{\sqrt{n}}{3n - 1}$$

By L'Hôpital's rule

$$\lim_{n \to +\infty} |u_n| = \lim_{n \to +\infty} \frac{\sqrt{n}}{3n - 1}$$

$$= \lim_{n \to +\infty} \frac{\frac{1}{2} n^{-1/2}}{3}$$

$$= 0$$

We have $|u_{n+1}| \leq |u_n|$ if and only if each of the following inequalities is satisfied:

$$\frac{\sqrt{n+1}}{3n+2} \leq \frac{\sqrt{n}}{3n-1}$$

$$\frac{n+1}{9n^2 + 12n + 4} \leq \frac{n}{9n^2 - 6n + 1}$$

$$(n+1)(9n^2 - 6n + 1) \leq n(9n^2 + 12n + 4)$$

$$9n^3 + 3n^2 - 5n + 1 \leq 9n^3 + 12n^2 + 4n$$

$$1 \leq 9n(n+1) \tag{1}$$

Because (1) is true for all positive integers n, we have $|u_{n+1}| \leq |u_n|$. Thus, by the alternating series test the given series is convergent.

12. Find the error if the sum of the first four terms is used as an approximation to the sum of the given infinite series.

$$\sum_{n=1}^{+\infty} (-1)^{n+1} \frac{1}{n^n}$$

SOLUTION:

$$\sum_{n=1}^{+\infty} (-1)^{n+1} \frac{1}{n^n} = 1 - \frac{1}{2^2} + \frac{1}{3^3} - \frac{1}{4^4} + \frac{1}{5^5} - \cdots \qquad (1)$$

We use Theorem 15.6.4 with

$$u_n = (-1)^{n+1} \frac{1}{n^n}$$

Because

$$\lim_{n \to +\infty} u_n = 0 \quad \text{and} \quad |u_{n+1}| < |u_n|$$

the error if the sum of the first four terms is used to approximate the sum in (1) is $|R_4|$ where $|R_4| < |u_5|$. Because

$$|u_5| = \frac{1}{5^5} = 0.00032$$

the error is less than 0.00032.

16. Find the sum of the given infinite series, accurate to three decimal places.

$$\sum_{n=1}^{+\infty} (-1)^n \frac{1}{(2n+1)^3}$$

SOLUTION:

$$\sum_{n=1}^{+\infty} (-1)^n \frac{1}{(2n+1)^3} = -\frac{1}{3^3} + \frac{1}{5^3} - \frac{1}{7^3} + \frac{1}{9^3} - \frac{1}{11^3} + \cdots$$

$$= -0.0370 + 0.0080 - 0.0029 + 0.0014 - 0.0008$$
$$+ 0.0005 - 0.0003 + \cdots \qquad (1)$$

Because the series is alternating, the terms are decreasing in absolute value, and the limit of the terms is zero, we may use Theorem 15.6.4. Thus, if the first six terms are used to approximate the sum in (1), the error is less than the absolute value of the seventh term; that is, the error is less than 0.0003. Adding the first six terms in (1), we obtain -0.031, which is the desired three-place approximation.

In Exercises 19–32 determine whether the given series is absolutely convergent, conditionally convergent, or divergent. Prove your answer.

20. $\displaystyle\sum_{n=1}^{+\infty} n\left(\frac{2}{3}\right)^n$

SOLUTION: We use the ratio test, Theorem 15.6.8. Because $u_n = n(\frac{2}{3})^n$, then

$$\lim_{n \to +\infty} \left| \frac{u_{n+1}}{u_n} \right| = \lim_{n \to +\infty} \left[(n+1)\left(\frac{2}{3}\right)^{n+1} \div n\left(\frac{2}{3}\right)^n \right]$$

$$= \lim_{n \to +\infty} \left(\frac{2}{3}\right) \frac{n+1}{n} = \frac{2}{3}$$

Because $\frac{2}{3} < 1$, we conclude by Theorem 15.6.8(i) that the series is absolutely convergent.

24. $\sum\limits_{n=1}^{+\infty} (-1)^{n+1} \dfrac{1}{n(n+2)}$

SOLUTION: Let

$$u_n = \frac{(-1)^{n+1}}{n(n+2)}$$

Then

$$|u_n| = \frac{1}{n^2 + 2n}$$

We use a comparison test with

$$v_n = \frac{1}{n^2}$$

We have

$$0 < |u_n| < v_n$$

for all positive integers n, and because $\sum\limits_{n=1}^{+\infty} v_n$ is a p-series with $p = 2 > 1$, then $\sum\limits_{n=1}^{+\infty} v_n$ is convergent. Therefore, $\sum\limits_{n=1}^{+\infty} |u_n|$ is convergent, and by Definition 15.6.5 we conclude that the given series is absolutely convergent.

28. $\sum\limits_{n=1}^{+\infty} (-1)^n \dfrac{n^2 + 1}{n^3}$

SOLUTION: We try the ratio test.

$$|u_n| = \frac{n^2 + 1}{n^3}$$

Thus,

$$\lim_{n \to +\infty} \left| \frac{u_{n+1}}{u_n} \right| = \lim_{n \to +\infty} \frac{\dfrac{(n+1)^2 + 1}{(n+1)^3}}{\dfrac{n^2 + 1}{n^3}}$$

$$= \lim_{n \to +\infty} \frac{n^5 + 2n^4 + 2n^3}{n^5 + 3n^4 + 4n^3 + 4n^2 + 3n + 1} = 1$$

Because the limit is 1, the ratio test fails. However, the series is an alternating series, and

$$\lim_{n \to +\infty} |u_n| = \lim_{n \to +\infty} \left(\frac{1}{n} + \frac{1}{n^3} \right) = 0$$

Furthermore, $|u_{n+1}| < |u_n|$ because

$$\frac{1}{n+1} + \frac{1}{(n+1)^3} < \frac{1}{n} + \frac{1}{n^3}$$

Thus, by the alternating series test, the series is convergent. To determine whether the series is absolutely convergent, we consider the following series.

$$\sum_{n=1}^{+\infty} |u_n| = \sum_{n=1}^{+\infty} \left(\frac{1}{n} + \frac{1}{n^3} \right) \tag{1}$$

Because $\sum_{n=1}^{+\infty} 1/n$ is a divergent harmonic series and $\sum_{n=1}^{+\infty} 1/n^3$ is a convergent p-series, by Theorem 15.3.9 we conclude that series (1) is divergent. Hence, the given series is conditionally convergent.

32. $\sum_{n=1}^{+\infty} \dfrac{1 \cdot 3 \cdot 5 \cdot \cdots \cdot (2n-1)}{1 \cdot 4 \cdot 7 \cdot \cdots \cdot (3n-2)}$

SOLUTION: Let

$$u_n = \frac{1 \cdot 3 \cdot 5 \cdot \cdots \cdot (2n-1)}{1 \cdot 4 \cdot 7 \cdot \cdots \cdot (3n-2)}$$

Then

$$u_1 = \frac{1}{1} \qquad u_2 = \frac{1 \cdot 3}{1 \cdot 4} \qquad u_3 = \frac{1 \cdot 3 \cdot 5}{1 \cdot 4 \cdot 7} \qquad \text{and so on}$$

We have

$$u_{n+1} = \frac{1 \cdot 3 \cdot 5 \cdot \cdots \cdot (2n-1)(2n+1)}{1 \cdot 4 \cdot 7 \cdot \cdots \cdot (3n-2)(3n+1)}$$

and

$$\frac{u_{n+1}}{u_n} = \frac{1 \cdot 3 \cdot 5 \cdot \cdots \cdot (2n-1)(2n+1)}{1 \cdot 4 \cdot 7 \cdot \cdots \cdot (3n-2)(3n+1)} \cdot \frac{1 \cdot 4 \cdot 7 \cdot \cdots \cdot (3n-2)}{1 \cdot 3 \cdot 5 \cdot \cdots \cdot (2n-1)}$$

$$= \frac{2n+1}{3n+1}$$

Thus,

$$\lim_{n \to +\infty} \left| \frac{u_{n+1}}{u_n} \right| = \lim_{n \to +\infty} \frac{2n+1}{3n+1} = \frac{2}{3}$$

By the ratio test, we conclude that the series converges absolutely.

36. Show by means of an example that the converse of Exercise 35 is not true.

SOLUTION: The converse of Exercise 35 is as follows: If $\sum_{n=1}^{+\infty} u_n^2$ is convergent, then $\sum_{n=1}^{+\infty} u_n$ is absolutely convergent. For a counterexample, let $u_n = 1/n$. Then

$$\sum_{n=1}^{+\infty} u_n^2 = \sum_{n=1}^{+\infty} \frac{1}{n^2}$$

is a convergent p-series because $p = 2 > 1$. However,

$$\sum_{n=1}^{+\infty} u_n = \sum_{n=1}^{+\infty} \frac{1}{n}$$

is a harmonic series, which is divergent.

15.7 POWER SERIES   Until now, we have considered infinite series in which each term is a constant. We now consider infinite series that contain terms that are variables. The series

$$\sum_{n=0}^{+\infty} c_n x^n = c_0 + c_1 x + c_2 x^2 + \cdots + c_n x^n + \cdots \tag{1}$$

is called a *power series* in x. Sometimes a power series in x converges for certain replacements of x and diverges for other replacements of x. It depends on how the constants c_0, c_1, c_2, etc., are chosen. We have the following.

15.7.4 Theorem Let $\sum\limits_{n=0}^{+\infty} c_n x^n$ be a given power series. Then exactly one of the following conditions holds:

(i) The series converges only when $x = 0$.

(ii) The series is absolutely convergent for all values of x.

(iii) There exists a number $R > 0$ such that the series is absolutely convergent for all values of x for which $|x| < R$ and is divergent for all values of x for which $|x| > R$.

The number R in Theorem 15.7.4(iii) is called the *radius of convergence* of the power series. If condition (i) holds, we take $R = 0$, and if condition (ii) holds, we write $R = +\infty$. The set of all x for which the power series converges is called the *interval of convergence*. The following steps are used to find the interval of convergence of a power series in x that is represented by $\sum\limits_{n=0}^{+\infty} u_n$ where $u_n = c_n x^n$.

---

1. Find L, where $L = \lim\limits_{n \to +\infty} \left| \dfrac{u_{n+1}}{u_n} \right|$.

   a. If $L = 0$, the series converges absolutely for all x.

   b. If $L = +\infty$ (except when $x = 0$), the series converges only when $x = 0$.

   c. If $L < 1$ for $x_1 < x < x_2$, the series converges absolutely for all x in the interval (x_1, x_2).

2. If (x_1, x_2) is the interval found in step 1(c), replace x by x_1 in the given power series and use any test except the ratio test (which fails because $L = 1$) to determine whether the series converges. Repeat this for $x = x_2$.

3. The interval of convergence is the union of the set of all x found in steps 1 and 2 for which the series converges.

If x is replaced by $(x - a)$ in power series (1), we have

$$\sum_{n=0}^{+\infty} c_n(x - a)^n = c_0 + c_1(x - a) + c_2(x - a)^2 + \cdots + c_n(x - a)^n + \cdots$$

which is a power series in $x - a$. For such a series, exactly one of the following conditions holds.

(i) The series converges only when $x = a$.

(ii) The series is absolutely convergent for all x.

(iii) There exists a number $R > 0$ such that the series is absolutely convergent for all x for which $|x - a| < R$ and is divergent for all x for which $|x - a| > R$.

If condition (i) holds, we take $R = 0$. If condition (ii) holds, we write $R = +\infty$. Otherwise, the number R in condition (iii) is the radius of convergence. To find the interval of convergence, we follow the steps given for a power series in x.

Exercises 15.7

In Exercises 1–28 find the interval of convergence of the given power series.

4. $\sum\limits_{n=0}^{+\infty} \dfrac{n^2 x^n}{2^n}$

SOLUTION: Let

$$u_n = \frac{n^2 x^n}{2^n}$$

Then

$$u_{n+1} = \frac{(n+1)^2 x^{n+1}}{2^{n+1}}$$

and

$$\lim_{n \to +\infty} \left| \frac{u_{n+1}}{u_n} \right| = \lim_{n \to +\infty} \left| \frac{(n+1)^2 x^{n+1}}{2^{n+1}} \div \frac{n^2 x^n}{2^n} \right|$$

$$= \lim_{n \to +\infty} \left| \frac{(n+1)^2 x^{n+1}}{2^{n+1}} \cdot \frac{2^n}{n^2 x^n} \right|$$

$$= \lim_{n \to +\infty} \left| \frac{(n+1)^2 x}{2n^2} \right|$$

$$= \left| \frac{x}{2} \right| \lim_{n \to +\infty} \left(1 + \frac{1}{n} \right)^2 = \left| \frac{x}{2} \right|$$

If

$$\left| \frac{x}{2} \right| < 1$$

then

$$|x| < 2$$
$$-2 < x < 2$$

Thus, by the ratio test we conclude that the given series converges absolutely for all x in the open interval $(-2, 2)$. We test each endpoint of the interval for convergence. If $x = 2$, the given power series becomes

$$\sum_{n=0}^{+\infty} \frac{n^2 2^n}{2^n} = \sum_{n=0}^{+\infty} n^2$$

which is divergent because $\lim\limits_{n \to +\infty} n^2 \neq 0$.

If $x = -2$, the given power series becomes

$$\sum_{n=0}^{+\infty} \frac{n^2 (-2)^n}{2^n} = \sum_{n=0}^{+\infty} (-1)^n n^2$$

which is divergent because $\lim\limits_{n \to +\infty} (-1)^n n^2 \neq 0$. Hence, the interval of convergence for the given power series is the open interval $(-2, 2)$.

8. $\sum\limits_{n=1}^{+\infty} (-1)^n \dfrac{x^{2n}}{(2n)!}$

SOLUTION: This is a power series in x^2, and we may proceed as for a power series in x.

$$\lim_{n \to +\infty} \left| \frac{u_{n+1}}{u_n} \right| = \lim_{n \to +\infty} \left| \frac{x^{2n+2}}{(2n+2)!} \cdot \frac{(2n)!}{x^{2n}} \right|$$

$$= x^2 \lim_{n \to +\infty} \frac{(2n)!}{(2n+2)(2n+1)(2n)!}$$

$$= x^2 \lim_{n \to +\infty} \frac{1}{(2n+2)(2n+1)} = 0$$

Because the limit is 0 for all x, by the ratio test we conclude that the series is absolutely convergent for all x. The interval of convergence is $(-\infty, +\infty)$.

12. $\displaystyle\sum_{n=0}^{+\infty} \frac{x^n}{(n+1)5^n}$

SOLUTION:

$$\lim_{n \to +\infty} \left| \frac{u_{n+1}}{u_n} \right| = \lim_{n \to +\infty} \left| \frac{x^{n+1}}{(n+2)5^{n+1}} \cdot \frac{(n+1)5^n}{x^n} \right|$$

$$= \left| \frac{x}{5} \right| \lim_{n \to +\infty} \frac{n+1}{n+2} = \left| \frac{x}{5} \right|$$

If

$$\left| \frac{x}{5} \right| < 1$$

then

$$-5 < x < 5$$

Thus, the series converges absolutely for $-5 < x < 5$. We test each endpoint. If $x = -5$, the given power series becomes

$$\sum_{n=0}^{+\infty} \frac{(-5)^n}{(n+1)5^n} = \sum_{n=0}^{+\infty} \frac{(-1)^n}{n+1}$$

which converges by the alternating series test. If $x = 5$, the given power series becomes

$$\sum_{n=0}^{+\infty} \frac{5^n}{(n+1)5^n} = \sum_{n=0}^{+\infty} \frac{1}{n+1}$$

which diverges because it is the harmonic series. Hence, the interval of convergence is $[-5, 5)$.

16. $\displaystyle\sum_{n=1}^{+\infty} \frac{(x+2)^n}{(n+1)2^n}$

SOLUTION:

$$\lim_{n \to +\infty} \left| \frac{u_{n+1}}{u_n} \right| = \lim_{n \to +\infty} \left| \frac{(x+2)^{n+1}}{(n+2)2^{n+1}} \cdot \frac{(n+1)2^n}{(x+2)^n} \right|$$

$$= \left| \frac{x+2}{2} \right| \lim_{n \to +\infty} \frac{n+1}{n+2} = \left| \frac{x+2}{2} \right|$$

If

$$\left| \frac{x+2}{2} \right| < 1$$

then

$$-1 < \frac{x+2}{2} < 1$$

$$-2 < x+2 < 2$$

$$-4 < x < 0$$

Thus, the series converges absolutely for all x in $(-4, 0)$. If $x = -4$, the given power series is

$$\sum_{n=1}^{+\infty} \frac{(-2)^n}{(n+1)2^n} = \sum_{n=1}^{+\infty} \frac{(-1)^n}{n+1}.$$

which converges by the alternating series test. If $x = 0$, the given series is

$$\sum_{n=1}^{+\infty} \frac{2^n}{(n+1)2^n} = \sum_{n=1}^{+\infty} \frac{1}{n+1}$$

which diverges because it is the harmonic series with the first term omitted. Thus, the interval of convergence of the given power series is $[-4, 0)$.

20. $\displaystyle\sum_{n=1}^{+\infty} \frac{(x+5)^{n-1}}{n^2}$

SOLUTION:

$$\lim_{n \to +\infty} \left| \frac{u_{n+1}}{u_n} \right| = \lim_{n \to +\infty} \left| \frac{(x+5)^n}{(n+1)^2} \cdot \frac{n^2}{(x+5)^{n-1}} \right|$$

$$= |x+5| \lim_{n \to +\infty} \frac{n^2}{(n+1)^2} = |x+5|$$

If $|x+5| < 1$, then

$$-1 < x+5 < 1$$
$$-6 < x < -4$$

Thus, the power series converges absolutely for all x in $(-6, -4)$. If $x = -6$, the given series is

$$\sum_{n=1}^{+\infty} \frac{(-1)^{n-1}}{n^2}.$$

which converges by the alternating series test. If $x = -4$, the given series is

$$\sum_{n=1}^{+\infty} \frac{1}{n^2}$$

which converges because it is a p-series with $p = 2$. Thus, the interval of convergence of the given power series is $[-6, -4]$.

24. $\displaystyle\sum_{n=1}^{+\infty} \frac{x^n}{n^n}$

SOLUTION:

$$\lim_{n \to +\infty} \left| \frac{u_{n+1}}{u_n} \right| = \lim_{n \to +\infty} \left| \frac{x^{n+1}}{(n+1)^{n+1}} \cdot \frac{n^n}{x^n} \right|$$

$$= |x| \lim_{n \to +\infty} \frac{n^n}{(n+1)(n+1)^n}$$

$$= |x| \lim_{n \to +\infty} \left(\frac{1}{n+1} \right) \lim_{n \to +\infty} \left(\frac{n}{n+1} \right)^n \qquad (1)$$

To find the limit in (1), we let $z = 1/n$. Then $n \to +\infty$ is equivalent to $z \to 0^+$. Thus,

$$\lim_{n \to +\infty} \left(\frac{n}{n+1}\right)^n = \lim_{z \to 0^+} \left(\frac{\dfrac{1}{z}}{\dfrac{1}{z}+1}\right)^{1/z}$$

$$= \lim_{z \to 0^+} \frac{1}{(1+z)^{1/z}}$$

$$= \frac{1}{e} \qquad \text{[By Eq. (23) in Section 8.5]} \tag{2}$$

Substituting from (2) into (1), we obtain

$$\lim_{n \to +\infty} \left|\frac{u_{n+1}}{u_n}\right| = |x| \lim_{n \to +\infty} \left(\frac{1}{n+1}\right) \cdot \frac{1}{e} = 0$$

Therefore, the given series converges absolutely for all x. The interval of convergence is $(-\infty, +\infty)$.

28. $\displaystyle\sum_{n=1}^{+\infty} \frac{(-1)^{n+1} \, 1 \cdot 3 \cdot 5 \cdot \cdots \cdot (2n-1)}{2 \cdot 4 \cdot 6 \cdot \cdots \cdot (2n)} x^n$

SOLUTION:

$$\lim_{n \to +\infty} \left|\frac{u_{n+1}}{u_n}\right| = \lim_{n \to +\infty} \left|\frac{1 \cdot 3 \cdot 5 \cdot \cdots \cdot (2n-1)(2n+1)x^{n+1}}{2 \cdot 4 \cdot 6 \cdot \cdots \cdot (2n)(2n+2)} \cdot \frac{2 \cdot 4 \cdot 6 \cdot \cdots \cdot (2n)}{1 \cdot 3 \cdot 5 \cdot \cdots \cdot (2n-1)x^n}\right|$$

$$= |x| \lim_{n \to +\infty} \frac{2n+1}{2n+2} = |x|$$

If $|x| < 1$, then $-1 < x < 1$. Thus, the series converges absolutely for all x in $(-1, 1)$. If $x = -1$, the given series becomes

$$\sum_{n=1}^{+\infty} \frac{(-1)^{2n+1} \, 1 \cdot 3 \cdot 5 \cdot \cdots \cdot (2n-1)}{2 \cdot 4 \cdot 6 \cdot \cdots \cdot (2n)} = -\sum_{n=1}^{+\infty} \frac{1 \cdot 3 \cdot 5 \cdot \cdots \cdot (2n-1)}{2 \cdot 4 \cdot 6 \cdot \cdots \cdot (2n)} \tag{1}$$

We use a comparison test to show that the series in (1) is divergent. Let

$$u_n = \frac{1 \cdot 3 \cdot 5 \cdot \cdots \cdot (2n-1)}{2 \cdot 4 \cdot 6 \cdot \cdots \cdot (2n)} \quad \text{and} \quad v_n = \frac{1}{2n+1}$$

then

$$\frac{u_n}{v_n} = \frac{3 \cdot 5 \cdot 7 \cdot \cdots \cdot (2n-1)(2n+1)}{2 \cdot 4 \cdot 6 \cdot \cdots \cdot (2n-2)(2n)} \tag{2}$$

Each factor in the numerator of (2) is larger than the corresponding factor in the denominator; that is, $2n + 1 > 2n$ for $n = 1, 2, \ldots$. Thus,

$$u_n > v_n \qquad \text{for all } n$$

Furthermore,

$$\sum_{n=1}^{+\infty} v_n = \sum_{n=1}^{+\infty} \frac{1}{2n+1} \tag{3}$$

The series in (3) is divergent by a limit comparison test with the harmonic series. Thus, the series in (1) is also divergent.

Next, we replace x by 1 in the given series. The result is

$$\sum_{n=1}^{+\infty} \frac{(-1)^{n+1} \, 1 \cdot 3 \cdot 5 \cdot \cdots \cdot (2n-1)}{2 \cdot 4 \cdot 6 \cdot \cdots \cdot (2n)} \tag{4}$$

We use the alternating series test to show that (4) is convergent. As shown when applying the ratio test above,

$$\left| \frac{u_{n+1}}{u_n} \right| = \frac{2n+1}{2n+2} < 1$$

Thus, $|u_{n+1}| < |u_n|$. To show that $\lim_{n \to +\infty} |u_n| = 0$, we use a special "trick." We have

$$|u_n| = \frac{1}{2} \cdot \frac{3}{4} \cdot \frac{5}{6} \cdot \cdots \cdot \frac{2n-1}{2n} \tag{5}$$

If we add 1 to each factor in both the numerator and denominator of u_n, the result is

$$v_n = \frac{2}{3} \cdot \frac{4}{5} \cdot \frac{6}{7} \cdot \cdots \cdot \frac{2n}{2n+1} \tag{6}$$

Each fraction in (5) is smaller than the corresponding fraction in (6). That is, for all n

$$\frac{2n-1}{2n} < \frac{2n}{2n+1}$$

because

$$4n^2 - 1 < 4n^2$$

Thus,

$$|u_n| < v_n \tag{7}$$

Furthermore,

$$v_n = \frac{2 \cdot 4 \cdot 6 \cdot \cdots \cdot (2n)}{1 \cdot 3 \cdot 5 \cdot \cdots \cdot (2n-1)} \cdot \frac{1}{2n+1}$$

$$= \frac{1}{|u_n|} \cdot \frac{1}{2n+1} \tag{8}$$

Substituting from (8) into (7), we obtain

$$|u_n| < \frac{1}{|u_n|} \cdot \frac{1}{2n+1}$$

$$0 < |u_n|^2 < \frac{1}{2n+1} \tag{9}$$

Because

$$\lim_{n \to +\infty} \frac{1}{2n+1} = 0$$

By (9) and the squeeze theorem, we have

$$\lim_{n \to +\infty} |u_n|^2 = 0$$

Thus,

$$\lim_{n \to +\infty} |u_n| = 0$$

By the alternating series test, the given power series is convergent if $x = 1$. Hence, the interval of convergence of the power series is $(-1, 1]$.

32. Prove that if $\lim\limits_{n \to +\infty} \sqrt[n]{|u_n|} = L \ (L \neq 0)$, then the radius of convergence of the power series $\sum\limits_{n=1}^{+\infty} u_n x^n$ is $1/L$.

SOLUTION: Because $\lim\limits_{n \to +\infty} \sqrt[n]{|u_n|} = L$, then

$$\left[\lim_{n \to +\infty} \sqrt[n]{|u_n|} \right]^n = L^n$$

$$\lim_{n \to +\infty} \left[\sqrt[n]{|u_n|} \right]^n = L^n$$

$$\lim_{n \to +\infty} |u_n| = L^n \tag{1}$$

Replacing n by $(n + 1)$ in Eq. (1), we have

$$\lim_{n \to +\infty} |u_{n+1}| = L^{n+1} \tag{2}$$

We use the ratio test for the series $\sum\limits_{n=1}^{+\infty} u_n x^n$. We have

$$\lim_{n \to +\infty} \left| \frac{u_{n+1} x^{n+1}}{u_n x^n} \right| = |x| \frac{\lim\limits_{n \to +\infty} |u_{n+1}|}{\lim\limits_{n \to +\infty} |u_n|} \tag{3}$$

Substituting from (1) and (2) into (3), we obtain

$$\lim_{n \to +\infty} \left| \frac{u_{n+1} x^{n+1}}{u_n x^n} \right| = |x| \frac{L^{n+1}}{L^n} = |x| L$$

Thus, the series is absolutely convergent if

$$|x| L < 1$$

$$|x| < \frac{1}{L}$$

15.8 DIFFERENTIATION OF POWER SERIES

15.8.3 Theorem Let $\sum\limits_{n=0}^{+\infty} c_n x^n$ be a power series whose radius of convergence is $R > 0$. Then if f is the function defined by

$$f(x) = \sum_{n=0}^{+\infty} c_n x^n$$

$f'(x)$ exists for every x in the open interval $(-R, R)$, and it is given by

$$f'(x) = \sum_{n=1}^{+\infty} n c_n x^{n-1}$$

Thus, if a function f is defined by a power series, we may use term-by-term differentiation on the power series to find a power-series representation of f'.

The result of Example 3 is often stated as a theorem.

Theorem

$$e^x = \sum_{n=0}^{+\infty} \frac{x^n}{n!}$$

$$= 1 + x + \frac{x^2}{2!} + \frac{x^3}{3!} + \cdots$$

And we often use the following power series, which is a geometric series.

Theorem

$$\frac{1}{1-x} = \sum_{n=0}^{+\infty} x^n \qquad \text{if } |x| < 1$$

$$= 1 + x + x^2 + \cdots \qquad \text{if } |x| < 1$$

In Table 15.8a we show a flowchart that can be used to program a computer to calculate the partial sum $S_k = \sum_{n=0}^{k} u_n$, which is an approximation for the power series $S = \sum_{n=0}^{+\infty} u_n$, provided the series converges and provided k is chosen to be large enough. We assume that $u_n \neq 0$ for $n = 0, 1, 2, \ldots, k$. Thus, $S_0 = u_0$, and for $n = 1, 2, 3, \ldots, k$, we have

$$S_n = S_{n-1} + u_n$$

and

$$u_n = u_{n-1} \cdot \frac{u_n}{u_{n-1}}$$

We hand calculate the first term u_0 and the ratio u_n/u_{n-1}, which we reduce to simplest form and use to define Q. We read in the numbers x and k. The computer printout gives the partial sum. In Sections 15.9–15.11 we show how to program the computer to determine k so that when S_k is used to approximate S, the error is less than any $\epsilon > 0$.

In Table 15.8b we show an actual computer program written in BASIC that can be used to approximate e^x by using the partial sum S_k. With the data shown, the program calculates an approximation for $e^{-2.5}$ using $k = 15$. We use the power series

$$e^x = \sum_{n=0}^{+\infty} \frac{x^n}{n!}$$

Table 15.8a

Calculation of $S_k = \sum_{n=0}^{k} u_n$

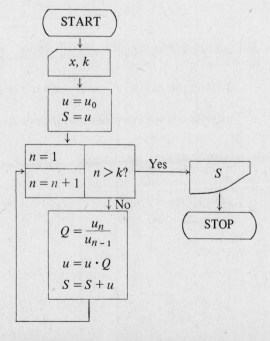

Table 15.8b

Approximation for e^x **using** $\displaystyle\sum_{n=0}^{k} \frac{x^n}{n!}$

```
10 READ X,K
20 DATA −2.5,15
30 LET S=U=1
40 FOR N=1 TO K
50 LET Q=X/N
60 LET U=U*Q
70 LET S=S+U
80 NEXT N
90 PRINT "S = ";S
999 END

RUN
RIEMAN

S = 8.20848E−02
```

Thus,

$$u_n = \frac{x^n}{n!}$$

Hence we take $u_0 = 1$ in step 30 of the program. Because

$$\frac{u_n}{u_{n-1}} = \frac{x^n}{n!} \div \frac{x^{n-1}}{(n-1)!} = \frac{x^n}{n!} \cdot \frac{(n-1)!}{x^{n-1}} = \frac{x}{n}$$

we take

$$Q = \frac{x}{n}$$

in step 50 of the program.

Exercises 15.8

In Exercises 1–10 a function f is defined by a power series. In each exercise do the following:

    (a) Find the radius of convergence of the given power series and the domain of f.

    (b) Write the power series that defines the function f' and find its radius of convergence by using methods of Section 15.7 (thus verifying Theorem 15.8.1).

    (c) Find the domain of f'.

4. $f(x) = \displaystyle\sum_{n=2}^{+\infty} \frac{(x-2)^n}{\sqrt{n-1}}$

SOLUTION:

    (a) We use the ratio test to find the radius of convergence.

$$\lim_{n \to +\infty} \left| \frac{u_{n+1}}{u_n} \right| = \lim_{n \to +\infty} \left| \frac{(x-2)^{n+1}}{\sqrt{n}} \cdot \frac{\sqrt{n-1}}{(x-2)^n} \right|$$

$$= |x - 2| \lim_{n \to +\infty} \sqrt{1 - \frac{1}{n}} = |x - 2|$$

Because the power series converges for all x such that $|x - 2| < 1$ and diverges for all x with $|x - 2| > 1$, the radius of convergence is $R = 1$. If $|x - 2| < 1$, then $1 < x < 3$. Thus, the domain of f contains all x in the open interval $(1, 3)$. We test the endpoints. We have

$$f(1) = \sum_{n=2}^{+\infty} \frac{(-1)^n}{\sqrt{n - 1}} \tag{1}$$

The series in (1) converges by the alternating series test. Hence, 1 is in the domain of f. Moreover,

$$f(3) = \sum_{n=2}^{+\infty} \frac{1}{\sqrt{n - 1}} \tag{2}$$

The series in (2) diverges by comparison with the p-series with $p = \frac{1}{2}$. Thus, 3 is not in the domain of f. We conclude that the domain of f is $[1, 3)$.

(b) Differentiating f, we obtain

$$f'(x) = \sum_{n=2}^{+\infty} \frac{n(x - 2)^{n-1}}{\sqrt{n - 1}}$$

We find the radius of convergence of f'. We have

$$\lim_{n \to +\infty} \left| \frac{u_{n+1}}{u_n} \right| = \lim_{n \to +\infty} \left| \frac{(n + 1)(x - 2)^n}{\sqrt{n}} \cdot \frac{\sqrt{n - 1}}{n(x - 2)^{n-1}} \right|$$

$$= |x - 2| \lim_{n \to +\infty} \sqrt{1 - \frac{1}{n} \left(1 + \frac{1}{n} \right)} = |x - 2|$$

Because the series converges for all x with $|x - 2| < 1$ and diverges for all x with $|x - 2| > 1$, then $R = 1$.

(c) Because $|x - 2| < 1$ if $1 < x < 3$, the domain of f' contains all numbers in $(1, 3)$. Testing the endpoints, we have

$$f'(1) = \sum_{n=1}^{+\infty} \frac{n(-1)^{n-1}}{\sqrt{n - 1}} \tag{3}$$

Because

$$\lim_{n \to +\infty} \frac{n}{\sqrt{n - 1}} = \lim_{n \to +\infty} \frac{1}{\sqrt{\frac{1}{n} - \frac{1}{n^2}}} = +\infty$$

the series in (3) diverges. Thus $f'(1)$ is not defined. Furthermore,

$$f'(3) = \sum_{n=1}^{+\infty} \frac{n}{\sqrt{n - 1}} \tag{4}$$

As before, the series in (4) is divergent. Thus, $f'(3)$ is not defined. Hence, the domain of f' is $(1, 3)$.

8. $f(x) = \sum_{n=1}^{+\infty} \frac{x^{2n-2}}{(2n - 2)!}$

SOLUTION:

(a) We use the ratio test to find R.

$$\lim_{n \to +\infty} \left| \frac{u_{n+1}}{u_n} \right| = \lim_{n \to +\infty} \left| \frac{x^{2n}}{(2n)!} \cdot \frac{(2n-2)!}{x^{2n-2}} \right|$$

$$= x^2 \lim_{n \to +\infty} \frac{(2n-2)!}{(2n)(2n-1)(2n-2)!}$$

$$= x^2 \lim_{n \to +\infty} \frac{1}{(2n)(2n-1)} = 0$$

Thus, the series converges absolutely for all x. We have $R = +\infty$ and the domain of f is $(-\infty, +\infty)$.

(b) Differentiating f, we have

$$f'(x) = \sum_{n=1}^{+\infty} \frac{(2n-2)x^{2n-3}}{(2n-2)!}$$

If $n = 1$, then $2n - 2 = 0$ and $(2n - 2)! = 0! = 1$. Thus, we begin the sum for $f'(x)$ with $n = 2$. That is,

$$f'(x) = \sum_{n=2}^{+\infty} \frac{(2n-2)x^{2n-3}}{(2n-2)(2n-3)!}$$

$$= \sum_{n=2}^{+\infty} \frac{x^{2n-3}}{(2n-3)!}$$

$$\lim_{n \to +\infty} \left| \frac{u_{n+1}}{u_n} \right| = \lim_{n \to +\infty} \left| \frac{x^{2n-1}}{(2n-1)!} \cdot \frac{(2n-3)!}{x^{2n-3}} \right|$$

$$= x^2 \lim_{n \to +\infty} \frac{(2n-3)!}{(2n-1)(2n-2)(2n-3)!}$$

$$= x^2 \lim \frac{1}{(2n-1)(2n-2)} = 0$$

Thus, the series converges for all x, and hence $R = +\infty$.

(c) Because the power series for f' converges for all x, the domain of f' is $(-\infty, +\infty)$.

---

12. Use the result of Example 3 (in the text) to find a power-series representation of $e^{\sqrt{x}}$.

SOLUTION: In Example 3 we are given that

$$e^x = \sum_{n=0}^{+\infty} \frac{x^n}{n!} \qquad \text{for all } x \in R$$

If $x \geq 0$ we may replace x by $\sqrt{x}$ and thus obtain

$$e^{\sqrt{x}} = \sum_{n=0}^{+\infty} \frac{(\sqrt{x})^n}{n!}$$

$$= \sum_{n=0}^{+\infty} \frac{x^{n/2}}{n!} \qquad \text{if } x \geq 0$$

which is the required power-series representation.

---

16. (a) Use series (2) to find a power-series representation for $1/(1 + x^3)$.

(b) Differentiate term by term the series found in part (a) to find a power-series representation for $-3x^2/(1 + x^3)^2$.

SOLUTION:

(a) Series (2) is

$$1 - x + x^2 - x^3 + \cdots + (-1)^n x^n + \cdots = \frac{1}{1+x} \quad \text{if } |x| < 1$$

If we replace x by x^3, the result is

$$1 - x^3 + x^6 - x^9 + \cdots + (-1)^n x^{3n} + \cdots = \frac{1}{1+x^3} \quad \text{if } |x| < 1$$

(b) Differentiating with respect to x on both sides, we obtain

$$-3x^2 + 6x^5 - 9x^8 + \cdots + (-1)^n (3n) x^{3n-1} + \cdots = \frac{-3x^2}{(1+x^3)^2} \quad \text{if } |x| < 1$$

which is equivalent to

$$3 \sum_{n=1}^{+\infty} (-1)^n n x^{3n-1} = \frac{-3x^2}{(1+x^3)^2} \quad \text{if } |x| < 1$$

20. If

$$f(x) = \sum_{n=0}^{+\infty} (-1)^n \frac{x^{2n}}{3^n}$$

find $f'(\tfrac{1}{2})$ correct to four decimal places.

SOLUTION: Differentiating f and disregarding the term with value zero, we have

$$f'(x) = \sum_{n=0}^{+\infty} (-1)^n \frac{2n x^{2n-1}}{3^n}$$

$$= \sum_{n=1}^{+\infty} (-1)^n \frac{2n x^{2n-1}}{3^n}$$

Thus,

$$f'\left(\frac{1}{2}\right) = \sum_{n=1}^{+\infty} (-1)^n \frac{2n(\tfrac{1}{2})^{2n-1}}{3^n}$$

$$= \sum_{n=1}^{+\infty} (-1)^n \frac{n}{3^n 2^{2n-2}}$$

$$= -\frac{1}{3} + \frac{2}{3^2 \cdot 2^2} - \frac{3}{3^3 \cdot 2^4} + \frac{4}{3^4 \cdot 2^6} - \frac{5}{3^5 \cdot 2^8} + \cdots$$

$$= -0.33333 + 0.05556 - 0.00694 + 0.00077 - 0.00008 + \cdots \tag{1}$$

Because series (1) is an alternating series that satisfies the hypothesis of the alternating series test, if the first five terms of (1) are used to approximate the sum, the absolute value of the error is less than the absolute value of the sixth term. Thus, the first five terms give an approximation that is accurate to four decimal places. Adding the first five terms in (1), we obtain -0.28402, which we round off to -0.2840. Thus, $f'(\tfrac{1}{2}) = -0.2840$.

24. (a) Find a power-series representation for $(e^x - 1)/x$.
(b) By differentiating term by term the power series in part (a), show that

$$\sum_{n=1}^{+\infty} \frac{n}{(n+1)!} = 1$$

SOLUTION:

(a) From Example 3 of the text, we have for all x

$$e^x = 1 + x + \frac{x^2}{2!} + \frac{x^3}{3!} + \cdots$$

Thus,

$$e^x - 1 = x + \frac{x^2}{2!} + \frac{x^3}{3!} + \cdots$$

$$= \sum_{n=1}^{+\infty} \frac{x^n}{n!}$$

Dividing on both sides by x, we obtain

$$\frac{e^x - 1}{x} = \sum_{n=1}^{+\infty} \frac{x^{n-1}}{n!} \tag{1}$$

which is the required power-series representation.

(b) Differentiating on both sides of (1), we obtain

$$\frac{xe^x - e^x + 1}{x^2} = \sum_{n=1}^{+\infty} \frac{(n-1)x^{n-2}}{n!}$$

$$= \sum_{n=2}^{+\infty} \frac{(n-1)x^{n-2}}{n!} \tag{2}$$

Replacing n by $n + 1$ in (2), we obtain

$$\frac{xe^x - e^x + 1}{x^2} = \sum_{n=1}^{+\infty} \frac{nx^{x-1}}{(n+1)!} \tag{3}$$

If $x = 1$ in (3), we obtain

$$1 = \sum_{n=1}^{+\infty} \frac{n}{(n+1)!}$$

28. (a) Use only properties of power series to find a power-series representation of the function f for which $f(x) > 0$ and $f'(x) = 2xf(x)$ for all x, and $f(0) = 1$.

(b) Verify your result in part (a) by solving the differential equation $D_x y = 2xy$ with the boundary condition $y = 1$ when $x = 0$.

SOLUTION:

(a) Let a power-series representation of f be given by

$$f(x) = c_0 + c_1 x + c_2 x^2 + c_3 x^3 + c_4 x^4 + c_5 x^5 + \cdots + c_n x^n + \cdots \tag{1}$$

We find the constants c_i, $i = 0, 1, 2, \ldots$. Because $f(0) = 1$, by substituting in (1) we have $c_0 = 1$. Differentiating on both sides we obtain

$$f'(x) = c_1 + 2c_2 x + 3c_3 x^2 + 4c_4 x^3 + 5c_5 x^4 + 6c_6 x^5 + \cdots + nc_n x^{n-1} + \cdots \tag{2}$$

Multiplying on both sides of (1) by $2x$ with $c_0 = 1$, we obtain

$$2xf(x) = 2x + 2c_1 x^2 + 2c_2 x^3 + 2c_3 x^4 + 2c_4 x^5 + \cdots + 2c_{n-2}x^{n-1} + \cdots \tag{3}$$

Because $f'(x) = 2x f(x)$ for all x, coefficients of like powers of x in (2) and (3) must be equal. First, we consider the coefficients of even powers of x. Equating corresponding coefficients in (2) and (3), by induction we have

$$c_1 = 0$$
$$c_3 = \tfrac{2}{3} c_1 = 0$$
$$c_5 = \tfrac{2}{5} c_3 = 0$$

·

·

·

$$c_{2n+1} = 0$$

Next, we consider the coefficients of odd powers of x. Equating corresponding coefficients in (2) and (3), by induction we obtain

$$c_2 = 1$$

$$c_4 = \frac{1}{2} c_2 = \frac{1}{2} \cdot 1 = \frac{1}{2!}$$

$$c_6 = \frac{1}{3} c_4 = \frac{1}{3} \cdot \frac{1}{2!} = \frac{1}{3!}$$

·

·

·

$$c_{2n} = \frac{1}{n!}$$

Substituting the values for c_i, $i = 0, 1, 2, \ldots$ in (1), we obtain

$$f(x) = 1 + \frac{x^2}{1!} + \frac{x^4}{2!} + \frac{x^6}{3!} + \cdots + \frac{x^{2n}}{n!} + \cdots$$

which is the desired power-series representation of f.

(b) If $y = f(x) > 0$ and if $f'(x) = 2x f(x)$, we have

$$D_x y = 2xy$$

$$\frac{dy}{dx} = 2xy$$

$$\frac{dy}{y} = 2x \, dx$$

$$\int \frac{dy}{y} = \int 2x \, dx$$

$$\ln y = x^2 + C \tag{4}$$

Because $y = 1$ when $x = 0$, from (4) we have $C = 0$. Thus,

$$\ln y = x^2$$
$$y = e^{x^2}$$
$$f(x) = e^{x^2} \tag{5}$$

From Example 3 we have

$$e^x = 1 + x + \frac{x^2}{2!} + \frac{x^3}{3!} + \cdots + \frac{x^n}{n!} + \cdots$$

Replacing x by x^2, we get

$$e^{x^2} = 1 + x^2 + \frac{x^4}{2!} + \frac{x^6}{3!} + \cdots + \frac{x^{2n}}{n!} + \cdots \tag{6}$$

Substituting from (6) into (5), we have

$$f(x) = 1 + x^2 + \frac{x^4}{2!} + \frac{x^6}{3!} + \cdots + \frac{x^{2n}}{n!} + \cdots$$

which agrees with the result in (a).

15.9 INTEGRATION OF POWER SERIES

15.9.1 Theorem   Let $\sum\limits_{n=0}^{+\infty} c_n x^n$ be a power series whose radius of convergence is $R > 0$. Then if f is the function defined by

$$f(x) = \sum_{n=0}^{+\infty} c_n x^n$$

f is integrable on every closed subinterval of $(-R, R)$, and we evaluate the integral of f by integrating the given power series term by term; that is, if x is in $(-R, R)$, then

$$\int_0^x f(t)\, dt = \sum_{n=0}^{+\infty} \frac{c_n}{n+1} x^{n+1}$$

Furthermore, R is the radius of convergence of the resulting series.

Thus, if a function f is defined by a power series, we may use term-by-term integration on the power series to find a power-series representation of $F(x) = \int_0^x f(t)\, dt$. And if x_1 is a constant that is in the interval of convergence of the power-series representation of $F(x)$, we may use the first n terms of this power series to approximate the definite integral $\int_0^{x_1} f(t)\, dt$.

We have the following power-series representations which are stated as theorems.

Theorem $\ln(1 + x) = \sum\limits_{n=1}^{+\infty} (-1)^{n-1} \dfrac{x^n}{n!}$ if $|x| < 1$

$$= x - \frac{x^2}{2} + \frac{x^3}{3} - \frac{x^4}{4} + \cdots \qquad \text{if } |x| < 1$$

Theorem $\ln \dfrac{1 + x}{1 - x} = 2 \sum\limits_{n=1}^{+\infty} \dfrac{x^{2n-1}}{2n - 1}$ if $|x| < 1$

$$= 2\left(x + \frac{x^3}{3} + \frac{x^5}{5} + \cdots \right) \qquad \text{if } |x| < 1$$

Theorem $\tan^{-1} x = \sum\limits_{n=0}^{+\infty} (-1)^n \dfrac{x^{2n+1}}{2n + 1}$ if $|x| \leq 1$

$$= x - \frac{x^3}{3} + \frac{x^5}{5} - \cdots \qquad \text{if } |x| \leq 1$$

In Table 15.9a we show a flowchart that can be used to approximate the alternating series $S = \sum\limits_{n=0}^{+\infty} u_n$ with error less than $\epsilon > 0$, provided S converges by the alternating series test. By Theorem 15.6.4, if the partial sum $\sum\limits_{n=0}^{k} u_n$ is used to approximate S, then the error is less than $|u_{k+1}|$. Thus, in Table 15.9a we terminate the sum when $|u| < \epsilon$. We must hand calculate the first term u_0 and the ratio u_n/u_{n-1} used to define the function Q.

Table 15.9a
Alternating Series

Approximation for $S = \sum\limits_{n=0}^{+\infty} u_n$ with error less than ϵ if $\dfrac{u_n}{u_{n-1}} < 0$ for $n = 1, 2,$ 3, . . .

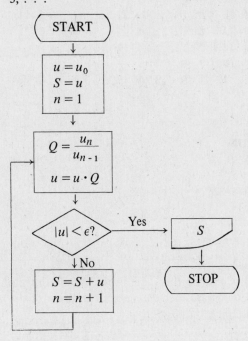

In Table 15.9b we show an actual computer program written in BASIC that can be used to calculate $\int_0^2 e^{-x^2} \, dx$ with error less than 10^{-4}. We use the power-series representation of Example 1.

$$\int_0^x e^{-t^2} \, dt = \sum_{n=0}^{+\infty} (-1)^n \frac{x^{2n+1}}{n!(2n+1)}$$

Thus,

$$\int_0^2 e^{-x^2} \, dx = \sum_{n=0}^{+\infty} \frac{(-1)^n 2^{2n+1}}{n!(2n+1)}$$

We have

$$u_n = \frac{(-1)^n 2^{2n+1}}{n!(2n+1)}$$

Thus, $u_0 = 2$ in step 10 of the program. Because

$$\frac{u_n}{u_{n-1}} = \frac{(-1)^n 2^{2n+1}}{n!(2n+1)} \cdot \frac{(n-1)!(2n-1)}{(-1)^{n-1} 2^{2n-1}}$$

$$= \frac{-2^2 (2n-1)}{n(2n+1)}$$

$$= \frac{4 - 8n}{n(2n+1)}$$

We take

$$Q = \frac{4 - 8n}{n(2n+1)}$$

in step 30 of the program.

Table 15.9b

```
1Ø  LET  S=U=2
2Ø  LET  N=1
3Ø  LET  Q=(4-8*N)/N/(2*N+1)
4Ø  LET  U=U*Q
5Ø  IF  ABS(U)<.0001  THEN  9Ø
6Ø  LET  S=S+U
7Ø  LET  N=N+1
8Ø  GOTO  3Ø
9Ø  PRINT "S=";S,"N = ";N
999  END
```

RUN
15.9B

$S = .882124 \qquad N = 15$

Exercises 15.9

4. Compute the value of the definite integral to three decimal places by two methods: **(a)** the fundamental theorem of the calculus and **(b)** a power series. Compare the results.

$$\int_0^{1/3} \ln(1 + x)\, dx$$

(a) We use integration by parts to evaluate the integral. First, let $z = 1 + x$ and $dz = dx$. Then

$$\int_0^{1/3} \ln(1 + x)\, dx = \int_1^{4/3} \ln z\, dz$$

Now let $u = \ln z$ and $dv = dz$. Then $du = z^{-1}\, dz$ and $v = z$, so

$$\int_1^{4/3} \ln z\, dz = z \ln z - z \Big]_1^{4/3}$$

$$= \left(\frac{4}{3} \ln \frac{4}{3} - \frac{4}{3}\right) - (-1) = 0.050$$

(b) In Example 3 of the text a power-series representation of $\ln(1 + x)$ is obtained. From this result we have

$$\ln(1 + t) = \sum_{n=1}^{+\infty} (-1)^{n-1} \frac{t^n}{n} \qquad \text{if } |t| < 0$$

By Theorem 15.9.1,

$$\int_0^x \ln(1 + t)\, dt = \sum_{n=1}^{+\infty} (-1)^{n-1} \frac{x^{n+1}}{n(n + 1)}$$

If $x = \frac{1}{3}$, we have

$$\int_0^{1/3} \ln(1 + t)\, dt = \sum_{n=1}^{+\infty} (-1)^{n-1} \frac{\left(\frac{1}{3}\right)^{n+1}}{n(n + 1)}$$

$$= \frac{1}{1(2)(3^2)} - \frac{1}{2(3)(3^3)} + \frac{1}{3(4)(3^4)} - \cdots$$

$$= 0.0556 - 0.0062 + 0.0010 - \cdots$$

$$= 0.050$$

In Exercises 5–16 compute the value of the given integral, accurate to three decimal places, by using series.

8. $\displaystyle\int_0^{1/3} \frac{dx}{1 + x^4}$

SOLUTION: First, we find a power-series representation of the function f such that

$$f(x) = \frac{1}{1 + t^4}$$

We have the geometric series

$$\frac{1}{1 - x} = \sum_{n=0}^{+\infty} x^n \qquad \text{if } |x| < 1$$

Replacing x by $-t^4$, we obtain

$$\frac{1}{1 + t^4} = \sum_{n=0}^{+\infty} (-1)^n t^{4n} \qquad \text{if } |t| < 1 \tag{1}$$

By Theorem 15.9.1 and (1) we have

$$\int_0^x \frac{dt}{1 + t^4} = \sum_{n=0}^{+\infty} \int_0^x (-1)^n t^{4n} \qquad \text{if } |x| < 1$$

$$= \sum_{n=0}^{+\infty} \frac{(-1)^n x^{4n+1}}{4n + 1} \qquad \text{if } |x| < 1 \tag{2}$$

If $x = \frac{1}{3}$ in (2), we have

$$\int_0^{1/3} \frac{dt}{1 + t^4} = \sum_{n=0}^{+\infty} \frac{(-1)^n}{(4n + 1)3^{4n+1}}$$

$$= \frac{1}{3} - \frac{1}{5 \cdot 3^5} + \frac{1}{9 \cdot 3^9} - \cdots$$

$$= 0.33333 - 0.00082 + 0.00001 - \cdots = 0.3325$$

Because (3) is an alternating series, the error in the approximation is less than 0.00001, the absolute value of the last term considered. Therefore, the approximation is accurate to four decimal places.

12. $\displaystyle\int_0^1 h(x)\, dx \qquad$ where $h(x) = \begin{cases} \dfrac{\sinh x}{x} & \text{if } x \neq 0 \\ 1 & \text{if } x = 0 \end{cases}$

SOLUTION:

$$h(x) = \begin{cases} \dfrac{e^x - e^{-x}}{2x} & \text{if } x \neq 0 \\ 1 & \text{if } x = 0 \end{cases}$$

We have

$$e^x = 1 + x + \frac{x^2}{2!} + \frac{x^3}{3!} + \frac{x^4}{4!} + \cdots \tag{1}$$

Thus,

$$e^{-x} = 1 - x + \frac{x^2}{2!} - \frac{x^3}{3!} + \frac{x^4}{4!} - \cdots \tag{2}$$

Subtracting the members of (2) from the members of (1), we obtain

$$e^x - e^{-x} = 2x + 2 \cdot \frac{x^3}{3!} + 2 \cdot \frac{x^5}{5!} + \cdots$$

Dividing on both sides by $2x$, we get

$$\frac{e^x - e^{-x}}{2x} = 1 + \frac{x^2}{3!} + \frac{x^4}{5!} + \cdots$$

Thus, for all x we have

$$h(x) = 1 + \frac{x^2}{3!} + \frac{x^4}{5!} + \cdots \tag{3}$$

Integrating term by term in (3), we obtain

$$\int_0^x h(t) \, dt = x + \frac{x^3}{3 \cdot 3!} + \frac{x^5}{5 \cdot 5!} + \cdots$$

Thus,

$$\int_0^1 h(t) \, dt = 1 + \frac{1}{3 \cdot 3!} + \frac{1}{5 \cdot 5!} + \frac{1}{7 \cdot 7!} + \cdots$$

$$= 1.00000 + 0.05556 + 0.00167 + 0.00003 + \cdots = 1.0573$$

16. $\displaystyle\int_0^1 g(x) \, dx$ where $g(x) = \begin{cases} \dfrac{\cosh x - 1}{x} & \text{if } x \neq 0 \\ 0 & \text{if } x = 0 \end{cases}$

SOLUTION:

$$g(x) = \begin{cases} \dfrac{\frac{1}{2}(e^x + e^{-x}) - 1}{x} & \text{if } x \neq 0 \\ 0 & \text{if } x = 0 \end{cases}$$

Because

$$e^x = 1 + x + \frac{x^2}{2!} + \frac{x^3}{3!} + \frac{x^4}{4!} + \cdots$$

and

$$e^{-x} = 1 - x + \frac{x^2}{2!} - \frac{x^3}{3!} + \frac{x^4}{4!} - \cdots$$

then by adding corresponding terms we obtain

$$e^x + e^{-x} = 2 + 2 \cdot \frac{x^2}{2!} + 2 \cdot \frac{x^4}{4!} + \cdots$$

Hence,

$$\frac{1}{2}(e^x + e^{-x}) = 1 + \frac{x^2}{2!} + \frac{x^4}{4!} + \cdots$$

$$\frac{1}{2}(e^x + e^{-x}) - 1 = \frac{x^2}{2!} + \frac{x^4}{4!} + \frac{x^6}{6!} + \cdots$$

$$\frac{\frac{1}{2}(e^x + e^{-x}) - 1}{x} = \frac{x}{2!} + \frac{x^3}{4!} + \frac{x^5}{6!} + \cdots$$

Thus, for all t we have

$$g(t) = \frac{t}{2!} + \frac{t^3}{4!} + \frac{t^5}{6!} + \cdots$$

Integrating term by term, we obtain

$$\int_0^x g(t)\, dt = \frac{x^2}{2(2!)} + \frac{x^4}{4(4!)} + \frac{x^6}{6(6!)} + \cdots$$

and if $x = 1$, the result is

$$\int_0^1 g(t)\, dt = \frac{1}{2(2!)} + \frac{1}{4(4!)} + \frac{1}{6(6!)} + \cdots$$

$$= 0.2500 + 0.0104 + 0.0002 = 0.261$$

20. Use the power series in Eq. (8) to compute ln 3 accurate to four decimal places.

SOLUTION: From Eq. (8) we have

$$\ln \frac{1+x}{1-x} = 2\left(x + \frac{x^3}{3} + \frac{x^5}{5} + \cdots + \frac{x^{2n-1}}{2n-1} + \cdots\right) \quad \text{if } |x| < 1 \tag{1}$$

To find x, we let

$$\frac{1+x}{1-x} = 3$$

Solving for x, we obtain $x = \frac{1}{2}$. Replacing x by $\frac{1}{2}$ in (1), we obtain

$$\ln 3 = 2\left(\frac{1}{2} + \frac{1}{3 \cdot 2^3} + \frac{1}{5 \cdot 2^5} + \frac{1}{7 \cdot 2^7} + \frac{1}{9 \cdot 2^9} + \frac{1}{11 \cdot 2^{11}} + \cdots\right)$$

$$= 2(0.50000 + 0.04167 + 0.00625 + 0.00112 + 0.00022 + 0.00004 + \cdots)$$

$$= 1.0986$$

24. Find a power series for xe^x by multiplying the series for e^x by x, and then integrate the resulting series term by term from 0 to 1 and show that

$$\sum_{n=1}^{+\infty} \frac{1}{n!(n+2)} = \frac{1}{2}$$

SOLUTION: First, we use integration by parts to obtain

$$\int_0^1 te^t\, dt = \left[te^t - e^t\right]_0^1 = 1 \tag{1}$$

In the power-series representation, we have

$$e^x = \sum_{n=0}^{+\infty} \frac{x^n}{n!}$$

$$xe^x = \sum_{n=0}^{+\infty} \frac{x^{n+1}}{n!}$$

Thus,

$$\int_0^x te^t \, dt = \sum_{n=0}^{+\infty} \int_0^x \frac{t^{n+1}}{n!}$$

$$= \sum_{n=0}^{+\infty} \frac{x^{n+2}}{n!(n+2)} \qquad (2)$$

If $x = 1$ in (2), we have

$$\int_0^1 te^t \, dt = \sum_{n=0}^{+\infty} \frac{1}{n!(n+2)}$$

$$= \frac{1}{2} + \sum_{n=1}^{+\infty} \frac{1}{n!(n+2)} \qquad (3)$$

Substituting from (1) into (3), we have

$$1 = \frac{1}{2} + \sum_{n=1}^{+\infty} \frac{1}{n!(n+2)}$$

$$\sum_{n=1}^{+\infty} \frac{1}{n!(n+2)} = \frac{1}{2}$$

28. Show that the interval of convergence of the power series in Eq. (9) is $[-1, 1]$ and that the power series is a representation of $\tan^{-1}x$ for all x in its interval of convergence.

SOLUTION: From Eq. (9) we have

$$\tan^{-1}x = \sum_{n=0}^{+\infty} (-1)^n \frac{x^{2n+1}}{2n+1} \qquad \text{if } |x| < 1 \qquad (1)$$

If we let $x = 1$ in the series in (1), we obtain

$$\sum_{n=0}^{+\infty} (-1)^n \frac{1}{2n+1}$$

which converges by the alternating series test. And if we let $x = -1$ in the series of (1), we get

$$\sum_{n=0}^{+\infty} (-1)^n \frac{(-1)^{2n+1}}{2n+1} = \sum_{n=0}^{+\infty} \frac{(-1)^{2n}(-1)^{n+1}}{2n+1}$$

$$= \sum_{n=0}^{+\infty} \frac{(-1)^{n+1}}{2n+1}$$

which also converges by the alternating series test. Thus, the interval of convergence for series (1) is $[-1, 1]$. In Example 4 of the text, we showed that the power series is a representation of $\tan^{-1}x$ for all x in $(-1, 1)$. We consider the endpoints of the interval. For $x = 1$ we wish to prove that

$$\tan^{-1} 1 = \sum_{n=0}^{+\infty} \frac{(-1)^n}{2n+1}$$

$$= \sum_{n=1}^{+\infty} \frac{(-1)^{n-1}}{2n-1}$$

For this infinite series, the nth partial sum is

$$S_n = 1 - \frac{1}{3} + \frac{1}{5} - \frac{1}{7} + \cdots + (-1)^{n-1} \frac{1}{2n-1} \qquad (2)$$

If we show that $\lim\limits_{n \to +\infty} S_n = \tan^{-1} 1$, we will have proved that the sum of the series is $\tan^{-1} 1$. Consider the formula for the sum of a geometric progression:

$$a + ar + ar^2 + ar^3 + \cdots + ar^{n-1} = \frac{a - ar^n}{1 - r}$$

Let $a = 1$ and $r = -t^2$. We have

$$1 - t^2 + t^4 - t^6 + \cdots + (-t^2)^{n-1} = \frac{1 - (-t^2)^n}{1 + t^2}$$

$$= \frac{1}{1 + t^2} + (-1)^{n+1} \frac{t^{2n}}{1 + t^2}$$

Integrating from 0 to 1, we have

$$\int_0^1 [1 - t^2 + t^4 - t^6 + \cdots + (-1)^{n-1} t^{2n-2}]\, dt = \int_0^1 \frac{dt}{1 + t^2} + (-1)^{n+1} \int_0^1 \frac{t^{2n}}{1 + t^2}\, dt$$

$$\left[1 - \frac{1}{3} + \frac{1}{5} - \frac{1}{7} + \cdots + (-1)^{n-1} \frac{1}{2n - 1} \right] = \tan^{-1} 1 + (-1)^{n+1} \int_0^1 \frac{t^{2n}}{1 + t^2}\, dt \quad (3)$$

Substituting from (2) into (3), we obtain

$$S_n = \tan^{-1} 1 + (-1)^{n+1} \int_0^1 \frac{t^{2n}}{1 + t^2}\, dt \qquad (4)$$

Let

$$R_n = (-1)^{n+1} \int_0^1 \frac{t^{2n}}{1 + t^2}\, dt$$

Thus, Eq. (4) can be written as

$$S_n = \tan^{-1} 1 + R_n \qquad (5)$$

Because

$$\frac{t^{2n}}{1 + t^2} \leq t^{2n}$$

then

$$\int_0^1 \frac{t^{2n}}{1 + t^2}\, dt \leq \int_0^1 t^{2n}\, dt$$

Thus,

$$0 \leq |R_n| = \int_0^1 \frac{t^{2n}}{1 + t^2}\, dt \leq \int_0^1 t^{2n}\, dt = \frac{1}{2n + 1} \qquad (6)$$

Because

$$\lim_{n \to +\infty} \frac{1}{2n + 1} = 0$$

by the squeeze theorem and inequality (6), it follows that $\lim\limits_{n \to +\infty} R_n = 0$. Thus, from (5) we have

$$\lim_{n \to +\infty} S_n = \tan^{-1} 1 + \lim_{n \to +\infty} R_n$$

$$= \tan^{-1} 1$$

Therefore, the sum of the series is $\tan^{-1} 1$. By a similar method, we prove that Eq. (1) also holds when $x = -1$.

15.10 TAYLOR SERIES The *Taylor series* of the function f at the number a is given by

$$\sum_{n=0}^{+\infty} \frac{f^{(n)}(a)}{n!}(x-a)^n = f(a) + f'(a)(x-a) + \frac{f''(a)}{2!}(x-a)^2 + \cdots$$

$$+ \frac{f^{(n)}(a)}{n!}(x-a)^n + \cdots$$

If $a = 0$ in the Taylor series of f at a, we have the special case called the *Maclaurin series* given by

$$\sum_{n=0}^{+\infty} \frac{f^{(n)}(0)}{n!} x^n = f(0) + f'(0)x + \frac{f''(0)}{2!}x^2 + \cdots + \frac{f^{(n)}(0)}{n!}x^n + \cdots$$

The power series given in Section 15.8 that represent e^x and $1/(1-x)$ are Maclaurin series, and the power series given in Section 15.9 that represent $\ln(1 + x)$, $\ln[(1 + x)/(1 - x)]$, and $\tan^{-1} x$ are also Maclaurin series. We also have the following Maclaurin series:

Theorem
$$\sin x = \sum_{n=0}^{+\infty} \frac{(-1)^n x^{2n+1}}{(2n+1)!} = x - \frac{x^3}{3!} + \frac{x^5}{5!} - \frac{x^7}{7!} + \cdots$$

Theorem
$$\cos x = \sum_{n=0}^{+\infty} \frac{(-1)^n x^{2n}}{(2n)!} = 1 - \frac{x^2}{2!} + \frac{x^4}{4!} - \frac{x^6}{6!} + \cdots$$

Theorem
$$\sinh x = \sum_{n=0}^{+\infty} \frac{x^{2n+1}}{(2n+1)!} = x + \frac{x^3}{3!} + \frac{x^5}{5!} + \frac{x^7}{7!} + \cdots$$

Theorem
$$\cosh x = \sum_{n=0}^{+\infty} \frac{x^{2n}}{(2n)!} = 1 + \frac{x^2}{2!} + \frac{x^4}{4!} + \frac{x^6}{6!} + \cdots$$

The following theorem provides a test for determining whether a function is represented by its Taylor series.

15.10.1 Theorem Let f be a function such that f and all of its derivatives exist in some interval $(a - r, a + r)$. Then the function is represented by its Taylor series

$$\sum_{n=0}^{+\infty} \frac{f^{(n)}(a)}{n!}(x-a)^n$$

for all x such that $|x - a| < r$ if and only if

$$\lim_{n \to +\infty} R_n(x) = \lim_{n \to +\infty} \frac{f^{(n+1)}(\xi_n)}{(n+1)!}(x-a)^{n+1} = 0$$

where each ξ_n is between x and a.

Let $S = \sum_{n=0}^{+\infty} u_n$ be a series of positive terms and let the partial sum $\sum_{n=0}^{k} u_n$ be used to approximate S. If there is a constant R with $0 < R < 1$ such that

$$\frac{u_n}{u_{n-1}} \leq R \qquad \text{for } n \geq k+1$$

then the error in the approximation is E_k where

$$E_k \leq \frac{u_{k+1}}{1 - R}$$

In Table 15.10a, we show a flowchart that can be used to approximate the series $S = \sum_{n=0}^{+\infty} u_n$ with error less than $\epsilon > 0$. We must hand calculate the first term u_0 and the ratio u_n/u_{n-1} used to define Q. The way in which we define R depends on the behavior of the series. If

Table 15.10a
Series of Positive Terms

Approximation for $S = \sum_{n=0}^{+\infty} u_n$ with error less than ϵ if

(i) $u_n > 0$ for $n = 0, 1, 2, \ldots$
(ii) for each positive integer k

$$\frac{u_n}{u_{n-1}} \leq R < 1 \qquad \text{for } n \geq k+1$$

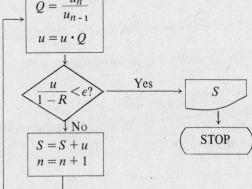

$$a_n = \frac{u_n}{u_{n-1}}$$

and $\{a_n\}$ is a decreasing sequence, then for any positive integer k we have $a_n \leq a_{k+1}$ for $n \geq k + 1$. Thus, if $0 < a_{k+1} < 1$ we may choose $R = a_{k+1}$.

In Table 15.10b we show an actual computer program written in BASIC that can be used to approximate $\int_0^2 \cosh \sqrt{x}\, dx$ with error less than 10^{-4}. We use the Maclaurin series.

$$\cosh x = \sum_{n=0}^{+\infty} \frac{x^{2n}}{(2n)!}$$

to obtain

$$\cosh \sqrt{x} = \sum_{n=0}^{+\infty} \frac{x^n}{(2n)!}$$

Integrating term by term, we have

$$\int_0^2 \cosh \sqrt{x}\, dx = \sum_{n=0}^{+\infty} \frac{x^{n+1}}{(n+1)(2n)!} \Bigg]_0^2$$

$$= \sum_{n=0}^{+\infty} \frac{2^{n+1}}{(n+1)(2n)!}$$

Thus,

$$u_n = \frac{2^{n+1}}{(n+1)(2n)!}$$

We have $u_0 = 2$ in step 10 of the program. Because

$$\frac{u_n}{u_{n-1}} = \frac{2^{n+1}}{(n+1)(2n)!} \cdot \frac{n(2n-2)!}{2^n}$$

$$= \frac{2n}{(n+1)(2n)(2n-1)}$$

$$= \frac{1}{(n+1)(2n-1)}$$

we take

$$Q = \frac{1}{(n+1)(2n-1)}$$

in step 30 of the program. Let

$$a_n = \frac{1}{(n+1)(2n-1)}$$

Because $\{a_n\}$ is a decreasing sequence, we take

$$R = \frac{1}{(n+1)(2n-1)}$$

Thus, $R = Q$ in the flowchart of Table 15.10a, and we use this to write step 50 in the program.

Table 15.10b

```
10   LET  S=U=2
20   LET  N=1
30   LET  Q=1/(N+1)/(2*N-1)
40   LET  U=U*Q
50   LET  R=Q
60   IF  ABS(U/(1-R))<.0001 THEN 100
70   LET  S=S+U
80   LET  N=N+1
90   GOTO 30
100  PRINT "S= ";S,"N = ";N
999  END
```

RUN
15.10B

$S = 3.11683 \qquad N = 5$

Exercises 15.10

4. Obtain the Maclaurin series for the cosine function by differentiating the Maclaurin series for the sine function. Also obtain the Maclaurin series for the sine function by differentiating the one for the cosine function.

SOLUTION: The Maclaurin series for the sine function is

$$\sin x = \sum_{n=0}^{+\infty} \frac{(-1)^n x^{2n+1}}{(2n+1)!} \qquad (1)$$

$$= x - \frac{x^3}{3!} + \frac{x^5}{5!} - \frac{x^7}{7!} + \cdots$$

Differentiating on both sides of (1) with respect to x, we obtain

$$\cos x = \sum_{n=0}^{+\infty} \frac{(-1)^n (2n+1) x^{2n}}{(2n+1)!}$$

$$= \sum_{n=0}^{+\infty} \frac{(-1)^n x^{2n}}{(2n)!} \qquad (2)$$

$$= 1 - \frac{x^2}{2!} + \frac{x^4}{4!} - \frac{x^6}{6!} + \cdots$$

which is the Maclaurin series for the cosine function. Because the first term of series (2) is the constant 1, we may write (2) as follows.

$$\cos x = 1 + \sum_{n=1}^{+\infty} \frac{(-1)^n x^{2n}}{(2n)!}$$

Differentiating on both sides with respect to x gives

$$-\sin x = \sum_{n=1}^{+\infty} \frac{(-1)^n (2n) x^{2n-1}}{(2n)!}$$

$$= \sum_{n=1}^{+\infty} \frac{(-1)^n x^{2n-1}}{(2n-1)!}$$

Replacing n with $k+1$, we have

$$-\sin x = \sum_{k+1=1}^{+\infty} \frac{(-1)^{k+1} x^{2k+1}}{(2k+1)!}$$

$$= -\sum_{k=0}^{+\infty} \frac{(-1)^k x^{2k+1}}{(2k+1)!}$$

And multiplying on both sides by -1, we obtain

$$\sin x = \sum_{k=0}^{+\infty} \frac{(-1)^k x^{2k+1}}{(2k+1)!}$$

$$= x - \frac{x^3}{3!} + \frac{x^5}{5!} - \frac{x^7}{7!} + \cdots$$

which is the Maclaurin series for the sine function.

8. Given $\ln 2 = 0.6931$, use the series obtained in Exercise 7 to find $\ln 3$ accurate to four decimal places.

SOLUTION: In Exercise 7 we obtained the Taylor series for $\ln x$ at 2. It is

$$\ln x = \ln 2 + \sum_{n=1}^{+\infty} (-1)^{n-1} \frac{(x-2)^n}{n(2^n)}$$

If $x = 3$, we have

$$\ln 3 = \ln 2 + \sum_{n=1}^{+\infty} (-1)^{n-1} \frac{1}{n(2^n)}$$

$$= \ln 2 + \frac{1}{2} - \frac{1}{2(2^2)} + \frac{1}{3(2^3)} - \frac{1}{4(2^4)} + \frac{1}{5(2^5)} - \frac{1}{6(2^6)}$$

$$+ \frac{1}{7(2^7)} - \frac{1}{8(2^8)} + \frac{1}{9(2^9)} - \frac{1}{10(2^{10})} + \cdots$$

$$= 0.69315 + 0.50000 - 0.12500 + 0.04167 - 0.01563 + 0.00625$$
$$- 0.00260 + 0.00112 - 0.00049 + 0.00022 - 0.00010 + 0.00004$$

$$= 1.0986$$

12. Find a power-series representation for the given function at the number a and determine its radius of convergence.

$$f(x) = \frac{1}{x}; \quad a = 1$$

SOLUTION: We have the geometric series

$$\frac{1}{1-x} = \sum_{n=0}^{+\infty} x^n \quad \text{if } |x| < 1$$

Replacing x by $-x$, we obtain

$$\frac{1}{1+x} = \sum_{n=0}^{+\infty} (-1)^n x^n \quad \text{if } |x| < 1 \tag{1}$$

Because

$$\frac{1}{x} = \frac{1}{1-(x-1)}$$

we replace x by $x - 1$ in (1). Thus,

$$\frac{1}{1+(x-1)} = \sum_{n=0}^{+\infty} (-1)^n (x-1)^n \quad \text{if } |x-1| < 1$$

or, equivalently,

$$\frac{1}{x} = \sum_{n=0}^{+\infty} (-1)^n (x-1)^n \quad \text{if } |x-1| < 1$$

The radius of convergence is $R = 1$.

16. Find the Maclaurin series for $\cos^2 x$. (*Hint:* Use $\cos^2 x = \frac{1}{2}(1 + \cos 2x)$.)

SOLUTION: We have for all x

$$\cos x = \sum_{n=0}^{+\infty} \frac{(-1)^n x^{2n}}{(2n)!}$$

Replacing x by $2x$, we obtain

$$\cos 2x = \sum_{n=0}^{+\infty} \frac{(-1)^n (2x)^{2n}}{(2n)!}$$

$$= \sum_{n=0}^{+\infty} \frac{(-1)^n 2^{2n} x^{2n}}{(2n)!}$$

$$= 1 + \sum_{n=1}^{+\infty} \frac{(-1)^n 2^{2n} x^{2n}}{(2n)!}$$

Adding 1 to both sides and multiplying both sides by $\frac{1}{2}$, we obtain

$$\frac{1}{2}(1 + \cos 2x) = \frac{1}{2}\left[2 + \sum_{n=1}^{+\infty} \frac{(-1)^n 2^{2n} x^{2n}}{(2n)!}\right]$$

$$\cos^2 x = 1 + \sum_{n=1}^{+\infty} \frac{(-1)^n 2^{2n-1} x^{2n}}{(2n)!}$$

20. Use a power series to compute the value of $\sqrt[5]{e}$ to four decimal places.

SOLUTION: We have for all x

$$e^x = \sum_{n=0}^{+\infty} \frac{x^n}{n!}$$

Replacing x by $\frac{1}{5}$, we have

$$\sqrt[5]{e} = \sum_{n=0}^{+\infty} \frac{1}{5^n n!} \tag{1}$$

$$= 1 + \frac{1}{5} + \frac{1}{5^2(2!)} + \frac{1}{5^3(3!)} + \frac{1}{5^4(4!)} + \cdots$$

$$= 1 + 0.20000 + 0.02000 + 0.00133 + 0.00007 + \cdots = 1.2214 \tag{2}$$

We show that the approximation (2) is accurate to four decimal places. From (1) we have

$$u_n = \frac{1}{5^n n!} \tag{3}$$

Let

$$a_n = \frac{u_n}{u_{n-1}}$$

$$= \frac{1}{5n} \tag{4}$$

Thus, $\{a_n\}$ is a decreasing sequence, and we may take $R = a_{k+1}$ in the error formula

$$E_k \le \frac{u_{k+1}}{1 - R} \tag{5}$$

Because the approximation (2) is found by using the partial sum $\sum_{n=0}^{3} u_n$, the error is E_3. From (5) we have

$$E_3 \le \frac{u_4}{1 - R}$$

where from (3) and (4) we have

$$u_4 = 0.00007 \quad \text{and} \quad R = a_4 = \tfrac{1}{20}$$

Thus,

$$E_3 \le \frac{0.00007}{1 - \frac{1}{20}} = 0.000074$$

Because $E_3 < 10^{-4}$, the approximation is accurate to four decimal places.

24. Use a power series to compute $\sqrt[3]{29}$ to three decimal places.

SOLUTION: We apply Definition 8.5.2.

$$a^x = \exp[x \ln a]$$

and the power series for e^x and $\ln(1 + x)$. Because 27 is the perfect cube nearest to 29, we write

$$\sqrt[3]{29} = \sqrt[3]{27} \cdot \sqrt[3]{\frac{29}{27}} = 3\sqrt[3]{1.0741} \tag{1}$$

From Definition 8.5.2 we have

$$\sqrt[3]{1.0741} = \exp[\tfrac{1}{3} \ln(1.0741)] \tag{2}$$

Because

$$\ln(1 + x) = x - \frac{x^2}{2} + \frac{x^3}{3} - \frac{x^4}{4} + \cdots \qquad \text{if } |x| < 1$$

we may let $x = 0.0741$, and obtain

$$\ln(1.0741) = 0.0741 - \frac{0.0741^2}{2} + \frac{0.0741^3}{3} - \cdots$$

$$= 0.0741 - 0.0027 + 0.0001 - \cdots$$

$$= 0.0715 \tag{3}$$

Substituting from (3) into (2), we have

$$\sqrt[3]{1.0741} = \exp[\tfrac{1}{3}(0.0715)] = \exp(0.0238) \tag{4}$$

Because

$$e^x = 1 + x + \frac{x^2}{2!} + \frac{x^3}{3!} + \cdots$$

we may let $x = 0.0238$ and obtain

$$\exp(0.0238) = 1 + 0.0238 + \frac{0.0238^2}{2} + \frac{0.0238^3}{3} + \cdots$$

$$= 1 + 0.0238 + 0.0003 + 0.000004 + \cdots$$

$$= 1.0241 \tag{5}$$

substituting from (5) into (4) and then from (4) into (1), we obtain

$$\sqrt[3]{29} = 3(1.0241) = 3.072$$

28. Compute the value of the definite integral accurate to three decimal places.

$$\int_0^1 \cos \sqrt{x}\, dx$$

SOLUTION: We have

$$\cos x = \sum_{n=0}^{+\infty} \frac{(-1)^n x^{2n}}{(2n)!}$$

Thus,

$$\cos \sqrt{x} = \sum_{n=0}^{+\infty} \frac{(-1)^n x^n}{(2n)!}$$

$$\int_0^1 \cos \sqrt{x}\ dx = \sum_{n=0}^{+\infty} \int_0^1 \frac{(-1)^n x^n}{(2n)!}\ dx$$

$$= \sum_{n=0}^{+\infty} \frac{(-1)^n}{(n+1)(2n)!}$$

$$= 1 - \frac{1}{2 \cdot 2!} + \frac{1}{3 \cdot 4!} - \frac{1}{4 \cdot 6!} + \frac{1}{5 \cdot 8!}$$

$$= 1 - 0.25000 + 0.01389 - 0.00035 + 0.000005 + \cdots$$

$$= 0.7635$$

Because we have an alternating series, the error is less than the absolute value of the last term considered, namely $5 \cdot 10^{-6}$.

32. The function E defined by

$$E(x) = \frac{2}{\sqrt{\pi}} \int_0^x e^{-t^2}\ dt$$

is called the *error function,* and it is important in mathematical statistics. Find the Maclaurin series for the error function.

SOLUTION: We have

$$e^x = \sum_{n=0}^{+\infty} \frac{x^n}{n!}$$

Thus,

$$e^{-t^2} = \sum_{n=0}^{+\infty} \frac{(-t^2)^n}{n!}$$

$$= \sum_{n=0}^{+\infty} \frac{(-1)^n t^{2n}}{n!}$$

Integrating, we have

$$\int_0^x e^{-t^2}\ dt = \sum_{n=0}^{+\infty} \frac{(-1)^n x^{2n+1}}{(2n+1)n!}$$

Hence,

$$\frac{2}{\sqrt{\pi}} \int_0^x e^{-t^2}\ dt = \frac{2}{\sqrt{\pi}} \sum_{n=0}^{+\infty} \frac{(-1)^n x^{2n+1}}{(2n+1)n!}$$

15.11 THE BINOMIAL SERIES

15.11.1 Theorem (Binomial Theorem) If m is any real number, then

$$(1+x)^m = 1 + \sum_{n=1}^{+\infty} \frac{m(m-1)(m-2) \cdots (m-n+1)}{n!} x^n$$

for all values of x such that $|x| < 1$.

If $0 < x < 1$ in Theorem 15.11.1, then the binomial series is an alternating series for all terms with $n > m$, and we may use the flowchart in Table 15.9a to approximate the series with error less than any $\epsilon > 0$. If $-1 < x < 0$, then either all terms with $n > m$ are positive or all terms with $n > m$ are negative. If $m \le -1$ and $-1 < x < 0$, then all terms are positive. If

$$a_n = \frac{u_n}{u_{n-1}}$$

then $\{a_n\}$ is a decreasing sequence. Thus, we may use the flowchart in Table 15.10a with $R = Q$, as in the program of Table 15.10b, to approximate the series with error less than any $\epsilon > 0$. If $m > -1$ and $-1 < x < 0$, then for all terms with $n > m + 1$, $\{a_n\}$ is an increasing sequence. Because $\lim\limits_{n \to +\infty} a_n = |x|$, we have $0 < a_n \le |x|$ for all $n > m + 1$. Thus, we may use the flowchart in Table 15.10a with $R = |x|$ to approximate the series with error less than any $\epsilon > 0$.

Suppose we wish to compute the value of the following definite integral accurate to four decimal places.

$$\int_0^{0.9} \sqrt{1 - x^3}\, dx$$

We use Theorem 15.11.1 with $m = \frac{1}{2}$ to obtain

$$(1 + x)^{1/2} = 1 + \sum_{n=1}^{+\infty} \frac{(\frac{1}{2})(\frac{1}{2} - 1)(\frac{1}{2} - 2) \cdots (\frac{1}{2} - n + 1)x^n}{n!} \qquad \text{if } |x| < 1$$

Replacing x by $-x^3$, we get

$$(1 - x^3)^{1/2} = 1 + \sum_{n=1}^{+\infty} \frac{(\frac{1}{2})(-\frac{1}{2})(-\frac{3}{2}) \cdots (\frac{3}{2} - n)(-1)^n x^{3n}}{n!} \qquad \text{if } |x| < 1$$

Integrating both sides, we obtain

$$\int_0^{0.9} \sqrt{1 - x^3}\, dx = \left[x + \sum_{n=1}^{+\infty} \frac{(-1)^n(\frac{1}{2})(-\frac{1}{2})(-\frac{3}{2}) \cdots (\frac{3}{2} - n)x^{3n+1}}{(3n+1)n!} \right]_0^{0.9}$$

$$= 0.9 + \sum_{n=1}^{+\infty} \frac{(-1)^n(\frac{1}{2})(-\frac{1}{2})(-\frac{3}{2}) \cdots (\frac{3}{2} - n)(0.9)^{3n+1}}{(3n+1)n!}$$

Thus, we have $u_0 = 0.9$ in step 10 of Table 15.11.

Because

$$u_n = \frac{(-1)^n(\frac{1}{2})(-\frac{1}{2}) \cdots (\frac{3}{2} - n)(0.9)^{3n+1}}{(3n+1)n!}$$

in step 30 of Table 15.11, we take

$$Q = \frac{u_n}{u_{n-1}}$$

$$= \frac{(-1)^n(\frac{1}{2}) \cdots (\frac{3}{2} - n)(0.9)^{3n+1}}{(3n+1)n!} \cdot \frac{(3n-2)(n-1)!}{(-1)^{n-1}(\frac{1}{2})(-\frac{1}{2}) \cdots (\frac{5}{2} - n)(0.9)^{3n-2}}$$

$$= -\frac{(\frac{3}{2} - n)(3n - 2)(0.9)^3}{(3n+1)n}$$

$$= \frac{(2n - 3)(3n - 2)(0.9)^3}{2n(3n + 1)}$$

Let $a_n = Q$. Because $\{a_n\}$ is an increasing sequence and

$$\lim_{n \to +\infty} a_n = \lim_{n \to +\infty} \frac{(2n - 3)(3n - 2)(0.9)^3}{2n(3n + 1)} = (0.9)^3$$

we have $a_n \le (0.9)^3$ for all n. Thus, we take $R = (0.9)^3$ in step 50 of Table 15.11.

Table 15.11

```
10   LET  S=U=.9
20   LET  N=1
30   LET  Q=(2*N-3)*(3*N-2)*.9↑3/2/N/(3*N+1)
40   LET  U=U*Q
50   LET  R=.9↑3
60   IF  ABS(U/(1-R))<.0001  THEN  100
70   LET  S=S+U
80   LET  N=N+1
90   GOTO  30
100  PRINT "S = ";S,"N = ";N
999  END
```

RUN
15.110

$$S = .805931 \qquad N = 9$$

Exercises 15.11

In Exercises 1–10 use a binomial series to find the Maclaurin series for the given function. Determine the radius of convergence of the resulting series.

4. $f(x) = \sqrt[3]{8 + x}$

SOLUTION: We have

$$\sqrt[3]{8 + x} = 2(1 + \tfrac{1}{8}x)^{1/3} \tag{1}$$

By Theorem 15.11.1, with $m = \tfrac{1}{3}$, we have

$$(1 + x)^{1/3} = 1 + \sum_{n=1}^{+\infty} \frac{\tfrac{1}{3}(\tfrac{1}{3} - 1)(\tfrac{1}{3} - 2) \cdots (\tfrac{1}{3} - n + 1)}{n!} x^n \qquad \text{if } |x| < 1$$

$$= 1 + \sum_{n=1}^{+\infty} \frac{\tfrac{1}{3}(-\tfrac{2}{3})(-\tfrac{5}{3}) \cdots (\tfrac{4}{3} - n)}{n!} x^n \qquad \text{if } |x| < 1$$

$$= 1 + \sum_{n=1}^{+\infty} \frac{(-1)^n(-1)(2)(5) \cdots (3n - 4)}{3^n n!} x^n \qquad \text{if } |x| < 1$$

Replacing x by $\tfrac{1}{8}x$, we obtain

$$\left(1 + \frac{1}{8}x\right)^{1/3} = 1 + \sum_{n=1}^{+\infty} \frac{(-1)^n(-1)(2)(5) \cdots (3n - 4)}{3^n n!}\left(\frac{1}{8}x\right)^n \qquad \text{if } \left|\frac{1}{8}x\right| < 1$$

$$= 1 + \sum_{n=1}^{+\infty} \frac{(-1)^n(-1)(2)(5) \cdots (3n - 4)x^n}{3^n 2^{3n} n!} \qquad \text{if } |x| < 8$$

Multiplying both sides by 2, we obtain

$$\sqrt[3]{8 + x} = 2 + \sum_{n=1}^{+\infty} \frac{(-1)^n(-1)(2)(5) \cdots (3n - 4)x^n}{3^n 2^{3n-1} n!} \qquad \text{if } |x| < 8$$

The radius of convergence is 8.

8. $f(x) = \dfrac{x}{\sqrt{1 - x}}$

SOLUTION: $f(x) = x(1 - x)^{-1/2}$

We apply Theorem 15.11.1 with $m = -\frac{1}{2}$ to obtain

$$(1 + x)^{-1/2} = 1 + \sum_{n=1}^{+\infty} \frac{(-\frac{1}{2})(-\frac{1}{2} - 1)(-\frac{1}{2} - 2) \cdots (-\frac{1}{2} - n + 1)}{n!} x^n \qquad \text{if } |x| < 1$$

$$= 1 + \sum_{n=1}^{+\infty} \frac{(-1)^n \left(\frac{1}{2}\right)\left(\frac{3}{2}\right)\left(\frac{5}{2}\right) \cdots \left(\frac{2n-1}{2}\right)}{n!} x^n \qquad \text{if } |x| < 1$$

$$= 1 + \sum_{n=1}^{+\infty} \frac{(-1)^n \cdot 1 \cdot 3 \cdot 5 \cdot \cdots \cdot (2n-1) x^n}{2^n(n!)} \qquad \text{if } |x| < 1$$

Replacing x with $-x$, we have

$$(1 - x)^{-1/2} = 1 + \sum_{n=1}^{+\infty} \frac{(-1)^n \cdot 1 \cdot 3 \cdot 5 \cdot \cdots \cdot (2n-1)(-x)^n}{2^n(n!)} \qquad \text{if } |x| < 1$$

$$= 1 + \sum_{n=1}^{+\infty} \frac{(-1)^{2n} \cdot 1 \cdot 3 \cdot 5 \cdot \cdots \cdot (2n-1) x^n}{2^n(n!)} \qquad \text{if } |x| < 1$$

$$= 1 + \sum_{n=1}^{+\infty} \frac{1 \cdot 3 \cdot 5 \cdot \cdots \cdot (2n-1) x^n}{2^n(n!)} \qquad \text{if } |x| < 1$$

Multiplying on both sides by x, we obtain

$$x(1 - x)^{-1/2} = x + \sum_{n=1}^{+\infty} \frac{1 \cdot 3 \cdot 5 \cdot \cdots \cdot (2n-1) x^{n+1}}{2^n(n!)} \qquad \text{if } |x| < 1$$

Thus,

$$f(x) = x + \sum_{n=1}^{+\infty} \frac{1 \cdot 3 \cdot 5 \cdot \cdots \cdot (2n-1) x^{n+1}}{2^n(n!)} \qquad \text{if } |x| < 1$$

In Exercises 11–16 compute the value of the given quantity accurate to three decimal places by using a binomial series.

12. $\sqrt{51}$

SOLUTION: Because 49 is the perfect square that is nearest to 51, we first write

$$\sqrt{51} = \sqrt{49} \cdot \sqrt{\frac{51}{49}} = 7\sqrt{1.0408} \qquad (1)$$

We use the binomial series with $m = \frac{1}{2}$. Thus,

$$(1 + x)^{1/2} = 1 + \frac{1}{2} x + \left(\frac{1}{2}\right)\left(-\frac{1}{2}\right) \frac{x^2}{2!} + \left(\frac{1}{2}\right)\left(-\frac{1}{2}\right)\left(-\frac{3}{2}\right) \frac{x^3}{3!} + \cdots$$

$$= 1 + \frac{x}{2} - \frac{x^2}{8} + \frac{x^3}{16} - \cdots$$

If $x = 0.0408$, we have

$$(1.0408)^{1/2} = 1 + \frac{0.0408}{2} - \frac{0.0408^2}{8} + \frac{0.0408^3}{16} - \cdots$$

$$= 1 + 0.0204 - 0.00021 + \cdots = 1.0202 \qquad (2)$$

Substituting from (2) into (1), we obtain

$$\sqrt{51} = 7(1.0202) = 7.141$$

16. Compute the value of the given quantity accurate to three decimal places by using a binomial series.

$$\frac{1}{\sqrt[5]{31}}$$

SOLUTION: Because $2^5 = 32$ is the closest fifth power of an integer, we take

$$\frac{1}{\sqrt[5]{31}} = 31^{-1/5}$$

$$= \left(32 \cdot \frac{31}{32}\right)^{-1/5}$$

$$= \frac{1}{2}\left(1 - \frac{1}{32}\right)^{-1/5} \tag{1}$$

We use the binomial theorem with $m = -\frac{1}{5}$. Thus,

$$(1+x)^{-1/5} = 1 + \left(-\frac{1}{5}\right)x + \left(-\frac{1}{5}\right)\left(-\frac{6}{5}\right)\frac{x^2}{2!} + \cdots \qquad \text{if } |x| < 1 \tag{2}$$

If $x = -\frac{1}{32}$ in (2), we get

$$\left(1 - \frac{1}{32}\right)^{-1/5} = 1 + \left(-\frac{1}{5}\right)\left(-\frac{1}{32}\right) + \left(-\frac{1}{5}\right)\left(-\frac{6}{5}\right)\left(\frac{1}{2!}\right)\left(-\frac{1}{32}\right)^2 + \cdots$$

$$= 1.000 + 0.0063 + 0.0001 + \cdots = 1.006 \tag{3}$$

Substituting from (3) into (1), we obtain

$$\frac{1}{\sqrt[5]{31}} = \frac{1}{2}(1.006) = 0.503 \tag{4}$$

Because $m > -1$, then for all n we have

$$0 < \frac{u_n}{u_{n-1}} < |x|$$

Thus, we may take $R = |x| = \frac{1}{32}$ in the error formula

$$E_k \le \frac{u_{k+1}}{1 - R}$$

we have

$$E_1 \le \frac{u_2}{1 - R}$$

$$= \frac{0.0001}{1 - \frac{1}{32}} = 0.000103$$

Therefore, the error in (4) is $\frac{1}{2}E_1 = 0.000052 < 10^{-4}$. Hence, the approximation is accurate to three decimal places.

20. Compute the value of the given definite integral accurate to four decimal places.

$$\int_0^{1/2} \sqrt{1 - x^3}\, dx$$

SOLUTION: By Theorem 15.1.1 with $m = \frac{1}{2}$ we have for all $|x| < 1$

$$(1+x)^{1/2} = 1 + \sum_{n=1}^{+\infty} \frac{\left(\frac{1}{2}\right)\left(-\frac{1}{2}\right)\left(-\frac{3}{2}\right) \cdots \left(\frac{3}{2} - n\right)x^n}{n!}$$

$$= 1 + \frac{x}{2} + \sum_{n=2}^{+\infty} \frac{(-1)^{n-1} \cdot 1 \cdot 3 \cdot 5 \cdot \cdots \cdot (2n-3)x^n}{2^n(n!)}$$

Replacing x by $-x^3$, we obtain for all $|x| < 1$

$$(1 - x^3)^{1/2} = 1 - \frac{x^3}{2} + \sum_{n=2}^{+\infty} \frac{(-1)^{n-1} 1 \cdot 3 \cdot 5 \cdot \cdots \cdot (2n-3)(-x^3)^n}{2^n(n!)}$$

$$= 1 - \frac{x^3}{2} + \sum_{n=2}^{+\infty} \frac{(-1)^{2n-1} \cdot 1 \cdot 3 \cdot 5 \cdot \cdots \cdot (2n-3)x^{3n}}{2^n(n!)}$$

$$= 1 - \frac{x^3}{2} - \sum_{n=2}^{+\infty} \frac{1 \cdot 3 \cdot 5 \cdot \cdots \cdot (2n-3)x^{3n}}{2^n(n!)}$$

Therefore, by integrating on both sides, we obtain

$$\int_0^{1/2} \sqrt{1 - x^3} \, dx = \left[x - \frac{x^4}{8} - \sum_{n=2}^{+\infty} \frac{1 \cdot 3 \cdot 5 \cdot \cdots \cdot (2n-3)x^{3n+1}}{(3n+1)(2^n)(n!)} \right]_0^{1/2}$$

$$= \frac{1}{2} - \frac{1}{8(2^4)} - \frac{1}{(7)(2^2)(2!)(2^7)} - \frac{1 \cdot 3}{(10)(2^3)(3!)(2^{10})} - \cdots$$

$$= 0.50000 - 0.00781 - 0.00014 - 0.000006 - \cdots$$

$$= 0.4920$$

24. Compute the value of the definite integral accurate to four decimal places.

$$\int_0^{1/2} f(x) \, dx \qquad \text{where } f(x) = \begin{cases} \dfrac{\sin^{-1} x}{x} & \text{if } x \neq 0 \\ 1 & \text{if } x = 0 \end{cases}$$

SOLUTION: We use the series of Example 2.

$$\sin^{-1} x = x + \sum_{n=1}^{+\infty} \frac{1 \cdot 3 \cdot 5 \cdot \cdots \cdot (2n-1)}{2^n n!} \cdot \frac{x^{2n+1}}{2n+1} \qquad \text{for } |x| < 1$$

Thus,

$$\frac{\sin^{-1} x}{x} = 1 + \sum_{n=1}^{+\infty} \frac{1 \cdot 3 \cdot 5 \cdot \cdots \cdot (2n-1)}{2^n n!} \cdot \frac{x^{2n}}{2n+1} \qquad \text{if } 0 < |x| < 1$$

Because this sum has value 1 when $x = 0$, we have

$$f(x) = 1 + \sum_{n=1}^{+\infty} \frac{1 \cdot 3 \cdot 5 \cdot \cdots \cdot (2n-1)}{2^n n!} \cdot \frac{x^{2n}}{2n+1} \qquad \text{if } |x| < 1$$

Thus,

$$\int_0^{1/2} f(x) \, dx = \left[x + \sum_{n=1}^{+\infty} \frac{1 \cdot 3 \cdot 5 \cdot \cdots \cdot (2n-1)x^{2n+1}}{2^n n!(2n+1)^2} \right]_0^{1/2}$$

$$= \frac{1}{2} + \sum_{n=1}^{+\infty} \frac{1 \cdot 3 \cdot 5 \cdot \cdots \cdot (2n-1)}{2^n n!(2n+1)^2 2^{2n+1}}$$

$$= \frac{1}{2} + \sum_{n=1}^{+\infty} \frac{1 \cdot 3 \cdot 5 \cdot \cdots \cdot (2n-1)}{2^{3n+1} n!(2n+1)^2} \qquad (1)$$

$$= \frac{1}{2} + \frac{1}{2^4 \cdot 3^2} + \frac{1 \cdot 3}{2^7(2!)5^2} + \frac{1 \cdot 3 \cdot 5}{2^{10}(3!)7^2} + \cdots$$

$$= 0.50000 + 0.00694 + 0.00047 + 0.00005 + \cdots$$

$$= 0.5075$$

We show that the error is less than 10^{-4}.

From (1) we have

$$u_n = \frac{1 \cdot 3 \cdot 5 \cdot \cdots \cdot (2n-1)}{2^{3n+1} n! (2n+1)^2}$$

Let

$$a_n = \frac{u_n}{u_{n-1}}$$

$$= \frac{1 \cdot 3 \cdot 5 \cdot \cdots \cdot (2n-1)}{2^{3n+1} n! (2n+1)^2} \cdot \frac{2^{3n-2}(n-1)!(2n-1)^2}{1 \cdot 3 \cdot 5 \cdot \cdots \cdot (2n-3)}$$

$$= \frac{(2n-1)^3}{2^3 n (2n+1)^2}$$

Because $\{a_n\}$ is an increasing sequence and

$$\lim_{n \to +\infty} a_n = \lim_{n \to +\infty} \frac{(2n-1)^3}{2^3 n (2n+1)^2} = \frac{1}{4}$$

for all n, we have

$$0 < \frac{u_n}{u_{n-1}} \le \frac{1}{4}$$

Thus, we may use the error formula of Section 15.10 with $R = \frac{1}{4}$. We have

$$E_k \le \frac{u_{k+1}}{1-R}$$

Thus,

$$E_2 \le \frac{0.00005}{1 - \frac{1}{4}} < 0.00007 < 10^{-4}$$

Review Exercises for Chapter 15

In Exercises 1–8 write the first four numbers of the sequence and find the limit of the sequence, if it exists.

4. $\left\{ \dfrac{n + 3n^2}{4 + 2n^3} \right\}$

SOLUTION: Let

$$a_n = \frac{n + 3n^2}{4 + 2n^3}$$

Then

$$a_1 = \tfrac{2}{3} \qquad a_2 = \tfrac{7}{10} \qquad a_3 = \tfrac{15}{29} \qquad a_4 = \tfrac{13}{33}$$

$$\lim_{n \to +\infty} a_n = \lim_{n \to +\infty} \frac{n + 3n^2}{4 + 2n^3}$$

$$= \lim_{n \to +\infty} \frac{\dfrac{1}{n^2} + \dfrac{3}{n}}{\dfrac{4}{n^3} + 2} = 0$$

8. $\left\{ \dfrac{(n+2)^2}{n+4} - \dfrac{(n+2)^2}{n} \right\}$

SOLUTION: Let

$$a_n = \frac{(n+2)^2}{n+4} - \frac{(n+2)^2}{n}$$

$$= \frac{-4(n+2)^2}{n(n+4)}$$

The first four terms are

$$a_1 = -\tfrac{36}{5} \qquad a_2 = -\tfrac{16}{3} \qquad a_3 = -\tfrac{100}{21} \qquad a_4 = -\tfrac{9}{2}$$

and

$$\lim_{n \to +\infty} a_n = \lim_{n \to +\infty} \frac{-4(n+2)^2}{n(n+4)}$$

$$= -4 \lim_{n \to +\infty} \frac{n^2 + 4n + 4}{n^2 + 4n}$$

$$= -4 \lim_{n \to +\infty} \frac{1 + \dfrac{4}{n} + \dfrac{4}{n^2}}{1 + \dfrac{4}{n}}$$

$$= (-4)(1) = -4$$

In Exercises 11–17 determine whether the series is convergent or divergent. If the series is convergent, find its sum.

12. $\displaystyle\sum_{n=1}^{+\infty} e^{-2n}$

SOLUTION:

$$\sum_{n=1}^{+\infty} e^{-2n} = \sum_{n=1}^{+\infty} \left(\frac{1}{e^2} \right)^n$$

$$= \sum_{n=1}^{+\infty} \left(\frac{1}{e^2} \right)\left(\frac{1}{e^2} \right)^{n-1} \tag{1}$$

Because (1) is a geometric series with first term $a = e^{-2}$ and ratio $r = e^{-2}$, then

$$\sum_{n=1}^{+\infty} e^{-2n} = \frac{a}{1-r} = \frac{e^{-2}}{1 - e^{-2}} = \frac{1}{e^2 - 1}$$

16. $\displaystyle\sum_{n=0}^{+\infty} \cos^n \frac{1}{3} \pi$

SOLUTION: Because $\cos \tfrac{1}{3}\pi = \tfrac{1}{2}$, then

$$\sum_{n=0}^{+\infty} \cos^n \frac{1}{3} \pi = \sum_{n=0}^{+\infty} \left(\frac{1}{2} \right)^n \tag{1}$$

Because (1) is a geometric series with first term $a = (\tfrac{1}{2})^0 = 1$ and ratio $r = \tfrac{1}{2}$, we have

$$\sum_{n=0}^{+\infty} \cos^n \frac{1}{3}\pi = \frac{a}{1-r} = \frac{1}{1-\frac{1}{2}} = 2$$

In Exercises 18–32 determine if the series is convergent or divergent.

20. $\displaystyle\sum_{n=1}^{+\infty} \frac{3 + \sin n}{n^2}$

SOLUTION: Because $-1 \leq \sin n < 1$, then

$$\frac{2}{n^2} \leq \frac{3 + \sin n}{n^2} \leq \frac{4}{n^2} \tag{1}$$

Because $\displaystyle\sum_{n=1}^{+\infty} 1/n^2$ is a p-series with $p = 2$, it is convergent. Thus, $\displaystyle\sum_{n=1}^{+\infty} 4/n^2$ is convergent, and from (1) and the comparison test we conclude that the given series is convergent.

24. $\displaystyle\sum_{n=1}^{+\infty} \frac{(-1)^{n+1}}{1 + \sqrt{n}}$

SOLUTION: We apply the alternating series test (15.6.2). Let

$$u_n = \frac{(-1)^{n+1}}{1 + \sqrt{n}}$$

Then

$$|u_n| = \frac{1}{1 + \sqrt{n}}$$

and

$$|u_{n+1}| = \frac{1}{1 + \sqrt{n+1}}$$

Thus, we have $|u_{n+1}| < |u_n|$ for all positive integers n. Furthermore,

$$\lim_{n \to +\infty} |u_n| = \lim_{n \to +\infty} \frac{1}{1 + \sqrt{n}} = 0$$

Hence

$$\lim_{n \to +\infty} u_n = 0$$

so the hypothesis of the alternating series test is satisfied. Therefore, the given series is convergent.

28. $\displaystyle\sum_{n=0}^{+\infty} \frac{n!}{10^n}$

SOLUTION: We have

$$\left| \frac{u_{n+1}}{u_n} \right| = \frac{(n+1)!}{10^{n+1}} \cdot \frac{10^n}{n!} = \frac{n+1}{10}$$

Thus,

$$\lim_{n \to +\infty} \left| \frac{u_{n+1}}{u_n} \right| = +\infty$$

By the ratio test, the series is divergent.

32. $\sum\limits_{n=1}^{+\infty} n(3^{-n^2})$

SOLUTION: We apply the integral test (15.5.1). Let

$$f(x) = x(3^{-x^2})$$

Because

$$f'(x) = x(3^{-x^2})(-2x)(\ln 3) + 3^{-x^2}$$
$$= 3^{-x^2}[-2x^2(\ln 3) + 1]$$

then $f'(x) < 0$ for all $x \geq 1$. Thus, f is continuous, decreasing, and positive valued for all $x \geq 1$, and so the hypothesis of the integral test is satisfied. Furthermore,

$$\int_1^{+\infty} f(x)\, dx = \int_1^{+\infty} x(3^{-x^2})\, dx$$

$$= \lim_{b \to +\infty} \int_1^b x(3^{-x^2})\, dx \qquad (1)$$

Let $u = 3^{-x^2}$ and $du = -2x(\ln 3)(3^{-x^2})\, dx$. Thus,

$$\int x(3^{-x^2})\, dx = \int \frac{-du}{2 \ln 3}$$

$$= -\frac{u}{2 \ln 3} + C$$

$$= \frac{-3^{-x^2}}{2 \ln 3} + C \qquad (2)$$

Substituting from (2) into (1), we have

$$\int_1^{+\infty} f(x)\, dx = \lim_{b \to +\infty} \left[\frac{-3^{-x^2}}{2 \ln 3} \right]_1^b$$

$$= \frac{-1}{2 \ln 3} \lim_{b \to +\infty} \left[\frac{1}{3^{b^2}} - \frac{1}{3} \right]$$

$$= \frac{1}{6 \ln 3}$$

Therefore, the given series is convergent.

In Exercises 35–42 determine if the given series is absolutely convergent, conditionally convergent, or divergent. Prove your answer.

36. $\sum\limits_{n=0}^{+\infty} (-1)^n \dfrac{5^{2n+1}}{(2n+1)!}$

SOLUTION: We apply the ratio test. Let

$$u_n = (-1)^n \frac{5^{2n+1}}{(2n+1)!}$$

Then

$$\lim_{n \to +\infty} \left| \frac{u_{n+1}}{u_n} \right| = \lim_{n \to +\infty} \left| \frac{5^{2n+3}}{(2n+3)!} \cdot \frac{(2n+1)!}{5^{2n+1}} \right|$$

$$= \lim_{n \to +\infty} \frac{5^2(2n+1)!}{(2n+3)(2n+2)(2n+1)!}$$

$$= \lim_{n \to +\infty} \frac{25}{(2n+3)(2n+2)} = 0$$

Therefore, the series is absolutely convergent.

40. $\displaystyle\sum_{n=1}^{+\infty} (-1)^n \frac{\sqrt{2n-1}}{n}$

SOLUTION: We have an alternating series with

$$|u_n| = \frac{\sqrt{2n-1}}{n}$$

The inequality

$$|u_{n+1}| \le |u_n| \tag{1}$$

is equivalent to

$$\frac{\sqrt{2n+1}}{n+1} \le \frac{\sqrt{2n-1}}{n}$$

$$\frac{2n+1}{(n+1)^2} \le \frac{2n-1}{n^2}$$

$$n^2(2n+1) \le (n+1)^2(2n-1)$$

$$2n^3 + 2n^2 \le 2n^3 + 3n^2 - 1$$

$$n^2 \ge 1 \tag{2}$$

Because (2) is true, we conclude that (1) is true. Furthermore,

$$\lim_{n \to +\infty} |u_n| = \lim_{n \to +\infty} \frac{\sqrt{2n-1}}{n}$$

$$= \lim_{n \to +\infty} \sqrt{\frac{2}{n} - \frac{1}{n^2}} = 0$$

By the alternating series test, the given series is convergent. To test for absolute convergence, we consider the series

$$\sum_{n=1}^{+\infty} |u_n| = \sum_{n=1}^{+\infty} \frac{\sqrt{2n-1}}{n}$$

We make a limit comparison test with

$$u_n = \frac{\sqrt{2n-1}}{n} \quad \text{and} \quad v_n = \frac{1}{\sqrt{n}}$$

Thus,

$$\lim_{n \to +\infty} \frac{u_n}{v_n} = \lim_{n \to +\infty} \frac{\sqrt{2n-1}}{n} \cdot \frac{\sqrt{n}}{1}$$

$$= \lim_{n \to +\infty} \sqrt{2 - \frac{1}{n}} = \sqrt{2} \tag{3}$$

The series

$$\sum_{n=1}^{+\infty} v_n = \sum_{n=1}^{+\infty} \frac{1}{\sqrt{n}}$$

is divergent, because it is a p-series with $p = \frac{1}{2}$. By (3) and Theorem 15.4.3(i), the series $\sum\limits_{n=1}^{+\infty} |u_n|$ is divergent. Thus, the given series is conditionally convergent.

In Exercises 43–52 find the interval of convergence of the given power series.

44. $\sum\limits_{n=1}^{+\infty} \dfrac{(x-2)^n}{n}$

SOLUTION:

$$\lim_{n \to +\infty} \left| \frac{u_{n+1}}{u_n} \right| = \lim_{n \to +\infty} \left| \frac{(x-2)^{n+1}}{n+1} \cdot \frac{n}{(x-2)^n} \right|$$

$$= |x-2| \lim_{n \to +\infty} \left(1 + \frac{1}{n} \right) = |x-2|$$

If $|x-2| < 1$ or, equivalently, $1 < x < 3$, the series converges absolutely. For $x = 3$, the given series is $\sum\limits_{n=1}^{+\infty} 1/n$ which is the divergent harmonic series. For $x = 1$, the given series is $\sum\limits_{n=1}^{+\infty} (-1)^n/n$ which converges by the alternating series test. Thus, the interval of convergence of the given series is $[1, 3)$.

48. $\sum\limits_{n=1}^{+\infty} \dfrac{(-1)^{n-1}x^{2n-1}}{(2n-1)!}$

SOLUTION:

$$\lim_{n \to +\infty} \left| \frac{u_{n+1}}{u_n} \right| = \lim_{n \to +\infty} \left| \frac{(-1)^n x^{2n+1}}{(2n+1)!} \cdot \frac{(2n-1)!}{(-1)^{n-1}x^{2n-1}} \right|$$

$$= x^2 \lim_{n \to +\infty} \frac{1}{(2n+1)(2n)} = 0$$

Thus, the power series converges for all x in $(-\infty, +\infty)$.

52. $\sum\limits_{n=1}^{+\infty} n^n x^n$

SOLUTION:

$$\lim_{n \to +\infty} \left| \frac{u_{n+1}}{u_n} \right| = \lim_{n \to +\infty} \left| \frac{(n+1)^{n+1}x^{n+1}}{n^n x^n} \right|$$

$$= |x| \lim_{n \to +\infty} \frac{(n+1)^n (n+1)}{n^n}$$

$$= |x| \lim_{n \to +\infty} \left(1 + \frac{1}{n} \right)^n (n+1)$$

$$= |x| e \lim_{n \to +\infty} (n+1) = +\infty$$

Thus, the given series is divergent for all $x \neq 0$.

In Exercises 53–62 use a power series to compute the value of the given quantity accurate to four decimal places.

56. $\sin^{-1} 1$

SOLUTION: From Example 2 of Section 15.11 we have

$$\sin^{-1} x = x + \sum_{n=1}^{+\infty} \frac{1 \cdot 3 \cdot 5 \cdot \cdots \cdot (2n-1)x^{2n+1}}{2^n n!(2n+1)} \qquad \text{if } |x| < 1 \qquad (1)$$

We cannot use (1) when $x = 1$, but since $\sin^{-1} 1 = \frac{1}{2}\pi$ and $\sin^{-1} \frac{1}{2} = \frac{1}{6}\pi$, then $\sin^{-1} 1 = 3 \sin^{-1} \frac{1}{2}$. Thus, we use (1) to find $\sin^{-1} \frac{1}{2}$. With $x = \frac{1}{2}$, (1) becomes

$$\sin^{-1} \frac{1}{2} = \frac{1}{2} + \sum_{n=1}^{+\infty} \frac{1 \cdot 3 \cdot 5 \cdot \cdots \cdot (2n-1)}{2^{3n+1} n!(2n+1)}$$

$$= \frac{1}{2} + \frac{1}{2^4 \cdot 3} + \frac{1 \cdot 3}{2^7 \cdot 2! \cdot 5} + \frac{1 \cdot 3 \cdot 5}{2^{10} \cdot 3! \cdot 7}$$

$$+ \frac{1 \cdot 3 \cdot 5 \cdot 7}{2^{13} \cdot 4! \cdot 9} + \frac{1 \cdot 3 \cdot 5 \cdot 7 \cdot 9}{2^{16} \cdot 5! \cdot 11} + \cdots$$

$$= 0.50000 + 0.02083 + 0.00234 + 0.00035 + 0.00006 + 0.00001 + \cdots$$

$$= 0.52359$$

Multiplying both sides by 3, we have

$$3 \sin^{-1} \tfrac{1}{2} = 1.5708$$

Thus,

$$\sin^{-1} 1 = 1.5708$$

60. $\displaystyle\int_0^1 \cos x^3 \, dx$

SOLUTION: We use the Maclaurin series for $\cos x$. Thus,

$$\cos x = \sum_{n=0}^{+\infty} \frac{(-1)^n x^{2n}}{(2n)!}$$

$$\cos x^3 = \sum_{n=0}^{+\infty} \frac{(-1)^n x^{6n}}{(2n)!}$$

$$\int_0^1 \cos x^3 \, dx = \sum_{n=0}^{+\infty} \frac{(-1)^n x^{6n+1}}{(6n+1)(2n)!} \Big]_0^1$$

$$= \sum_{n=0}^{+\infty} \frac{(-1)^n}{(6n+1)(2n)!}$$

$$= 1 - \frac{1}{7 \cdot 2!} + \frac{1}{13 \cdot 4!} - \frac{1}{19 \cdot 6!} + \cdots$$

$$= 1 - 0.07143 + 0.00321 - 0.00007 = 0.9317$$

64. Find the Maclaurin series for the given function and find its interval of convergence.

$$f(x) = \frac{1}{2-x}$$

SOLUTION: Because

$$\frac{1}{2-x} = \frac{\frac{1}{2}}{1 - \frac{1}{2}x}$$

we use the geometric series

$$\frac{1}{1-x} = \sum_{n=0}^{+\infty} x^n \qquad \text{if } |x| < 1$$

We replace x by $\frac{1}{2}x$. Thus,

$$\frac{1}{1-\frac{1}{2}x} = \sum_{n=0}^{+\infty} \left(\frac{x}{2}\right)^n \qquad \text{if } \left|\frac{1}{2}x\right| < 1$$

Multiplying both sides by $\frac{1}{2}$, we have

$$\frac{\frac{1}{2}}{1-\frac{1}{2}x} = \frac{1}{2} \sum_{n=0}^{+\infty} \left(\frac{x}{2}\right)^n \qquad \text{if } |x| < 2$$

or, equivalently,

$$\frac{1}{2-x} = \sum_{n=0}^{+\infty} \frac{x^n}{2^{n+1}} \qquad \text{if } |x| < 2$$

68. Find the Taylor series for the given function at the given number.

$$f(x) = \frac{1}{x}; \text{ at } 2$$

SOLUTION: Because

$$\frac{1}{x} = \frac{\frac{1}{2}}{1+\frac{1}{2}(x-2)}$$

we use the geometric series

$$\frac{1}{1+x} = \sum_{n=0}^{+\infty} (-1)^n x^n \qquad \text{if } |x| < 1$$

Replacing x by $\frac{1}{2}(x-2)$, we obtain

$$\frac{1}{1+\frac{1}{2}(x-2)} = \sum_{n=0}^{+\infty} \frac{(-1)^n(x-2)^n}{2^n} \qquad \text{if } |\frac{1}{2}(x-2)| < 1$$

Multiplying both sides by $\frac{1}{2}$, we get

$$\frac{\frac{1}{2}}{1+\frac{1}{2}(x-2)} = \sum_{n=0}^{+\infty} \frac{(-1)^n(x-2)^n}{2^{n+1}} \qquad \text{if } |\frac{1}{2}(x-2)| < 1$$

or, equivalently,

$$\frac{1}{x} = \sum_{n=0}^{+\infty} \frac{(-1)^n(x-2)^n}{2^{n+1}} \qquad \text{if } 0 < x < 4$$

72. Show that $y = J_0(x)$ is a solution of the differential equation.

$$x\frac{d^2y}{dx^2} + \frac{dy}{dx} + xy = 0$$

SOLUTION: We have

$$y = J_0(x) = \sum_{n=0}^{+\infty} (-1)^n \frac{x^{2n}}{n!\, n!\, 2^{2n}} = 1 + \sum_{n=1}^{+\infty} \frac{(-1)^n x^{2n}}{n!\, n!\, 2^{2n}} \qquad (1)$$

Thus, by differentiating (1) successively we have

$$\frac{dy}{dx} = \sum_{n=1}^{+\infty} \frac{(-1)^n 2n x^{2n-1}}{n!\, n!\, 2^{2n}} \qquad (2)$$

$$\frac{d^2y}{dx^2} = \sum_{n=1}^{+\infty} \frac{(-1)^n 2n(2n-1)x^{2n-2}}{n!\, n!\, 2^{2n}}$$

Multiplying both sides by x, we obtain

$$x\frac{d^2y}{dx^2} = \sum_{n=1}^{+\infty} \frac{(-1)^n 2n(2n-1)x^{2n-1}}{n!\, n!\, 2^{2n}} \tag{3}$$

Adding the corresponding members of (2) and (3), we obtain

$$x\frac{d^2y}{dx^2} + \frac{dy}{dx} = \sum_{n=1}^{+\infty} \frac{(-1)^n (2n)x^{2n-1}[(2n-1)+1]}{n!\, n!\, 2^{2n}}$$

$$= \sum_{n=1}^{+\infty} \frac{(-1)^n x^{2n-1}}{(n-1)!(n-1)!2^{2n-2}} \tag{4}$$

Replacing n by $n+1$ in (4), we obtain

$$x\frac{d^2y}{dx^2} + \frac{dy}{dx} = \sum_{n=0}^{+\infty} \frac{(-1)^{n+1} x^{2n+1}}{n!\, n!\, 2^{2n}}$$

$$= -\sum_{n=0}^{+\infty} \frac{(-1)^n x^{2n+1}}{n!\, n!\, 2^{2n}} \tag{5}$$

Multiplying both sides of (1), by x, we get

$$xy = \sum_{n=0}^{+\infty} \frac{(-1)^n x^{2n+1}}{n!\, n!\, 2^{2n}} \tag{6}$$

Adding the corresponding members of (5) and (6), we get

$$x\frac{d^2y}{dx^2} + \frac{dy}{dx} + xy = 0$$

Appendix

A.1 CHAPTER TESTS   After taking each test with the time limit indicated, turn to the page on which the solutions are given and correct your paper. It may be most beneficial to you if you try to simulate the conditions under which an actual test might be given for your class. Thus, do not refer to the examples in the text or to your notes unless you will be allowed to do this in class.

Test for Chapter 8   **(65 minutes)**   **Solutions on page 763.**

1. Find $f'(x)$ and simplify your answer, if possible.

   (a) $f(x) = \ln \sqrt{4 - x^2}$ **(b)** $f(x) = 2^{3x}$

   (c) $f(x) = \dfrac{e^{2x}}{x}$ **(d)** $f(x) = x^{\ln x}$

2. Find the indefinite integral

   (a) $\displaystyle\int \frac{x^2 \, dx}{x - 1}$ **(b)** $\displaystyle\int \frac{e^x \, dx}{\sqrt{1 + e^x}}$

3. Calculate the definite integral and simplify your answer.

   $$\int_{-2}^{10} \frac{dx}{2x + 7}$$

4. Find the relative extrema of f.

   $$f(x) = x^2 e^x$$

5. Use implicit differentiation to find $D_x y$.

$$y = \ln(xy)$$

6. Find the area of the region bounded by the curve $y = e^{2x}$, the line $y = 4$, and the y-axis.

7. For the function f, defined below,
   (a) Show that f^{-1} exists.
   (b) Find $f^{-1}(x)$.
   (c) Find the domain of f^{-1}.

$$f(x) = \frac{2x + 3}{x + 1}; 0 \le x \le 1$$

8. (a) Write the equation that defines $\ln x$.
   (b) What are the domain and range of the function defined by $f(x) = \ln(2x - 1)$?
   (c) Write the equation that defines a^x if $a > 0$ and x is any real number.
   (d) Express e as the limit of a function of x as x approaches zero.

9. The rate at which a body changes temperature is proportional to the difference between its temperature and that of the surrounding medium. Suppose that a body with temperature 100° is placed in air with temperature 60°, and the body cools to 80° after 3 minutes. How much longer will it take for the body to cool to 65°?

Test for Chapter 9   (60 minutes)   Solutions on page 765.

1. Find the derivative of the given function. Simplify your answer, if possible.

   (a) $f(x) = \sec^3 2x \tan 2x$

   (b) $f(x) = \dfrac{\sinh x}{1 + \cosh x}$

   (c) $f(x) = x \cosh^{-1} x - \sqrt{x^2 - 1}$

2. Find the indefinite integral.

   (a) $\displaystyle\int \cot^5 x\, dx$

   (b) $\displaystyle\int \frac{dx}{1 + \tan^2 4x}$

   (c) $\displaystyle\int \frac{(x + 1)\, dx}{x^2 + 16}$

3. Calculate the definite integral and simplify your answer.

   (a) $\displaystyle\int_{-1}^{\sqrt{2}} \frac{dx}{\sqrt{4 - x^2}}$ (b) $\displaystyle\int_{\pi/3}^{2\pi/3} \tan\left(\frac{1}{2} x\right) dx$

4. Find the acute angle between the curves $y = x^2$ and $x = y^3$ at the point $(1, 1)$.

5. Show that $5\sqrt{5}$ is the absolute minimum value of the function f on the interval $(0, \frac{1}{2}\pi)$.

   $$f(x) = \sec x + 8 \csc x$$

6. A picture 2 feet high is placed on a wall with its base 1 foot above the eye level of a man who is walking toward the wall at the rate of 5 ft/sec. Find the rate of change of the radian measure of the angle subtended by the picture when the man is 3 ft from the wall.

Test for Chapter 10   (110 minutes)   Solutions on page 767.

1. Find the indicated indefinite integral.

(a) $\displaystyle\int x^2 \tan^{-1} x \, dx$

(b) $\displaystyle\int e^x \cos 2x \, dx$

(c) $\displaystyle\int \frac{x^2 \, dx}{(16 - x^2)^{3/2}}$

(d) $\displaystyle\int \frac{x^2 - 2x + 4}{x^3 + 2x} \, dx$

(e) $\displaystyle\int \frac{(x^2 + 2) \, dx}{x^4 + 2x^2 + 1}$

(f) $\displaystyle\int \frac{dx}{x(1 + \sqrt[3]{x})}$

2. Calculate the indicated definite integral.

(a) $\displaystyle\int_4^5 \frac{\sqrt{x^2 - 16}}{x^2} \, dx$

(b) $\displaystyle\int_0^{\pi/2} \frac{dx}{4 + 5 \cos x}$

3. Let R be the region bounded by the curve $y = x \ln x$, the x-axis, and the line $x = 2$. Use a definite integral with the cylindrical shell method to find the volume of the solid of revolution if R is revolved about the y-axis.

4. Express the indefinite integral in terms of an inverse hyperbolic function and also in terms of a natural logarithm.

$$\int \frac{dx}{\sqrt{x^2 - 2x + 5}}$$

5. Evaluate the integral and express the answer in terms of a natural logarithm.

$$\int_2^3 \frac{dx}{\sqrt{x^2 + 6x}}$$

Test for Chapter 11   (50 minutes)   Solutions on page 771.

1. The linear density of a 2-ft rod is $\sqrt{4x + 1}$ slugs/ft, where x is the number of feet from the end of the rod. Find the total mass of the rod and the center of mass.

2. Find the centroid of the region in the first quadrant that is bounded by the curve $y = x^3$ and the line $y = x$.

3. Find the centroid of the solid of revolution if the region bounded by the curve $y = x^2$ and the line $y = 4$ is revolved about the y-axis.

4. A dam with vertical face has a gate that is in the shape of an isosceles triangle 3 ft wide at the top and 2 ft high, and the upper edge of the gate is 10 ft below the surface of the water. Find the total force on the gate due to water pressure.

5. Use the theorem of Pappus to find the centroid of a right triangular region with base b units and altitude h units.

Test for Chapter 12   **(55 minutes)**   **Solutions on page 773.**

1. (a) Find a polar equation of the graph having the Cartesian equation
$y^2 - x^2 = 4$.

(b) Find a Cartesian equation of the graph having the polar equation

$$r = \frac{2}{1 - \sin \theta}$$

2. Draw a sketch of the graph of the polar equation

$$r = \sqrt{3} + 2 \sin \theta$$

3. Find all points of intersection of the graphs of the given polar equations and draw a sketch of the graphs using the same pole and polar axis.

$$\begin{cases} r = -3 \cos \theta \\ r = 1 - \cos \theta \end{cases}$$

4. Find $\tan \beta$ where β is the acute angle between the curves $r = \sin \theta$ and $r = \cos 2\theta$ at the point where $\theta = \frac{1}{6}\pi$.

5. Find the area of the region bounded by one loop of the graph of $r = \cos 3\theta$.

6. Find the area of the region that is the intersection of the region in the first quadrant that is bounded by the curves $r^2 = \cos 2\theta$ and $r = \sqrt{2} \sin \theta$.

Test for Chapter 13   **(90 minutes)**   **Solutions on page 775.**

1. Find the vertex, the focus, an equation of the directrix, and the length of the latus rectum of the parabola whose equation is as follows. Draw a sketch of the graph.

$$8x - 4y - y^2 - 12 = 0$$

2. Find the center, the vertices, the foci, and equations of the asymptotes of the hyperbola whose equation is as follows. Draw a sketch of the graph.

$$4x^2 - y^2 - 8x + 6y - 1 = 0$$

3. Find a Cartesian equation of the ellipse with center at the origin, one vertex at the point $(3, 0)$ and with eccentricity $\frac{2}{3}$.

4. Find the eccentricity, idenfity the conic, write an equation of the directrix that is nearest the pole, and draw a sketch of the graph of the conic whose polar equation is as follows.

$$r = \frac{6}{3 + 2 \sin \theta}$$

5. Use the definition of a parabola to find an equation of the parabola with focus at the point $(1, -3)$ and directrix the line $x - y = 2$. Simplify your answer.

6. The graph of each of the following equations is a conic section. Use a theorem to determine whether the graph is a parabola, an ellipse, or a hyperbola without actually drawing the graph.

(a) $x^2 - 4xy + 4y^2 + x = 1$
(b) $x^2 + 3xy + y^2 = 4$

7. A parabolic arch is 20 feet wide at the base and 8 feet high at its center. Find the distance from the base of the arch to its focus.

8. For the given equation, do the following:

(i) Find $\cos \alpha$ and $\sin \alpha$ where α is the angle with least positive measure such that the xy term is eliminated when the axes are rotated through the angle α.
(ii) Find an equation with variables $\bar{x}$ and $\bar{y}$ for the given curve if the axes are rotated through the angle α.
(iii) Find the x and y coordinates of the vertices of the given curve and draw a sketch that shows the original x and y axes, the new $\bar{x}$ and $\bar{y}$ axes, and the curve.

$$21x^2 - 16xy + 9y^2 = 25$$

Test for Chapter 14 **(55 minutes)** **Solutions on page 777.**

1. Use L'Hôpital's rule to find the limit.

(a) $\displaystyle\lim_{x \to 0} \frac{\sin x - x}{x^3}$

(b) $\displaystyle\lim_{x \to 0} \tan x \ln x$

(c) $\displaystyle\lim_{x \to +\infty} \left(1 - \frac{2}{x}\right)^x$

2. Show whether or not the integral is convergent, and find the limit if it exists.

(a) $\displaystyle\int_{-\infty}^{0} \sqrt{e^{3x}}\, dx$

(b) $\displaystyle\int_{-1}^{+\infty} x^{-4/3}\, dx$

(c) $\displaystyle\int_{1}^{+\infty} xe^{-x}\, dx$

3. Let $f(x) = x^{1/2}$
(a) Find the third-degree Taylor polynomial about $a = 1$ for $f(x)$.
(b) Use the above polynomial to calculate an approximation for $\sqrt{2}$.
(c) Use the Lagrange form of the remainder to find an upper bound for the absolute value of the error in part (b).

Test for Chapter 15 **(100 minutes)** **Solutions on page 779.**

1. Let $u_n = \dfrac{n!}{1 \cdot 3 \cdot 5 \cdot \, \cdots \, \cdot (2n - 1)}$

(a) Show that the sequence $\{u_n\}$ is monotonic.
(b) Show that the sequence $\{u_n\}$ is convergent.

2. For each of the following do the following:

    (i) Determine whether the series is a p-series or a geometric series.
    (ii) Find p for each p-series and find r for each geometric series.
    (iii) Indicate whether the series is convergent or divergent.

    (a) $\displaystyle\sum_{n=1}^{+\infty} \frac{1}{n\sqrt{n}}$ **(b)** $\displaystyle\sum_{n=1}^{+\infty} \left(\frac{5}{2}\right)^{-n}$

    (c) $\displaystyle\sum_{n=1}^{+\infty} \frac{(-2)^n}{3}$ **(d)** $\displaystyle\sum_{n=1}^{+\infty} \frac{n}{n^{4/3}}$

3. Consider the series

$$\sum_{n=1}^{+\infty} \frac{(-1)^{n+1}\sqrt{n}}{n+1}$$

    (a) Use the alternating series test to show that the series is convergent.
    (b) Use a limit comparison test to determine whether the series is absolutely convergent or conditionally convergent.

4. Find the interval of convergence, including the endpoints if they exist, of the following power series.

$$\sum_{n=1}^{+\infty} \frac{2^n(x-1)^{n-1}}{n}$$

5. Show that the hypothesis for the integral test is satisfied and use the test to determine whether or not the following series is convergent.

$$\sum_{n=1}^{+\infty} n \cdot e^{-n^2}$$

6. Let $\{u_n\}$ and $\{S_n\}$ be sequence functions with $S_n = \displaystyle\sum_{k=1}^{n} u_k$. For each of the following answer true if the statement is always true and answer false otherwise.

    (a) If the sequence $\{u_n\}$ has an upper bound, it must have a least upper bound.
    (b) If the sequence $\{u_n\}$ is monotonic, it must be bounded.
    (c) If the sequence $\{u_n\}$ is bounded, it must be convergent.
    (d) If $\displaystyle\lim_{n\to+\infty} u_n = 0$, then the sequence $\{S_n\}$ must be convergent.
    (e) If $\displaystyle\lim_{n\to+\infty} u_n = \frac{1}{2}$, then the sequence $\{S_n\}$ must be divergent.
    (f) If $u_n \geq 0$ for all n, the sequence $\{S_n\}$ must be monotonic.

7. Use the Maclaurin series for $\cos x$ to find an infinite series whose sum is

$$\int_0^1 \cos\sqrt{x}\, dx$$

8. Use the binomial theorem to find the Maclaurin series for $1/\sqrt{1+x^3}$, if $|x| < 1$. Write your answer in the form $1 + \displaystyle\sum_{n=1}^{+\infty} u_n$, where u_n is reduced to a simple fraction.

9. Let $f(x) = \displaystyle\int_0^x e^{t^2}\, dt$. Find the Maclaurin series $\displaystyle\sum_{n=0}^{+\infty} u_n$ for $f(x)$ and find the simplest form of the ratio u_n/u_{n-1} for the series.

10. Let

$$f(x) = \sum_{n=0}^{+\infty} x^n = \frac{1}{1-x} \quad \text{if } |x| < 1$$

Use the power series for $f'(x)$ to show that

$$\sum_{n=1}^{+\infty} \frac{n}{3^n} = \frac{3}{4}$$

A.2 SOLUTIONS FOR CHAPTER TESTS
Solutions for Test 8

1. (a) $f(x) = \frac{1}{2}\ln(4 - x^2)$

$$f'(x) = \frac{1}{2} \cdot \frac{-2x}{4 - x^2} = \frac{x}{x^2 - 4}$$

(b) $f'(x) = 2^{3x}(3)\ln 2$

(c) $f'(x) = \frac{2xe^{2x} - e^{2x}}{x^2} = \frac{e^{2x}(2x - 1)}{x^2}$

(d) $\quad y = x^{\ln x}; \; x > 0$

$$\ln y = \ln x^{\ln x} = (\ln x)(\ln x) = \ln^2 x$$

$$\frac{1}{y} D_x y = (2 \ln x)\left(\frac{1}{x}\right)$$

$$D_x y = \left(\frac{2}{x}\ln x\right) x^{\ln x} = (2 \ln x)x^{-1 + \ln x}$$

2. (a)

$$\begin{array}{r} x + 1 \\ x - 1 \overline{)\; x^2 } \\ \underline{x^2 - x} \\ x \\ \underline{x - 1} \\ 1 \end{array}$$

$$\int \frac{x^2 \, dx}{x - 1} = \int (x + 1) \, dx + \int \frac{dx}{x - 1}$$

$$= \frac{1}{2}x^2 + x + \ln|x - 1| + C$$

(b) $u = 1 + e^x; \; du = e^x \, dx$

$$\int \frac{e^x \, dx}{\sqrt{1 + e^x}} = \int u^{-1/2} \, du$$

$$= 2u^{1/2} + C$$

$$= 2\sqrt{1 + e^x} + C$$

3. $\displaystyle\int_{-2}^{10} \frac{dx}{2x + 7} = \frac{1}{2}\int_{-2}^{10} \frac{d(2x + 7)}{2x + 7}$

$$= \frac{1}{2} \ln|2x + 7| \, \Big]_{-2}^{10}$$

$$= \frac{1}{2}[\ln 27 - \ln 3]$$

$$= \frac{1}{2} \ln 9 = \ln 3$$

4. $f'(x) = x^2 e^x + e^x(2x) = xe^x(x+2)$

$$\begin{array}{c|ccc} & -2 & & 0 \\ \hline f'(x) & + & - & + \end{array}$$

$f(0) = 0$, and thus 0 is a relative minimum value. $f(-2) = 4e^{-2}$, and thus $4e^{-2}$ is a relative maximum value.

5. $y = \ln x + \ln y$

$$D_x y = \frac{1}{x} + \frac{1}{y} D_x y$$

$$xy D_x y = y + x D_x y$$

$$D_x y = \frac{y}{xy - x}$$

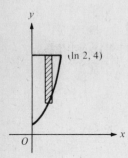

Figure 8. 6T

6. When $y = 4$, $e^{2x} = 4$, or $e^x = 2$ and $x = \ln 2$. (See Fig. 8.6T.)

$$A = \int_0^{\ln 2} (4 - e^{2x})\, dx$$

$$= 4x - \frac{1}{2} e^{2x}\Big]_0^{\ln 2}$$

$$= (4 \ln 2 - \tfrac{1}{2} e^{2 \ln 2}) - (0 - \tfrac{1}{2} e^0)$$

$$= 4 \ln 2 - \tfrac{3}{2}$$

7. (a) $f'(x) = \dfrac{(x+1)(2) - (2x+3)}{(x+1)^2} = \dfrac{-1}{(x+1)^2}$

Because $f'(x) < 0$ for $0 \le x \le 1$, then f is decreasing on $[0, 1]$, and thus f^{-1} exists.

(b) $y = \dfrac{2x+3}{x+1}$

Interchange x and y.

$$x = \frac{2y+3}{y+1}$$

$$xy + x = 2y + 3$$

$$y(x - 2) = -x + 3$$

$$y = \frac{-x+3}{x-2}$$

Thus,

$$f^{-1}(x) = \frac{-x+3}{x-2}$$

(c) $f(0) = 3$ and $f(1) = \frac{5}{2}$. Thus, the range of f is $[\frac{5}{2}, 3]$, which is also the domain of f^{-1}.

8. (a) $\ln x = \displaystyle\int_1^x \frac{dt}{t}$

(b) $2x - 1 > 0$

$$x > \tfrac{1}{2}$$

Domain is $(\frac{1}{2}, +\infty)$ and range is $(-\infty, +\infty)$.

(c) $a^x = \exp(x \ln a) = e^{x \ln a}$

(d) $e = \lim_{x \to 0} (1 + x)^{1/x}$

9. Let x degrees be the temperature after t minutes.

$$\frac{dx}{dt} = k(x - 60)$$

$$\int \frac{dx}{x - 60} = \int k \, dt$$

$$\ln|x - 60| = kt + C$$

$$x - 60 = A e^{kt}$$

Because $x = 100$ when $t = 0$, then $A = 40$. Thus,

$$x - 60 = 40 e^{kt}$$

Because $x = 80$ when $t = 3$, then

$$80 - 60 = 40 e^{3k}$$

$$\tfrac{1}{2} = (e^k)^3$$
$$e^k = (\tfrac{1}{2})^{1/3} = 2^{-1/3}$$

Thus,

$$x - 60 = 40 \cdot 2^{-t/3}$$

Let $x = 65$. Then

$$65 - 60 = 40 \cdot 2^{-t/3}$$
$$2^{-3} = 2^{-t/3}$$
$$t = 9$$

Thus, it takes 6 minutes longer.

Solutions for Test 9

1. (a) $f'(x) = \sec^3(2x) \sec^2(2x)(2) + \tan(2x)(3) \sec^2(2x) \sec(2x) \tan(2x)(2)$
$$= 2 \sec^5(2x) + 6 \sec^3(2x) \tan^2(2x)$$
$$= 2 \sec^5(2x) + 6 \sec^3(2x)[\sec^2(2x) - 1]$$
$$= 8 \sec^5(2x) - 6 \sec^3(2x)$$

(b) $f'(x) = \dfrac{(1 + \cosh x)\cosh x - \sinh x \sinh x}{(1 + \cosh x)^2}$

$$= \frac{\cosh x + \cosh^2 x - \sinh^2 x}{(1 + \cosh x)^2}$$

$$= \frac{\cosh x + 1}{(1 + \cosh x)^2} = \frac{1}{1 + \cosh x}$$

(c) $f'(x) = x \dfrac{1}{\sqrt{x^2 - 1}} + \cosh^{-1} x - \dfrac{1}{2}(x^2 - 1)^{-1/2}(2x)$

$$= \frac{x}{\sqrt{x^2 - 1}} + \cosh^{-1} x - \frac{x}{\sqrt{x^2 - 1}} = \cosh^{-1} x$$

2. (a) $\displaystyle\int \cot^5 x \, dx = \int (\csc^2 x - 1)\cot^3 x \, dx$

$$= \int [\cot^3 x \csc^2 x - (\csc^2 x - 1)\cot x] \, dx$$

$$= \int (\cot^3 x - \cot x)\csc^2 x \, dx + \int \cot x \, dx$$

$$= -\frac{1}{4} \cot^4 x + \frac{1}{2} \cot^2 x + \ln|\sin x| + C$$

(b) $\displaystyle\int \frac{dx}{1 + \tan^2 (4x)} = \int \frac{dx}{\sec^2 4x}$

$$= \int \cos^2 4x \, dx$$

$$= \frac{1}{2} \int (1 + \cos 8x) \, dx$$

$$= \frac{1}{2}(x + \frac{1}{8} \sin 8x) + C$$

(c) $\displaystyle\int \frac{(x+1) \, dx}{x^2 + 16} = \int \frac{x \, dx}{x^2 + 16} + \int \frac{dx}{x^2 + 16}$

$$= \frac{1}{2} \ln(x^2 + 16) + \frac{1}{4} \tan^{-1} (\frac{1}{4} x) + C$$

3. (a) $\displaystyle\int_{-1}^{\sqrt{2}} \frac{dx}{\sqrt{4 - x^2}} = \sin^{-1}\left(\frac{1}{2}x\right)\bigg]_{-1}^{\sqrt{2}}$

$$= \sin^{-1}\left(\frac{1}{2}\sqrt{2}\right) - \sin^{-1}\left(-\frac{1}{2}\right)$$

$$= \frac{1}{4}\pi - \left(-\frac{1}{6}\pi\right) = \frac{5}{12}\pi$$

(b) $\displaystyle\int_{\pi/3}^{2\pi/3} \tan\left(\frac{1}{2}x\right) dx = 2 \ln \sec\left(\frac{1}{2}x\right)\bigg]_{\pi/3}^{2\pi/3}$

$$= 2\left[\ln \sec\left(\frac{1}{3}\pi\right) - \ln \sec\left(\frac{1}{6}\pi\right)\right]$$

$$= 2\left[\ln 2 - \ln \frac{2}{3}\sqrt{3}\right]$$

$$= 2 \ln \sqrt{3} = \ln 3$$

4. Let $f(x) = x^2$ and $g(x) = x^{1/3}$. Then

$$f'(x) = 2x \qquad g'(x) = \frac{1}{3} x^{-2/3}$$
$$f'(1) = 2 \qquad g'(1) = \frac{1}{3}$$

$$\tan \theta = \frac{m_2 - m_1}{1 + m_2 m_1} = \frac{2 - \frac{1}{3}}{1 + 2(\frac{1}{3})} = 1$$

Thus, $\theta = \frac{1}{4}\pi$.

5. $f'(x) = \sec x \tan x - 8 \csc x \cot x$

$$= \frac{\sin x}{\cos^2 x} - \frac{8 \cos x}{\sin^2 x}$$

If $f'(x) = 0$, then

$$\frac{\sin x}{\cos^2 x} = \frac{8 \cos x}{\sin^2 x}$$

Figure 9. 5T

$$\sin^3 x = 8 \cos^3 x$$

$$\tan^3 x = 8$$

$$\tan x = 2$$

If $\tan x = 2$ (see Fig. 9.5T), then

$$f(x) = \sec x + 8 \csc x$$
$$= \sqrt{5} + 8(\tfrac{1}{2}\sqrt{5}) = 5\sqrt{5}$$

Furthermore,

$$f''(x) = \sec x(\sec^2 x) + \tan x(\sec x \tan x) - 8[\csc x(-\csc^2 x)$$
$$+ \cot x(-\csc x \cot x)]$$
$$= \sec^3 x + \sec x \tan^2 x + 8 \csc^3 x + 8 \csc x \cot^2 x$$

Because $f''(x) > 0$ for $0 < x < \tfrac{1}{2}\pi$, then f has a relative minimum value at the point where $f'(x) = 0$. Because there is only one critical number in $(0, \tfrac{1}{2}\pi)$, the relative minimum value is also an absolute minimum value.

6. $\theta = \cot^{-1}(\tfrac{1}{3}x) - \cot^{-1}x$ (See Fig. 9.6T)

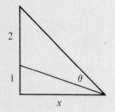

Figure 9. 6T

$$D_t\theta = -\frac{\tfrac{1}{3}D_t x}{1 + \tfrac{1}{9}x^2} + \frac{D_t x}{1 + x^2}$$

Let $D_t x = -5$ and $x = 3$.

$$D_t\theta = \frac{\tfrac{5}{3}}{1 + 1} + \frac{-5}{10} = \frac{5}{6} - \frac{3}{6} = \frac{1}{3}$$

Thus, the angle is increasing at the rate of $\tfrac{1}{3}$ radians per second.

Solutions for Test 10   **1. (a)** $\displaystyle\int x^2 \tan^{-1} x\, dx$

$$u = \tan^{-1} x \qquad dv = x^2\, dx$$

$$du = \frac{dx}{1 + x^2} \qquad v = \frac{1}{3}x^3$$

$$\int x^2 \tan^{-1} x\, dx = \frac{1}{3}x^3 \tan^{-1}x - \frac{1}{3}\int \frac{x^3\, dx}{1 + x^2}$$

$$= \frac{1}{3}x^3 \tan^{-1}x - \frac{1}{3}\left[\int x\, dx - \int \frac{x\, dx}{x^2 + 1}\right]$$

$$= \frac{1}{3}x^3 \tan^{-1}x - \frac{1}{6}x^2 + \frac{1}{6}\ln(x^2 + 1) + C$$

(b) $\displaystyle\int e^x \cos 2x\, dx$

$$u = e^x \qquad dv = \cos 2x\, dx$$

$$du = e^x\, dx \qquad v = \tfrac{1}{2}\sin 2x$$

$$\int e^x \cos 2x\, dx = \frac{1}{2}e^x \sin 2x - \frac{1}{2}\int e^x \sin 2x\, dx$$

$$\bar{u} = e^x \qquad d\bar{v} = \sin 2x\, dx$$

$$d\bar{u} = e^x\, dx \qquad \bar{v} = -\tfrac{1}{2}\cos 2x$$

$$\int e^x \cos 2x\, dx = \frac{1}{2}e^x \sin 2x - \frac{1}{2}\left[-\frac{1}{2}e^x \cos 2x + \frac{1}{2}\int e^x \cos 2x\, dx\right]$$

$$\int e^x \cos 2x \, dx = \frac{1}{2} e^x \sin 2x + \frac{1}{4} e^x \cos 2x - \frac{1}{4} \int e^x \cos 2x \, dx$$

$$\frac{5}{4} \int e^x \cos 2x \, dx = \frac{1}{4} e^x (2 \sin 2x + \cos 2x)$$

$$\int e^x \cos 2x \, dx = \frac{1}{5} e^x (2 \sin 2x + \cos 2x) + C$$

(c) $\displaystyle\int \frac{x^2 \, dx}{(16 - x^2)^{3/2}}$

$$\theta = \sin^{-1} \frac{x}{4}$$

$$x = 4 \sin \theta$$

$$dx = 4 \cos \theta \, d\theta$$

$$(16 - x^2)^{3/2} = [16(1 - \sin^2 \theta)]^{3/2} = 64 \cos^3 \theta$$

$$\int \frac{x^2 \, dx}{(16 - x^2)^{3/2}} = \int \frac{(16 \sin^2 \theta)(4 \cos \theta \, d\theta)}{64 \cos^3 \theta}$$

$$= \int \tan^2 \theta \, d\theta$$

$$= \int (\sec^2 \theta - 1) \, d\theta$$

$$= \tan \theta - \theta + C$$

$$= \frac{x}{\sqrt{16 - x^2}} - \sin^{-1}\left(\frac{x}{4}\right) + C$$

(d) $\displaystyle\int \frac{x^2 - 2x + 4}{x^3 + 2x} \, dx$

$$\frac{x^2 - 2x + 4}{x(x^2 + 2)} = \frac{A}{x} + \frac{Bx + C}{x^2 + 2}$$

$$x^2 - 2x + 4 = A(x^2 + 2) + (Bx + C)x$$

$$= (A + B)x^2 + Cx + 2A$$

$$A + B = 1$$

$$C = -2$$

$$2A = 4$$

$$A = 2 \qquad B = -1 \qquad C = -2$$

$$\int \frac{x^2 - 2x + 4}{x^3 + 2x} \, dx = \int \frac{2 \, dx}{x} - \int \frac{(x + 2) \, dx}{x^2 + 2}$$

$$= 2 \ln|x| - \frac{1}{2} \ln|x^2 + 2| - \frac{2}{\sqrt{2}} \tan^{-1}\left(\frac{x}{\sqrt{2}}\right) + C$$

$$= \ln\left(\frac{x^2}{\sqrt{x^2 + 2}}\right) - \sqrt{2} \tan^{-1}\left(\frac{x}{\sqrt{2}}\right) + C$$

(e) $\displaystyle\int \frac{x^2 + 2}{x^4 + 2x^2 + 1} \, dx$

$$\frac{x^2 + 2}{(x^2 + 1)^2} = \frac{Ax + B}{x^2 + 1} + \frac{Cx + D}{(x^2 + 1)^2}$$

$$x^2 + 2 = (Ax + B)(x^2 + 1) + Cx + D$$

$$= Ax^3 + Bx^2 + (A + C)x + (B + D)$$

$$A = 0 \qquad B = 1 \qquad A + C = 0 \qquad B + D = 2$$
$$C = 0 \qquad\qquad D = 1$$

$$\int \frac{x^2 + 2}{(x^2 + 1)^2} \, dx = \int \frac{dx}{x^2 + 1} + \int \frac{dx}{(x^2 + 1)^2}$$

$$\theta = \tan^{-1} x$$
$$x = \tan \theta$$
$$dx = \sec^2 \theta \, d\theta$$
$$(x^2 + 1)^2 = \sec^4 \theta$$

$$\int \frac{x^2 + 2}{(x^2 + 1)^2} \, dx = \tan^{-1} x + \int \frac{\sec^2 \theta \, d\theta}{\sec^4 \theta}$$

$$= \tan^{-1} x + \int \cos^2 \theta \, d\theta$$

$$= \tan^{-1} x + \frac{1}{2} \int (1 + \cos 2\theta) \, d\theta$$

$$= \tan^{-1} x + \frac{1}{2} \left(\theta + \frac{1}{2} \sin 2\theta \right) + C$$

$$= \tan^{-1} x + \frac{1}{2} (\theta + \sin \theta \cos \theta) + C$$

$$= \tan^{-1} x + \frac{1}{2} \tan^{-1} x + \frac{1}{2} \cdot \frac{x}{\sqrt{x^2 + 1}} \cdot \frac{1}{\sqrt{x^2 + 1}} + C$$

$$= \frac{3}{2} \tan^{-1} x + \frac{x}{2(x^2 + 1)} + C$$

(f) $\displaystyle \int \frac{dx}{x(1 + \sqrt[3]{x})}$

$$u = \sqrt[3]{x} \qquad x = u^3 \qquad dx = 3u^2 \, du$$

$$\int \frac{dx}{x(1 + \sqrt[3]{x})} = \int \frac{3u^2 \, du}{u^3(1 + u)}$$

$$= \int \frac{3 \, du}{u(1 + u)}$$

$$\frac{3}{u(1 + u)} = \frac{A}{u} + \frac{B}{1 + u}$$
$$3 = A(1 + u) + Bu$$

$$u = 0; \quad 3 = A$$

$$u = -1; \quad 3 = -B$$

$$\int \frac{3 \, du}{u(1 + u)} = \int \frac{3 \, du}{u} - \int \frac{3 \, du}{1 + u}$$

$$= 3(\ln|u| - \ln|1 + u|) + C$$

$$= 3 \ln \left| \frac{x^{1/3}}{1 + x^{1/3}} \right| + C$$

2. (a) $\displaystyle \int_4^5 \frac{\sqrt{x^2 - 16}}{x^2} \, dx$

$$\theta = \sec^{-1}(\tfrac{1}{4}x)$$

$$x = 4 \sec \theta$$

$$dx = 4 \sec \theta \tan \theta \, d\theta$$

$$\sqrt{x^2 - 16} = \sqrt{16(\sec^2 \theta - 1)} = 4 \tan \theta$$

$$\int_4^5 \frac{\sqrt{x^2 - 16}}{x^2} \, dx = \int_0^{\sec^{-1}(5/4)} \frac{(4 \tan \theta)(4 \sec \theta \tan \theta \, d\theta)}{16 \sec^2 \theta}$$

$$= \int_0^{\sec^{-1}(5/4)} \frac{\tan^2 \theta}{\sec \theta} \, d\theta$$

$$= \int_0^{\sec^{-1}(5/4)} \frac{\sec^2 \theta - 1}{\sec \theta} \, d\theta$$

$$= \int_0^{\sec^{-1}(5/4)} (\sec \theta - \cos \theta) \, d\theta$$

$$= \ln|\sec \theta + \tan \theta| - \sin \theta \Big]_0^{\sec^{-1}(5/4)}$$

$$= \ln \left| \frac{5}{4} + \frac{3}{4} \right| - \frac{3}{5} = \ln 2 - \frac{3}{5}$$

(b) $\displaystyle\int_0^{\pi/2} \frac{dx}{4 + 5 \cos x}$

$$z = \tan \tfrac{1}{2} x \qquad dx = \frac{2 \, dz}{1 + z^2} \qquad \cos x = \frac{1 - z^2}{1 + z^2}$$

$$\int_0^{\pi/2} \frac{dx}{4 + 5 \cos x} = \int_0^1 \frac{\dfrac{2 \, dz}{1 + z^2}}{4 + 5 \cdot \dfrac{1 - z^2}{1 + z^2}}$$

$$= \int_0^1 \frac{2 \, dz}{9 - z^2}$$

$$= \frac{1}{3} \ln \left| \frac{3 + z}{3 - z} \right| \, \Big]_0^1 = \frac{1}{3} \ln 2$$

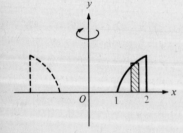

Figure 10. 3T

3. (See Fig. 10.3T.)

$$\Delta_i V = 2\pi \bar{r}_i h_i \, \Delta_i r$$

$$= 2\pi \bar{x}_i (\bar{x}_i \ln \bar{x}_i) \, \Delta_i x$$

$$V = 2\pi \int_1^2 x^2 \ln x \, dx$$

$$u = \ln x \qquad dv = x^2 \, dx$$

$$du = \frac{dx}{x} \qquad v = \frac{1}{3} x^3$$

$$V = 2\pi \left[\frac{1}{3} x^3 \ln x - \int \frac{1}{3} x^2 \, dx \right]_1^2$$

$$= 2\pi \left[\frac{1}{3} x^3 \ln x - \frac{1}{9} x^3 \right]_1^2$$

$$= \frac{2}{9} \pi (24 \ln 2 - 7)$$

4. $\displaystyle\int \frac{dx}{\sqrt{x^2 - 2x + 5}} = \int \frac{dx}{\sqrt{(x-1)^2 + 4}}$

$$= \sinh^{-1}\left(\frac{x-1}{2}\right) + C$$

$$= \ln(x - 1 + \sqrt{x^2 - 2x + 5}) + C$$

5. $\displaystyle\int_2^3 \frac{dx}{\sqrt{x^2 + 6x}} = \int_2^3 \frac{dx}{\sqrt{(x+3)^2 - 9}}$

$$= \ln(x + 3 + \sqrt{x^2 + 6x})\Big]_2^3$$

$$= \ln(6 + \sqrt{27}) - \ln(5 + \sqrt{16})$$

$$= \ln\left(\frac{6 + 3\sqrt{3}}{9}\right)$$

$$= \ln\left(\frac{2 + \sqrt{3}}{3}\right)$$

Solutions for Test 11   **1.** $M = \displaystyle\int_0^2 \sqrt{4x + 1}\ dx$

$$= \frac{1}{4}\int_0^2 (4x + 1)^{1/2}(4\ dx)$$

$$= \tfrac{1}{6}(4x + 1)^{3/2}\Big]_0^2 = \tfrac{13}{3}$$

Total mass is $\frac{13}{3}$ slugs.

$$m \cdot \bar{x} = \int_0^2 x\sqrt{4x + 1}\ dx$$

$$u = 4x + 1$$

$$du = 4\ dx$$

$$x = \frac{1}{4}(u - 1)$$

$$m \cdot \bar{x} = \int_1^9 \frac{1}{4}(u - 1)u^{1/2} \cdot \frac{1}{4}\ du$$

$$= \frac{1}{16}\int_1^9 (u^{3/2} - u^{1/2})\ du$$

$$= \frac{1}{16}\left[\frac{2}{5}u^{5/2} - \frac{2}{3}u^{3/2}\right]_1^9$$

$$= \frac{1}{16}\left[\frac{2}{5}(3^5 - 1) - \frac{2}{3}(3^3 - 1)\right]$$

$$= \frac{1}{16}\left(\frac{486}{5} - \frac{52}{3}\right) = \frac{1198}{240}$$

$$\bar{x} = \frac{1198}{240} \cdot \frac{3}{13} = 1.15$$

Center of mass is 1.15 ft from the end.

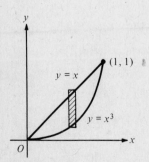

Figure 11. 2T

2. See Fig. 11.2T.

$$A = \int_0^1 (x - x^3)\, dx$$

$$= \frac{1}{2}x^2 - \frac{1}{4}x^4 \Big]_0^1 = \frac{1}{4}$$

$$A \cdot \bar{x} = \int_0^1 (x - x^3)x\, dx$$

$$= \frac{1}{3}x^3 - \frac{1}{5}x^5 \Big]_0^1 = \frac{2}{15}$$

$$A \cdot \bar{y} = \frac{1}{2} \int_0^1 (x - x^3)(x + x^3)\, dx$$

$$= \frac{1}{2}\left[\frac{1}{3}x^3 - \frac{1}{7}x^7\right]_0^1 = \frac{2}{21}$$

Thus, $\bar{x} = \frac{2}{15} \div \frac{1}{4} = \frac{8}{15}$ and $\bar{y} = \frac{2}{21} \div \frac{1}{4} = \frac{8}{21}$. Centroid is $(\frac{8}{15}, \frac{8}{21})$.

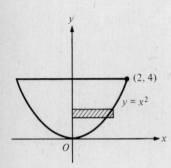

Figure 11. 3T

3. See Fig. 11.3T.

$$V = \pi \int_0^4 y\, dy$$

$$= \frac{1}{2}\pi y^2 \Big]_0^4 = 8\pi$$

$$V \cdot \bar{y} = \pi \int_0^4 y^2\, dy$$

$$= \frac{1}{3}\pi y^3 \Big]_0^4 = \frac{64}{3}\pi$$

Thus,

$$\bar{y} = \frac{64}{3}\pi \div 8\pi = \frac{8}{3}$$

The centroid is $(0, \frac{8}{3}, 0)$.

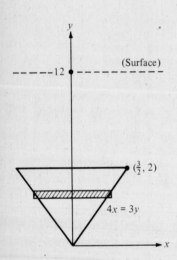

Figure 11. 4T

4. See Fig. 11.4T.

$$\Delta A = 2x\, \Delta y = \frac{3}{2}y\, \Delta y$$
$$h = 12 - y$$

$$F = \frac{3}{2}\rho \int_0^2 (12 - y)y\, dy$$

$$= \frac{3}{2}\rho\left[6y^2 - \frac{1}{3}y^3\right]_0^2$$

$$= \frac{3}{2}\rho[24 - \frac{8}{3}] = 32\rho$$

Force is 32ρ pounds.

5. See Fig. 11.5T.

Area of triangle: $A = \frac{1}{2}bh$
Volume of solid of revolution about y-axis: $V = \frac{1}{3}\pi b^2 h$ (cone)
Theorem of Pappus: $V = 2\pi \bar{x} A$
Therefore,

$$\frac{1}{3}\pi b^2 h = 2\pi \bar{x}(\frac{1}{2}bh)$$

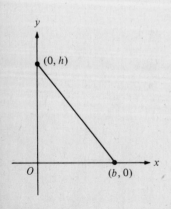

Figure 11. 5T

$$\frac{b}{3} = \bar{x}$$

Volume of solid of revolution about x-axis: $V = \frac{1}{3}\pi bh^2$

Theorem of Pappus: $V = 2\pi \bar{y}A$

Therefore,

$$\frac{1}{3}\pi bh^2 = 2\pi \bar{y}(\frac{1}{2}bh)$$

$$\frac{h}{3} = \bar{y}$$

Centroid is $(\frac{1}{3}b, \frac{1}{3}h)$.

Solutions for Test 12

1. (a) $r^2 \sin^2 \theta - r^2 \cos^2 \theta = 4$

$$-r^2(\cos^2 \theta - \sin^2 \theta) = 4$$
$$-r^2 \cos 2\theta = 4$$

(b) $r - r \sin \theta = 2$

$$r = 2 + r \sin \theta$$
$$\pm\sqrt{x^2 + y^2} = 2 + y$$
$$x^2 + y^2 = 4 + 4y + y^2$$
$$x^2 = 4 + 4y$$

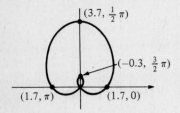

$(3.7, \frac{1}{2}\pi)$

$(-0.3, \frac{3}{2}\pi)$

$(1.7, \pi)$ $(1.7, 0)$

*Figure 12. 2*T

2.

θ	0	$\frac{1}{2}\pi$	π	$\frac{4}{3}\pi$	$\frac{3}{2}\pi$	$\frac{5}{3}\pi$	2π
r	1.7	3.7	1.7	0	-0.3	0	1.7

The graph is shown in Fig. 12.2T.

3. Each graph contains the pole. Furthermore, if

$$-3 \cos \theta = 1 - \cos \theta$$
$$-2 \cos \theta = 1$$
$$\cos \theta = -\frac{1}{2}$$
$$\theta = \pm\frac{2}{3}\pi$$

Intersection points are $(\frac{3}{2}, \frac{2}{3}\pi)$, $(\frac{3}{2}, -\frac{2}{3}\pi)$ and the pole. The graph is shown in Fig. 12.3T.

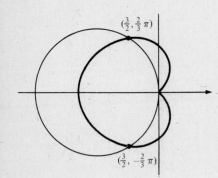

$(\frac{3}{2}, \frac{2}{3}\pi)$

$(\frac{3}{2}, -\frac{2}{3}\pi)$

*Figure 12. 3*T

4. $r = \sin \theta$ $\qquad$ $r = \cos 2\theta$

$$\frac{dr}{d\theta} = \cos \theta \qquad \frac{dr}{d\theta} = -2 \sin 2\theta$$

$$\frac{r}{\dfrac{dr}{d\theta}} = \frac{\sin \theta}{\cos \theta} \qquad \frac{r}{\dfrac{dr}{d\theta}} = \frac{\cos 2\theta}{-2 \sin 2\theta}$$

$$= \tan \theta \qquad = -\frac{1}{2} \cot 2\theta$$

$$\tan \chi_1 = \tan \frac{1}{6}\pi \qquad \tan \chi_2 = -\frac{1}{2} \cot \frac{1}{3}\pi$$

$$= \frac{1}{\sqrt{3}} \qquad = -\frac{1}{2\sqrt{3}}$$

$$\tan \beta = \left| \frac{\tan \chi_1 - \tan \chi_2}{1 + \tan \chi_1 \tan \chi_2} \right|$$

$$= \left| \frac{\dfrac{1}{\sqrt{3}} + \dfrac{1}{2\sqrt{3}}}{1 - \dfrac{1}{\sqrt{3}} \cdot \dfrac{1}{2\sqrt{3}}} \right|$$

$$= \frac{\dfrac{3}{2\sqrt{3}}}{1 - \dfrac{1}{6}} = \tfrac{3}{5}\sqrt{3}$$

5. If $r = 0$, then $\cos 3\theta = 0$ and $3\theta = \pm\tfrac{1}{2}\pi$, so $\theta = \pm\tfrac{1}{6}\pi$. Thus,

$$A = \frac{1}{2}\int_{-\pi/6}^{\pi/6} \cos^2 3\theta \ d\theta$$

$$= \frac{1}{4}\int_{-\pi/6}^{\pi/6} (1 + \cos 6\theta) \ d\theta$$

$$= \frac{1}{4}\left[\theta + \frac{1}{6}\sin 6\theta\right]_{-\pi/6}^{\pi/6}$$

$$= \frac{1}{4}\left[\frac{1}{6}\pi - \left(-\frac{1}{6}\pi\right)\right] = \frac{1}{12}\pi$$

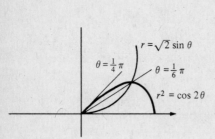

$r = \sqrt{2}\sin\theta$

$\theta = \tfrac{1}{4}\pi$ $\theta = \tfrac{1}{6}\pi$

$r^2 = \cos 2\theta$

Figure 12. 6T

6. See Fig. 12.6T. To find the point of intersection let

$$\cos 2\theta = (\sqrt{2}\sin\theta)^2$$
$$1 - 2\sin^2\theta = 2\sin^2\theta$$
$$1 = 4\sin^2\theta$$
$$\sin\theta = \tfrac{1}{2}$$
$$\theta = \tfrac{1}{6}\pi$$

Thus,

$$A_1 = \frac{1}{2}\int_0^{\pi/6} (\sqrt{2}\sin\theta)^2 \ d\theta$$

$$= \int_0^{\pi/6} \frac{1}{2}(1 - \cos 2\theta) \ d\theta$$

$$= \frac{1}{2}\left[\theta - \frac{1}{2}\sin 2\theta\right]_0^{\pi/6}$$

$$= \frac{1}{2}\left[\frac{1}{6}\pi - \frac{1}{2}\cdot\frac{1}{2}\sqrt{3}\right]$$

$$= \frac{1}{12}\pi - \frac{1}{8}\sqrt{3}$$

If $\cos 2\theta = 0$, then $\theta = \tfrac{1}{4}\pi$. Thus,

$$A_2 = \frac{1}{2}\int_{\pi/6}^{\pi/4} \cos(2\theta) \ d\theta$$

$$= \frac{1}{4}\sin 2\theta\bigg]_{\pi/6}^{\pi/4}$$

$$= \frac{1}{4}\left[1 - \frac{1}{2}\sqrt{3}\right]$$

$$= \frac{1}{4} - \frac{1}{8}\sqrt{3}$$

Thus,

$$A = A_1 + A_2 = \tfrac{1}{12}\pi + \tfrac{1}{4} - \tfrac{1}{4}\sqrt{3}$$

Solutions for Test 13

1. $8x - 12 = y^2 + 4y$

$\qquad 8x - 8 = y^2 + 4y + 4$

$\qquad 8(x - 1) = (y + 2)^2$

The vertex is $(1, -2)$. The axis is horizontal. Because $4p = 8$, then $p = 2$. Thus, the focus is $(3, -2)$, and the directrix is the line $x = -1$. The length of the latus rectum is 8. See Fig. 13.1T.

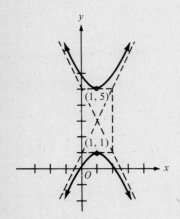

*Figure 13. 1*T

2.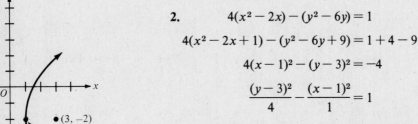

$$4(x^2 - 2x) - (y^2 - 6y) = 1$$

$$4(x^2 - 2x + 1) - (y^2 - 6y + 9) = 1 + 4 - 9$$

$$4(x - 1)^2 - (y - 3)^2 = -4$$

$$\frac{(y - 3)^2}{4} - \frac{(x - 1)^2}{1} = 1$$

The center is at $(1, 3)$. The transverse axis is vertical. Because $a^2 = 4$, then $a = 2$. Thus, the vertices are at $(1, 5)$ and $(1, 1)$. Because $b = 1$, then

$$b = a\sqrt{e^2 - 1}$$

$$1 = 2\sqrt{e^2 - 1}$$

$$1 = 4e^2 - 4$$

$$e = \tfrac{1}{2}\sqrt{5}$$

Thus $ae = \sqrt{5}$, and so the foci are at $(1, 3 + \sqrt{5})$ and $(1, 3 - \sqrt{5})$. Equations of the asymptotes are

$$\frac{y - 3}{2} = \pm \frac{x - 1}{1}$$

$$y - 3 = \pm 2(x - 1)$$

$$y = 2x + 1 \quad \text{and} \quad y = -2x + 5$$

See Fig. 13.2T.

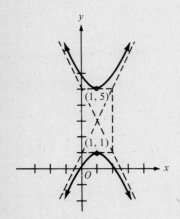

*Figure 13. 2*T

3. Because the center is at the origin and a vertex is at $(3, 0)$, the major axis is horizontal. Thus

$$\frac{x^2}{a^2} + \frac{y^2}{b^2} = 1$$

Furthermore $a = 3$, and $b = a\sqrt{1 - e^2}$. Thus,

$$b = 3\sqrt{1 - \tfrac{4}{9}} = \sqrt{5}$$

Thus,

$$\frac{x^2}{9} + \frac{y^2}{5} = 1$$

4. $r = \dfrac{6}{3 + 2\sin\theta} = \dfrac{2}{1 + \tfrac{2}{3}\sin\theta}$

Thus, $e = \tfrac{2}{3}$ and the curve is an ellipse. Because $ed = 2$, then $d = 3$. Therefore, an equation of the directrix nearest the pole is $r \sin\theta = 3$.

π	0	$\frac{1}{2}\pi$	π	$\frac{3}{2}\pi$
r	2	1.2	2	6

See Fig. 13.4T.

5. Let (x, y) be any point on the parabola. Because the distance between (x, y) and the focus $(1, -3)$ equals the distance between (x, y) and the directrix $x - y - 2 = 0$, we have

$$\sqrt{(x-1)^2 + (y+3)^2} = \frac{|x-y-2|}{\sqrt{2}}$$

$$x^2 - 2x + 1 + y^2 + 6y + 9 = \frac{x^2 + y^2 + 4 - 2xy - 4x + 4y}{2}$$

$$2x^2 - 4x + 2 + 2y^2 + 12y + 18 = x^2 + y^2 + 4 - 2xy - 4x + 4y$$

$$x^2 + 2xy + y^2 + 8y + 16 = 0$$

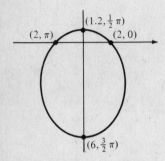

$(1.2, \frac{1}{2}\pi)$
$(2, \pi)$ $(2, 0)$
$(6, \frac{3}{2}\pi)$

*Figure 13. 4*T

6. (a) $B^2 - 4AC = (-4)^2 - 4 \cdot 1 \cdot 4 = 0$
   The graph is a parabola.
   (b) $B^2 - 4AC = 3^2 - 4 \cdot 1 \cdot 1 = 5 > 0$
   The graph is a hyperbola.

7. See Fig. 13.7T. We have

$$x^2 = 4py$$
$$10^2 = (4p)(-8)$$
$$p = -\frac{25}{8}$$

The focus is at the point $(0, -\frac{25}{8})$. Because $8 - \frac{25}{8} = \frac{39}{8}$, the focus is $\frac{39}{8}$ ft above the base.

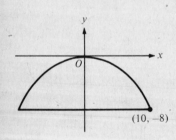

$(10, -8)$

*Figure 13. 7*T

8. (i) $\cot 2\alpha = \dfrac{A - C}{B} = \dfrac{21 - 9}{-16} = -\dfrac{3}{4}$

$$\tan 2\alpha = -\frac{4}{3}$$

$$\cos^2 2\alpha = \frac{1}{1 + \tan^2 2\alpha} = \frac{1}{1 + (-\frac{4}{3})^2} = \frac{9}{25}$$

$$\cos 2\alpha = -\sqrt{\cos^2 2\alpha} = -\tfrac{3}{5}$$

$$\sin \alpha = \sqrt{\frac{1 - \cos 2\alpha}{2}} = \sqrt{\frac{1 + \frac{3}{5}}{2}} = \frac{2}{\sqrt{5}}$$

$$\cos \alpha = \sqrt{\frac{1 + \cos 2\alpha}{2}} = \sqrt{\frac{1 - \frac{3}{5}}{2}} = \frac{1}{\sqrt{5}}$$

(ii) $x = \bar{x}\cos \alpha - \bar{y}\sin \alpha = \dfrac{\bar{x} - 2\bar{y}}{\sqrt{5}}$

$$y = \bar{x}\sin \alpha + \bar{y}\cos \alpha = \frac{2\bar{x} + \bar{y}}{\sqrt{5}}$$

$$x^2 = \frac{\bar{x}^2 - 4\bar{x}\bar{y} + 4\bar{y}^2}{5} \qquad xy = \frac{2\bar{x}^2 - 3\bar{x}\bar{y} - 2\bar{y}^2}{5} \qquad y^2 = \frac{4\bar{x}^2 + 4\bar{x}\bar{y} + \bar{y}^2}{5}$$

$$21x^2 - 16xy + 9y^2 = 25$$

$$\frac{21\bar{x}^2 - 84\bar{x}\bar{y} + 84\bar{y}^2}{5} + \frac{-32\bar{x}^2 + 48\bar{x}\bar{y} + 32\bar{y}^2}{5} + \frac{36\bar{x}^2 + 36\bar{x}\bar{y} + 9\bar{y}^2}{5} = 25$$

$$25\bar{x}^2 + 125\bar{y}^2 = 125$$

$$\frac{\bar{x}^2}{5} + \bar{y}^2 = 1$$

(iii) For the vertex, we have $\bar{x} = \pm\sqrt{5}$ and $\bar{y} = 0$. Thus,

$$x = \frac{\bar{x} - 2\bar{y}}{\sqrt{5}} = \pm 1$$

$$y = \frac{2\bar{x} + \bar{y}}{\sqrt{5}} = \pm 2$$

See Fig. 13.8T for the graph.

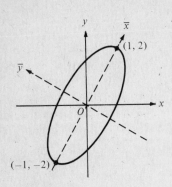

Figure 13. 8T

Solutions for Test 14 **1. (a)** $\displaystyle\lim_{x\to 0} \frac{\sin x - x}{x^3} = \lim_{x\to 0} \frac{\cos x - 1}{3x^2}$

$$= \lim_{x\to 0} \frac{-\sin x}{6x}$$

$$= \lim_{x\to 0} \frac{-\cos x}{6} = -\frac{1}{6}$$

(b) $\displaystyle\lim_{x\to 0} \tan x \ln x = \lim_{x\to 0} \frac{\ln x}{\cot x}$

$$= \lim_{x\to 0} \frac{\dfrac{1}{x}}{-\csc^2 x}$$

$$= \lim_{x\to 0} \frac{-\sin^2 x}{x}$$

$$= \lim_{x\to 0} \frac{-2\sin x \cos x}{1} = 0$$

(c) Let $y = \left(1 - \dfrac{2}{x}\right)^x$

$$\ln y = x \ln\left(1 - \frac{2}{x}\right)$$

$$\ln y = \frac{\ln(1 - 2x^{-1})}{x^{-1}}$$

$$\lim_{x\to +\infty} (\ln y) = \lim_{x\to +\infty} \frac{\dfrac{2x^{-2}}{1 - 2x^{-1}}}{-x^{-2}}$$

$$= \lim_{x\to +\infty} \frac{-2}{1 - 2x^{-1}} = -2$$

Thus,

$$\ln(\lim_{x\to +\infty} y) = -2$$

$$\lim_{x\to +\infty} y = e^{-2}$$

2. (a) $\int_{-\infty}^{0} \sqrt{e^{3x}} \, dx = \lim_{a \to -\infty} \int_{a}^{0} e^{3x/2} \, dx$

$$= \lim_{a \to -\infty} \left[\frac{2}{3} e^{3x/2} \right]_{a}^{0}$$

$$= \frac{2}{3} \lim_{a \to -\infty} [1 - e^{3a/2}] = \frac{2}{3}$$

(b) $\int_{-1}^{+\infty} x^{-4/3} \, dx = \int_{-1}^{0} x^{-4/3} \, dx + \int_{0}^{+\infty} x^{-4/3} \, dx$

Then

$$\int_{-1}^{0} x^{-4/3} \, dx = \lim_{\epsilon \to 0^-} \int_{-1}^{\epsilon} x^{-4/3} \, dx$$

$$= \lim_{\epsilon \to 0^-} \left[-3x^{-1/3} \right]_{-1}^{\epsilon}$$

$$= -3 \lim_{\epsilon \to 0^-} [\epsilon^{-1/3} + 1] = +\infty$$

Thus, the integral is divergent.

(c) $\int_{1}^{+\infty} xe^{-x} \, dx = \lim_{b \to +\infty} \int_{1}^{b} xe^{-x} \, dx$

Use integration by parts.

$$u = x \qquad dv = e^{-x} \, dx$$
$$du = dx \qquad v = -e^{-x}$$

$$\int xe^{-x} \, dx = -xe^{-x} + \int e^{-x} \, dx = -xe^{-x} - e^{-x}$$

Thus,

$$\int_{1}^{+\infty} xe^{-x} \, dx = \lim_{b \to +\infty} \left[-e^{-x}(x + 1) \right]_{1}^{b}$$

$$= \lim_{b \to +\infty} \frac{b + 1}{-e^{b}} + \frac{2}{e}$$

$$= \lim_{b \to +\infty} \left(\frac{1}{-e^{b}} \right) + \frac{2}{e} \qquad \text{(L'Hôpital's rule)}$$

$$= \frac{2}{e}$$

3. (a)
$$f(x) = x^{1/2} \qquad\qquad f(1) = 1$$
$$f'(x) = \tfrac{1}{2}x^{-1/2} \qquad\quad f'(1) = \tfrac{1}{2}$$
$$f''(x) = -\tfrac{1}{4}x^{-3/2} \qquad f''(1) = -\tfrac{1}{4}$$
$$f'''(x) = \tfrac{3}{8}x^{-5/2} \qquad\; f'''(1) = \tfrac{3}{8}$$

Thus,

$$P_3(x) = 1 + \tfrac{1}{2}(x - 1) - \frac{\tfrac{1}{4}(x - 1)^2}{2!} + \frac{\tfrac{3}{8}(x - 1)^3}{3!}$$

$$= 1 + \tfrac{1}{2}(x - 1) - \tfrac{1}{8}(x - 1)^2 + \tfrac{1}{16}(x - 1)^3$$

(b) $\sqrt{2} = f(2)$

$$P_3(2) = 1 + \tfrac{1}{2} - \tfrac{1}{8} + \tfrac{1}{16} = \tfrac{23}{16} = 1.44$$

(c) $\quad R_3(2) = \dfrac{f^{(iv)}(\xi)(2-1)^4}{4!} \quad$ with ξ between 1 and 2

$$f^{(iv)}(x) = -\tfrac{15}{16} x^{-7/2}$$

$$|f^{(iv)}(\xi)| \le \tfrac{15}{16} \cdot 1^{-7/2}$$

Thus,

$$|R_3(2)| \le \frac{15}{16} \cdot \frac{1^4}{4!} = \tfrac{5}{128} = 0.039$$

Solutions for Test 15

1. (a) $\dfrac{u_{n+1}}{u_n} = \dfrac{(n+1)!}{1 \cdot 3 \cdot 5 \cdot \cdots \cdot (2n-1)(2n+1)} \cdot \dfrac{1 \cdot 3 \cdot 5 \cdot \cdots \cdot (2n-1)}{n!}$

$$= \frac{n+1}{2n+1}$$

Because $n + 1 < 2n + 1$, then $u_{n+1} < u_n$, and thus the sequence $\{u_n\}$ is monotonic decreasing.

(b) Because $\{u_n\}$ is decreasing, then $u_1 = 1$ is an upper bound. Because $u_n > 0$ for all n, then 0 is a lower bound. Because $\{u_n\}$ is monotonic and bounded, then $\{u_n\}$ is convergent.

2. (a) $\displaystyle\sum_{n=1}^{+\infty} \frac{1}{n\sqrt{n}} = \sum_{n=1}^{+\infty} \frac{1}{n^{3/2}}$

Thus, the series is a p-series with $p = \tfrac{3}{2}$. Because $p > 1$, the series is convergent.

(b) $\displaystyle\sum_{n=1}^{+\infty} \left(\frac{5}{2}\right)^{-n} = \sum_{n=1}^{+\infty} \left(\frac{2}{5}\right)^n$

Thus, the series is a geometric series with $r = \tfrac{2}{5}$. Because $|r| < 1$, the series is convergent.

(c) $\displaystyle\sum_{n=1}^{+\infty} \frac{(-2)^n}{3} = \sum_{n=1}^{+\infty} \left(-\frac{2}{3}\right)(-2)^{n-1}$

Thus, the series is a geometric series with $r = -2$. Because $|r| > 1$, the series is divergent.

(d) $\displaystyle\sum_{n=1}^{+\infty} \frac{n}{n^{4/3}} = \sum_{n=1}^{+\infty} \frac{1}{n^{1/3}}$

Thus, the series is a p-series with $p = \tfrac{1}{3}$. Because $p < 1$, the series is divergent.

3. (a) $|u_n| = \dfrac{\sqrt{n}}{n+1}$

Thus, $|u_{n+1}| \le |u_n|$ if and only if

$$\frac{\sqrt{n+1}}{n+2} \le \frac{\sqrt{n}}{n+1}$$

$$\frac{n+1}{(n+2)^2} \le \frac{n}{(n+1)^2}$$

$$(n+1)^3 \le n(n+2)^2$$

$$n^3 + 3n^2 + 3n + 1 \leq n^3 + 4n^2 + 4n$$

$$1 \leq n^2 + n$$

Hence, $\{|u_n|\}$ is decreasing. Furthermore,

$$\lim_{n \to +\infty} |u_n| = \lim_{n \to +\infty} \frac{\sqrt{n}}{n+1}$$

$$= \lim_{n \to +\infty} \frac{\sqrt{\dfrac{1}{n}}}{1 + \dfrac{1}{n}} = 0$$

Thus, the alternating series $\{u_n\}$ is convergent by the alternating series test.

(b) Let $|u_n| = \sqrt{n}/(n+1)$ and let $v_n = 1/\sqrt{n}$.

$$\lim_{n \to +\infty} \frac{|u_n|}{v_n} = \lim_{n \to +\infty} \frac{n}{n+1} = 1$$

Furthermore,

$$\sum_{n=1}^{+\infty} v_n = \sum_{n=1}^{+\infty} \frac{1}{n^{1/2}}$$

is divergent because it is a p-series with $p = \frac{1}{2}$. Therefore, $\displaystyle\sum_{n=1}^{+\infty} |u_n| = \sum_{n=1}^{+\infty} \sqrt{n}/(n+1)$ is divergent. Hence, the given alternating series is conditionally convergent.

4.
$$\lim_{n \to +\infty} \left| \frac{u_{n+1}}{u_n} \right| = \lim_{n \to +\infty} \left| \frac{2^{n+1}(x-1)^n}{n+1} \cdot \frac{n}{2^n(x-1)^{n-1}} \right|$$

$$= \lim_{n \to +\infty} \frac{2n}{n+1} |x-1| = 2|x-1|$$

If $2|x-1| < 1$, then

$$|x-1| < \tfrac{1}{2}$$
$$-\tfrac{1}{2} < x - 1 < \tfrac{1}{2}$$
$$\tfrac{1}{2} < x < \tfrac{3}{2}$$

We test the endpoints. If $x = \frac{3}{2}$, then

$$\sum_{n=1}^{+\infty} \frac{2^n(x-1)^{n-1}}{n} = \sum_{n=1}^{+\infty} \frac{2^n(\frac{1}{2})^{n-1}}{n}$$

$$= 2 \sum_{n=1}^{+\infty} \frac{1}{n}$$

which is divergent because it is 2 times the harmonic series. Moreover, if $x = \frac{1}{2}$, then

$$\sum_{n=1}^{+\infty} \frac{2^n(x-1)^{n-1}}{n} = \sum_{n=1}^{+\infty} \frac{2^n(-\frac{1}{2})^{n-1}}{n} = 2 \sum_{n=1}^{+\infty} \frac{(-1)^{n-1}}{n}$$

which is convergent by the alternating series test. Thus, the interval of convergence is $\{x : \frac{1}{2} \leq x < \frac{3}{2}\} = [\frac{1}{2}, \frac{3}{2})$.

5. Let $f(x) = xe^{-x^2}$. Then

$$f'(x) = -2x^2 e^{-x^2} + e^{-x^2} = \frac{1 - 2x^2}{e^{x^2}}$$

Because $f'(x) < 0$ if $x \geq 1$, then f is decreasing and continuous on $[1, +\infty)$. Thus, the hypothesis of the integral test is satisfied. Moreover,

$$\int_1^{+\infty} xe^{-x^2}\,dx = \lim_{b \to +\infty} \left[-\frac{1}{2} e^{-x^2} \right]_1^b$$

$$= -\frac{1}{2} \lim_{b \to +\infty} [e^{-b^2} - e^{-1}] = \frac{1}{2} e^{-1}$$

Therefore, the given series is convergent.

6. (a) True **(b)** False **(c)** False **(d)** False **(e)** True **(f)** True

7.
$$\cos x = \sum_{n=0}^{+\infty} \frac{(-1)^n x^{2n}}{(2n)!}$$

$$\cos \sqrt{x} = \sum_{n=0}^{+\infty} \frac{(-1)^n x^n}{(2n)!}$$

$$\int_0^1 \cos \sqrt{x}\,dx = \left[\sum_{n=0}^{+\infty} \frac{(-1)^n x^{n+1}}{(2n)!(n+1)} \right]_0^1$$

$$= \sum_{n=0}^{+\infty} \frac{(-1)^n}{(n+1)(2n)!}$$

8. $\dfrac{1}{\sqrt{1 + x^3}} = (1 + x^3)^{-1/2}$

$$= 1 + \sum_{n=1}^{+\infty} \frac{(-\frac{1}{2})(-\frac{1}{2} - 1)(-\frac{1}{2} - 2) \cdots (-\frac{1}{2} - n + 1)}{n!} (x^3)^n \quad \text{if } |x^3| < 1$$

$$= 1 + \sum_{n=1}^{+\infty} \frac{(-1)^n 1 \cdot 3 \cdot 5 \cdots (2n-1) x^{3n}}{2^n n!} \quad \text{if } |x| < 1$$

9.
$$e^x = \sum_{n=0}^{+\infty} \frac{x^n}{n!}$$

$$e^{t^2} = \sum_{n=0}^{+\infty} \frac{t^{2n}}{n!}$$

$$\int_0^x e^{t^2}\,dt = \left[\sum_{n=0}^{+\infty} \frac{t^{2n+1}}{n!(2n+1)} \right]_0^x$$

$$= \sum_{n=0}^{+\infty} \frac{x^{2n+1}}{n!(2n+1)}$$

Furthermore,

$$\frac{u_n}{u_{n-1}} = \frac{x^{2n+1}}{n!(2n+1)} \cdot \frac{(n-1)!(2n-1)}{x^{2n-1}} = \frac{x^2(2n-1)}{n(2n+1)}$$

10. $f'(x) = \sum_{n=1}^{+\infty} nx^{n-1}$

$$f'\left(\frac{1}{3}\right) = \sum_{n=1}^{+\infty} \frac{n}{3^{n-1}} = \sum_{n=1}^{+\infty} \frac{3n}{3^n} = 3 \sum_{n=1}^{+\infty} \frac{n}{3^n}$$

Furthermore, because $f(x) = (1-x)^{-1}$, then

$$f'(x) = -(1-x)^{-2}(-1) = \frac{1}{(1-x)^2}$$

$$f'\left(\frac{1}{3}\right) = \frac{1}{(1-\frac{1}{3})^2} = \frac{9}{4}$$

Thus,

$$3\sum_{n=1}^{+\infty} \frac{n}{3^n} = \frac{9}{4}$$

$$\sum_{n=1}^{+\infty} \frac{n}{3^n} = \frac{3}{4}$$

A TABLE OF DERIVATIVES

1. $D_x(u^n) = nu^{n-1} D_x u$

2. $D_x(u + v) = D_x u + D_x v$

3. $D_x(uv) = u D_x v + v D_x u$

4. $D_x\left(\dfrac{u}{v}\right) = \dfrac{v D_x u - u D_x v}{v^2}$

5. $D_x(e^u) = e^u D_x u$

6. $D_x(a^u) = a^u \ln a \, D_x u$

7. $D_x(\ln u) = \dfrac{1}{u} D_x u$

8. $D_x(\sin u) = \cos u \, D_x u$

9. $D_x(\cos u) = -\sin u \, D_x u$

10. $D_x(\tan u) = \sec^2 u \, D_x u$

11. $D_x(\cot u) = -\csc^2 u \, D_x u$

12. $D_x(\sec u) = \sec u \tan u \, D_x u$

13. $D_x(\csc u) = -\csc u \cot u \, D_x u$

14. $D_x(\sin^{-1} u) = \dfrac{1}{\sqrt{1 - u^2}} D_x u$

15. $D_x(\cos^{-1} u) = \dfrac{-1}{\sqrt{1 - u^2}} D_x u$

16. $D_x(\tan^{-1} u) = \dfrac{1}{1 + u^2} D_x u$

17. $D_x(\cot^{-1} u) = \dfrac{-1}{1 + u^2} D_x u$

18. $D_x(\sec^{-1} u) = \dfrac{1}{|u| \sqrt{u^2 - 1}} D_x u$

19. $D_x(\csc^{-1} u) = \dfrac{-1}{|u| \sqrt{u^2 - 1}} D_x u$

20. $D_x(\sinh u) = \cosh u \, D_x u$

21. $D_x(\cosh u) = \sinh u \, D_x u$

22. $D_x(\tanh u) = \operatorname{sech}^2 u \, D_x u$

23. $D_x(\coth u) = -\operatorname{csch}^2 u \, D_x u$

24. $D_x(\operatorname{sech} u) = -\operatorname{sech} u \tanh u \, D_x u$

25. $D_x(\operatorname{csch} u) = -\operatorname{csch} u \coth u \, D_x u$

A TABLE OF INTEGRALS

Some Elementary Forms

1. $\displaystyle\int du = u + C$

2. $\displaystyle\int a \, du = au + C$

3. $\displaystyle\int [f(u) + g(u)]du = \int f(u)du + \int g(u)du$

4. $\displaystyle\int u^n \, du = \dfrac{u^{n+1}}{n + 1} + C \quad (n \neq -1)$

5. $\displaystyle\int \dfrac{du}{u} = \ln|u| + C$

Rational Forms Containing $a + bu$

6. $\displaystyle\int \dfrac{u \, du}{a + bu} = \dfrac{1}{b^2}[a + bu - a \ln|a + bu|] + C$

7. $\displaystyle\int \dfrac{u^2 \, du}{a + bu} = \dfrac{1}{b^3}\left[\dfrac{1}{2}(a + bu)^2 - 2a(a + bu) + a^2\ln|a + bu|\right] + C$

8. $\displaystyle\int \dfrac{u \, du}{(a + bu)^2} = \dfrac{1}{b^2}\left[\dfrac{a}{a + bu} + \ln|a + bu|\right] + C$

9. $\displaystyle\int \dfrac{u^2 \, du}{(a + bu)^2} = \dfrac{1}{b^3}\left[a + bu - \dfrac{a^2}{a + bu} - 2a \ln|a + bu|\right] + C$

10. $\displaystyle\int \dfrac{u \, du}{(a + bu)^3} = \dfrac{1}{b^2}\left[\dfrac{a}{2(a + bu)^2} - \dfrac{1}{a + bu}\right] + C$

11. $\displaystyle\int \dfrac{du}{u(a + bu)} = \dfrac{1}{a} \ln\left|\dfrac{u}{a + bu}\right| + C$

12. $\displaystyle\int \dfrac{du}{u^2(a + bu)} = -\dfrac{1}{au} + \dfrac{b}{a^2} \ln\left|\dfrac{a + bu}{u}\right| + C$

13. $\displaystyle\int \dfrac{du}{u(a + bu)^2} = \dfrac{1}{a(a + bu)} + \dfrac{1}{a^2} \ln\left|\dfrac{u}{a + bu}\right| + C$

Forms Containing $\sqrt{a + bu}$

14. $\displaystyle\int u\sqrt{a + bu} \, du = \dfrac{2}{15b^3}(3bu - 2a)(a + bu)^{3/2} + C$

15. $\displaystyle\int u^2\sqrt{a + bu} \, du = \dfrac{2}{105b^3}(15b^2u^2 - 12abu + 8a^2)(a + bu)^{3/2} + C$

16. $\displaystyle\int u^n\sqrt{a + bu} \, du = \dfrac{2u^n(a + bu)^{3/2}}{b(2n + 3)} - \dfrac{2an}{b(2n + 3)}\int u^{n-1}\sqrt{a + bu} \, du$

17. $\displaystyle\int \dfrac{u \, du}{\sqrt{a + bu}} = \dfrac{2}{3b^2}(bu - 2a)\sqrt{a + bu} + C$

18. $\displaystyle\int \dfrac{u^2 \, du}{\sqrt{a + bu}} = \dfrac{2}{15b^3}(3b^2u^2 - 4abu + 8a^2)\sqrt{a + bu} + C$

19. $\displaystyle\int \dfrac{u^n \, du}{\sqrt{a + bu}} = \dfrac{2u^n\sqrt{a + bu}}{b(2n + 1)} - \dfrac{2an}{b(2n + 1)}\int \dfrac{u^{n-1} \, du}{\sqrt{a + bu}}$

20. $\displaystyle\int \dfrac{du}{u\sqrt{a + bu}} = \begin{cases} \dfrac{1}{\sqrt{a}} \ln\left|\dfrac{\sqrt{a + bu} - \sqrt{a}}{\sqrt{a + bu} + \sqrt{a}}\right| + C & \text{if } a > 0 \\[2ex] \dfrac{2}{\sqrt{-a}} \tan^{-1}\sqrt{\dfrac{a + bu}{-a}} + C & \text{if } a < 0 \end{cases}$

21. $\displaystyle\int \dfrac{du}{u^n\sqrt{a + bu}} = -\dfrac{\sqrt{a + bu}}{a(n - 1)u^{n-1}} - \dfrac{b(2n - 3)}{2a(n - 1)}\int \dfrac{du}{u^{n-1}\sqrt{a + bu}}$

22. $\displaystyle\int \dfrac{\sqrt{a + bu} \, du}{u} = 2\sqrt{a + bu} + a\int \dfrac{du}{u\sqrt{a + bu}}$

23. $\displaystyle\int \dfrac{\sqrt{a + bu} \, du}{u^n} = -\dfrac{(a + bu)^{3/2}}{a(n - 1)u^{n-1}} - \dfrac{b(2n - 5)}{2a(n - 1)}\int \dfrac{\sqrt{a + bu} \, du}{u^{n-1}}$

Forms Containing $a^2 \pm u^2$

24. $\displaystyle\int \frac{du}{a^2 + u^2} = \frac{1}{a}\tan^{-1}\frac{u}{a} + C$

25. $\displaystyle\int \frac{du}{a^2 - u^2} = \frac{1}{2a}\ln\left|\frac{u+a}{u-a}\right| + C = \begin{cases} \dfrac{1}{a}\tanh^{-1}\dfrac{u}{a} + C & \text{if } |u| < a \\[2mm] \dfrac{1}{a}\coth^{-1}\dfrac{u}{a} + C & \text{if } |u| > a \end{cases}$

26. $\displaystyle\int \frac{du}{u^2 - a^2} = \frac{1}{2a}\ln\left|\frac{u-a}{u+a}\right| + C = \begin{cases} -\dfrac{1}{a}\tanh^{-1}\dfrac{u}{a} + C & \text{if } |u| < a \\[2mm] -\dfrac{1}{a}\coth^{-1}\dfrac{u}{a} + C & \text{if } |u| > a \end{cases}$

Forms Containing $\sqrt{u^2 \pm a^2}$

In formulas 27 through 38, we may replace

$\ln(u + \sqrt{u^2 + a^2})$ by $\sinh^{-1}\dfrac{u}{a}$

$\ln|u + \sqrt{u^2 - a^2}|$ by $\cosh^{-1}\dfrac{u}{a}$

$\ln\left|\dfrac{a + \sqrt{u^2 + a^2}}{u}\right|$ by $\sinh^{-1}\dfrac{a}{u}$

27. $\displaystyle\int \frac{du}{\sqrt{u^2 \pm a^2}} = \ln|u + \sqrt{u^2 \pm a^2}| + C$

28. $\displaystyle\int \sqrt{u^2 \pm a^2}\, du = \frac{u}{2}\sqrt{u^2 \pm a^2} \pm \frac{a^2}{2}\ln|u + \sqrt{u^2 \pm a^2}| + C$

29. $\displaystyle\int u^2 \sqrt{u^2 \pm a^2}\, du = \frac{u}{8}(2u^2 \pm a^2)\sqrt{u^2 \pm a^2}$
$\qquad\qquad\qquad\qquad\qquad -\dfrac{a^4}{8}\ln|u + \sqrt{u^2 \pm a^2}| + C$

30. $\displaystyle\int \frac{\sqrt{u^2 + a^2}\, du}{u} = \sqrt{u^2 + a^2} - a\ln\left|\frac{a + \sqrt{u^2 + a^2}}{u}\right| + C$

31. $\displaystyle\int \frac{\sqrt{u^2 - a^2}\, du}{u} = \sqrt{u^2 - a^2} - a\sec^{-1}\left|\frac{u}{a}\right| + C$

32. $\displaystyle\int \frac{\sqrt{u^2 \pm a^2}\, du}{u^2} = -\frac{\sqrt{u^2 \pm a^2}}{u} + \ln|u + \sqrt{u^2 \pm a^2}| + C$

33. $\displaystyle\int \frac{u^2\, du}{\sqrt{u^2 \pm a^2}} = \frac{u}{2}\sqrt{u^2 \pm a^2} - \frac{\pm a^2}{2}\ln|u + \sqrt{u^2 \pm a^2}| + C$

34. $\displaystyle\int \frac{du}{u\sqrt{u^2 + a^2}} = -\frac{1}{a}\ln\left|\frac{a + \sqrt{u^2 + a^2}}{u}\right| + C$

35. $\displaystyle\int \frac{du}{u\sqrt{u^2 - a^2}} = \frac{1}{a}\sec^{-1}\left|\frac{u}{a}\right| + C$

36. $\displaystyle\int \frac{du}{u^2\sqrt{u^2 \pm a^2}} = -\frac{\sqrt{u^2 \pm a^2}}{\pm a^2 u} + C$

37. $\displaystyle\int (u^2 \pm a^2)^{3/2}\, du = \frac{u}{8}(2u^2 \pm 5a^2)\sqrt{u^2 \pm a^2}$
$\qquad\qquad\qquad\qquad\qquad + \dfrac{3a^4}{8}\ln|u + \sqrt{u^2 \pm a^2}| + C$

38. $\displaystyle\int \frac{du}{(u^2 \pm a^2)^{3/2}} = \frac{u}{\pm a^2\sqrt{u^2 \pm a^2}} + C$

Forms Containing $\sqrt{a^2 - u^2}$

39. $\displaystyle\int \frac{du}{\sqrt{a^2 - u^2}} = \sin^{-1}\frac{u}{a} + C$

40. $\displaystyle\int \sqrt{a^2 - u^2}\, du = \frac{u}{2}\sqrt{a^2 - u^2} + \frac{a^2}{2}\sin^{-1}\frac{u}{a} + C$

41. $\displaystyle\int u^2\sqrt{a^2 - u^2}\, du = \frac{u}{8}(2u^2 - a^2)\sqrt{a^2 - u^2} + \frac{a^4}{8}\sin^{-1}\frac{u}{a} + C$

42. $\displaystyle\int \frac{\sqrt{a^2 - u^2}\, du}{u} = \sqrt{a^2 - u^2} - a\ln\left|\frac{a + \sqrt{a^2 - u^2}}{u}\right| + C$
$\qquad\qquad\qquad\quad = \sqrt{a^2 - u^2} - a\cosh^{-1}\dfrac{a}{u} + C$

43. $\displaystyle\int \frac{\sqrt{a^2 - u^2}\, du}{u^2} = -\frac{\sqrt{a^2 - u^2}}{u} - \sin^{-1}\frac{u}{a} + C$

44. $\displaystyle\int \frac{u^2\, du}{\sqrt{a^2 - u^2}} = -\frac{u}{2}\sqrt{a^2 - u^2} + \frac{a^2}{2}\sin^{-1}\frac{u}{a} + C$

45. $\displaystyle\int \frac{du}{u\sqrt{a^2 - u^2}} = -\frac{1}{a}\ln\left|\frac{a + \sqrt{a^2 - u^2}}{u}\right| + C$
$\qquad\qquad\qquad\quad = -\dfrac{1}{a}\cosh^{-1}\dfrac{a}{u} + C$

46. $\displaystyle\int \frac{du}{u^2\sqrt{a^2 - u^2}} = -\frac{\sqrt{a^2 - u^2}}{a^2 u} + C$

47. $\displaystyle\int (a^2 - u^2)^{3/2}\, du = -\frac{u}{8}(2u^2 - 5a^2)\sqrt{a^2 - u^2} + \frac{3a^4}{8}\sin^{-1}\frac{u}{a} + C$

48. $\displaystyle\int \frac{du}{(a^2 - u^2)^{3/2}} = \frac{u}{a^2\sqrt{a^2 - u^2}} + C$

Forms Containing $2au - u^2$

49. $\displaystyle\int \sqrt{2au - u^2}\, du = \frac{u - a}{2}\sqrt{2au - u^2} + \frac{a^2}{2}\cos^{-1}\left(1 - \frac{u}{a}\right) + C$

50. $\displaystyle\int u\sqrt{2au - u^2}\, du = \frac{2u^2 - au - 3a^2}{6}\sqrt{2au - u^2}$
$$+ \frac{a^3}{2}\cos^{-1}\left(1 - \frac{u}{a}\right) + C$$

51. $\displaystyle\int \frac{\sqrt{2au - u^2}\, du}{u} = \sqrt{2au - u^2} + a\cos^{-1}\left(1 - \frac{u}{a}\right) + C$

52. $\displaystyle\int \frac{\sqrt{2au - u^2}\, du}{u^2} = -\frac{2\sqrt{2au - u^2}}{u} - \cos^{-1}\left(1 - \frac{u}{a}\right) + C$

53. $\displaystyle\int \frac{du}{\sqrt{2au - u^2}} = \cos^{-1}\left(1 - \frac{u}{a}\right) + C$

54. $\displaystyle\int \frac{u\, du}{\sqrt{2au - u^2}} = -\sqrt{2au - u^2} + a\cos^{-1}\left(1 - \frac{u}{a}\right) + C$

55. $\displaystyle\int \frac{u^2\, du}{\sqrt{2au - u^2}} = -\frac{(u + 3a)}{2}\sqrt{2au - u^2} + \frac{3a^2}{2}\cos^{-1}\left(1 - \frac{u}{a}\right) + C$

56. $\displaystyle\int \frac{du}{u\sqrt{2au - u^2}} = -\frac{\sqrt{2au - u^2}}{au} + C$

57. $\displaystyle\int \frac{du}{(2au - u^2)^{3/2}} = \frac{u - a}{a^2\sqrt{2au - u^2}} + C$

58. $\displaystyle\int \frac{u\, du}{(2au - u^2)^{3/2}} = \frac{u}{a\sqrt{2au - u^2}} + C$

Forms Containing Trigonometric Functions

59. $\displaystyle\int \sin u\, du = -\cos u + C$

60. $\displaystyle\int \cos u\, du = \sin u + C$

61. $\displaystyle\int \tan u\, du = \ln|\sec u| + C$

62. $\displaystyle\int \cot u\, du = \ln|\sin u| + C$

63. $\displaystyle\int \sec u\, du = \ln|\sec u + \tan u| + C = \ln|\tan(\tfrac{1}{4}\pi + \tfrac{1}{2}u)| + C$

64. $\displaystyle\int \csc u\, du = \ln|\csc u - \cot u| + C = \ln|\tan \tfrac{1}{2}u| + C$

65. $\displaystyle\int \sec^2 u\, du = \tan u + C$

66. $\displaystyle\int \csc^2 u\, du = -\cot u + C$

67. $\displaystyle\int \sec u \tan u\, du = \sec u + C$

68. $\displaystyle\int \csc u \cot u\, du = -\csc u + C$

69. $\displaystyle\int \sin^2 u\, du = \tfrac{1}{2}u - \tfrac{1}{4}\sin 2u + C$

70. $\displaystyle\int \cos^2 u\, du = \tfrac{1}{2}u + \tfrac{1}{4}\sin 2u + C$

71. $\displaystyle\int \tan^2 u\, du = \tan u - u + C$

72. $\displaystyle\int \cot^2 u\, du = -\cot u - u + C$

73. $\displaystyle\int \sin^n u\, du = -\frac{1}{n}\sin^{n-1} u \cos u + \frac{n-1}{n}\int \sin^{n-2} u\, du$

74. $\displaystyle\int \cos^n u\, du = \frac{1}{n}\cos^{n-1} u \sin u + \frac{n-1}{n}\int \cos^{n-2} u\, du$

75. $\displaystyle\int \tan^n u\, du = \frac{1}{n-1}\tan^{n-1} u - \int \tan^{n-2} u\, du$

76. $\displaystyle\int \cot^n u\, du = -\frac{1}{n-1}\cot^{n-1} u - \int \cot^{n-2} u\, du$

77. $\displaystyle\int \sec^n u\, du = \frac{1}{n-1}\sec^{n-2} u \tan u + \frac{n-2}{n-1}\int \sec^{n-2} u\, du$

78. $\displaystyle\int \csc^n u\, du = -\frac{1}{n-1}\csc^{n-2} u \cot u + \frac{n-2}{n-1}\int \csc^{n-2} u\, du$

79. $\displaystyle\int \sin mu \sin nu\, du = -\frac{\sin(m+n)u}{2(m+n)} + \frac{\sin(m-n)u}{2(m-n)} + C$

80. $\displaystyle\int \cos mu \cos nu\, du = \frac{\sin(m+n)u}{2(m+n)} + \frac{\sin(m-n)u}{2(m-n)} + C$

81. $\displaystyle\int \sin mu \cos nu\, du = -\frac{\cos(m+n)u}{2(m+n)} - \frac{\cos(m-n)u}{2(m-n)} + C$

82. $\displaystyle\int u \sin u\, du = \sin u - u \cos u + C$

83. $\displaystyle\int u \cos u\, du = \cos u + u \sin u + C$

84. $\displaystyle\int u^2 \sin u\, du = 2u \sin u + (2 - u^2)\cos u + C$

85. $\displaystyle\int u^2 \cos u\, du = 2u \cos u + (u^2 - 2)\sin u + C$

86. $\displaystyle\int u^n \sin u\, du = -u^n \cos u + n\int u^{n-1}\cos u\, du$

87. $\displaystyle\int u^n \cos u\, du = u^n \sin u - n\int u^{n-1}\sin u\, du$

88. $\displaystyle\int \sin^m u \cos^n u\, du = -\frac{\sin^{m-1} u \cos^{n+1} u}{m+n} + \frac{m-1}{m+n}\int \sin^{m-2} u \cos^n u\, du$
$$= \frac{\sin^{m+1} u \cos^{n-1} u}{m+n} + \frac{n-1}{m+n}\int \sin^m u \cos^{n-2} u\, du$$

Forms Containing Inverse Trigonometric Functions

89. $\int \sin^{-1} u\, du = u \sin^{-1} u + \sqrt{1 - u^2} + C$

90. $\int \cos^{-1} u\, du = u \cos^{-1} u - \sqrt{1 - u^2} + C$

91. $\int \tan^{-1} u\, du = u \tan^{-1} u - \ln \sqrt{1 + u^2} + C$

92. $\int \cot^{-1} u\, du = u \cot^{-1} u + \ln \sqrt{1 + u^2} + C$

93. $\int \sec^{-1} u\, du = u \sec^{-1} u - \ln|u + \sqrt{u^2 - 1}| + C$
$$= u \sec^{-1} u - \cosh^{-1} u + C$$

94. $\int \csc^{-1} u\, du = u \csc^{-1} u + \ln|u + \sqrt{u^2 - 1}| + C$
$$= u \csc^{-1} u + \cosh^{-1} u + C$$

Forms Containing Exponential and Logarithmic Functions

95. $\int e^u\, du = e^u + C$

96. $\int a^u\, du = \dfrac{a^u}{\ln a} + C$

97. $\int u e^u\, du = e^u(u - 1) + C$

98. $\int u^n e^u\, du = u^n e^u - n \int u^{n-1} e^u\, du$

99. $\int u^n a^u\, du = \dfrac{u^n a^u}{\ln a} - \dfrac{n}{\ln a} \int u^{n-1} a^u\, du + C$

100. $\int \dfrac{e^u\, du}{u^n} = -\dfrac{e^u}{(n-1)u^{n-1}} + \dfrac{1}{n-1} \int \dfrac{e^u\, du}{u^{n-1}}$

101. $\int \dfrac{a^u\, du}{u^n} = -\dfrac{a^u}{(n-1)u^{n-1}} + \dfrac{\ln a}{n-1} \int \dfrac{a^u\, du}{u^{n-1}}$

102. $\int \ln u\, du = u \ln u - u + C$

103. $\int u^n \ln u\, du = \dfrac{u^{n+1}}{(n+1)^2} [(n+1) \ln u - 1] + C$

104. $\int \dfrac{du}{u \ln u} = \ln|\ln u| + C$

105. $\int e^{au} \sin nu\, du = \dfrac{e^{au}}{a^2 + n^2} (a \sin nu - n \cos nu) + C$

106. $\int e^{au} \cos nu\, du = \dfrac{e^{au}}{a^2 + n^2} (a \cos nu + n \sin nu) + C$

Forms Containing Hyperbolic Functions

107. $\int \sinh u\, du = \cosh u + C$

108. $\int \cosh u\, du = \sinh u + C$

109. $\int \tanh u\, du = \ln|\cosh u| + C$

110. $\int \coth u\, du = \ln|\sinh u| + C$

111. $\int \operatorname{sech} u\, du = \tan^{-1}(\sinh u) + C$

112. $\int \operatorname{csch} u\, du = \ln|\tanh \tfrac{1}{2}u| + C$

113. $\int \operatorname{sech}^2 u\, du = \tanh u + C$

114. $\int \operatorname{csch}^2 u\, du = -\coth u + C$

115. $\int \operatorname{sech} u \tanh u\, du = -\operatorname{sech} u + C$

116. $\int \operatorname{csch} u \coth u\, du = -\operatorname{csch} u + C$

117. $\int \sinh^2 u\, du = \tfrac{1}{4}\sinh 2u - \tfrac{1}{2}u + C$

118. $\int \cosh^2 u\, du = \tfrac{1}{4}\sinh 2u + \tfrac{1}{2}u + C$

119. $\int \tanh^2 u\, du = u - \tanh u + C$

120. $\int \coth^2 u\, du = u - \coth u + C$

121. $\int u \sinh u\, du = u \cosh u - \sinh u + C$

122. $\int u \cosh u\, du = u \sinh u - \cosh u + C$

123. $\int e^{au} \sinh nu\, du = \dfrac{e^{au}}{a^2 - n^2} (a \sinh nu - n \cosh nu) + C$

124. $\int e^{au} \cosh nu\, du = \dfrac{e^{au}}{a^2 - n^2} (a \cosh nu - n \sinh nu) + C$